Managing Measurement Risk in Building and Civil Engineering

Managing Measurement Risk in Building and Civil Engineering

Peter Williams

WILEY Blackwell

Registered Office
John Wiley & Sons, Ltd, The Atrium, Southern Gate, Chichester, West Sussex, PO19 8SQ, United Kingdom

Editorial Offices
9600 Garsington Road, Oxford, OX4 2DQ, United Kingdom
The Atrium, Southern Gate, Chichester, West Sussex, PO19 8SQ, United Kingdom

For details of our global editorial offices, for customer services and for information about how
to apply for permission to reuse the copyright material in this book please see our website at
www.wiley.com/wiley-blackwell.

Library of Congress Cataloging-in-Publication Data

Williams, Peter, 1947 November 20– author.
  Managing measurement risk in building and civil engineering / Peter Williams.
    pages   cm
  Includes bibliographical references and index.
  ISBN 978-1-118-56152-2 (pbk.)
1. Construction industry–Materials management.   2. Measurement–Risk assessment.
3. Civil engineering–Materials.   I. Title.
  TH437.W56 2015
  624.1′8–dc23
                            2015028049

A catalogue record for this book is available from the British Library.

Cover image reproduced with the permission of Kier Construction

Set in 9.5/11.5pt Sabon by SPi Global, Pondicherry, India
Printed and bound in Malaysia by Vivar Printing Sdn Bhd

1   2016

Contents

Preface

After a long, challenging and fascinating career – extremely varied, often high pressure and certainly never dull – I have decided to put down on paper my views on a subject that has been central to almost everything I have done in the construction industry.

Whilst working for various contractors and subcontractors, running a contracting business, lecturing both in higher education and at a professional level and undertaking various sorts of consultancy work, measurement has always played an important role in my life. It has permeated everything I have done in quantity surveying, estimating, financial management, contract administration and legal matters in my career in building and civil engineering.

Whilst I don't claim to be an expert in the subject, I have nonetheless always been an avid student of measurement, both theoretically and practically. I have owned and read many of the great standard textbooks on the subject, and I hold both the writers and the books themselves in the very highest regard. Some of them were instrumental in my own education, and some have been invaluable during my working career.

I have never been a professional quantity surveyor (PQS) as such – I wanted to be but, at the time that I qualified, contractors' quantity surveyors were excluded from membership of the RICS – the 'home' of the PQS in the City of Westminster, London (I sat the examinations of the Institute of Quantity Surveyors).

No, I have always been a 'contractor's man' at heart – happy to be ploughing through the mud, setting out with level and theodolite, doing physical measures on cold and windswept sites or 'arguing the toss' with main contractors who don't want to listen – and I was proud to be a fully qualified member of the sadly missed IQS, long since absorbed by the RICS.

Nevertheless, I have been involved in PQS-type work, both as a consultant to contracting firms, in loss adjusting for the insurance business and in undergraduate, postgraduate and practitioner training and education.

And so, it is from this background that this book has been written. Despite what some people may think, it is not just quantity surveyors who can measure – many engineers are involved in the measurement side of the industry and, increasingly, specialist subcontractors with no QS background find themselves preparing quantities as part of the bidding process. It is hoped, therefore, that this book will appeal to a broad 'church' of 'measurement practitioners', of whatever persuasion.

Whilst it is fervently hoped that this book will follow the traditions of the great measurement books, it is structured and written completely differently. The main theme of the book is 'risk', and so the chapters dealing with the various methods of measurement in particular focus on 'risk issues' that emanate from the measurement process or impact on it in some way. Such risk issues may relate to questions of interpretation of standard methods of measurement or may refer to risks arising from the relationship between measurement rules and standard forms of contract or procurement methods.

Inevitably over the last 25 years, computer technology has impacted the subject of construction measurement, and this is an important theme in this book. However, do not fear! The book has not been written by a computer boffin or rocket scientist. It has been written by a practitioner for practitioners (and would-be practitioners) using everyday language. Where 'technical' words are unavoidable, these are explained in a simple and understandable fashion (with apologies in advance to the 'computer buffs'!).

The UK construction industry is privileged to have been served by several outstanding quantity surveyors over the years, some of whom have achieved iconic status. The likes of 'Jim' Nisbet, 'Ted' Skoyles and Douglas Ferry et al. are part of quantity surveying history, but no more so than equally iconic personalities such as A.J. Willis and Ivor Seeley, who have contributed significantly in the area of construction measurement, both through their professional work and through their publications – particularly textbooks.

Measurement remains the core skill of the profession, and there will be few quantity surveyors anywhere who have not owned or studied a copy of 'Willis' or 'Seeley'. These books have played an immensely important role in helping aspiring quantity surveyors to master both the 'tools and techniques' and the 'art' of measurement.

The means by which an architectural or engineering design may be modelled financially is provided by measurement; this provides the framework within which such designs may be controlled and realised within defined cost parameters, to the satisfaction of the client. Measurement has a particular skill base, but it is elevated to an 'art' because the quantity surveyor is frequently called upon to interpret incomplete designs, in order to determine the precise quantitative and qualitative intentions of the designer, so that contractors may be fully informed when compiling their tenders.

The true art of measurement is undoubtedly the province of the 'professional' quantity surveyor, and a great deal of experience is required to fully master the subject. By definition, therefore, most construction professionals, and many quantity surveyors indeed, cannot be considered competent in measurement. They may be able to measure up to a point but, faced with a drawing chest full of AO drawings and a multiplicity of standard details and schedules, most would be daunted by the prospect of 'taking off' the quantities for a project of any size. Many wouldn't have the first idea where to begin.

This is still the case despite the huge advancements that have been made in IT-based measurement, but it does not mean that measurement is a 'closed book' or an inaccessible skill that might never be acquired. Nor does it mean that everyone involved in the construction process needs to be able to measure to the standard of a PQS. Not everyone who uses measurement needs to prepare quantities for the production of formal bills of quantities in the normally accepted PQS sense.

It is hoped, therefore, that this book will help those construction professionals, subcontractors and the like, who use measurement in their work, or deal with the output from the measurement process, to understand not only the 'ins and outs' of measuring construction work but also the relationship that measurement has with contracts, procurement, claims and post-contract control in construction. Measurement is part of a 'big picture' that extends well beyond the process of taking off quantities.

The views expressed in this book are mine, and mine only, but I apologise only for any errors there may be. Some may disagree with my line of thinking, which is fair enough, but the intention has been to write a practical, constructively critical and thought-provoking book about construction measurement, approached from a risk perspective. The observations made, and the risk issues raised, are also personal but are in no way meant to be authoritative.

Measurement has moved into a new and exciting era of on-screen quantification and BIM models, but this has changed nothing in terms of the basic principles underlying measurement – thoroughness, attention to detail, good organisation, making your work auditable and, above all, understanding the way building and engineering projects are designed and built. You must know the technology to be able to measure.

It is hoped that this book will help to give you the confidence to both 'measure' and understand measurement risk issues and to do so in the best traditions of the likes of Willis and Seeley to whom the industry owes a great debt of gratitude for their vision and expertise in the field of construction measurement.

Peter Williams
Chester
November 2014

Author Biography

Peter Williams began his studies in construction in the mid-1960s at the Liverpool College of Building, gaining a Higher National Diploma in Building for which he was awarded the top honour of the Chartered Institute of Building – the Silver Medal. His studies continued with a Master of Science Degree in Construction Management and Economics at the University of Aston in Birmingham. During this period, Peter also became a fully qualified Member of both the CIOB and the Institute of Quantity Surveyors, by examination, and thence became a Member of the RICS.

Peter's working career began as an assistant quantity surveyor and he then worked as a site engineer on a number of large building and civil engineering contracts. Several years as a building estimator and then civil engineering estimator followed and he then became a Senior Lecturer at Liverpool Polytechnic. During the 1980s and 90s, Peter was responsible for running a civil engineering and building contracting company and this period was followed by his appointment as a Principal Lecturer, and then Director of Quantity Surveying, at Liverpool John Moores University. Later, Peter became Head of Construction Management Development which involved the authorship of a distance learning MSc in Construction Health and Safety Management and tutoring on the Post-graduate Certificate in Construction Law. As well as lecturing on a wide range of construction management and quantity surveying subjects, Peter was responsible for the development and validation of the LJMU MSc in Construction Project Management.

Following retirement from the University, Peter has been engaged as a Consultant and Lecturer and has worked with a variety of contractors, subcontractors and client organisations in the fields of quantity surveying, construction law, health and safety management, delay analysis and claims.

Peter's writing career has included co-authorship of the best-selling *Construction Planning, Programming and Control* (with B. Cooke) and *Financial Management in Construction Contracting* (with A. Ross), both published by Wiley-Blackwell. His present interests are as a writer, researcher, lecturer and consultant with particular interests in contracts and finance, delay analysis and health and safety management.

Cooking, food and wine are among Peter's leisure interests. He is a keen DIY-er and enjoys sport, especially football. He follows his local football team, Chester FC, and his boyhood club, Wolverhampton Wanderers.

Acknowledgements

I have been privileged to have had the help and support of a number of people whilst writing this book, some of whom I have never met. Some have been very generous in supplying me with various resources, and everyone mentioned below has freely given the most valuable commodity of all – their time.

My sincere thanks, therefore, go to everyone I have been in contact with, both in the United Kingdom and overseas, all of who have taken the time and trouble to help me:

Andrew Bellerby	Managing Director, Tekla UK
Nicola Bingham	Ramboll UK Ltd
Tony Bolding	iSky Software Ltd (QSPro)
Joanna Chomeniuk	North West Construction Hub
Tim Cook	Causeway (CATO)
Greg Cooper	X-LAM Engineering Manager, B&K Structures
John Granville	Executive Director, New Zealand Institute of Quantity Surveyors
Patrick Hanlon	Director, BQH Quantity Surveyors, New Zealand
Martin Hodson-Walker	Commercial Manager, Roger Bullivant Limited
Michael Kirwan	BSS – Building Software Services (Buildsoft Cubit)
Hugh Mackie	WT Partnership, Brisbane, Australia
David Miller	Director, Rand Associates
Phil Vickers	Commercial Director, Kier Construction

I am particularly grateful to Madeleine Metcalfe and Harriet Konishi at Wiley-Blackwell, who have believed in me from the outset and who have been immensely supportive and patient during this project.

I would also like to thank an anonymous group of people – the Wiley-Blackwell book proposal reviewers – who all unknowingly contributed to the eventual outcome with their incisive comments, constructive criticism and professional guidance.

And last but not least, my gratitude goes to Paul Hodgkinson, who worked on many of the line drawings, and to Jaqueline, for being a good listener and for providing the moral support, good food and wine just when needed!

Glossary

3D BIM: The use of parametric design models and space programming tools to enable 3D visualisation, walk-throughs, clash detection and coordination, item scheduling, etc.

4D BIM: Sometimes referred to as *3D BIM plus time*,[1] 4D BIM is where 3D objects and assemblies are linked with the project programme and phasing strategy and where resources can be quantified and scheduled.

5D BIM: 5D BIM may be considered as 4D BIM plus cost[1] where the BIM design is linked to the cost planning, bill production and estimating functions of the construction process.

Activity schedule: A list of unquantified construction activities, usually prepared by the contractor, often, but not necessarily, linked to the contractor's programme.

Admeasurement: The act of ascertaining and apportioning in order to establish the difference between a final quantity and an original quantity of work, whether more or less.

Anding-on: Where a set of dimensions for one item is copied to another item description that has the same quantity.

Bill compiler: The person responsible for assembling a completed bill of quantities ready to issue with other tender documents.

Bill of quantities: A list of item descriptions, and firm or approximate quantities, based on a standard method of measurement.

BIM model: A 3D assembly of components of the same family that carry technical, geometric, measurement and other data.

BIM: An acronym used to describe the tools, processes and technologies that facilitate the digital representation of the physical and functional characteristics of a building or structure, thereby creating a shared knowledge resource of information that can be used for reliable decision making throughout its life cycle.

Builder's quantities: A list of quantities lacking the precision of measurement and description normally associated with a professional quantity surveyor measuring to a standard method of measurement.

CESMM: Civil Engineering Standard Method of Measurement.

Commercial opportunity: A strategy often employed by contractors at tender stage where a risk allowance is calculated on the basis of potential future gains should the contract be awarded that enables a lower initial tender bid to be made. Capitalising on buying gains for materials and subcontractors, profiting from errors in the tender documentation and tactical pricing of rates are some of the techniques employed.

Contract sum analysis: A breakdown of a contract sum, usually in design and build, used for post-contract administration.

Cost-value reconciliation: The process of matching cost and revenue at a common date by measuring the true value of work carried out.

DBFO: Design-Build-Finance-Operate procurement used for major projects where a consortium delivers a capital project (e.g. a tunnel or bridge) and operates the facility for a concession period (e.g. 25 years) in order to recover the initial investment.

Daywork: A method of measuring and valuing work on the basis of the resources expended rather than in relation to the quantities of work done.

Design cost control: The process of establishing a budget, deciding how to spend the money in order to satisfy the client's functional and aesthetic requirements and reconciling the cost limit with tenders received.

Design intent: Intended ambiguity in a completed design, which leaves the final design decisions to those undertaking the construction work.

Dim sheet: A specially ruled sheet of paper used to ensure that measured dimensions are recorded in the correct order and fashion needed to ensure clarity, accuracy and a visible audit trail.

Direct billing: A method of quantification where the dimensions, waste calculations, item descriptions, quantities and pricing columns are provided on the same page.

Dotting-on: A way of adding an additional number to a 'times-ing' calculation where a measured item possesses the same dimensions as an item previously measured.

Earthworks balance: Calculation of the volumes of excavation, filling and disposal to ensure that the required quantity of material is available for construction and any surplus is disposed of Levels may be realigned to ensure the optimum use of materials arising from the site and the minimisation of imported materials from off-site.

Extra over: The additional burden required to complete an item of work over and above that of a base item, where the two items of work have some dissimilar characteristics but are essentially of much the same nature.

Final account: The process of calculating the amount due or final payment owing on a contract.

Final account statement: A statement of the amount owing at the conclusion of a contract, calculated according to express terms of that contract.

Lump sum: A type of contract based on a contract sum which can only be adjusted if there are express terms in the contract to do so.

Mass-haul diagram: A diagram or computer model that identifies the quantities of excavation and fill arising on a site together with the movement of those materials required to achieve the optimum earthworks balance and the minimum requirement for imported material or off-site disposal.

Measure and value: A type of contract where the quantities (if any) are estimated and the difference between the original quantities and the final quantities determines the contract sum and whether any change is needed in the rates and prices to reflect the consequences of a change in quantities. 'Remeasurement' is used as a synonym but is not strictly correct.

Measurement: The action or an act of measuring or calculating a length, quantity, value, etc.

Measurement claim: A contractual, common law, *quantum meruit* or *ex gratia* claim submitted by contractors or subcontractors where there has been an error in a quantity, an error in an item description, a discrepancy between a pricing document (e.g. BQ and schedule of rates) and any other contract document(s), such as drawings or specifications, a departure from the rules of a method of measurement or an omission or alleged omission, to measure something required by a method of measurement.

MMHW: The Method of Measurement for Highway Works which is part of Volume 4 (Bills of Quantities for Highway Works) of the Manual of Contract Documents for Highway Works.

NRM1: New rules of measurement *for order of cost estimating and cost planning for capital building works.*

NRM2: New rules of measurement *for detailed measurement for building works.*

Order of cost estimate: An estimate of the possible cost of construction in the early stages of design which forms the basis for deciding on the cost limit and marks the beginning of the cost planning process.

Overbreak: The additional excavation, filling and disposal generated when ground conditions, or construction methods, result in excavation beyond the minimum limits stipulated in the contract.

Pain and gain: A system of risk and reward designed to encourage value-engineered solutions that result in shared cost savings.

Pareto principle: A 'rule of thumb' in management and business which states that, *for many events, roughly 80% of the effects come from 20% of the causes.*

Plug-in: A software application, such as an on-screen measurement tool, that supplements an existing software application, such as bill production software package, thereby enabling greater interactivity and customisation.

POM(I):   Principles of Measurement (International) for Works of Construction.

PQS: A professional quantity surveyor, engaged by the employer, whose duties may include giving cost advice, preparing cost plans, advising on contractor selection and procurement, the preparation of bills of quantities, tender reconciliation and dealing with pre- and post-contract issues on the employer's behalf.

Price list: A list of simple *ad hoc* items, with an accompanying unit of measurement, that may be used for small-scale construction projects and works orders.

Prime cost sum (PC sum):   A monetary allowance in a bill of quantities for the provision of specialist goods or services to be nominated by the employer which cannot be precisely defined or quantified at tender stage.

Provisional quantities: Estimated quantities of items of work in a lump sum contract where accurate quantities cannot be measured.

Provisional sums: Described, but not measured, items in a bill of quantities representing a sum of money to be expended, as required, by the contract administrator.

QS: An abbreviation for 'quantity surveyor' to be distinguished from the abbreviation PQS.

Rates and prices: Rates are normally multiplied by quantities to arrive at a total for each BQ item; prices are lump sums, such method-related charges or preliminaries. In NEC3, a rate multiplied by a quantity is a price.

Remeasurement: The measurement of something that has already been measured but is to be measured again, resulting in a fresh set of quantities that replace the original.

RICS: The Royal Institution of Chartered Surveyors which is the primary UK professional body for quantity surveyors.

Schedule of cost components: A statement of 'defined cost', employed by NEC3 forms of contract, used for valuing compensation events and for target and cost reimbursement contracts.

Schedule of rates: An unquantified list of item descriptions, not necessarily based on a standard method of measurement, used where the nature of the work required is known but not the extent.

Schedule of works: An unquantified list of work items, usually composite, that is not based on any particular standard method of measurement.

SMM: Standard Method of Measurement of Building Works.

Written short: A method of writing item descriptions, or item coverages, where one item description or item coverage relies on another item description or item coverage in order to convey the complete meaning of the item description or item coverage 'written short'.

Note

1. http://www.fgould.com/uk-europe/articles/5d-bim-explained/#sthash.ExMxUsfc.dpuf (accessed 7 April 2015).

Addendum: Infrastructure Conditions of Contract – *With Quantities Version*

Since the manuscript for this book was completed at the end of November 2014, a new form of contract has been published which merits inclusion in the text in so far as it concerns the subject matter of the book.

The ICC – *With Quantities Version* (2014) is a new addition to the suite of contracts published by the Association for Consultancy and Engineering (ACE) on behalf of ACE and the Civil Engineering Contractors Association (CECA). The ICC suite is largely a rebranded version of the former ICE Conditions of Contract which were adopted by ACE and CECA in 2011 following the decision of the Institution of Civil Engineers in 2010 to sponsor the New Engineering Contract (NEC) suite as its contract of preference.

A.1 Type of Contract

The ICC – *With Quantities Version* is a lump sum contract intended for engineer/consultant-designed projects, with the option for an element of contractor design if desired. Being a lump sum contract, the *With Quantities Version* presumes that the design is sufficiently developed at tender stage in order that a bill of quantities (BQ) with fixed (not approximate) quantities may be produced as a basis for inviting tenders.

This new contract contrasts sharply with the ICC – *Measurement Version*, a measure and value (admeasurement) contract, which has been widely used for civil engineering contracts for many years. The ICC – *With Quantities Version* provides an alternative to this arrangement in circumstances where quantities can be measured with more certainty at tender stage and the parties can thus enter into a lump sum contract.

A.1.1 Lump Sum versus Measure and Value

The essential difference between a lump sum and a measure and value contract is that the parties agree to a contract sum, that is, an accepted offer to carry out the works for a defined sum of money. This sum can only be adjusted if there are express terms written into the contract to do so. Grounds for adjusting the contract sum may include:

- Remeasurement of approximate/provisional quantities
- Variations ordered by the contract administrator

- Adjustment of provisional sums
- Adjustment of prime cost sums, etc.

In a measure and value contract, the total of the BQ provides a tender total (i.e. not a contract sum) which is purely a total by which the various tenders received may be compared.

A.1.2   Form of Tender

Surprisingly, in a lump sum contract, the Appendix to the ICC – *With Quantities Version* is identical to that in the *Measurement Version*. In both cases, the contractor offers to construct and complete the works *for such sum as may be ascertained*. This indicates a measure and value, as opposed to a lump sum, contract, because the priced BQ is brought to a total – the 'tender total' – and the contract sum is 'ascertained' when the works are complete.

The term 'tender total' is defined in Clause 1.1(w), but this differs from the *Measurement Version* definition in that the tender total *means the total of the Contractor's tender for the design, construction and completion of the Works*. This is less than clear especially when read in conjunction with Clause 1(d): *Contract Price* which is defined as *the sum to be ascertained and paid in accordance with the provisions of the Contract for the construction and completion of the Works*. Both definitions resonate more with a measure and value arrangement than a lump sum contract.

A.1.3   Contract Documents

Clause 1.2(b) of the ICC – *With Quantities Version* provides that the *Contract Documents* are those defined in Clause 1.1(b). This would appear to be a drafting error as it is Clause 1.1(c) that refers to the *Contract*.

Inexplicably, there is no reference to 'drawings' or 'specification' amongst the various documents referred to in Clause 1.1(c), and whilst the term *Works Data* is defined in Clause 1.1(y), this simply *includes, without limitation, the Employer's Requirements and the Contractor's Proposals*. Such wording is normally used when referring to contractor design, but the words *without limitation* may be intended to have wider implications.

Notwithstanding the definition of *Works Data* in Clause 1.1(y), Clause 4.6 states that the *Employer's design shall be contained in the Works Data*. This sits more comfortably with a lump sum contract based largely on an engineer/consultant design and with a BQ prepared on the basis that the design is complete. Consequently, albeit somewhat tortuously, it is clear that the drawings and specification, upon which the BQ has been prepared, are intended to be contract documents.

A.1.4   Risk

One of the features of the ICC – *With Quantities Version* is that *Risk* is accorded its own clause where *Contractors Risk, Employer's Risks*, and *Excepted Risks* are collected in one place. Gone, for instance, is the famous Clause 12 from the *Measurement Version* and in its place is Clause 8.5(a) which places the risk of *physical conditions … or artificial obstructions which … could not reasonably have been foreseen by an experienced contractor* squarely with the employer, as before.

However, the means whereby such risks are valued is to be found in several places. Clause 8.9 states that *the Engineer may … order a variation* in such circumstances and, if so, this will be valued in accordance with (the new) Clause 12.6. Where, however, the Contractor *intends to claim any additional payment or any allowance of additional time*, notice must be given in accordance with Clause 13.1, and *full and detailed particulars of the claim* must be submitted in due course as directed by the engineer.

A.1.5 Contractor-Designed Works

In the ICC – *With Quantities Version*, 'contractor-designed works' is defined as *the part or parts of the Permanent Works to be designed, constructed and completed by or on behalf of the Contractor* (Clause 1(f) refers). Such work could feasibly be measured in the BQ, or a single item could be included for the contractor to price as a lump sum.

Inclusion of suitable measured items for contractor-designed works in the BQ is not straight-forward as there are no protocols in the ICC – *With Quantities Version* nor does CESMM4 contain provisions for billing contractor-designed works. Presumably, therefore, a suitable method measurement should be chosen which provides for contractor design, or, alternatively, the bill compiler should give appropriate consideration to ensuring that such works are prop-erly described in the BQ by careful application of the additional description rules contained in Paragraphs 5.9–5.11 and 5.14 of CESMM4.

Where a contractor-designed item of work is to be included in the BQ as a single, non-measured item, the ICC – *With Quantities Version* provides for the inclusion of a milestone sum provided that Supplementary Clause 21 is incorporated in the Contract Agreement (see Section A.1.6).

A.1.6 Supplementary Clauses

Amongst the four supplementary clauses in the ICC – *With Quantities Version* is Clause 21, which deals with milestones.

Whilst not expressly stated in the form of contract, milestone sums would be included in the BQ for items of work to be designed by the contractor that are not to be measured in detail by the employer's quantity surveyor/measurement engineer. Milestones are sums of money, *identified as such in the Bill of Quantities*, which are not subject to remeasurement, and, therefore, risk for the accuracy of the quantities underpinning such lump sums lies entirely with the contractor.

Payments to the contractor in respect of milestone sums take place *only upon achievement of the criteria set out in the Works Data*. Therefore, interim payments for a bridge over a 9-month construction period would only be made when each stage of construction (e.g. substructure, superstructure, deck and finishings), identified in the Works Data, has been substantially com-pleted, subject to deduction of retention as per Clause 21.3(a).

It is unfortunate that the ICC – *With Quantities Version* is tied to the use of formal BQ as milestone sums, linked to activity schedules, could have lent added flexibility to the form of contract.

A.2 Method of Measurement

The default method of measurement for the ICC – *With Quantities Version* is the CESMM 4th edition unless another edition, or another method of measurement, is stated in the Appendix (Clause 11.1 refers).

The *With Quantities Version* differs from the *Measurement Version* in that the employer warrants that the BQ has been prepared in accordance with the Method of Measurement

(Clause 11.1 refers). This is to be contrasted with Clause 57 of the ICC – *Measurement Version*, which states that the BQ is *deemed to have been* prepared in accordance with the method of measurement.

A.2.1 Employer's Warranty

ICC – *With Quantities Version* Clause 11.1 (Bill of Quantities) provides a non-contractual remedy to the contractor should the provisions of the method of measurement not be correctly observed when preparing the BQ. This remedy is available by virtue of the employer's warranty. The legal remedies provided by this warranty may well impose additional risk on the employer should the BQ not comply with the rules of measurement.

Breach of the employer's warranty may entitle the contractor to sue for damages outside the contract for the financial consequences emanating from the breach. Whilst such a breach does not go to the root of the contract or signify that the contract may be repudiated or rescinded, a warranty is a promise that one party may rely upon what has been promised by the other party, and breach of the warranty is established purely as a matter of fact. Whether such a 'sledgehammer' solution to a simple matter of contract administration was intended by the drafting committee remains to be seen.

A.2.2 Items Not in Accordance with the Method of Measurement

Clause 11.2 simply states that where *any item in the Bill of Quantities is not in accordance with the Method of Measurement*, the offending item *shall be corrected and the Contractor shall be paid the corrected amount*. One implication to be drawn from this is that the *corrected amount* is purely a corrected quantity, but this seems overly simplistic in view of the Employer's Clause 11.1 warranty and because the deviation from the Method of Measurement could be a failure to:

- Measure an item that should have been measured.
- Correctly describe an item in accordance with the descriptive rules.
- Provide additional description as prescribed by the method of measurement.
- Distinguish an item that displays different characteristics to a similar item.
- Clarify where a similar item is to be carried out in dissimilar conditions or in a different location, etc.

A.3   Bill of quantities

The quantities in a measure and value contract are all estimated and are therefore subject to change according to whether more or less work has been carried out. The process of determining the difference between the original and final quantities is called admeasurement (i.e. not remeasurement) because the change in quantity is added to or deducted from the original billed quantity. Confusingly, measure and value contracts are often referred to as 'remeasurement contracts', and the words 'admeasurement' and 'remeasurement' are often taken to be synonymous.

The ICC – *With Quantities Version*, on the other hand, is written on the basis that a BQ forms part of the contract and that the quantities stated therein are *Fixed Quantities* (Clause 11.2). This means that the quantities are not subject to admeasurement and can only be remeasured if the engineer has issued a variation. However, if there are quantities in the BQ that are not 'fixed' for some reason – uncertain ground conditions, for instance – these must be *expressly identified therein as subject to remeasurement*.

Presumably, this means that the phrase *subject to remeasurement* must be used in the BQ and not the commonly accepted terms 'approximate' or 'provisional'.

A.3.1   Work Subject to Remeasurement

Where work is to be remeasured, Clause 11.3 provides that *the Engineer shall ascertain the quantities executed* and submit them to the contractor for agreement. Whilst there is no right for the contractor to attend or assist in the remeasurement process, in contrast with Clause 56(3) of the *Measurement Version*, the contractor does have a right to dispute the engineer's quantities and *shall provide fully substantiated details* of the quantities disputed.

In common with the ICC – *Measurement Version* Clause 7(1), there is provision in the *With Quantities Version* for the engineer to issue *further instructions, including drawings and specifications* (Clause 4.6). Consequently, it must be assumed that remeasurement is to be based upon revised drawings, failing which site measurement is the only alternative. In such circumstances, it may have been prudent to include provision for the contractor to be present on-site to 'hold the other end of the tape' and agree measurements at the time.

For remeasured quantities, Clause 11.2 states that the contractor shall be paid for the *actual quantities of work at the rates contained in the Bill of Quantities for those items*. For any particular item, this infers that the final account shall be computed by deducting the product of the *original quantity x BQ rate* from the contract sum and by adding the *remeasured quantity x the BQ rate*. This is normal practice for a lump sum contract.

However, where the original and final quantities differ significantly or where changes in the quantities impact the contractor's methods and/or sequence of working, there should be a means of valuing the work in order to adequately compensate the contractor. In some contracts, the ICC – *Measurement Version* Clause 55(2) for instance, the contract administrator has authority under the contract to adjust the BQ rates and prices according to the prevailing circumstances. In other contracts, a variation to the contract would be issued and the work would be valued in accordance with the variation procedure.

There are no provisions in the ICC – *With Quantities Version* for the engineer to adjust rates and prices nor are changes in quantities recognised as constituting a variation to the contract.

A.4   Protocols

Provision is made in the ICC – *Measurement Version* for the inclusion of a BIM Protocol in the contract, should BIM be employed for the design, and the protocol to be adopted shall be named in the Appendix – Part 1. The contractual obligation to comply with the protocol is provided in Clause 20 which also states that an Information Protocol may be included in the contract in order to underpin the BIM Protocol, if desired.

A.4.1   BIM Protocol

BIM protocols are intended to encourage collaborative working between members of the project team and to ensure that models are created at defined stages of a project, that common standards of working are adopted and that those using the model(s) have a legal right to do so.

The inclusion of such a protocol in the contract creates additional rights and obligations between the employer and the contractor, but not with other members of the project team, who may nevertheless have similar contractual arrangements with the employer. An example is the CIC BIM Protocol[1] which has been designed to be incorporated into all direct contracts between the employer and the project team members.

Whilst not specifically identified in the CIC BIM Protocol, there are obvious implications for the accuracy and exchange of quantities extracted from model(s), especially in terms of the employer's warranty in Clause 11.1 of the ICC – *With Quantities Version*, and for the correction of quantities where there are design clashes or changes as the models develop.

A.4.2   Information Protocol

The CIC BIM Protocol requires the employer to appoint an information manager which could be a project manager, lead designer, a contractor or a stand-alone appointment. Amongst the duties of the appointee will be management of the agreed processes and procedures for exchanging project information, involvement in the preparation of project outputs (data drops) and implementation of the BIM Protocol.

Whilst model clash detection and coordination duties belong to the BIM coordinator, the information manager would clearly be involved in ensuring that the correct procedures are followed for quantities extraction from models and for the exchange of appropriate information required for the preparation of BQ.

The information manager appointment requires its own scope of services. There is a detailed *Scope of Services for the Role of Information Management*[2] available from the CIC as well as a simpler version, which may be incorporated with any other form of appointment, such as the NEC3 Professional Services Contract.

Notes

1. Construction Industry Council (2013), Building Information Model (BIM) Protocol, http://www.bimtaskgroup.org/bim-protocol/ (accessed 5 July 2015)
2. http://www.bimtaskgroup.org/scope-of-services-for-information-management/ (accessed 5 July 2015)

PART 1

Measurement in Construction

Part 1

Managing Measurement Risk in Building and Civil Engineering, First Edition. Peter Williams.
© 2016 John Wiley & Sons, Ltd. Published 2016 by John Wiley & Sons, Ltd.

Chapter 1
The Role and Purpose of Measurement

Think of a number. Double it. Add 175. Double it again. Subtract 21.5. And the answer is... Who cares? Just fill in any number you like. If this sounds a bit like how you fill in your annual tax return, you might just have the makings of a career in quantity surveying.

The way that quantity surveyors can get away with getting it so wrong is an art form in itself.

John Crace, *The Guardian*, Saturday, 25 February 2006[1]

Perhaps Wembley Stadium is not the greatest endorsement of the quantity surveyor's skills, at least to 'outsiders', but it is more than likely that the project quantity surveyors had little control over this grotesque cost overrun.

Quantity surveyors work with information provided by others – clients, architects, engineers and so on – but it is probably true to say that the results they achieve often belie the paucity of the information they are provided with. The quantity surveyor's skill, attention to detail and ability to 'read between the lines' give the profession its individuality and special value.

1.1 Measurement

Part of the professional armoury of the quantity surveyor is the ability to measure.

Since its origins in the latter part of the nineteenth century, the quantity surveying profession has developed and grown and, as the years have rolled by, measurement has become pretty much the sole province of the quantity surveyor in the construction world.

Not everyone values what the quantity surveyor does (certainly not the *Guardian* reporter!), but this is more evident when it comes to the skill of measurement. Measurement lacks the esteem that it once had. Even some quantity surveyor academics, of the author's acquaintance, look down on measurement as being the province of 'technicians' and one eminent academic once told me that "you can teach monkeys to take-off quantities!".

This is not true, of course, and those who sneer at measurement are usually those who have been unable to master the subject themselves.

Managing Measurement Risk in Building and Civil Engineering, First Edition. Peter Williams.
© 2016 John Wiley & Sons, Ltd. Published 2016 by John Wiley & Sons, Ltd.

It is true that, measurement, once the 'be-all and end-all' of the quantity surveyor's work, is now very much the 'Cinderella' of the profession having been overtaken by sexier and more lucrative pursuits such as procurement advice, whole life costing and value management.

However, whilst the focus on measurement and bills of quantities is apparently waning, especially in higher education courses, Ashworth et al. (2013) observe that *instead of preparing bills of quantities for clients, quantity surveyors are preparing bills for contractors.*

1.1.1   Counting bricks

Many QSs must have had a similar experience to the author – in my case it was a conversation with a typical tennis club 'Hooray Henry'! When I was asked "and what do you do for a living?", the retort to "quantity surveyor" was:

"Ah … one of those chaps that counts the bricks?"

Quantity surveyors don't count bricks; of course, they measure the area of brickwork so that someone else can work out how many bricks are needed. For a m² of half brick thick wall in stretcher bond, this is 59 bricks, but no one actually counts them, do they?

- Builders need to know how many bricks are required, but they order by the 100.
- The contractor's buyer? No, they order by the 1000 and they arrive in pallets of 400/500 give or take.
- Bricklayers? No, they count the number of courses to work out how many bricks they have laid and they are paid by the 1000.

All these people are measuring the same thing but in a different way. The thing that distinguishes quantity surveyors is, firstly, that they habitually measure accurately and, secondly, usually do so following a code, or method of measurement, using well-established conventions for recording their work.

1.1.2   Definition

The noun 'measurement' is so much part of everyday life that most people would not bother to find out what it actually means. A number of definitions may be found:

The Oxford English Dictionary:

- *The action or an act of measuring or calculating a length, quantity, value, etc.*

Collins Dictionary:

- *The act or process of measuring.*
- *An amount, extent, or size determined by measuring.*
- *A system of measures based on a particular standard.*

Wikipedia:

- From the Old French word – *mesurement.*
- Represents the assignment of numbers to objects or events.
- All measurements consist of three parts: magnitude, dimensions (units) and uncertainty.[2]

1.1.3   Who measures?

Measurement is most commonly associated with the quantity surveyor, and the quantity surveyor is, in turn, normally thought of as a construction economist associated with 'working out costs', 'counting bricks (!)' and producing 'bills of quantities'. Whilst there is a lot more to the profession than this, measurement is the basis of much that the quantity surveyor does.

In the construction industry, however, measurement is conducted by all sorts of participants:

- Surveyors for conducting site surveys.
- Professional quantity surveyors for producing bills of quantities for tendering purposes.
- Main contractors for calculating a tender price for a design and build project.
- Main contractors pricing partial contractor design in traditional contracts.
- Subcontractors tendering for work on a drawings and specification basis.
- Buyers for calculating the quantity of materials to be ordered.
- Bonus surveyors for calculating incentive payments.
- Planning engineers for deciding on production outputs in order to arrive at the duration of activities on a construction programme.
- Engineers for setting out buildings and engineering structures.
- Construction workers as part of their trade.

In fact, it would be interesting to discover just how many retractable measuring tapes may be found on a typical construction site!

Individuals approach measurement in different ways:

- Work on-site may be measured with a tape or with digital surveying equipment.
- Quantities may be produced by hand using traditional procedures or various sorts of software may be employed.
- On-screen measurement may be done with 'high-end' library-based systems or with software that allows the measurer to create the descriptions.
- Item descriptions may be non-standard, bespoke, quasi-formal or formal SMM based.

Measurement is used by many people, in different ways, using different standards, and this book tries to reflect this by broadening the traditional approach taken by measurement books to include consideration of:

- The identification of risk in standard methods of measurement.
- The identification of risk in the measurement process.
- The production of 'formal' and 'informal' bills of quantities.
- The production of 'informal' schedules of quantities of different sorts.
- The physical measurement of completed or partially completed work on-site.
- The valuation of work in progress.
- The valuation of variations to the contract.
- The measurement of work in connection with provisional sums.
- The remeasurement of approximate quantities.
- Post-contract measurement and physical measurement on-site.
- The measurement aspects of claims and final accounts.

1.2   The end of measurement or a new beginning?

The construction industry, irrespective of country or culture, is, of course, a paradox, and any discussion about it is likely to polarise opinion.

Crotty (2012), in his excellent and thought-provoking book, considers the industry to be locked into *a craft-based mode of operation* despite the extent of factory-based production preferred in some countries.

The obverse side of this argument is put eloquently by Radosavljevic and Bennett (2012), who identify that construction is an industry capable of delivering buildings and infrastructure that *are the most complex things produced by* humans.

The anachronistic face of the industry is exemplified by the almost medieval construction crafts still practised, and yet this may be contrasted with leading edge Building Information Modelling (BIM)-based architectural and engineering design and construction, such as Renzo Piano's 1016 foot skyscraper – The Shard at London Bridge Quarter. The industry has a fabled history of disputes and litigation and yet is capable of the highest standards of collaborative working, a shining example being the Llanelli Scarlets Rugby Stadium project (Eastman et al., 2011) in South Wales, United Kingdom.

The industry is changing slowly. In some cases, old-fashioned 'data-poor' documentation, such as paper-based drawings and bills of quantities, are being replaced by 'data-rich' models that can be shared interactively by all members of the design and construction team. This is a gradual process but it is gaining momentum. BIM in construction is being incrementally adopted – and not just on signature projects – but will this signal the end of measurement? No, BIM is not a threat; it just poses a different question – 'how will measurement be done in the future?'.

The RICS has taken a step in this direction by publishing the New Rules of Measurement suite of documents, and traditional practices are being challenged by new technology in both the design and construction of projects.

1.2.1   Anyone can measure

On the face of it, measurement is pretty straightforward; most people can measure physical things, whether by using a ruler, a measuring tape, a digital measuring device or by physically pacing out.

As far as construction work is concerned, anyone could take an intelligent guess as to how such things might be measured – it's fairly obvious:

- Rainwater pipes and gutters, drain pipes – linear metre, linear foot.
- Brickwork, plastering, painting – square metre, square foot, square yard.
- Excavation, concrete – cubic metre, cubic foot, cubic yard.

Measuring from drawings is more complex – far more complex than measurement textbooks imply. Drawings – being a less than full size representation – introduce the idea of scale. Consequently, unless the drawing is fully dimensioned, a measuring device is required that will measure to the appropriate scale – traditionally, a scale rule.

The modern day equivalent to this is the digitiser tablet which facilitates tracing a line, or shape, with an electronic device which then conveys the measurements taken to a personal computer.

Currently, measurements may be taken from scaled drawings that have been converted into a digital file by using on-screen measurement software and a computer mouse.

1.2.2   Measurement in the lecture theatre

Over the last 20 years, measurement has assumed less and less importance in academic circles, and many university courses now place significantly more emphasis on procurement and risk management topics to the exclusion of measurement.

The reasons for this trend reflect changes in the quantity surveying profession – Wikipedia provides a long list of services that quantity surveyors provide[3] – and it is certainly a fact that the subject of measurement is not regarded as sufficiently challenging academically.

The result of all this is a much more superficial coverage of the subject in quantity surveying degree courses such that employers frequently complain that sandwich course students and graduates cannot measure, never mind do so with any degree of competence. This is a real shame because 'measurement' is central to the quantity surveyor's skill base and it is a subject that ties all the other subjects in the syllabus together. Measurement is, after all, the basis of virtually everything that the quantity surveyor is about, and it is a transferable skill that is jealously coveted by other professions.

McDonnell (2010), who observes that *measurement of quantities is a core skill which must be inherent in all graduates from Quantity Surveying courses*, also considers that students find measurement difficult to understand and that they lack the basic knowledge of construction technology in order to be able to measure competently. These frailties are often exposed when students and graduates enter the industry and have to measure 'for real' – they struggle and disappointed employers complain that they are unable to 'hit the ground running' as far as measurement goes.

Paradoxically, McDonnell (2010) reports that students recognise the importance of measurement in their studies and also that they consider traditional 'paper-based' methods of teaching measurement as still having an important role to play despite the attractions of the 'computer-based' measurement software packages that are used both in teaching and in the industry generally.

Other commentators agree with the students, and Lee et al. (2014), amongst others, suggest that mastery of traditional paper-based taking-off is the key to being able to measure competently and confidently. A counterargument to this is that measurement from 3D models is simpler than from paper drawings because models can be 'delayered' to show only those items being measured and this, with the benefit of 3D perspective, is easier for students to comprehend.

It is often argued that measurement smacks of 'training' rather than 'education', and there may be some truth in this point of view, but measurement, when considered in its broadest context, draws together many complex legal, financial and technical issues sufficient to challenge the most experienced practitioners, never mind undergraduates.

1.3   How's your Latin?

Cash flow is the lifeblood of the construction trade.

Lord Denning
Former Master of the Rolls

Some people would argue that 'information' should have equivalent status in our thoughts:

Bringing together the right information with the right people will dramatically improve a company's ability to develop and act on strategic business opportunities.

The most meaningful way to differentiate your company from your competition … is to do an outstanding job with information. How you gather, manage, and use information will determine whether you win or lose.

Being flooded with information doesn't mean we have the right information or that we're in touch with the right people.

Bill Gates

1.3.1   Information

Information is all about communication, the Latin word for which – *commūnicāre* – means to share.

Sharing – of information – is how the construction industry communicates with itself. Drawings beget specifications, which together beget bills of quantities, and all this information is passed on down the line, via a complex system of contracts and subcontracts, with the final result being the completed building, bridge or tunnel.

There is no shortage of information in construction, but whether this is the information needed by the recipient is questionable and whether the correct information gets to those doing the work is even more doubtful. BIM responds to this challenge as it is capable of modelling exactly what has to be fabricated off-site, and assembled on-site, to pinpoint precision, with quantities already determined.

However, the construction industry has special characteristics which create a complex communication environment. Being of a project-based nature, the industry is fragmented and dynamic with many stakeholders operating in frequently changing sets of relationships which are contractually driven (Hoezen, 2006).

The system has its critics, and Crotty (2012) describes it as *a way of doing business that would be quite recognisable to medieval builders and their clients*. They might struggle with the mobile technology and with the finer points of *e-tendering*, but Latham (1994) and Egan (1998) would no doubt agree!

Speaking of which, the UK construction industry has been the subject of seemingly innumerable official and semi-official reports. This started with the Simon Report – no, not the 1930 study of potential constitutional reform in India – the 1944 Simon Committee Report on *The Placing and Management of Building Contracts*!

This report was followed by lots of others, including the Emmerson, Banwell and Wood reports. In more recent times, there have been the Latham (1994) and Egan (1998) reports and finally, for now at least, the Wolstenholme Report (2009), inspired by Constructing Excellence which was set up post-Egan (originally named the Movement for Innovation). This report concentrated on the impact of the Egan Report with particular regard to the performance targets set for the industry by Egan.

Murray and Langford (2003) usefully summarise the important issues raised in the 1944–98 reports.

They note that the principal driving force behind these government-initiated reports was two groups of powerful clients who had their own agendas for wanting change in the way that the construction industry operated. Pre-1980, it was largely government and parastatal clients driving the desire for change, and post-1980, it was influential private clients and construction employers who wished to conduct their business with the state in a different way.

1.3.2 The Tavistock Report

Murray and Langford observed that the desire to change the relationships of the participants in the construction process, with a view to improving industry performance, was a consistent theme in all of the reports. A recurring theme in the most influential reports was that of procurement and, more particularly, the inefficiencies generated by the separation of design and construction in the industry.

Procurement is especially important in the context of the subject matter of this book, with particular regard to the quality of project information and the negative impact on industry performance of:

- The lack of integration within design teams.
- The disconnect between designers and constructors.
- The incomplete and uncoordinated design information that results.

This was the theme of the so-called Tavistock Report (there have been several), written by Higgin and Jessop (1965), which concerned organisational structures and the exchange of information in the construction industry. The report was published independently by the Tavistock Institute of Human Relations in 1965, a not-for-profit organisation that carries out, *inter alia*, research work in the social sciences, and was not government sponsored.

The significance of the Higgin and Jessop report may not be fully understood for some time yet, but some observers consider that the successful adoption of BIM in the UK construction industry will be dependent on the industry's ability to embrace a new era of cooperative working, driven by high-quality intelligent information.

There may be a need for new methods of procurement and conditions of contract in order to do this, but there will certainly have to be a lot of 'head scratching' about how measurement fits into the BIM-scene *vis-a-vis* current methods of measurement.

1.3.3   A new business model?

This Government's four year strategy for BIM implementation will change the dynamics and behaviours of the construction supply chain, unlocking new, more efficient and collaborative ways of working. This whole sector adoption of BIM will put us at the vanguard of a new digital construction era and position the UK to become the world leaders in BIM.

Francis Maude
Minister for the Cabinet Office

The Government Construction Strategy, published by the Cabinet Office on 31 May 2011, announced the Government's intention to require collaborative 3D BIM on its projects by 2016, with all project and asset information, documentation and data being electronic.[4]

This is an enormous challenge to the current business models employed in construction, but if it can be achieved, the BIM revolution promises to fundamentally change current practice.

At the risk of wearing out the Bill Gates quotes:

This is a fantastic time to be entering the business world, because business is going to change more in the next 10 years than it has in the last 50.

There are two key challenges:

- Firstly, to coalesce the various design inputs involved in construction projects.
- Secondly, to harmonise the inputs of all contributors to the construction process – clients, designers, contractors and users – so that the final outcome is smarter and more predictable.

There is nothing new about this 'model' of the construction industry, of course, and in many respects, this is how construction work was organised for many centuries.

The client or architect would hire individuals or teams of craftsmen, organise the work, control the money and supervise work on-site. It is a system that worked well and led to the sort of harmony between designers and constructors that present-day BIM methodologies seek to create; the architect produced the drawings and the craftsman added final details as 'designer of last resort'. In some cases, craftsmen also created models or prototypes prior to actual construction.

The construction boom in the mid-nineteenth century changed all this, largely driven by the Industrial Revolution and the need to improve public health and sanitation. Many clients demanded market prices for their projects, and this gave rise to the idea of competitive tendering that has been the mainstay of the industry for over a century.

At the same time, the structure of the industry was changing, and the concept of the general contractor was established. It was the general contractor that organised the work according to the architect's or engineer's design, and the designer's role became one of the monitoring standards and certification of payment, thereby driving a wedge between design and construction that hitherto did not exist.

The system of competitive tendering in the late nineteenth century, with general contractors such as William Joliffe, Sir Edward Banks and Thomas Cubitt competing for work, was originally based on architect/engineer design with quantities prepared by each tendering contractor. This business model led to such wildly inconsistent tender prices that tendering contractors eventually contributed to the cost of preparing a single set of quantities upon which their tenders could be based.

A natural extension to this model was for the client to engage a quantity surveyor, both to produce the quantities needed for tendering and to provide other professional services. Lack of clarity in the documentation led, as a natural consequence, to the need for standardisation of the way that this documentation was presented. The concept of the standard method of measurement was born.

Over the years, the main focus of measurement has been on the production of bills of quantities – an aspect of quantity surveying pretty much dominated by the professional quantity surveyor (PQS) – but nothing major has changed in relation to the measurement of quantities since standardised measurement was first introduced:

- Standard methods of measurement have become more sophisticated, and different versions have been developed for use in different industry sectors.
- More attention has been paid to the value of quantities, and the way they are measured and presented, as a means of communication.
- The role of measurement in the cost management of buildings has become prominent since the early 1950s.
- Measurement techniques and conventions have developed over time.
- The introduction of computerised and on-screen measurement has transformed quantity surveying from a paper-based to a technology-based profession.

However, none of these developments has fundamentally changed the way that quantities are measured or the role that measurement plays in the procurement of construction work.

1.4   Standardised measurement

Formalised measurement has a long history dating back to the Middle Ages. Records exist of Royal expenditure on building work at the time of Henry II (1154–1189), and detailed accounts may be found from the monarchy of Henry III (1216–1272).

It is likely that we would call these documents 'final accounts' rather than bills of quantities.

Records of what appear to resemble bills of quantities appeared in Ireland in 1750, but Seeley and Winfield (2009) point out that the first method of measurement was produced in Scotland in 1802. There appears to be no evidence, however, that this method of measurement was in widespread use, or that bills of quantities were prepared to any extent, but there are firm indications that the system of measure and value on completion was used for most of the nineteenth century.

Towards the end of the nineteenth century, the practice of employing one surveyor to measure the quantities developed, due to the growth of competitive tendering, but this only brought uniformity to tendering and tender prices and not to the measurement process.

Standardised measurement is a more recent, twentieth-century phenomenon, which differs from formalised measurement in the sense that the latter is 'measurement written down' but standardised measurement is 'measurement written down according to a set of rules'.

The industry has a variety of standard methods or rules of measurement to suit different circumstances, but they each take a different approach to measuring the same thing, as illustrated in Table 1.1.

The item descriptions in Table 1.1 impact on how the work item is communicated to the tenderer/contractor and how the estimator goes about pricing the work. Whether, and to what

Table 1.1 Comparison of standard methods of measurement.

NRM2	**Excavating and filling**
	Excavations
	Excavation
	  Bulk excavation; not
	  exceeding 2 m deep
		• Clear description
		• No precise sense of how deep the excavation is
		• Masks the fact that this could equally be an excavation to reduce levels or a basement excavation

CEMM4	**Earthworks** General excavation **Material other than topsoil, rock or artificial material** maximum depth: 1–2 m	• Clear description of what is to be excavated and the type of excavation it is • Depth is calculated from the commencing level and is therefore between 1 m and 2 m deep

MMHW	**600:Earthworks**
	Excavation
	  Excavation of acceptable
	  material excluding Class 5A in
	  cutting or other excavation
		• No depth category
		• The excavation could be in cuttings or elsewhere on-site
		• The term 'acceptable' means that it could be reused on-site if required

POM(I)	Excavation
	Generally
	  Reduce levels
	  any depth
		• The type of excavation is clear
		• No depth category is stated
		• Additional description may be advisable

extent, ancillary items such as working space and earthwork support are measured, together with the precision of the description, impact on the balance of risk in the item.

1.4.1   Non-UK standard methods of measurement

Many countries of the world have their own standard methods or codes of measurement. Some such as Australia and South Africa have more than one and the United Kingdom has several different methods of measurement in use at any one time for different purposes.

In a survey of 64 countries including the United Kingdom, the RICS (Building Cost Information Service, 2003) found that 14 of the 32 countries that responded had their own standard method or code of measurement and 5 countries used at least one UK standard method of measurement.

Clearly, in any book, it would be unrealistic to include coverage of all standard methods and codes of measurement, particularly as some standard methods of measurement used around the world, such as the Malaysian Standard Method of Measurement of Building Works and the Agreed Rules of Measurement (ARM4) in the Republic of Ireland, are, at least to some extent, modelled on UK standard methods and others bear a close resemblance to the Principles of Measurement (International) (POM(I)). In some cases, inclusion would be impractical due to the language barrier.

In Australia, the Australian Standard Method of Measurement (currently Fifth Edition), published by the AIQS, is widely used, but the Australian Standard AS 1181–1982 method of measurement of civil engineering works and associated building works is seemingly less popular.

New Zealand has a number of standards, Standard NZS4202 being a standard method of measurement of building works, but there is reportedly a trend away from using it. The New Zealand Standard NZS4224: 1983 Code of practice for measurement of civil engineering quantities is apparently not widely used. In the NZ building industry, bills of quantities are referred to as 'schedules of quantities'.

South Africa has a strong quantity surveying heritage, and the ASAQS publishes a Standard System of Measuring Building Work. This is not a national standard but it is very widely used.

In civil engineering, there are two national standards, SABS120 and SABS1200, for measurement of civil engineering works. These standards are very comprehensive compared with the Australian and NZ equivalents.

1.4.2   The Standard Method of Measurement of Building Works

The *Standard Method of Measurement of Building Works: Seventh Edition* is part of a long history of such documents stretching back to 1922 when the first edition was published. The aim of the SMM has always been to introduce uniformity into the production of bills of quantities which were, hitherto, prepared on an *ad hoc* basis variously by quantity surveyors engaged by the employer or jointly by contractors keen to reduce the costs of tendering.

Consequently, the early years of the twentieth century witnessed the two surveyors' bodies at the time – The Surveyors' Institution and the Quantity Surveyors' Association – cooperating in an endeavour to create a standard set of measurement rules in conjunction with representatives of the contractors' organisations (the NFBTE and the Institute of Builders).

The new SMM created a 'level playing field' for tenderers and ensured consistency in the preparation of bills of quantities which were previously conditioned by *local custom* and *the idiosyncrasies of individual surveyors*.

SMM7 was first published in 1988, with minor amendments being made in 1988, 1989 and 1992 as a result experience and use of the document in practice. Three further amendments were made in 2000 (amendments 1 and 2) and in 2009 (amendment 3), and this is why the latest version is sometimes referred to as SMM7C.

SMM7 heralded two further benefits of standardisation in the shape of two major changes from the previous SMM6:

- The adoption of a measurement system based on classification tables, as opposed to prose, which has the advantage of being more closely aligned with the use of standard phraseology and computerised measurement.
- The restructuring of the SMM into the Common Arrangement of Work Sections (CAWS) developed by the Co-ordinating Committee for Project Information (CCPI) in order to promote the standardisation and coordination of bills of quantities, drawings and specifications.

The benefits of coordinated project information propounded at the time SMM7 was introduced were that obvious omissions in the design would be picked up in the specification and billing of the work (e.g. builders' work in connection with mechanical and electrical services that might otherwise be overlooked). Anecdotal evidence suggests, however, that such benefits did not materialise in practice.

1.4.3   Common arrangement of work sections

Construction projects are complex entities that rely on lots of information of different sorts including product information and specifications, as well as the usual drawings and bills of quantities. This information needs to be classified so as to be available and accessible to a variety of users, just like *Yellow Pages*, and bill compilers need access to this information so that they can describe as well as measure work. This is the role of classification systems.

There is no one classification system in use in construction, and many countries have their own ways of specifying construction products and construction work.

Prior to the publication of SMM7, the CI/SfB system was in common use, and many architectural libraries still use it. However, a new classification system was introduced in 1987 by the CCPI which was formed as a result of a 1970s coordinated project information initiative. This led to the formation of the Construction Industry Project Information Committee (CPIC), which led in turn to the introduction of SMM7 in 1988.

As a consequence, the structure and layout of SMM7 was based on the Common Arrangement of Work Sections (CAWS), and it is CAWS which is responsible for the SMM7 alphanumeric numbering system. For instance, reinforcement for in situ concrete (reference E30) is to be found in Work Section E: *In situ concrete/Large precast concrete.*

UK construction industry classification systems have moved on since 1988, although the CAWS classification is still around. Uniclass, a new classification scheme for the construction industry, which was introduced in 1997 to implement BS ISO 12006-2 in the United Kingdom, incorporates the CAWS amongst its 15 classification tables – Table J (Work Sections for buildings).

However, due to deficiencies in its structuring and classification of information for BIM, Uniclass is not now considered fit for purpose and this has led to the development of Uniclass2, a fully searchable albeit not yet fully developed system.

1.4.4   RICS new rules of measurement

Since the 1 January 2013, SMM7 has been superseded by NRM2 and all indications are that this should now be used from this date on.

NRM2 has been issued as a guidance note, and as such, there is no compulsion to use it but not to do so may be considered to be contrary to good practice, and this may have implications as regards professional competence and professional negligence.

The JCT has issued an update to those of its 2011 contracts where measurement is relevant, including the SBC/Q, SBC/AQ and SBC/XQ, and this suggests that NRM2 *should be used instead of SMM7* for all relevant contracts entered into after the 1 January 2013. The RICS subscription online information portal, *isurv*,[5] also states that NRM2 has now replaced SMM7.

As part of the suite of documents that make up the RICS New Rules of Measurement, NRM1 was introduced a set of measurement rules for use during the design process. Its full title is *NRM1: Order of cost estimating and cost planning for capital building works*. It was first published in 2009 as NRM1: *Order of cost estimating and elemental cost planning* and is now in its second edition which became operative on 1 January 2013.

1.4.5   Civil Engineering Standard Method of Measurement

Standardised measurement of civil engineering work has a long history dating back to 1933 when the Institution of Civil Engineers published a report which provided, for the first time, a standard procedure for the drafting of bills of quantities for civil engineering work (Seeley and Murray, 2001).

This was revised in 1953 when the *Standard Method of Measurement of Civil Engineering Quantities* was published which, itself, was amended in 1963 and 1968 (metric version). In 1964, the ICE set up a committee to look into revising the existing standard method, and research work was initiated by the Construction Industry Research and Information Association (CIRIA). This culminated in Report No. 34 which was published in 1971 (Barnes and Thompson, 1971).

The work which led to the publication of the Civil Engineering Standard Method of Measurement (CESMM) was conducted for the ICE by Dr Martin Barnes under the guidance of a steering group of the Institution. Dr Barnes acknowledges the contribution of the CIRIA research, and the involvement of P. A. Thompson of the University of Manchester Institute of Science and Technology (UMIST), in the eventual successful launch of CESMM in 1976 (Barnes, 1977).

When the CESMM was first published, it represented a major shift away from the previous codes for measuring civil engineering works. The principal characteristics that made CESMM different included:

- A classification system for developing item descriptions.
- An item coding system to assist in sorting bills of quantities and for cost analysis.
- Greater standardisation of measured items and descriptions leading to more clarity and less need for interpretation of item descriptions.
- Greater pricing certainty and less confusion for tenderers.
- The recognition that method and time were determining factors in the pricing of civil engineering work and that these should be separated from the measured work items in the form of method-related charges so that construction method could be more accurately reflected in the contractor's prices.
- A method of measurement that enabled bills of quantities to be better used for programming, site management and post-contract control.

1.4.6   Method of Measurement for Highway Works

Since the dawn of formalised measurement in the early 1920s, a distinction has been made between the measurement rules for building work and those for civil engineering. Some would argue that this is a false distinction in the context of a single construction industry that uses transferable skills, resources and techniques, whilst others would make the point that building work is more detailed and intricate than civil engineering work and that there is therefore merit in the distinction.

Perhaps the wider adoption of BIM will force the industry to adopt a simpler and universal approach to measurement in the future, but in the meantime, the distinction prevails almost to the extent that building and civil engineering are seen as separate industries. The Method of Measurement for Highway Works (MMHW) makes the distinction between these different approaches to measurement even more pronounced in that it creates a distinction between the measurement of different types of work in the same industry sector – civil engineering.

The MMHW relates to a specific type of civil engineering work – major roadworks. This type of work involves earthworks, drainage, road pavements, structures – including bridges and associated geotechnical works – road lighting and communications installations, etc.

The origins of this method of measurement date back to 1971 with the introduction of the *Method of Measurement for Road and Bridge Works*. Prior to this, Hunter (1997) records for posterity that *Notes on the Preparation of the Bill of Quantities* were first published in 1951 along with the first edition of the *Specification for Road and Bridge Works*. In their third edition in 1963, the *Notes on the Preparation of the Bill of Quantities* were based on the *ICE Standard Method of Measurement of Civil Engineering Quantities* (1962).

The first edition of the MMHW, which, at the time, had the title *Method of Measurement for Road and Bridge Works*, coincided with the UK motorway boom of the 1970s but post-dated the first UK motorways; Preston Bypass and the first section of the M1 were opened in 1958 and 1959, respectively. The 1971, and subsequent, editions of the *Method of Measurement for Road and Bridge Works*, and the accompanying *Specification for Road and Bridge Works*, were printed in small handbooks that could fit in a 'duffle coat' pocket. The first and second editions of the *Method of Measurement for Road and Bridge Works* were grey and green, respectively. The various editions of the Specification were also distinctively coloured.

Despite the MMRB name change in 1987, the MMHW was accorded 'Third Edition' status. It was published in ring binders, as was the *Specification for Highway Works* which became known as the 'Brown Book' even though it comprised several booklets. The 'Brown Book' documents were published along with Notes for Guidance, a Library of Standard Item Descriptions and Highway Construction Details which were the forerunners of the current *Manual of Contract Documents for Highway Works* (MCHW), first published in 1991.

The current *MMHW*, notwithstanding amendments over the years, is the Fourth Edition.

The novel part of the *Method of Measurement for Road and Bridge Works*, and the current *MMHW*, is the idea that civil engineering work can be measured simply, with relatively few items, so long as the extent of the work covered by the items is clear and underpinned by an unequivocal specification of the work involved. The central feature of the method of measurement, therefore, is the 'item coverage' rules that accompany the measured work items in the bill of quantities, and these items depend heavily on the *Specification for Highway Works*; the SHW is the 'spine' of the *Manual of Contract Documents* which includes the *MMHW* and other documents.

1.4.7   Principles of Measurement (International)

In 2003, an RICS survey found that the most common standard method of measurement used internationally was POM(I), which was reportedly used in eight countries albeit that its use in the United Kingdom was reported to be 'minimal'. Of the countries where at least one standard method or code existed, 17 reported that their use was recommended or adopted by national bodies, and 15 said that such documents were regularly named in contract conditions. POM(I) was used for some years in the Republic of Ireland before the Agreed Rules of Measurement (ARM) were introduced.

In common with other standard methods of measurement in use internationally, such as the South African Standard System of Measuring Building Work (Sixth Edition) 1999, POM(I) is refreshingly simple and easy to follow. The measurement rules hark back to the earlier editions of the *Standard Method of Measurement of Building Work* and the *Standard Method of Measurement of Civil Engineering Quantities* in the United Kingdom in that there are no complex rules, no tables, no coverage rules and the like and there is no tendency to measure to the *n*th degree – thus following the precepts of the Pareto principle (see Chapter 3).

First published in 1979 by the RICS, and not revised since, POM(I) takes a significantly different approach compared to the standard methods commonly used nowadays in the United Kingdom. It is a model of simplicity such that the bill compiler is able to measure in a flexible way, building clear and comprehensive item descriptions but without the constraints of too many rules. Consequently, item descriptions can be developed within a recognisable framework whilst also allowing the skill and experience of the bill compiler to shine through.

To many contractors and subcontractors who are required to prepare bills of quantities for various reasons, POM(I) should be much more attractive than vaguely following a more complex method of measurement because a BQ can be prepared that complies to recognised rules that everyone can understand but also requires less time and resources to prepare the take-off.

1.4.8   Bills of quantities

The RICS Contracts in Use Survey (2011) reveals that the use of bills of quantities has been in decline for some years and that bills of quantities were used in only one in five contracts included in the survey due, in the main, to the growth in popularity of design and build procurement and the increased use of contracts based on drawings and specifications. The RICS Survey does not include the civil engineering sector wherein bills of quantities are still commonly used.

When drawing conclusions about the use of bills of quantities, a distinction needs to be made between bills of quantities prepared as a 'traditional' tender document (formal bills of quantities) and those prepared internally by contractors (informal bills of quantities) as a means of establishing the quantity of work in a project so that a price can be calculated for tendering purposes.

The key difference between the two is that:

- **Formal bills of quantities** are based on standardised measurement rules.
- **Informal bills of quantities** are based on the so-called builders' quantities determined by each of the individual tendering contractors.

Tenders based on builders' quantities introduce an aspect of risk and variability of tender prices that needs to be recognised, and quantities produced in this way have to be prepared in a considered, informed and consistent manner.

Thus, bills of quantities are not 'dead' but have been reinvented in a modern context for a variety of uses according to the procurement and contractual arrangements for any given project. In this regard, it is significant that the need for a more modern approach to measurement and bills of quantities has been recognised in the recent publication of the RICS NRM suite and the contract neutral CESMM4 documents.

1.5   Measurement: skill or art?

If construction measurement was simply concerned with determining lengths, areas and volumes, almost anyone could do it, but it isn't. The ability to measure to a professional standard is not easily acquired, and not all quantity surveyors, even, are competent in measurement or fully understand the subject.

Measurement demands two key skills:

1. The ability to measure construction work, in accordance with generally accepted quantity surveying practice and procedure and in accordance with a specific standard method of measurement.
2. The ability to produce formal bills of quantities with the aim of inviting tenders capable of being converted into a formal contract or subcontract.

To be able to measure to this standard requires an in-depth knowledge of technology and a thorough understanding of how buildings and engineering structures are designed and built. This is underpinned by the ability to set down dimensions in a professional way, thereby creating an audit trail that may be followed by others. Added to these skills is the need for a sound knowledge of procurement and contracts, experience of industry practices, knowledge of standard methods of measurement, reasoning ability and cognitive intuition.

Modern computer software packages have, to some extent, made measurement much less of a chore, but the fact remains that 'skill' is required to obtain accurate results.

The 'art' of measurement is in recognising what is missing from the information provided, whether drawing or model, and making sure that appropriate allowances are made in quantities taken and the documentation produced.

1.5.1   Uses of measurement

There is much more to measurement than how to 'take off' quantities from drawings using a standard method of measurement, and it is a subject that is no longer solely the province of the PQS engaged in producing traditional bills of quantities.

Measurement is used in a variety of ways by many participants in the construction process including:

- Quantity surveyors, engineers, building surveyors and others who are involved in the preparation of pre-contract documentation and post-contract control for main contracts and subcontracts.

- Quantity surveyors, working for contractors and subcontractors, who are involved with interim valuations and payment applications, the handling of subcontract accounts and remeasures and dealing with the internal cost-value reconciliation process.
- Site engineers, trades foremen and the like who are required to record and measure events on-site with a view to establishing entitlement to variations, extras and contractual claims.
- Contractors' planning engineers who need to use measurement when compiling the pre-tender programme, the master programme or short-term programmes, for updating programmes for progress so that projects may be controlled effectively and for determining the causes and effects of delay and disruption to the programme in order that claims for extensions of time and loss and expense may be pursued.
- Quantity surveyors and engineers who need to be able to check that main contractors' payment applications have been submitted in accordance with the contract documents generally and with the appropriate standard method of measurement in particular.
- Those involved in the submission and checking of applications for extensions of time and contractual money entitlements who may need to measure physical and financial progress and interpret this information in relation to the contract documents in general and the contract bills of quantities in particular.
- Those involved with making payment applications on measure and value contracts or where schedules are used or for valuing variations on design and build and other contracts.
- Subcontractors who are not quantity surveyors but nevertheless need to understand a particular method of measurement when pricing tenders, submitting payment applications and dealing with variations, daywork and the like.
- Students of quantity surveying, construction management, civil engineering and related courses who need an understanding of the principles and practice of measurement, an appreciation of various standard methods of measurement and a sound grasp of the role of measurement in construction contracts and its relationship with a variety of procurement methodologies, contractual arrangements and standard forms of contract.

It is important to recognise that measurement is used by different people in different ways and is an indispensable facet of many aspects of a construction project; indeed, this book has been specifically designed and written with this fact in mind.

1.5.2   Risk management

Prima facie 'measurement' is concerned with quantifying the work required to realise a proposed construction project with a view to obtaining an acceptable price from a contractor, which will then enable a civil contract to be drawn up so as to facilitate construction and completion of the project.

However, there is much more to the subject of 'measurement' than this:

- Firstly, each standard method of measurement has different measurement and item coverage rules, and this imposes a different balance of risk upon the contracting parties.
- Secondly, such risks will be viewed differently as between client/professional advisers, main contractor and subcontractors, and this will influence their respective attitudes at pre-contract, contract and post-contract stages.
- A third factor is that the extent of risk will be conditioned by the procurement option chosen for a specific project and by the conditions of contract employed for that project.
- The risks posed by the project in question will impact technically, financially and in project management terms, and such risks need to be managed appropriately by each of the participants in the construction process in turn.

Measurement has many uses and applications, and its centrality in the day-to-day management of the construction process is undervalued. There is an important link between methods of measurement, conditions of contract and procurement, and the link is **risk**.

This is a strong theme of this book.

1.5.3   The author's objectives

The objectives of writing this book were:

- To present the subject of measurement in a modern context with a risk management emphasis.
- To recognise the interrelationship of measurement with the complex web of procurement and contractual issues that characterise construction contracting.
- To emphasise the role of measurement in the entirety of the contracting process.
- To convey the basic principles of measurement and to put them in the modern context of on-screen measurement and BIM model quantity take-off.
- To incorporate consideration of five common methods of measurement currently used in construction.
- To widen the accessibility of measurement beyond the province of the professional quantity surveyor.
- To recognise the measurement risk issues facing quantity surveyors, contractors, subcontractors and others in the context of construction work when tendering, applying for payment, controlling cost and value and pursuing legal entitlement.

Much of the above has not been attempted in a measurement book before.

1.5.4   Rationale

The idea behind this book was to write something different and to try, as best as the author is able, to provide a text for practitioners who are able to measure but, perhaps, need a reference to confirm their instincts about measurement issues that are not dealt with in traditional measurement books.

There are many practitioners in the industry who are engaged in both building and civil engineering projects, or have to juggle more than one method of measurement on the same project, and this book is written with them in mind.

The book is structured in four Parts:

Part 1 – Measurement in construction establishes the connection between measurement and design with particular reference to the RIBA Plan of Work 2013 process model that depicts industry procedures. The importance of measurement in a BIM environment is emphasised, and quantity surveying conventions are explained in the context of measurement software applications.

Formal bills of quantities are recognised as only one of many outputs from the measurement process, and the measurement of construction work where no particular standard method of measurement is used is covered in detail.

This includes informal bills of quantities, activity schedules under the EEC and JCT standard forms of contract and schedules of rates and schedules of works for term maintenance and refurbishment contracts.

Part 2 – Measurement risk is a unique feature of the book as it provides a detailed examination of five current, commonly used, standard methods (or rules) of measurement – NRM1, NRM2, CESMM4, MMHW and POM(I).

The theme in Part 2 is the identification of risk, and each method of measurement is examined in detail, both critically and practically. Numerous practical examples of the use and application of each method of measurement are provided. Shortcomings and errors in the documents are highlighted, and there is a comprehensive guide to the transition from CESMM3 to CESMM4.

Part 3 – Measurement risk in contract control deals with measurement risk in the post-contract control of construction work.

This includes coverage of physical measurement and the function of measurement in the valuation of variations and work in progress. Measurement claims and final accounts are considered in the context of the JCT, ECC, ICC and FIDIC forms of contract and with respect to the role that measurement plays in these post-contract processes.

Part 4 – Measurement case studies is devoted to a series of worked examples relating to each of the standard methods of measurement covered in the book and also to the preparation of builders' quantities.

There was never any intention to cover every 'dot and comma' of the methods of measurement referred to in the book nor could it possibly include extensive worked examples of quantity take-off and dimension sheets as is traditional in measurement books. More important than 'how to measure this or that' are the risk issues arising from the methods of measurement we use and the link with the various procurement methods and conditions of contract that are commonly employed in the construction industry.

The aim is to provide sufficient insight and guidance, at a strategic measurement level, such that the reader will be more than capable of delving into whatever corner of any of these documents suits their purpose. If in some small way this helps the reader to feel confident enough to explore the wider issues arising from the measurement and valuation of building and civil engineering work, then the effort writing the book will have been worthwhile.

Notes

1. http://www.theguardian.com/money/2006/feb/25/careers.work1 (accessed 29 March 2015).
2. http://en.wikipedia.org/wiki/Measurement (accessed 29 March 2015).
3. http://en.wikipedia.org/wiki/Quantity_surveyor (accessed 29 March 2015).
4. http://www.bimtaskgroup.org/ (accessed 29 March 2015).
5. http://www.isurv.com/ (accessed 29 March 2015).

References

Ashworth, A., Hogg, K. and Higgs, C. (2013) *Willis's Practice and Procedure for the Quantity Surveyor*, Wiley-Blackwell, Oxford.

Barnes, M. (1977) *Measurement in Contract Control*, Thomas Telford, London.

Barnes, N.M.L. and Thompson, P.A. (1971) *Civil Engineering Bills of Quantities*, Report 34, Construction Industry Research and Information Association, London.

Building Cost Information Service (2003) *International Survey 2003: Standard Methods of Measurement in Current Use*, RICS Construction Faculty.

Crotty, R. (2012) *The Impact of Building Information Modelling: Transforming Construction*, Spon Press, London.

Davis Langdon (AECOM) (2011) *Contracts in Use, A Survey of Building Contracts in Use During* 2010, RICS, London.

Eastman, C., Teicholz, P., Sacks, R. and Liston, K. (2011) *BIM Handbook – A Guide to Building Information Modelling*, Second Edition, Wiley, Hoboken.

Egan, J. (1998) *Rethinking Construction*, Construction Task Force, Department of Trade and Industry, London

Higgin, G. and Jessop, N. (1965) *Communications in the Building Industry: The Report of a Pilot Study*, Tavistock, London.

Hoezen, M.E.L. (2006) *The Problem of Communication in Construction*, University of Twente, Twente.

Hunter, R.N. (1997) *Claims on Highway Contracts*, Thomas Telford, London.

Latham, M., Sir (1994) *Constructing the Team*, HMSO, London.

Lee, S., William, T., Willis, A. (2014) *Willis's Elements of Quantity Surveying*, 12th Edition, Wiley-Blackwell, Oxford.

McDonnell, F. (2010) the relevance of teaching traditional measurement techniques to undergraduate quantity surveying students [paper]. *Journal for Education in the Built Environment*.

Murray, M. and Langford, D. (eds) (2003) *Construction Reports, 1944–98*. Blackwell Science, Oxford.

Radosavljevic, M. and Bennett, J. (2012) *Construction Management Strategies: A Theory of Construction Management*, Wiley-Blackwell, Oxford.

Seeley, I.H. and Murray, G.P. (2001) *Civil Engineering Quantities*, 6th Edition, Palgrave Macmillan, New York.

Seeley, I.H. and Winfield, R. (2009) *Building Quantities Explained*, Palgrave Macmillan, Basingstoke.

Wolstenholme, A. (2009) *Never Waste a Good Crisis: A review of progress since Rethinking Construction and Thoughts for our Future*, Constructing Excellence.

Chapter 2
Measurement and Design

For which of you, intending to build a tower, sitteth not down first, and counteth the cost, whether he have sufficient to finish it?

Luke 14.28

The Holy Bible: King James Version (1769)

Suppose one of you wants to build a tower. He will first sit down and estimate the cost to see whether he has enough money to finish it, won't he?

Luke 14.28

The Holy Bible: International Standard Version

This oft misquoted parable is allegorical, of course, and has nothing to do with building towers, but it is illustrative of the long association between design and measurement in the cost management of construction.

The need to control the cost of building and infrastructure projects is not likely to diminish in the future, but in the digital age that the construction industry is beginning to embrace, the system of doing so will undoubtedly be revolutionised.

In the United Kingdom, for example, the Government Construction Strategy (2011) makes a clear statement that efficiency, innovation, cost reduction, value for money and new procurement methods are prerequisites when it comes to spending public money and that the industry must provide integrated solutions to required outcomes and do so in a collaborative rather than adversarial manner.

This strategy document challenges established industry cultures, business models and practices and demands that cost reduction and innovation shall be achieved by focussing on the supply chain rather than on the bidding process. The UK government has also challenged the industry by mandating a minimum requirement for Level 2 BIM on all public sector projects by 2016.

Managing Measurement Risk in Building and Civil Engineering, First Edition. Peter Williams.
© 2016 John Wiley & Sons, Ltd. Published 2016 by John Wiley & Sons, Ltd.

2.1 Introduction

History provides many well-documented examples of the formal quantification of materials and labour as a means of establishing the cost of building medieval castles and the like.

In 1378, for example, a contract was entered into to carry out building works at Bolton Castle in Yorkshire whereby *John Lewyn agrees to build for Sir Richard Scrape, at Bolton:*

The design	Comments
• *A kitchen tower, 10 ells by 8 ells and 50 feet high, with walls 2 ells thick* • *Between this and the gate, a vaulted room with three storeys over it, 12 ells by 5½ ells and 40 feet high, the outer walls being 2 ells and the inner 4 feet thick* • *A tower 50 feet high, containing a gateway, with three storeys over it* • *In this tower, south of the gate, a four-storey building* • *West of the gate, a vaulted room with another over it and a chamber above, 10 ells by 5½ ells and 40 feet in height* • *All the rooms having doors, windows, fireplaces, privies, etc.* • *Also a vice in the kitchen tower and two in the gate tower* • *All inner walls to be 3 or 4 feet thick*	• No *form* of construction is mentioned so this must have been well understood and fairly prescriptive in the fourteenth century • *Various* units of measurements are used • *Ells* to delineate areas • *Feet* for wall thicknesses and heights

The contract (or Works Information if the ECC had been used!)

• *Sir Richard shall provide carriage for the materials timber for scaffolding and wood for the limekiln*	• Risk undertaken by the party best able to bear it
• *For all this work, John Lewyn shall have 50 marks and 100 s. for every perch, of 20 feet, of construction* (Salzman, 1968)	• Clear payment terms stated • Interim payment determined by measuring progress to date • The dictionary understanding of 'perch' as a unit of measurement approximates to 16½ feet and not 20 feet as stated in the payment terms

Notes:
1 ell = 1.25 yards = 1.14 m.
A perch (also called a pole or rod) was a measure of length, especially for land, equal to a quarter of a chain or 5½ yards (~5.029 m) (see Oxford Dictionary).

2.2 Design

2.2.1 Design process

Traditionally, design develops from an initial concept through to final production drawings and the ability to take measurements follows this development. Process models illustrate how this development progresses over time, the best known of which is the RIBA Plan of Work.

There are other models, the OGC Gateway and Construction Industry Council models, for example, but the RIBA model depicts not only the design and construction process but also the contributions made to the process by various participants and the exchange of information between them.

By understanding the design process, it is easier to understand where measurement fits into the picture and how models that seek to measure or quantify the design develop as the design develops. Different procurement methods dictate when, in the design process, measurement takes place and who performs the measurement function. In some cases, contractors undertake measurement tasks that were once the sole province of the professional quantity surveyor.

2.2.2   RIBA Plan of Work

The RIBA Plan of Work was first published in 1964 and has been held by many as the exemplar of standard practice over the ensuing 50 years.

In more recent times, the RIBA Plan of Work 'process model' has undergone a number of revisions. It originally comprised the RIBA Outline Plan of Work and the Work Stage Procedures which expanded the model in great detail. The RIBA Plan of Work was substantially reworked in 1998, and the 2007 edition was amended in 2008. In order to help the industry to accommodate the latest in Building Information Modelling (BIM) thinking, the 2007 edition (revised 2008) had a BIM Overlay and an additional stage (Stage M), concerned with Model Maintenance and Development.

Following a period of consultation, the RIBA published the RIBA Plan of Work 2013 (RIBA, 2013), which is substantially different from the outgoing model. The 2013 version has redefined the project stages and emphasised the role of procurement, programme and information exchange in the process model.

The RIBA Plan of Work 2013 stresses that procurement strategy does not fundamentally alter the progression of the design but that the information exchange points will vary according to the method of procurement chosen. Now that the Plan of Work is available digitally, bespoke models can be created for individual practices and individual projects so that the model more closely resembles the specific work in hand.

Figure 2.1 illustrates a two-stage tender process, where the preferred bidder is appointed during Work Stage 3, who then contributes to the development of the cost plan; this culminates in a bill of quantities being measured, priced and agreed prior to the contractor's appointment for the construction stage.

2.2.3   Design intent

In the days when buildings were constructed with simple 'natural' materials (stone, timber, slate, etc.), it was easy for architects to convey what they wanted to the builder. The architect understood the materials and the construction process, and the builder knew the materials and the construction process intimately, and enough about design, to understand what the architect wanted (Crotty, 2012).

In modern times, life is more complex. Buildings are largely constructed from manufactured materials, components and specialist equipment and, therefore, the job of the designer is much more conceptual. Great reliance is placed by designers on communicating the design to the builder via technical literature, such as product catalogues and data sheets, and the builder, or specialist installer, interprets this information in the context of the designer's drawings.

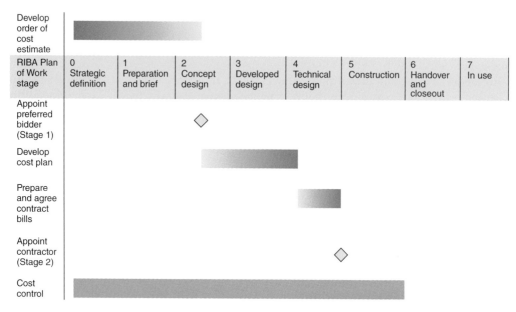

Figure 2.1   Cost control: two-stage tender.

This is not to imply that designs are not developed to a technical level of detail, far from it, but even at Work Stage 4 (Technical Design), the design is incomplete. The design may even be so incomplete that elements of the design are passed over to the contractor in a partial contractor design arrangement. This is common practice now.

As a result of all of these, the design communicated to the contractor is complete enough to be built from, but imprecise in terms of how this should be done, and building components are specified but not detailed to sufficiently accurate tolerances and specifications to be manufactured and fabricated. 2D CAD, for instance, is capable of conveying a design but, beyond the lines and circles on the drawings, the geometry of the design does not hold sufficient design information for construction purposes.

What the builder receives, therefore, is the 'design intent'.

In fact, the RIBA Plan of Work for Employer's Requirements (design and build) specifically recognises 'design intent' as an integral part of the make-up of the Employer's Requirement document.

There may be some purposefulness, or 'intended ambiguity', in a completed design which leaves the final construction decision-making to the contractor and specialist contractors. This might be to limit liability should something go wrong during construction or may be a way to tap into the collective knowledge of builders, specialists, manufacturers and tradespeople by involving them in the design process.

Examples of design intent:

- Where the standard of materials and workmanship are to be to the satisfaction of the architect, they are often stated in the contract to be to his *reasonable satisfaction.*
- In a performance specification, where the designer describes the effect that the contractor is required to achieve (such as xm^3 of air flow per minute), but not the ways and means by which this is to be achieved.
- Where a structural engineering design for roll-formed roof sheeting or engineered building products, such as prestressed concrete units, is stated in the contract as being expressed in detailed shop drawings and method statements to be provided by a specialist contractor.

All of these are examples of intended ambiguity.

Where there is purposeful ambiguity in a design, this influences the design information available (whether analogue or digital) and hence influences what is to be measured, how it is to be measured and what risk is attached to the measurements.

Consequently, the architect may conceptualise a complex roof but leave the design of that element to the contractor. This might be done through a partial contractor design arrangement whereby the tenderer prices a single bill of quantities item inclusive of the design of that item. This way of billing the item means that the tenderer has to design the roof, measure the quantities, price the work and undertake responsibility for the design.

The process of conveying a 'design intent' for the roof effectively transfers risk to the contractor whose design liability, unless qualified in the contract, is the 'fitness for purpose' liability of a contractor–designer and not the lesser standard of an architect–designer.

2.2.4 Design cost control

The problem with the biblical definition of cost control and, indeed, the historic cost accounting methods employed since medieval times is the lack of a procedure to ensure that the out-turn cost of a project is equal to or less than the amount that can be afforded by the client. The ability to do this with precision requires a system that controls cost as the design develops and, at the same time, enables the designer to make informed decisions that do not compromise the design concept.

The development of **elemental cost planning** – part of a system of cost control – enabled this to happen.

The concept was introduced in 1951 by James Nisbet (1951) and has been developed and refined over the intervening years both by private quantity surveying practices and by the Building Cost Information Service (BCIS) of the RICS.

The basic idea was to establish a budget, or cost limit, and then decide how to spend the money. This introduced a different concept of measuring buildings than hitherto.

In the initial stages of design, measurements are expressed in m² of gross internal floor area (GIFA) or, alternatively, as simple units of occupancy, such as hospital beds, theatre seats and school places. As the design develops, more complex measurements are produced, this time expressed as elements and sub-elements, but still with a unit of measurement of m² of GIFA.

More recently, the RICS initiative to introduce the *New Rules of Measurement* (NRM1, 2012) has led to a rethink about the process of design cost control such that budgets or cost limits are now referred to as *order of cost estimates* and cost plans are measured in more detail as shown in Table 2.1.

The BCIS 'Elemental Standard Form of Cost Analysis (NRM Edition)' 2012 has been developed by BCIS to work specifically with NRM1, and a 'component' level has been added to reflect the same *user-defined* component level in NRM1. It can only be supposed that this has been done to line up with the 'components' that make up BIM models.

Table 2.1 NRM1 elements.

Ref.	Group element	Ref.	Element	Ref.	Sub-element	Ref.	Component
2	Superstructure	2.5	External walls	2.5.3	Solar/rain screening	2.5.3.1	Overcladding

2.2.5   Measurement

Measurement and design are closely connected.

Measurement is a by-product of the design process, and the stage to which the design has been developed determines what measurements can be taken both in extent and in detail. In the early stages of design, little information is available from which to measure, but as the design develops, more becomes known about the designer's intentions, representations of the design are produced in the form of sketches, drawings and/or models and consequently more detailed measurements can be taken.

Measurement is inextricably linked to the work of designers. For instance, the volume of concrete in a reinforced concrete column cannot be determined without:

- Firstly, identifying the cross-sectional dimensions from a plan.
- Secondly, finding the height of the column by reference to a cross section of the building.

Not only does this process require the use of more than one drawing, but it may also require human judgement and interpretation.

If, for example, there are no dimensions on the drawings, they must be:

- Calculated from other dimensions.
- Measured using a traditional scale rule.
- Measured using one of several types of electronic digitiser.
- Measured from a true scale drawing on a computer screen.

This effort is required because the lines on the drawings are 'dumb' and do not contain information, so their length must be determined from written dimensions or by scaling or otherwise measuring them physically, usually to a predetermined scale.

It is only at the level of BIM-based 3D models where smart digital information is available such that little or no human interpretation is required in order to establish dimensions for measurement purposes.

2.3   BIM

We live in a world of acronyms – SMM, NRM, CESMM, MMHW, POM(I), PDF, CAD, etc. – and now we have BIM! This stands for Building Information Modelling, albeit there is no universal acceptance of this.

Unlike other acronyms, BIM offers a new world of amazing information technology that will (eventually) transform construction, as it has in other industries (such as aerospace, auto manufacturing, etc.) and measurement.

There is no question that BIM presents enormous challenges and opportunities for the construction industry and that the industry is facing a period of major change in its working practices over the next decade.

There is a great deal of mystique surrounding BIM, however. What is it? How does it work? Who is using it? How will it develop? What are the implications for measurement and quantity surveying?

Adding to the mystique is the technical jargon employed by those involved with BIM which, to most non-IT people, is confusing at best and alien and something of a 'turn-off' at worst.

In fact, it takes a fair amount of 'delving', research and surfing to really get to grips with BIM, to appreciate what it means for the construction industry and to understand the answers to the questions posed at a construction practitioner, as opposed to a 'computer-buff' level. Thankfully, there are some excellent and up-to-date reference books available on the subject but none of these focus to any great extent on measurement or, in BIMspeak, 'quantity take-off'.

2.3.1   Definitions

There are a number of definitions of 'BIM' on the Internet and in the books referenced to in this chapter. Whilst researching this book, however, the author looked high and low for a clear, concise and understandable definition of BIM. Typical, though by no means the most unintelligible, of those available is:

> [BIM] is a *digital representation of physical and functional characteristics of a facility creating a shared knowledge resource for information about it and forming a reliable basis for decisions during its life cycle, from earliest conception to demolition*[1].

To a non-designer and non-IT anorak, this definition has little meaning. Searching further, one definition stood out. It came from a local architect in Liverpool:

> *BIM is basically two things: Building an intelligent 3D model and team collaboration.*

This architect clearly understood what BIM meant at a practical level. He appreciated that BIM can create a design that readily exchanges data between contributors to the design and construction process and the client and also that, just like drawings, a BIM model is only as good as those who contribute to it and to the information upon which the model is based.

In the discussion, we did discuss the accuracy of the quantities in a BIM model, as it is open to question as to who inputs the quantities into the components in the model and how accurate they are. The impression seemed to be that the quantities are very accurate but whether this is at a QS level of accuracy remains to be seen. QSs tend to measure on centre lines and girths but it is unlikely that software people will do the same thing. One architect suggested, 'Get the QS to do a traditional take-off as a check against the BIM quants'. Good idea!

On the face of it, 'BIM' stands for Building Information Modelling; however, Race (2012), amongst others, suggests that it could equally mean Building Information Management. This view is quite persuasive because BIM models are built on 'information', and it is this information that creates the basis of a common language which can be used to exchange data, and facilitate communication, between all participants in a project that has never been seen before.

Whichever definition of BIM is preferred, the 'building' bit of the acronym is somewhat misleading because BIM models are not exclusive to building projects. They can and are created for civil engineering projects or, indeed, any construction project envisaged. The only limitation is the existence of suitable 'components' which are the building blocks of all current BIM packages.[2]

In the United States, a variety of terms and acronyms have been devised to distinguish between vertical construction (i.e. building) and horizontal construction (e.g. infrastructure) including Civil BIM, CIM, BIM for infrastructure, Heavy BIM, etc.

Whatever the label, however, the result is the same: intelligent, data-rich models in three or more dimensions that greatly enhance the quality of design, construction, collaboration and communication in the industry (The Business Value of BIM for Infrastructure, McGraw-Hill Construction SmartMarket Report, 2012).

2.3.2   BIM benefits

Once you have seen a BIM model, you will quickly appreciate the benefits to be gained from their use. Amongst the many are:

- Visualisation of the structure is 'lifelike'.
- Walking through the model is like 'being here'.

- Layers can be removed and replaced at will.
- Design becomes faster, highly customisable and flexible.
- Optimisation of schedule and cost.
- Coordination and collaboration is seamless.
- Design conflicts can be readily detected and risk mitigated.
- Easy maintenance of Building Life Cycle.

David Scott[3] believes that the benefits of BIM can be multiplied by earlier contractor engagement on a project.

Put simply, he argues, *if a project can be built virtually in a BIM world, it is possible to build it in the real world, offering guarantees on price and programme much earlier by significantly managing out risk.*

2.3.3   BIM levels

It is a little confusing, but necessary to appreciate, that BIM is not just about intelligent models. There are steps, or levels, that lead to 'full' BIM – the BIM that is slowly emerging in the industry but is not universally adopted as yet.

At this point, most articles about BIM leap to the Bew–Richards BIM Maturity Wedge[4] to visualise what the BIM levels are. Frankly, as a non-designer, the wedge 'doesn't do it' for this author and a simpler approach is preferred, as illustrated in Table 2.2.

Table 2.2   BIM levels.

Level 0	• Two-dimensional computer-aided design (2D CAD) • Replaced the old drawing board and *Rotring Rapidograph* pens	Analogue design process
Level 1	• 2D CAD with 3D conceptual models for visualising the finished product • Used by a single designer or trade contractor and thus referred to as lonely BIM • Design liability clearly with the designer • Traditional contractual and insurance arrangements	
Level 2	• 3D models produced by all key contributors of the design team • Not necessarily a single model • May eventually result in a combined (or 'federated') model • Design liability and contractual and insurance implications start to become 'blurred' • Project roles and design responsibilities need to be clarified • Design outputs (information exchanges) need to be carefully defined • Greater collaboration between the design team and designing contractors	The boundary between analogue and digital designs
Level 3	• Full digital design within a federated model • An intelligent model enabling rich data exchange • A fully collaborative design and construction process • Complex design liability and contractual and insurance issues to resolve	Digital design

2.3.4   BIM awareness

In the National BIM Report (NBS, 2012), an online survey of a variety of construction disciplines and businesses found that 21% of the 1000 respondents were neither aware of nor using BIM. A further 48% reported that they were 'just aware' of BIM, and the balance (31%) said that they were aware of and currently using BIM.

In the same report, David Philp wrote that *2011, with its cornucopia of BIM seminars, workshops and related articles, helped trigger a 'light bulb moment' for many in the UK construction industry.* This doesn't quite square with Richard Waterhouse's Introduction to the 2012 NBS Report who, whilst acknowledging the twofold increase in BIM awareness on the previous year, also made it clear that a certain degree of inertia exists in the industry despite the launch of the UK government's BIM strategy in 2011.

To those who know and understand how the construction industry works, the results of the NBS survey will not come as much of a surprise.

Construction is notoriously reactionary and Crotty (2012) is not expecting BIM to lead to the same sort of dramatic change in construction as has happened in the manufacturing industries over the last 20 years.

Based on no evidence whatsoever – simply personal instinct – it is more than likely that the NBS survey is a pretty accurate reflection of the state of affairs in the industry and that BIM is probably something of a 'black box' to many people in construction. Considering the 'hype' surrounding BIM, the confusing technical jargon used by BIM evangelists and the lack of clear and simple explanations as to what BIM is, it is no wonder that the industry has been somewhat slow to take BIM on board.

As part of the research for this book, some informal interviews were held with local architects – not huge practices – in order to learn more about BIM and try to understand how the world of architecture was dealing with this complex subject. All were 'aware' and some were using BIM, but only up to a point. Listed below are some of the observations received:

- Some practices were at the interface between Levels 1 and 2.
- Others could not justify the expense of the leap from Level 2 to Level 3.
- Some practices used BIM, but only up to planning permission stage as clients were not prepared to pay the fees for full BIM design.
- For smallish projects, the architect, as lead consultant, could/would be the project manager for the model, but for larger jobs, there would need to be a project manager who would be in charge of the model.
- Ideally, a central model would be available in cyberspace (the Cloud), but this needs a fast fibre-optic Internet connection which is not always available to all. The other option is to provide a copy of the model for others to work on. This raises an issue of syncing where changes in the copy model have to be integrated into the central model which could then detect clashes and other problems.
- Some saw problems with the legal aspects of federated models and the 'questionable' quality of input from other contributors.
- There may be limitations in existing software such as Revit (Autodesk) because model properties do not necessarily exhibit UK standards such as Building Regulations and thermal insulation standards.
- It might be difficult to measure a stepped DPC; this would be shown on the model as a single line but the detailed changes of direction would not appear as the model would become too complicated and require too much computing power.
- It was suggested that interfaces may pose a problem from a measurement point of view, for example, wall to roof at eaves, as they may not be adequately detailed in the model.

Part 1

2.3.5   Glossary of terms

BIM has a vocabulary, if not a language, of its own, much of which has its origins in the United States. To English language 'purists', this presents yet another example of the 'Americanisation' of the 'mother tongue', but to pragmatists, this is a horse that bolted a long time ago!

Some words, phrases and acronyms are alien to the UK construction industry, and others are simply an offence to the English language (e.g. interoperability)!

BIM component

Components are the 'objects' that build the model (e.g. precast concrete wall panels, acoustic panels). They are 'parametric', being members of the same 'family' (e.g. Industry Foundation Classes (IFC)), but might have variable properties.

The components are 'smart' in that each component in the model may have all sorts of structural, thermal, sonic and physical property attributes, such as mass, dimensions, areas and volumes.

The designer chooses the desired components and drops them into the model which then takes shape and form.

IFC

IFC is a class, or family, of objects or components for BIM models that is not controlled by any particular software vendor.

IFCs have been developed by buildingSMART[5] specifically to ease the interoperability problems between software platforms. They are 'platform neutral' so it is not necessary to buy a specific piece of software to use the components in the model. Any suitable software will do the job.

Interoperability

This is where one piece of software doesn't speak the same language as another.

If one tries to open a Revit drawing in Microsoft Word, it doesn't work. In some QS software, Revit and PDF files won't open, and another piece of software (a plug-in) is needed to open and read the drawing.

Rendering

BIM models can be produced with different visual effects, and rendering is the process of turning a visually uninteresting model into something rich and lifelike.

Think of adding colour and shade to a normal paper drawing – it's similar in effect but much more impressive in the models which, when fully rendered, have astounding visual appeal and reality.

3D BIM

This is a model of a building or structure that has all the hallmarks of the 'real thing'. The model can be viewed from any angle, components can be turned off and on again, and 'walk-throughs' can be performed. Each component in the model has all sorts of attributes attached to it, including quantities.

Note the phrase 'each component in the model'. Models are like drawings. If it hasn't been designed, it's not on the drawings and it's the same with models. There is no point spending lots of time and money developing a model to the nth degree of detail for a planning application.

Models develop over time – from concept to technical design – but if something is not in the model, there will be no quantities available. Even when the model is well advanced, there may be aspects of the design that haven't been included in the model – just like the paper-based drawing. The QS needs to be as intuitive and inventive as ever!

4D BIM

This is a module or bolt-on to project management software, such as Asta Powerproject.

A BIM model is uploaded to the Cloud, and this can then be accessed by the software. The model is based on IFC components. The software draws IFC data from the Cloud, and the project activities in Powerproject can then be linked to the components in the model. When the model develops more detail or is revised, the changes can be seen in the Cloud and the programme revised accordingly.

5D BIM

This is to do with quantities, where a BIM model is accessed to extract quantitative data.

In some instances, it is necessary to have the architectural software (e.g. Revit) to access the quantities, but if the QS software used has a BIM reader that can import and read a BIM model, the quantities can be extracted within the software.

There is a bit of 'smoke and mirrors' here because, whilst the BIM reader can select quantities from the model and return them to the QS software, or to a spreadsheet, it isn't possible to import the quantities into an SMM-based electronic dim sheet just yet.

2.3.6   BIM systems

There are numerous individual BIM systems which can be used for a variety of purposes including architectural design, structural design and the design of engineering services such as heating and ventilating, fire protection and air conditioning.

An individual BIM system enables the designer to model the design in virtual reality using 'intelligent' components which, Crotty (2012) explains, are *exactly analogous to building components in the physical world*. Consequently, the designer can 'build' the design just as it would be built on-site in real life. Such models also have the capacity to be viewed from all sorts of perspectives, in many different ways and by showing or hiding features depending on the complexity or clarity desired.

Race (2012), however, suggests that the idea of a single, fully interactive and responsive BIM platform does not match reality. It is tempting to suppose that a BIM model will allow changes to be made to the design which are then reflected in all parts of the model. For example, a change in the floor–ceiling height not only will be reflected in the architectural design but also will be seen in the structural design, MEP, scheduling and quantity take-off. Race informs us that no such platform exists at the present time.

From a measurement point of view, the capability to hide all details, except the items to be measured, allows much greater clarity and accuracy when it comes to taking-off quantities. 2D and 3D CAD packages have the same capability, but the significant difference with BIM models is the richness of the information contained within the model, generated by its intelligent components. 2D and 3D CAD do not possess this capability.

The possibility that this technology can create 'fault-free' designs that can be built on-site without recourse to the usual circus of Requests for Information (RFIs), Architect's Instructions (AIs) and loss and expense claims is truly beguiling.

Crotty (2012) reports that there were only 10 *structure-related RFIs* on the £20 million Norwich Open Academy project and *no field RFIs* on the £20 million Llanelli Scarlets Rugby Stadium project, both of which were 'BIM projects'.

However convincing these examples are, reality persuades us that construction involves much activity that is not 'risk-free', such as excavating below ground level, tunnelling, installation of temporary works and so on. It is, therefore, unlikely that construction projects are capable of being completed on the basis of 'perfect' information, and without any variations to the design, no matter the extent of the intelligence contained within BIM models.

2.4   BIM quantities

As two thirds of vendors at 2013 BIM Show Live were dealing with measurement/surveying, it is tempting to imagine that a BQ can be produced at the push of a button with BIM. However, full automation is not possible with BIM models just yet, and according to Eastman et al. (2011), there is no BIM tool available that will prepare a quantity take-off suitable for preparing accurate estimates of construction work.

Additionally, quantity surveyors have to interpret BIM models just as they have to interpret drawings. If an item is not included in the model, but is clearly to be quantified, the QS will have to find another way of including relevant quantities in the BQ or schedule.

Consequently, professional insight and judgement is needed when producing quantities from models, and the traditional skill set and interpretative abilities of the QS are still required as they always have been. There is no less requirement for the QS to understand the technology of construction just because a model is available, and collaboration doesn't mean that there is no need to pick up the telephone to talk to designers.

2.4.1   Limitations

The quantities produced by models are not calibrated to methods of measurement and, therefore, whilst considerable measurement effort is saved, the BIM quantities still have to be manipulated into recognisable units and descriptions. BIM models can generate their own plans, elevations and sections, which is just as well because there is considerable need for 2D data when extracting quantities for measurement. Also, some items are not shown in models, just as they were not shown on the drawings:

- Off-site disposal.
- Earthwork support.
- Formwork.
- Reinforcement is often not modelled, but it can be.
- Subcontract attendances.

Benge (2014) suggests that *BIM is intended to address issues of process management and data retention* and that *NRM1 is linked to this, enabling the consistent collection of construction cost data that is synchronised with the design data*.

This seems to be somewhat at odds with the findings of a recent RICS Research Report which indicates that the information generated by BIM models is out of step with the NRM1 standard for order of cost estimating and elemental cost planning (RICS, 2014).

2.4.2   A word of caution

The ability to take quantitative data directly from a BIM model is an attractive proposition but the old adage must be remembered – G.I.G.O. ('Garbage In–Garbage Out').

Quantity surveyors are trained to pay attention to detail and, by their nature, tend to be careful people. Any self-respecting QS will wish to be sure where quantity-based information comes from, and most professionals will tell you to always use figured dimensions, or dimensions calculated from figured dimensions, rather than relying on scaling.

Quantities abstracted from a BIM model need to be treated with caution because the quantity surveyor cannot verify the source of the information or who has generated it. The data will have been entered in the model by someone and that someone will be a designer or an IT person and not a quantity surveyor.

Professional quantity surveyors are legally liable for the accuracy of their work, and quantities taken from BIM models should not be used out of hand without running some checks and balances to be sure that the figures 'stack up'.

2.4.3   On-screen measurement

Measurement and billing of construction work using on-screen measurement software is now fairly commonplace in the industry, and both SMM-based packages and non-SMM-based packages allow PDF and 2D and 3D CAD files to be imported into an on-screen measurement environment. Some other software packages use plug-ins that facilitate quantity take-off on-screen from imported PDF or CAD files which can then be exported to the bill production software.

In some ways, on-screen measurement software has transformed the way in which construction work is quantified, and it has certainly made the process much quicker and cost-effective both for QS practices and on the contracting side of the industry.

However, the step-up from CAD measurement to full BIM measurement is a significant one and requires a different mind-set compared with both traditional quantity take-off and BQ production and on-screen measurement from CAD files. Quantity take-off using BIM models is conceptually quite different to that from traditional or even on-screen measurement for a number of reasons:

- It is the model that generates the quantity data as opposed to the software when using on-screen measurement.
- Quantities from models are not necessarily in SMM units of measurement.
- There are no SMM distinctions (e.g. classifications of concrete thicknesses).
- Quantities generated by the model must be exported into suitable software in order to create bills of quantities or schedules.
- The output from CAD files is units of finished work whether SMM based or not.
- The output from BIM models is object based.

The difference is significant because BIM models consist of 'intelligent' objects that carry lots of data, including measurement data, whereas CAD files do not. As a consequence of this difference, quantity data is extracted from the model and then used, manipulated or imported into a suitable software package, whereas quantity data from CAD files has to be generated by the measurer within a software package that supports the importation of, and measurement from, CAD files. This is the case whether or not the software is SMM based.

2.4.4   Software issues

Some software packages purport to be BIM-measurement packages but this is something of a fallacy because there are no software packages available yet that can measure from full BIM models. Claims of BIM-measurement capability are not intended to be misleading, however, as there are different levels of BIM, and both 2D and 3D CAD are staging posts on the way to full BIM as illustrated in the Bew–Richards Maturity Wedge diagram.

Using quantity take-off data generated from a model may be an alien concept to some quantity surveyors, who are more used to generating the quantities themselves, either from drawings or from CAD files on-screen. Quantity surveyors may also feel a degree of scepticism regarding the reliability of quantities generated from a model because:

- The quantities produced by the model cannot be controlled by the measurer/estimator.
- The model may not generate all the quantities that the measurer/estimator needs (e.g. the quantity of concrete may be generated but not formwork or rebar).
- The measurer/estimator does not know how the quantities have been measured (e.g. does the model measure on the centre line as a QS would?).
- Objects within the model display their properties, including their dimensions, but do changes in the model generate an audit trail of change as would a change in quantity taken from a drawing or CAD file?

Having established that the BIM tool generates quantities from the model, the question arises as to how to overcome the lack of capability within the BIM tool to use and manipulate the quantity take-off for producing bills of quantities or schedules for pricing. Eastman et al. (2011) suggest three viable options:

1. Export the quantities from the BIM tool into a suitable software package (e.g. MS Excel spreadsheet).
2. Link the BIM tool with a suitable software package via a plug-in (e.g. Autodesk Quantity Take-off with Revit).
3. Import the model from the BIM tool into a specialised measurement package (e.g. Causeway BIMMeasure or Buildsoft Cubit).

The disadvantage of options 1 and 2 is that suitable software, or a plug-in, must be acquired as well as a copy of the BIM tool or software used to create the model. This may not be cost-effective in some instances. Consequently, option 3, using a software package such as Causeway BIMMeasure or Buildsoft Cubit, might be more attractive, as each allows models to be imported into the package for quantity take-off purposes.

2.4.5   Example

Figure 2.2 illustrates a model imported into Causeway BIMMeasure together with the various windows used for extracting and presenting quantities. A short and simple explanation of how the software works is provided at http://www.youtube.com/watch?v=ur2jcCPw6Ag by CATO's Tim Cook, whose YouTube channel is well worth visiting.

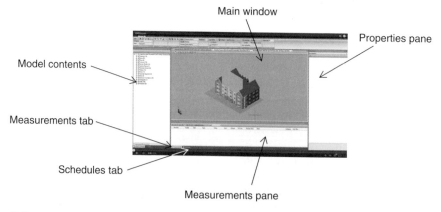

Figure 2.2   Causeway BIMMeasure (1).

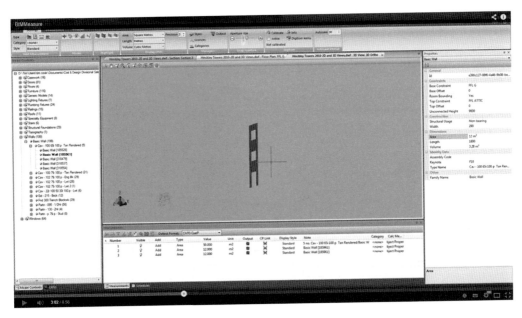

Figure 2.3   Causeway BIMMeasure (2).

In this demonstration, Tim explains four ways of producing quantities from the model:

1. Drag and drop the quantities from the list of Model Contents into the Measurements pane of the BIMMeasure window.
2. Highlight a model object in the main window and drag and drop its quantities into the Measurements pane.
3. Drag and drop the quantities for the object from the Properties pane of the BIMMeasure window.
4. Schedule the object(s) into a list by highlighting the object(s) in the main window and clicking the Schedules tab at the bottom left of the main window.

This is illustrated in Figure 2.3 which shows a model object and the selection cursor as well as object properties and measurement details.

2.4.6   Using 5D BIM

Causeway BIMMeasure allows quantities to be exported to MS Excel or they may be exported to Causeway's CATO suite. If an elemental cost plan is required, the BIM quantity take-off could be exported to the CATO Cost Planning tool.

Buildsoft Cubit, amongst whose many features are on-screen measurement and billing, also has a BIM reader in which models can be opened and quantities extracted. These quantities can then be exported from the BIM reader into Buildsoft Cubit in order to produce a bill of quantities, schedule or estimate (https://www.youtube.com/watch?v=l0IHqNZTth0).

As the Buildsoft Cubit measurement pane is basically a spreadsheet, this works very interactively, whereas, in other SMM-based take-off packages, there is no functionality between the BIM reader and the take-off dimension sheet.

Buildsoft Cubit also exports to MS Excel and the functionality of an NRM2 library is planned.

For a non-SMM-based bill of quantities, MS Excel could be used, though this is a bit long-winded compared to some other software that is available. Buildsoft Cubit is excellent for

Part 1

producing generic or non-SMM bills/schedules of quantities, and it can also extract BIM quantities to produce any style of bespoke measurement output.

SMM-based bills of quantities, or schedules, may be produced by deploying the CATO Take-off and Bills tool or, alternatively, with QSPro and the Bluebeam on-screen measurement plug-in.

One of the many attractions of BIM quantity take-off is the speed at which schedules of quantities can be produced. Such quantities are, however, generic and have no synergy with any of the standard methods of measurement used either in the United Kingdom or elsewhere. Consequently, further manipulation of the data is needed to produce the required output if formal bills of quantities or schedules are to be produced.

For informal use within contracting organisations and for the production of activity schedules, BIM quantity take-off is ideal. It can quickly provide the 'base' quantities required and a further benefit is that project management software such as Asta Powerproject can be linked to a 3D model in order to create a 4D programme, which can then be linked to the 5D BIM model (e.g. activity schedule) if desired.

BIM quantities are, however, generic and there is no standard method of measurement for use with BIM. Furthermore, it would appear that a common international standard method of measurement, compatible with all BIM software, is some way from being a reality (National Building Specification). Considering how long it has taken for mobile phone manufacturers to agree on a universal charging connector and notwithstanding the publication of the revised CESMM4 and the 'new' NRM2, a step in the right direction might be to move away from measurement driven by standard methods altogether and towards a more simplified approach to measurement.

Crotty (2012) subscribes to the view that a 'smart' bill of quantities, generated from a BIM model, would be able to retain all of the detailed information about a structure at the component level that is compressed in a traditional bill of quantities by the aggregation of individual items demanded by standard methods of measurement.

Perhaps Ted Skoyles was not too far wrong after all with his Operational Bills approach to measurement in the 1960s!

Notes

1. Construction Project Information Committee (CPIC).
2. Ibid.
3. Structural Engineering Discipline Lead, Laing O'Rourke, Engineering Excellence Group.
4. http://www.thenbs.com/topics/BIM/articles/whatIsCOBie.asp (accessed 20 April 2015).
5. http://www.buildingsmart.org.uk/ (accessed 20 April 2015).

References

Benge, D.P. (2014) *NRM1 Cost Management Handbook*, Routledge, Abingdon.

Cabinet Office, Government Construction Strategy (May 2011) https://www.gov.uk/government/publications/government-construction-strategy (accessed 24 April 2015).

Crotty, R. (2012) *The Impact of Building Information Modelling: Transforming Construction*, Spon Press, Oxford.

David, P., Head of BIM Implementation at the Cabinet Office and BIM Programme Director at Balfour Beatty, National BIM Report 2012, NBS.

Eastman, C., Teicholz, P., Sacks, R. and Liston, K. (2011) *BIM Handbook*, John Wiley & Sons, Inc., Hoboken.

National Building Specification. http://www.thenbs.com/bim/what-is-bim.asp (accessed 20 April 2015).

NBS (2012) National BIM Report 2012, NBS. http://www.thenbs.com/topics/bim/articles/nbsNational BimSurvey_2012.asp (accessed 24 April 2015).

Nisbet, J. (1951) Ministry of Education Building Bulletin No 4. HMSO.

NRM1 (2012) *RICS New Rules of Measurement*, NRM1: Order of cost estimating and cost planning for capital works, 2nd edition, RICS, Coventry.

Race, S. (2012) *BIM Demystified*, RIBA Publishing, London.

RIBA (2013) *Guide to Using the RIBA Plan of Work 2013*, RIBA Publishing, London. http://www.architecture.com/RIBA/Professionalsupport/RIBAOutlinePlanofWork2013.aspx (accessed 1 April 2015).

Richard Waterhouse, Chief Executive, RIBA Enterprises, National BIM Report 2012, NBS.

RICS (2014) *How Can Building Information Modelling [BIM] Support the New Rules of Measurement [NRM1]*, RICS, www.rics.org/research (accessed 1 April 2015).

Salzman, L.F. (1968) *Building in England Down to 1540: A Documentary History*, Oxford University Press, Oxford, pp. 454–455.

The Business Value of BIM for Infrastructure, McGraw-Hill Construction SmartMarket Report, 2012. www.construction.com/market_research (retrieved 8 December 2012; accessed 20 April 2015).

Chapter 3
Measurement Conventions

"The quantity surveyor should always use a pen to enter dimensions – use of a pencil and eraser shows a lack of moral fibre."

These were the words of my measurement lecturer back in the 1960s – and quite right too!

The essence of his message was that taking off quantities is meant to be a thorough and precise process which, if followed correctly, should give the measurer the confidence to enter dimensions in ink and not in pencil.

The process starts with a close scrutiny of the drawings, the preparation of a 'take-off list' and the careful consideration of what is to be measured and how it should be set down on the page. This includes paying particular attention to how side-cast calculations are to be set out, making sure that there is plenty of room between item descriptions and ensuring that, when mistakes are made, they are corrected in the proper manner using established and accepted practice.

Taking-off is also done in conjunction with a specification that informs the description of items to be measured and establishes either the standard of materials and workmanship required (prescriptive specification) or the end result required from the work to be carried out (performance specification).

3.1 Traditional conventions

Edexcel Level 3 Unit 9: *Measuring, Estimating and Tendering Processes in Construction and the Built Environment* dated January 2010 includes 'traditional' and 'cut and shuffle' bill production methods in the unit content.

This reflects the *raison d'être* of measurement as a means of producing quantities of construction work, fixed in place, so that bills of quantities can be produced, thereby enabling tenders to be invited from interested contractors.

Whilst the RICS has indicated a significant decline in the use of formal bills of quantities since 1987, 24.5% of all projects (18.8% by value) nevertheless seem to be procured in the traditional way, and, therefore, traditional quantity surveying services are still in demand (Davis Langdon (AECOM), 2011). However, with the marked increase in the popularity of design and build/design and construct and drawings and specification methods of procurement, informal bills of quantities are commonly prepared by contractors and/or subcontractors for tendering purposes and for work package procurement.

Consequently, measurement still figures strongly in the activities of the industry, and it makes sense, therefore, to embrace all that modern computer software can offer to practitioners involved in producing and/or using quantities.

3.1.1 Taking off quantities

Measurement (aka taking-off, quantity take-off) is the process of obtaining dimensions and calculating quantities from drawings; this may be done physically on-site, from paper-based drawings or from digital drawings, images or models.

In the construction industry, the process of measurement is mainly associated with quantity surveyors who, over the years, have developed conventions, or methods of working, that bring order, consistency, clarity and auditability to the measurement process.

Quantity surveying conventions are well described in traditional textbooks such as Willis and Seeley, but it bears repeating here that the conventions for entering dimensions are equally applicable to on-screen measurement as paper-based take-off.

The essential discipline for competent measurement is following the correct order for entering the **dimensions** so that it is easy to revisit the dim sheet to see where the quantities have come from. For this reason, it is important to enter length, followed by width followed by depth. On paper, it is important that the dimensions are entered vertically on the correct paper; otherwise, the measurer will quickly end up in a mess. On-screen, it doesn't matter whether this is done vertically or horizontally, as the software will dictate how the dims are entered. Some software give a choice of vertical or horizontal dimension entry; some don't.

The dimension sheet is laid out as illustrated in Table 3.1 where it can be seen that dimensions are entered in a strict order, on specially ruled paper, according to convention. Also illustrated is a typical item description with side-cast (or waste) calculations that make up the principal dimension entered in the dimension column. This is to be contrasted with an on-screen equivalent which shows the same conventional build-up of a side cast but with a different method of presentation.

Traditional conventions also include a means of multiplying a set of dimensions by another number. In Table 3.2, it can be seen that there are 3 nr items possessing the same dimensions and that **times-ing** is a way of reducing repetition in the take-off. The on-screen version is identical.

Often, in a take-off, items are measured which possess the same dimensions as an item previously measured but which has already been 'times-ed'. A way of adding an additional number to the calculation is by the convention of **dotting-on**. This is illustrated in Table 3.3 where it can be seen that the on-screen version is a close representation of the traditional method except that the 'dot' is replaced by a 'D'.

It will be noted that the **incorrect dimensions** shown in Table 3.4 are deleted correctly as per my old lecturer's instructions!

Dotting-on and times-ing are relevant in the context of computer software packages, which provide the facility to do the calculation but have a different way of entering the figures (e.g. in CATO 3 dot-on 5 is entered 3D5, as the computer would take a full stop as a decimal point).

Table 3.5 shows grouped dimensions relating to the same item with a bracket keeping them together.

This is not necessary in CATO or QSPro because the dimensions relating to individual items are retained in their individual dim sheet.

The **anding**-on of one set of dimensions to another item description shown in Table 3.5 is replicated in the QS software, but CATO and QSPro achieve this in slightly different ways. In effect, though, the dims are copied and pasted from one dim sheet to another.

Table 3.1 Dimension sheet.

Dimensions

The dimension paper is laid out in two equal columns each identical to the example shown. These are numbered as 'pages' Each of the pages is subdivided into four columns:

1. One column for times-ing
2. One for dimensions
3. One for squaring
4. A final column for the item description and any waste or side-cast calculations

X-ing	Dims	Squaring	Description	Sidecast
ENTERING DIMENSIONS				
	3.00		Cube	
	2.00			
	4.00	24.00		
			Superficial	
	3.00			
	2.00	6.00		
			Linear	
	3.00	3.00		
			Number	
	4			
	Item		Item	

Item descriptions and side casts

X-ing	Dims	Squaring	Description	Sidecast
DIMENSION SHEET				
				Centre line
			Half brick thick facings (P.C. £900/1000); cement mortar (1:4); in skin of hollow wall; pointed one side	2/49.60 99.200
				2/32.10 64.200
	162.95			163.400
	4.66	759.36		less 4/112 0.448
				162.952

Causeway TOAB - Project : 111 (SMM7)

Measurements and Descriptions for 111 Dim File BBB - Superstructure (Bill Order)

Part 1

Table 3.2 Times-ing.

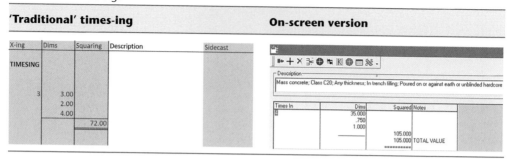

Table 3.3 Dotting-on.

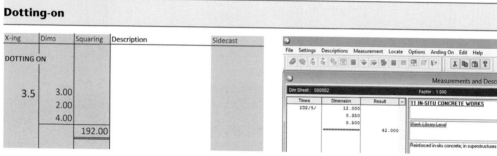

Dotting-on is a means of adding two or more times-ings together in order to multiply a set of dimensions by the factor desired

In the above example, it can be seen that the dimensions 3.00 × 2.00 × 4.00 = 24 are multiplied by 8 (=5 + 3 or 5 dot-on 3) giving a total of 192

Dotting-on to a times-ing on-screen:

- 2 + (or dot-on) 2 × 5 = 20
- 20 × 12 × 0.35 × 0.5 = 42.00

Table 3.4 Incorrect dimensions.

X-ing	Dims	Squaring	Description	Side cast
ALTERATIONS				
15	3.00			
	2.00			
	4.00			
		360.00		

Whilst there is undoubtedly still a role for traditional taking-off using dimension paper, or site 'dim books', measurement has moved on significantly since the 1980s, and computer software for measurement now plays a significant role in the quantity surveyor's work.

However, there are occasions when written down dimensions are preferable to computer-generated dims, physical measurement on-site being a prime example. It is also

Table 3.5   Grouping and anding-on dimensions.

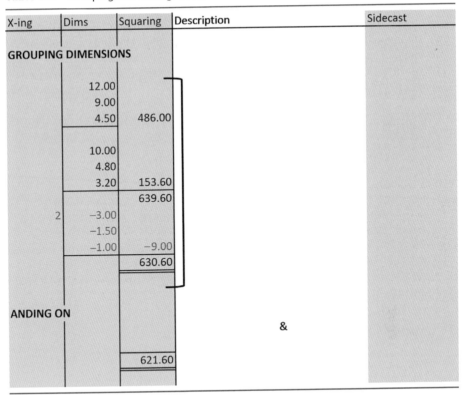

X-ing	Dims	Squaring	Description	Sidecast
GROUPING DIMENSIONS				
	12.00			
	9.00			
	4.50	486.00		
	10.00			
	4.80			
	3.20	153.60		
		639.60		
2	−3.00			
	−1.50			
	−1.00	−9.00		
		630.60		
ANDING ON			&	
		621.60		

probable that many small subcontractors prefer to write dimensions down as they feel that they cannot justify the expense of measurement software or the time to learn how to use it (they would be wrong on both counts!).

Nonetheless, the decline in the use of formal bills of quantities and the increased use of drawings and specification procurement are key drivers for investing in some sort of software in order to ensure that quantities can be taken off quickly and efficiency in order to meet the demands of the tendering process.

Top-end measurement packages are admittedly pricey, but some software systems such as QSPro are very reasonably priced. Microsoft Excel© offers a 'cheap and cheerful' alternative, but a great deal of repetitive typing and entering of formulae is necessary. This is not 'smart' for significant measurement tasks.

3.1.2   Cut and shuffle

'Cut and shuffle' was developed in the early 1960s and, by the 1970s, was perhaps the most popular and widely used method of bill production in QS practices. It was a form of pre-computer age 'automation' for the take-off and billing process. In common with the 'traditional' method of taking-off, there is no empirical evidence that cut and shuffle is still used to any great extent, but, equally, there is no reason why it shouldn't be.

Unlike the traditional method, there is no standardised paper or rulings for the cut and shuffle system, and according to Spain (1995), some professional offices even used different sorts of paper for different purposes.

The system uses a specially coated or 'sensitised' two-sheet paper that produces a copy of the dimensions which are then 'squared' and totalled. At this stage, the 'top' copy of the take-off is 'cut' (some QSs used specially perforated paper) and 'shuffled' into BQ order. All sheets of the same item are pinned together and brought to a total for billing. The 'bottom' copy is filed.

Cut and shuffle pages are laid out in A4 landscape and are split into four or five sections, each section representing a dimension page. Each cut and shuffle 'parent slip' contains only one item description, and any subsequent 'children' slips or pages contain only dimensions or side-cast calculations relating to that item. All pages are numerated and referenced back to the 'parent' item containing the item description. Once the slips are in BQ order, further slips may be introduced for headings, collections and summary pages as necessary.

The principal characteristics of cut and shuffle are:

- It is supposedly quicker than the 'traditional' method.
- It avoids separate abstracting and billing.
- Descriptions are written only once.
- Mistakes are avoided because there is no need to transfer squared dimensions to an abstract sheet.
- Taking-off may well be more time consuming due to the need for extensive referencing of the slips.
- The system is seemingly less useful for post-contract control purposes as measured items have to be searched for in all the take-off slips.

3.1.3   Direct billing

Direct billing is a neat and practical alternative to the traditional 'take-off, abstract and bill' system where dimensions, waste calculations, item descriptions, quantities and pricing columns are provided on the same page.

The system is particularly useful for contractors and subcontractors who simply require an 'internal' quantity take-off and pricing arrangement that is not intended for publication as a formal tender document.

Depending on the size and complexity of the project in question, direct billing can have a number of disadvantages:

- There is no means of collecting dimensions for similar items together unless like items are measured at the same time.
- Deductions for openings must be taken with the relevant item which can result in a great deal of repetition, for example:
  - Deductions for doors and windows must be taken with the appropriate masonry, plastering and painting items and cannot be 'anded-on'.
- The eventual bill of quantities can be 'bitty' because like items are not kept together as a consequence of not measuring in a strict 'take-off list' order.
- The direct bill of quantities will be much longer and bulkier than a traditional paper-based BQ.

Direct billing is particularly advantageous for taking off quantities for 'composite' items in bills of quantities, for example:

- Excavations where earthwork support and working space is included in the item coverage.
- Concreting items where formwork and reinforcement items are not measured separately.
- Underground drainage for manholes and gullies.

The modern equivalent of 'direct billing' is direct-entry or on-screen measurement software, but the big advantage of the software is that like items can be collected together and a professional-looking bill of quantities can be printed.

3.1.4   Item descriptions

Most standard textbooks on measurement rely on 2D paper drawings and traditional dimension paper in order to illustrate the basic principles of measurement, and it is probable that most quantity surveyors learn this way. It is a good way to develop an understanding of QS conventions and to gain confidence and the ability to 'read' drawings – presupposing, of course, a sound knowledge of construction technology and a basic grounding in arithmetic and geometry.

Perhaps the most difficult aspect of measurement is developing the ability to write clear item descriptions, and this is confirmed by Lee et al. (2014) who explain the need for long experience and the ability to write clearly yet briefly so as to avoid verbose and misleading descriptions.

The item description is the key to helping the contractor's estimator to understand the measured item and to subsequently price the item accurately and competitively. In the absence of a clear and precise description, there will always be a doubt as to the true meaning of the work described, and it is likely that each competing contractor will interpret the item in a different way. This leads to inconsistency in pricing, tenders that are not 'like for like' and the potential for arguments and claims once the project gets underway.

3.1.5   Standard phraseology

Of considerable help with regard to writing BQ item descriptions was the publication in 1965 of the Fletcher–Moore (Fletcher and Moore, 1965) system of standard phraseology. The system consisted of paper-based reference manuals which were referred to by the QS in order to build up suitable item descriptions. Effectively, the Fletcher–Moore library was a database of item descriptions taken from the prevailing standard method of measurement.

Some estimators at the time felt that the Fletcher–Moore item descriptions were somewhat 'clipped' in that they were very brief and lacked the personality of item descriptions phrased by experienced quantity surveyors, who had the skill to both comply with the rules of the standard method of measurement and to add nuances of information that helped the estimator to price the job.

3.1.6   2D paper-based drawings

There are two ways of producing quantities using measurement software in conjunction with paper-based drawings:

- Using figured dimensions or by scaling from the drawings using a traditional scale rule.
- Using a digitiser or graphics tablet which enables data to be captured electronically:
  - ○ By tracing lines manually.
  - ○ By plotting the corners of linear polylines[1] or shapes.

3.1.7   Figured dimensions/scaling

Using dimensions shown on the drawings is the traditional way of taking dimensions for quantities; it is still used in the industry and perhaps will always have a role to play especially for small take-offs, or where the quantities required are simple, or for measuring variations.

The main problem with this method is that drawings are never fully dimensioned and therefore some scaling is inevitable. When dimensions have to be scaled, accuracy is always questionable, and even when drawings are full scale (i.e. not reduced or photocopied), there are always inaccuracies built in by the drawing reproduction/printing process, and there is always a bit of guesswork involved when using a scale rule. Straight lines are not so bad, but curves represent a degree of difficulty both in scaling and with regard to accuracy.

A further issue that often crops up is that figured dimensions can sometimes be stated incorrectly (yes, even architects and engineers make mistakes!), and an overall total figured dimension may not equate to the figured dimensions that make it up.

Anecdotal evidence would suggest that there is a certain degree of indiscipline amongst practitioners when using paper-based taking-off. Non-QS-trained personnel invariably write down dimensions horizontally on non-QS paper, as opposed to vertically on dimension paper as a QS would; the dimensions are usually haphazardly organised with no audit trail as to where dimensions have come from or how a composite measurement (such as a centre line girth) has been calculated.

This leads to problems of clarity, accuracy and difficulties (often some time later) when dimensions have to be referred to (e.g. where there has been a variation or where the work has to be remeasured). Happily, using measurement software helps to overcome some of these problems.

3.1.8   Digitised measurement

Digitising tablets may be corded or uncorded, and they come in a variety of sizes; some are solid stand-alone work stations (like an architect's drawing board on a stand), some are solid but smaller for use on the desktop, and some are of the 'roll-up' variety for better portability. The data is digitised using either a stylus (a bit like a pen) or a puck; this is like a computer mouse but with an optical viewer and for precise targeting of points on the drawing.

Useful Website
http://www.visualprecision.co.uk

Measurement by digitiser has certain advantages compared to using traditional scale rules:

- Data capture is quicker and more 'fun'.
- Accuracy is significantly better.
- Drawings that have been reduced or are otherwise not to scale can be accommodated.
- Photocopied drawings can also be measured accurately.

Out of scale drawings can be measured by simply calibrating the drawing. This is done by clicking on the end points of a line of known dimension (it could be a figured dimension on the drawing or the scale indicator itself), and the software will work out the scale automatically.

Telling the software that a particular line represents (say) 10.34 m enables it to calibrate the scale of the drawing which is then applied to other digitised lines. It is a good idea to calibrate two lines as a double check.

3.2   Modern conventions

Modern quantity surveying conventions, whilst respecting their traditional heritage, are very much determined by the computer software that many quantity surveyors now use. Dimension entry, side casts and other protocols are still part of the quantity surveyor's work, but they may be done differently due to the way that the software is designed and operated. Different ways of working are now necessary too, as QSs frequently work in a variety of media such as 2D and 3D CAD, PDF files, 3D models and BIM models.

Ashworth (2010) notes that computers were first used for processing bills of quantities back in the 1960s, but the computers and associated software were a far cry from their modern-day equivalents. Masterbill Micro Systems Ltd claim to have developed the world's first bill production software for the personal computer in 1981, and now their Masterbill Elite© and Masterbill QSCad© are widely recognised as two of the leading software packages available.

There are, however, a number of well-known – and some not so well-known – software packages available to the quantity surveyor or subcontractor, with origins both in the United Kingdom and in other countries. The QS software used in this book is CATO and QSPro from the United Kingdom and the excellent Australian product, Buildsoft Cubit.

Computer-based measurement systems have taken the Fletcher–Moore idea of a library or database of standard item descriptions to a new level, so much so that some of the measurement software packages available provide several databases within the same software package. Consequently, some software houses provide a CESMM3/4, MMHW4 database as well as SMM7/NRM/ARM4 which enables users to switch database seamlessly according to whether the project is building, civil engineering or highway work.

3.2.1   Measurement software

There are basically three types of measurement software packages:

- SMM-based packages.
- Non-SMM-based packages.
- Hybrid packages.

SMM-based packages

SMM-based packages rely on a database of item descriptions for building up complete bills of quantities. Some software systems include a number of databases in the software package that can be accessed depending upon the type of work being measured.

Common SMM databases include:

United Kingdom
- The RICS New Rules of Measurement – NRM2.
- The Civil Engineering Standard Method of Measurement – CESMM4.
- The Method of Measurement for Highway Works – MMHW4.

Republic of Ireland
- The Agreed Rules of Measurement – ARM4.

Australia
- Australian Standard Method of Measurement of Building Works 5th Edition.

Republic of South Africa
- Standard System of Measuring Building Work 6th Edition.
- The Civil Engineering Standard Method of Measurement (South African Edition) – CESMM3.

International
- Principles of Measurement (International) – POM(I).

Examples of SMM-based software are Causeway CATO Take-Off and Bills, QSPro for Windows and Vector by Snape Software. Each of these has extensive libraries of a variety of standard methods of measurement including SMM7, NRM1 and 2, CESMM and MMHW, etc. CATO and Vector also have the POM(I) library.

All SMM-based software works in much the same way by selecting item descriptions from the library and then entering dimensions, either manually or from digital drawings. The features that distinguish software packages one from another – flexibility, multiple sorting capability,

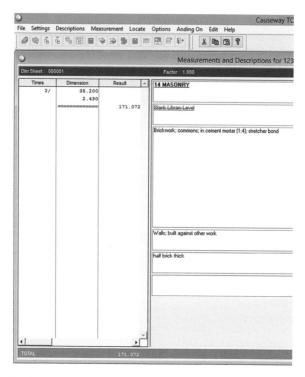

Figure 3.1   CATO dim sheets.

ease of use, library availability, estimating and cost planning facility, report functionality, afford-ability, etc. – must be assessed by the individual user.

CATO is based on a system of traditional 'dim sheets' which the user may group together conveniently (e.g. into elements such as substructure, superstructure, finishes and so on). Item descriptions are chosen from the library, edited as necessary, and dimensions are entered into the dim sheet as illustrated in Figure 3.1.

This is a comparable system to traditional quantity take-off on dimension paper.

QSPro similarly relies on a comprehensive library of item descriptions, but this is accessed differently than CATO.

As illustrated in Figure 3.2, the window for building the item descriptions is split, with the library in the lower pane (NRM2) and the item description above. Quantities are entered directly into the quantity 'box' or built up, with side casts and 'anding-on' if required, in a separate window as shown.

QSPro is flexible enough to offer two-dimensional layouts, a horizontal display as illustrated in Figure 3.2 and the traditional vertical method as shown in Figure 3.3.

Non-SMM-based packages

Non-SMM-based packages, such as Buildsoft Cubit, require item descriptions to be developed by the compiler.

Buildsoft Cubit enables bills, or schedules, of quantities to be produced, either for a single trade or for an entire project. This may be done:

- From scratch.
- From a previous project imported into the current project.
- From a 'template'.

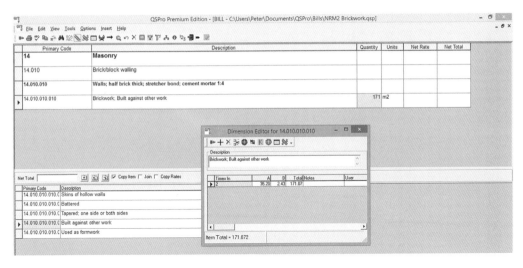

Figure 3.2   QSPro dim sheet.

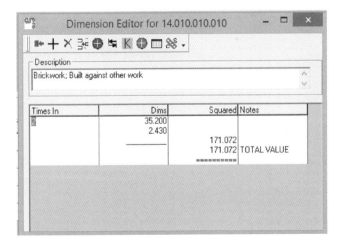

Figure 3.3   QSPro – vertical dimensions.

Templates may be developed very easily in Buildsoft Cubit by nominating a previous project and naming it as a 'template'. The software saves the item descriptions, units and all the formatting but does not save quantities or the drawings that the quantities were based on. The template is then ready to work on in the new project.

Projects are split up into 'jobs' which is where the quantities for individual parts of a project reside. 'Jobs' are user defined which can be elemental, work section, trade or any other classification.

Quantities are entered either manually or by on-screen measurement in Buildsoft Cubit, and on-screen take-off is aided by a 2D to 3D toggle which enable walls, columns, beams, etc. to be viewed in three dimensions rather than simply on plan. This is especially useful in that deductions for doors and windows can be clearly seen and walls of different heights and configurations can be viewed in perspective.

Items are created in the left-hand 'estimate' sheet, with the detailed calculations, including deductions, that make up the quantities being performed in the lower left 'calculation' sheet.

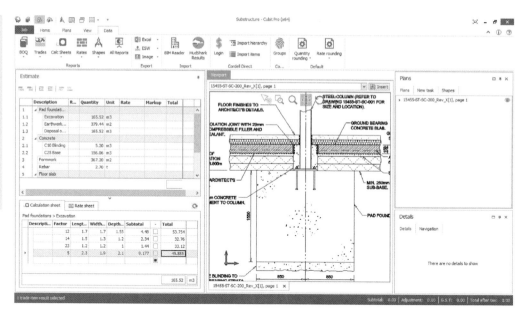

Figure 3.4   Buildsoft Cubit.

	Description	R...	Quantity	Unit	Rate	Markup	Total
1	⊿ Pad foundations						
1.1	Excavation		165.52	m3			
1.2	Earthwork support		379.44	m2			
1.3	Disposal off site		165.52	m3			
2	⊿ Concrete						
2.1	C10 Blinding		5.20	m3			
2.2	C25 Base		156.06	m3			
3	Formwork		367.20	m2			
4	Rebar		2.70	t			
5	⊿ Floor slab						
5.1	Sub-base		148.50	m3			
5.2	Sand blinding		29.70	m3			
5.3	A193 Mesh		594.00	m2			
5.4	C30 Concrete slab		118.80	m3			

Figure 3.5   Buildsoft Cubit hierarchy.

Drawings are imported into the central 'Viewport', and it is in this pane that on-screen measurement is performed. To do this, an item is created, a unit is chosen, and the drawing is then measured by mouse clicking from point to point.

Some of the functionality of Buildsoft Cubit is illustrated in Figure 3.4.

In Figure 3.5, it can be seen that Buildsoft Cubit can easily produce a hierarchical schedule of quantities with non-SMM-based item descriptions which can then be printed as a trade report, or bill of quantities, within Cubit or exported to MS Excel.

Other non-SMM-based measurement software includes Autodesk Quantity Take-off, Exactal CostX, Swiftplan and Trimble VICO.

Hybrid packages

Some software packages offer the best of both worlds – a choice of SMM-based standard libraries and the option to create bespoke libraries. The latter might be based loosely on a specific standard method of measurement, or the item descriptions might be drafted 'in-house'.

For example, at the time of writing (November 2014), the additional functionality of an NRM2 Library is in its latter stages of development which will effectively make Buildsoft Cubit a 'hybrid' software package.

This is an important development because the addition of a standard library of descriptions to an essentially non-SMM-based package is attractive. Some QS practices use two software packages – one SMM based and the other not, as it is then possible to measure and bill items that do not appear in the standard libraries, such as partial contractor design elements and provisional sums that would normally have to be word processed.

Use of non-SMM-based software for BQ preparation is a matter of risk perception.

When a standard method of measurement is used, everyone understands (or should understand):

- What is to be included in the rate for an item and what is not.
- The balance of risk in the employer–contractor or contractor–subcontractor relationship.
- The legal situation should there be a deviation from the SMM (e.g. as regards unfair contract terms).

This is not the case with bills of quantities containing non-SMM item descriptions, and this can lead to difficulties, even disputes, about what is included and excluded from the item coverage.

However, there can also be significant advantages in using non-SMM item descriptions:

- Item descriptions can be kept brief.
- A bill of quantities can be produced quickly.
- Less attention need be paid to measuring minor items.
- Any configuration of BQ/schedule is possible.
- Activity schedules, price lists and contract sum analyses can be produced.
- The estimator can price the tender far quicker provided the item coverage is clearly understood.

Microsoft excel

Strictly speaking, Excel is not a type of measurement software, but, to all intents and purposes, it can be classified as such because of its widespread use for measurement and related tasks particularly by site-based quantity surveyors and measurement engineers.

The big disadvantage of Excel compared with library-based software is that there is no database of standard descriptions readily available. This can be developed 'in-house', but a great deal of time and effort is required to set up the database in the first instance. Another difficulty with Excel is that each calculation requires an individual formula, whereas measurement software packages perform all calculations automatically.

Remembering that there are a number of standard methods of measurement for different types of construction work, Excel can only realistically be used for creating builders' quantities for individual projects where the number of item descriptions is limited or for subcontracts or for site remeasures and for the measurement of variations and extra work.

A major advantage of Excel is that some measurement software packages, and 2D and 3D CAD systems, can link directly to Excel enabling quantities to be automatically generated. As most subcontractors will not own a specialist measurement package, this enables bills of quantities/schedules to be prepared for trade and subcontract enquiries, and exported electronically, in a language that most businesses have on their desktop. An Excel export from QSPro is shown in Figure 3.6 where it can be seen that the full formatting of the bill of quantities is retained within the Excel spreadsheet environment. The full functionality of Excel is also retained.

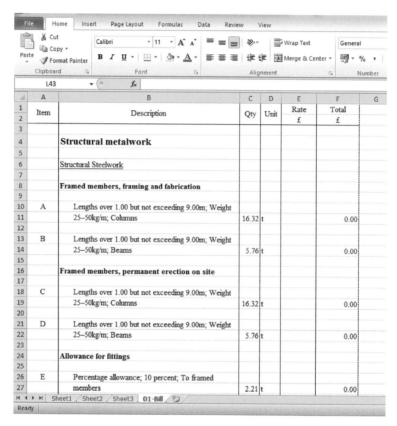

Figure 3.6   QSPro export to MS Excel.

3.2.2   Electronic communications

It is now commonplace in the construction industry for drawings to be transmitted electronically. This can be done is several ways:

- PDF.
- Web interface, for example, Eleco BIMCloud.
- Online file-transferring platform, for example, WeTransfer Plus.
- File hosting service, cloud storage service and online file storage provider, for example, Dropbox Business.
- Web-based portals such as DocElite, which have access restricted by user name and password.

Whether communicating between design team members, or to contractors or by main contractors to subcontractors, drawings frequently arrive by email attachment for review, measurement, tendering or subcontract quotations. The days of a pile of folded paper drawings in a brown envelope landing on the door mat are pretty well over!

Most people working in construction, however, are used to reading paper-based drawings and, therefore, the temptation when receiving drawings that have been produced electronically is to print off a hard copy.

This may well be fine for some purposes but can be problematic from a measurement perspective. The paper printout from a desktop printer is unlikely to be to a true scale, and frequently printouts are A4 or A3 at best and are too small to read comfortably.

Even with the time and expense of a true to scale printout, the paper drawings so produced will not be completely reliable due to processing inaccuracies. More importantly, printing a hard copy will limit the extent to which measurement software can be used unless it is intended to measure using a digitiser. The whole point about electronic drawings is to make the most of the speed, accuracy and efficiency of computerised taking-off.

Quantities can be extracted from Vector or Raster plans and drawings.

Raster plans are single-layer, flattened images such as PDF, JPG, BMP, PNG and TIFF. They require a drawn scale and pixelate when zoomed in. This may be contrasted with Vector plans which possess 'layer' data and more accurate lines for taking-off. The 'layer' data in Vector plans includes outlines, walls, room names and text which can be selected with suitable software, such as Exactal CostX, Buildsoft Cubit, CATO and Snape Vector, and used to create schedules or bills of quantities. Layer data can be turned off or on.

Whether taking-off quantities using measurement software or manually, the starting point for all measurement is the information provided by the designer in the shape of project drawings, schedules and standard details.

This design information forms the basis of the eventual bills of quantities which depends for its completeness on the stage that has been reached in the development of the design. Irrespective as to who receives this information – be it PQS, contractor or subcontractor – the information may arrive in several forms, including 2D paper-based drawings, as described in Section 3.1.6. Modern convention, however, is to transmit design information electronically, either with drawings or digital images.

3.2.3   2D or 3D CAD drawings

Electronically produced drawings and other documents may arrive in one or more file formats which are signified by the file ending. Just as a Microsoft Word document may have a file ending of (*.doc) or (*.docx), for example, so a CAD drawing file will have its own file ending. There are lots of file types used in computer-aided design and for document exchange, but some of the most commonly encountered ones are:

- *.dwg – DraWinG
- *.dgn – DesiGN
- *.dwf – Design Web Format
- *.dxf – Drawing eXchange Format
- *.pdf – Portable Document Format

DWG files

*.dwg is an intelligent file source and is the native format for several CAD packages. Autodesk products, such as AutoCAD and Autodesk Revit, use DWG files for storing 2D and 3D design data, and it is reputedly the most popular file format for CAD drawings. Plans with *.dwg file endings are Vector plans.

DWG technology facilitates a variety of methods for navigating around drawings including zooming in/out and panning and also enables 'layers' of the drawing to be turned on/off. This is a great feature for measurement because the complexities of reading a paper-based drawing can be overcome by selecting a particular layer to view (e.g. partitions, windows, plumbing appliances). As well as a variety of design and annotation capabilities, DWG files facilitate simple measurement.

The DWG file format is the native format for some other CAD packages as well, such as IntelliCAD and Caddie, and DWG files are also supported non-natively by other CAD applications.

Useful Website
http://www.autodesk.co.uk/

DGN files

DGN is the native CAD file format supported by Bentley Systems. Bentley CAD software products, such as MicroStation, are used on a wide variety of construction projects for both 2D and 3D design and drafting. The DGN file format is not as widely used as the DWG format preferred by Autodesk.

DGN files display similar functionality to DWG files.

DWF files

The DWF file format was developed by Autodesk to facilitate the exchange of data-rich design files between design team members in order that they may be viewed, reviewed or printed. The highly compressed nature of DWF files means that they are smaller and faster to transmit than design files, and, because of their functionality, they can be used to transmit complex CAD drawings in a single file as well as 3D models from most Autodesk Design applications.

DWF files are not intended to replace AutoCAD drawings (DWG) or other native CAD formats but may be converted into DWG files with a suitable DWF–DWG converter. Files other than Autodesk Design Web Format files may use the DWF file extension.

DXF files

DXF is a CAD data file format, developed by Autodesk, which facilitates data transfer (interoperability) between AutoCAD and other software programs. AutoCAD and some other CAD software programs are capable of creating, opening and editing DWF files. The Autodesk Design Review program can view, print and mark up DWF files. Files with *.dxf endings contain Vector plans that have the capability of 'layer' access that can be readily enabled/disabled.

PDF files

Drawings are frequently transmitted in the PDF file format – especially to contractors and subcontractors for tendering purpose – as the files are easy to open. PDF files – both drawings and other documents – do not rely on specific application software, computer operating systems or computer hardware. All that is required to view a PDF file is a suitable 'reader' such as Adobe Acrobat Reader or Nitro PDF Reader which are available online as free downloads.

PDF is the most commonly used file type. They can be flat Raster plans (i.e. a scanned image) or a smart layered Vector plan produced from CAD. Unlike CAD plans, which have the correct scale embedded in the file, PDF plans need to be scaled because they can be up to 3% out of scale. This can be simply overcome by calibrating the drawing as described in Section 3.2.6.

3.2.4 Digital images

Images taken with a digital camera, or mobile phone, can be measured using appropriate software such as PhotoModeler.

The software extracts 3D measurements and models from photographs taken with an ordinary camera and can produce accurate lengths, areas, volumes and cross sections from almost any photograph.

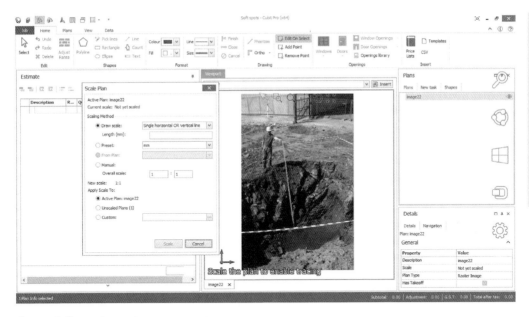

Figure 3.7   Scaling soft spot excavation.

Measurement may also be done from image files using proprietary measurement software such CATO or Buildsoft Cubit.

The image must be in a file type that the software can read, and there must be a dimension, or an object representing a dimension (e.g. measuring staff) in the image, which is capable of being scaled.

Figures 3.7 and 3.8 show Buildsoft Cubit goes about the measurement of a soft spot from a site photograph.

A JPEG image is first imported into the Viewport, and the software asks for the image to be scaled (see Figure 3.7). Note the measuring staff acting as the scale.

The excavation can then be measured on-screen using the polyline tool, and a volume for the item is calculated in the Estimate window as shown in Figure 3.8.

3.2.5   Viewing electronic files

Appropriate software is needed to be able to read electronic files, but it is not necessary to have a CAD package on your computer to be able to view CAD files.

Bentley Systems offer a free viewer download called Bentley View into which a variety of file formats can be imported and then viewed. A large number of file formats are supported by the viewer including DGN, DWG and DXF, but PDF is not a recognised format.

Useful Website
http://www.bentley.com/bentleyview

Alternatively, Autodesk's AutoCAD® WS is a free cloud-based web and mobile app that facilitates viewing, editing and sharing DWG files. The software works in a web browser or on a smart phone or other mobile device. AutoCAD® WS converts to PDF and DWF and allows DWG, DWF and DXF files to be opened directly from an email.

Useful Website
https://www.autocadws.com/web

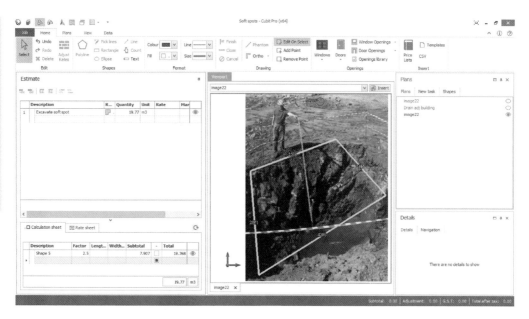

Figure 3.8   Calculating the soft spot volume.

Autodesk® DWG TrueView™ is a free file viewing and conversion software that enables DWG and DXF files to be opened as they would be in AutoCAD. Also from Autodesk, the free Autodesk® Design Review software download enables DWG files to be opened, viewed and printed, and changes to Autodesk 2D and 3D design files can also be tracked. All this is possible without the original design software.

Useful Website

http://www.autodesk.co.uk/designreview

If you have access to measurement software, it is important to verify which file types the software package is capable of reading in order to be able to access the drawing or other information. Should the software be unable to read certain file types, it is possible to download a file converter from the Internet. File converters can convert several file types including:

- DWG files to DXF or DWF
- DWG files to PDF
- PDF to DXF
- DWG files to an image file such as JPG/TIF/BMP

Acquiring a file converter may necessitate a purchase, but some file converter software is available on a free trial.

Useful Websites

http://www.autodwg.com/
http://www.teklabimsight.com/

3.2.6   Calibrating electronic drawings

CAD files may be imported into on-screen measurement software without calibration, but good practice suggests that they should, nevertheless, be checked before taking off quantities. Care has to be taken with PDF drawings, however, to make sure that the scale is true and any measurements taken will be accurate. To do this:

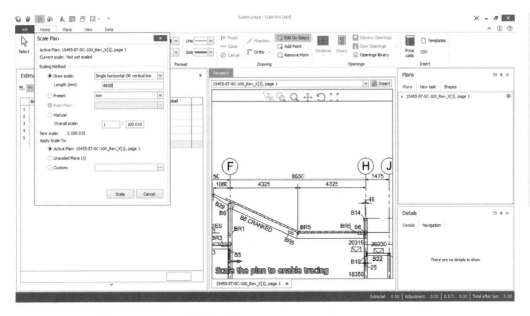

Figure 3.9   Calibrating PDF in Buildsoft Cubit.

1. Open the file in the measurement software.
2. On the horizontal x-axis, click on each end of a line with a known dimension.
3. Type the dimension into the pop-up box.
4. Press the 'enter' key.
5. The drawing is now calibrated.
6. Repeat the process on the y-axis as a double check.

Good practice when calibrating the scale of a drawing is to use the largest known line on the drawing and then to check another known length. Calibration is a common process with all on-screen measurement packages, and this is illustrated in Figure 3.9 using Buildsoft Cubit.

3.3   BIM conventions

BIM models are a completely different proposition compared to 2D and 3D CAD. CAD drawings are based on lines and polylines, whereas BIM models consist of parametric objects that carry technical, geometric and other data. BIM models are 'intelligent', whereas CAD drawings are 'dumb' in the sense that the lines that make them up do not carry any such data.

Consequently, the components in a BIM model may (but not always) possess measurement data that can be extracted and imported into a variety of software packages for quantity take-off, estimating and project management purposes. Take-offs, counts and measurements can be generated directly from the underlying model.

However, the nature and content of BIM models is different to 2D and 3D CAD, and their use requires a different mindset and different protocols compared with traditional quantity surveying conventions or their modern interpretations to date. The quantity surveyor's workflow is different to that demanded by traditional or even modern conventions, because it is the model that generates the quantities and not the quantity surveyor.

This doesn't mean that the quantity surveyor is redundant in the 'BIM-space' environment – far from it – it just means that the same analytical and critical skills, allied to professional judgement, knowledge of construction processes and procurement methods and the ability to ferret out detail, must be applied in a different way.

3.3.1   BIM models

BIM is object-based modelling where designs are derived by selecting objects from a library, as opposed to drawing lines. A window, for example, is a component that will have its own properties, but it also understands and can interact with other components in the model. This means that a window can be introduced into an external wall, and the wall object reacts by creating a suitable opening.

The properties possessed by the objects in a model may include quantities that can be extracted very quickly and may then be imported into measurement or estimating software for manipulation as required. This data can be used for a variety of purposes, including cost planning and detailed estimating, using the dimensions associated with the various object types contained within the design model.

This sounds like a perfect world for designers where quantity surveyors are no longer needed and bills or schedules of quantities may be obtained from the model at the push of a button. This can be done, but the problem is that the quantities will not be in an intelligible form, and they will invariably be wrong. The cryptic descriptions attached to model components are often not clear, and this makes it difficult to identify just what particular components are.

Additionally, the quantities generated from models will not be complete because BIM models do not model everything. A door may be modelled, but the architraves and door stops won't be.

3.3.2   BIM quantities

Quantities can be extracted from BIM models – they are embedded in the various components or objects that make up the model. However, unless a designer understands the quantity surveyor's way of thinking, inaccuracies can be quickly accumulated because, for instance, Revit default settings are set to zero decimal places, and considerable rounding errors could occur on a large contract. IFC files do not generate quantities by default, and this has to be accommodated by checking the 'include base quantities' box.

Models can also be used to derive quantities. In this case, model components, such as reinforced concrete columns, can be used to calculate the quantities of associated items that are not modelled, such as formwork and rebar.

Other quantities may have to be generated from 2D representations of the model. Therefore, expansion joints in a concrete floor slab will have to be measured by scaling from paper printouts or by on-screen measurement once their positioning and specification have been determined.

The quantity surveyor's skill is expressed by following traditional conventions in a modern context, and the 'art' of the profession is in validating the content of the model to establish what has been modelled and what has not.

3.3.3   Quantity extraction

Because reliable quantities cannot be generated automatically by BIM models, quantities have to be extracted using a suitable tool so that the data can then be manipulated as required. Software such as CATO, Buildsoft Cubit and Exactal CostX can do this, but it is only feasible to import the data into a spreadsheet-type workbook or Excel file. These are non-SMM-based media, and, therefore, further manipulation is needed to import the data into a BQ production package.

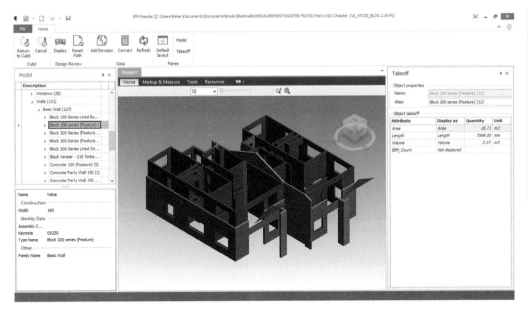

Figure 3.10   BIM component 'standouts'. Reproduced by kind permission of Building Software Services (BSS).

Care must be exercised when extracting quantities from a model because objects that appear to be fully detailed may only have reached an early stage of design resolution. Rendered models are impressive, and it is beguiling to imagine that the level of design development is consistent with a fully rendered level of design detail. This is unlikely to be the case in some instances, and the degree of reliability of the quantity data must be established by interrogating the model and talking to designers.

The beauty of BIM models, however, is their transparency. As one architect recounted, "there is no hiding place with BIM", and models can be interrogated to see whether a particular connection, joint, fitting or connecting plate has been designed or not. Designers cannot get away with inserting a 'cloud' to represent an unresolved design issue as they would do on a CAD drawing.

Depending on the software, quantities can be extracted from a model in several ways:

- From the model description window on the left which lists the contents of the model and the number count of each component.
- From the take-off (or properties) window on the right which displays the attributes or properties of the selected component.
- By dragging and dropping component data into an integrated spreadsheet direct from the model on-screen.
- By scheduling quantities in the integrated spreadsheet by dragging and dropping from the properties window.

In Figure 3.10, also from Buildsoft Cubit, the walls have been made to standout by hiding other components – this makes it easier to read the model and extract the desired information. It can be seen that the walls are hierarchically categorised in the model contents window and that 'Block 200 series' walls are displayed in the take-off panel together with the properties of this category of wall. The attributes displayed give the area, length and volume of the 'Block 200 series' walls which can then be extracted to the conventional Cubit take-off via the Return to Cubit menu button (top left).

3.3.4   Validation of quantities

Quantity surveyors, contractors and subcontractors, who are used to producing quantities traditionally or using on-screen measurement, need to adopt different protocols or conventions when extracting and using quantities from BIM models.

Of primary concern is to find out what has been modelled and what has not and whether further detail has been added by the designer in 2D or 3D CAD. Several sources of design information need to be looked at in order to get to all the quantities, and 2D drawings may be required to see whether there has been any duplication of quantities in the model because of overlapping components.

It is also important to understand how quantities in a model have been calculated because this will not have been done as a QS would do it. Pipe fittings are objects, and therefore, the quantities given for pipe runs (drainage, plumbing, etc.) are measured from fitting to fitting (there is no convention of 'extra over' in BIM). It may not be clear whether model quantities are given net of deductions for openings or pile caps in ground floor slabs, and there may be some degree of unreliability when it comes to the quantification of irregular areas in models.

Architectural models do not contain all the detail needed to produce a complete bill of quantities, and even structural engineers' models contain only the principal structural elements. Details of fittings and connections would only be derived from the fabricator's model, and thus, a number of different models would have to be interrogated to determine a complete 'picture'.

3.3.5   Reading BIM files

BIM modelling platforms are clearly commercial software systems. They come with their own highly developed libraries of parametric objects, and the output files carry native file extensions such as *.RVT (Autodesk Revit) and *.DB1 (Tekla Structures). If you do not possess a legitimate copy of the software, the files cannot be read in their native form.

BIM models can, nevertheless, be 'read' with a variety of free downloadable BIM investigation and review tools such as Autodesk Design Review, Solibri Model Checker and Tekla BIMSight. This software is fine for interrogating models to see what is there and what isn't modelled, but they have little or no capacity to extract quantities.

BIM models can be exported from their original platforms into DWF or DWFx file formats, but these are not native files. They can, nonetheless, be read by some measurement software via an integrated BIM reader. The non-native BIM file is imported into the reader, and the software can extract the desired properties – that is, quantities – from the model. BIM models can also be exported to IFC format which, being non-proprietary, can also be read by BIM Readers.

Figure 3.11 illustrates a DWFx file (BIM model) that has been imported into the Viewport of Buildsoft Cubit BIM Reader. It can be seen that the model has 12 nr columns, 26 nr floors and 5 nr Roofs.

3.3.6   BIM model output

It will be observed that the quantities provided by the model illustrated, whilst in recognisable units, are not expressed as a quantity surveyor would express them:

- They are not in SMM format.
- They are not in a loosely framed descriptive language that a contractor would be able to price.

They are, however:

- Presented in a hierarchical breakdown structure.
- Capable of being exported to other software that can add the descriptive and other details that are lacking.

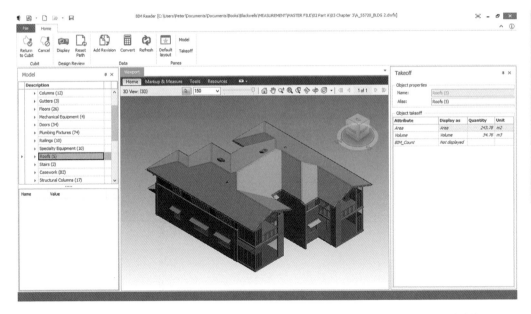

Figure 3.11   BIM model. Reproduced by kind permission of Building Software Services (BSS).

BIM quantities must be treated with caution. Where the quantities come from, who created them, how have they been calculated and their accuracy or reliability are doubts that a quantity surveyor may have, until persuaded otherwise.

It is quite clear that 5D BIM is not fully developed at the moment, and there are a number of interoperability and other issues to resolve before it can be considered established practice. There are no agreed rules of measurement as yet, there are no conventions as there are in 'orthodox' measurement, and there is no certainty that traditional practice will translate well into the BIM world of construction measurement.

If there is one BIM measurement convention that is paramount, it is the validation of quantities through thorough and meticulous detective work.

Note

1.   A group of interconnecting lines or curves in two or three dimensions; also called polygonal chain.

References

Ashworth, A. (2010) *Cost Studies of Buildings*, 5th Edition, Prentice Hall, Harlow.

Davis Langdon (AECOM) (2011) *Contracts in Use, A Survey of Building Contracts in Use During* 2010, RICS London, RICS Survey of Contracts in Use.

Fletcher, L. and Moore, T. (1965) *Standard Phraseology for Bills of Quantities*, George Godwin Ltd, London.

Lee, S., Williams, T., Willis, A. (2014) *Willis's Elements of Quantity Surveying*, 12th Edition, Wiley-Blackwell, Oxford.

Spain, B. (1995) *Taking off Quantities: Civil Engineering*, edited by Tweeds, Taylor & Francis, Boca Raton.

Chapter 4
Approaches to Measurement

There is no question that bills of quantities have been the 'big beast' of the UK construction industry for well over a century. Whilst it is true that many projects do not merit or require such detailed measurement, it is also true that bills of quantities, in some form, have underpinned the traditional methods of procurement used in the United Kingdom both prior to and since the advent of formalised measurement in 1922 (Royal Institution of Chartered Surveyors, 1922).

History shows that bills of quantities have not, in fact, changed radically for over 500 years, and it remains true today that a contractor cannot price anything but the smallest of projects without preparing some sort of quantified 'list' of the work required. The supremacy of bills of quantities was challenged briefly in the 1960s, unsuccessfully as it turned out, by the operational bill, but this failed to catch on, predominantly due to the lack of computing power available at that time.

The UK industry 'fixation' with bills of quantities is probably due to the omnipresence of 'formal' bills of quantities prepared by professional quantity surveyors (PQSs) as part of the traditional tendering process. This 'fixation' is also the likely reason why the use of bills of quantities is reported to be in decline. Both the RICS Survey of Contracts in Use 2010 (Davis Langdon (AECOM), 2011) and the NBS National Construction Contracts and Law Survey 2013 appear to indicate that changes in procurement methods and, especially, the growth of design and build (DB) are behind this shift in emphasis, and it is also probably true that the use of activity schedules popularised by the Engineering and Construction Contract has also displaced the 'formal' BQ to some extent.

However, Hibberd (2014) usefully points out that the RICS survey *does not look at 'traditional procurement' as such but considers procurement in terms of the use of firm or approximate bills, specification and drawings, design and build, target cost, etc.* This is, therefore, more indicative of pricing mechanisms than procurement methods. Hibberd also concludes that 'traditional procurement' is still dominant in the marketplace if the use of firm bills of quantities and drawings and specification is assumed to represent 'traditional procurement'. Both the RICS survey (76% of projects) and the NBS survey (72%) confirm this to be the case albeit that there has been a steady decline in traditional methods since the 90% levels reported in 1987.

It would be a mistake, however, to assume that the decline of firm bills of quantities also means a decline in the need to measure and prepare quantities. Everything that had to be measured

Managing Measurement Risk in Building and Civil Engineering, First Edition. Peter Williams.
© 2016 John Wiley & Sons, Ltd. Published 2016 by John Wiley & Sons, Ltd.

'formally' still has to be measured in order to prepare a price, and there is no 'magic' way around this fact of life. Drawings and specification-type contracts, for instance, need to be quantified in order to arrive at a tender price, and thus, the void caused by the decline in 'formal' bills of quantities has been filled by 'informal' bills of quantities (or similar) whether prepared by contractors, subcontractors or otherwise. This chapter focuses on the various approaches to measuring construction work available.

4.1 Measurement skills

At one time, the acquisition of measurement skills, though hard won, was a fairly straightforward process:

- Understand the technology of construction.
- Learn how to read a drawing.
- Drill yourself in the practices and conventions of the quantity surveyor.
- Acquire the ability to use a scale rule.
- The ability to write accurate and concise item descriptions would come with practice and experience.

Nowadays, the process is much more technological with the availability of digitised measurement, 2D and 3D computer-aided design, electronic drawings, BIM models and the like. Despite these advances, Lee et al. (2011) and Ostrowski (2013), amongst others, still believe that it is crucial to understand what to measure and how to measure it and to do so by grasping the basic principles of measurement in a traditional way, that is, by learning to measure using worked examples presented in 2D line drawing format and by setting out the resulting side-casts and dimensions in the traditional handwritten form.

It is hard to disagree with this point of view, but equally, computer-generated drawings and models can help to make measurement much easier, more fun and less prone to error than taking off from 2D paper-based drawings. Some design packages, for example, have the facility to remove or hide various 'layers' of the design in order to reveal only those items that are to be measured. In this way, internal partitions, doors, windows, plumbing appliances and the like can be visualised in isolation from the rest of the design in a way that was previously not possible with paper-based drawings.

4.2 Uses of measurement

As discussed earlier in this book, measurement is used for many purposes in the construction industry. Most people, however, would probably say that measurement is principally used as a means of quantifying the work required in a construction contract. This provides data that can be used for the pre- and post-contract administration of a project, including:

- Pricing tenders whether it be for a lump sum, measure and value contract or target-type contract.
- Preparation of the pretender programme of works.
- The valuation of work in progress for the purpose of interim payments.
- Valuation of variations and additional work.
- Remeasurement of completed work where required by the contract.

The construction industry uses a number of ways of communicating the work required in a contract according to the preferred method of procurement and the form of contract used.

Risk issue

Irrespective of the choices available, it would be unwise to enter into a contract without some sort of analysis or breakdown of the contractor's price or without a list of prices or rates.

Without such information, even for relatively small projects, the employer would be exposed to the risk of exploitation by the contractor especially as regards post-contract administration.

The choice of pricing document is a matter for the employer according to the level of detail that is thought necessary for a particular contract.

4.3   Pareto principle

The Pareto principle, or 80/20 rule, is a widely recognised rule of thumb in management and business circles, which states that:

- *For many events, roughly 80% of the effects come from 20% of the causes.*[1]

A typical example is where a small painting and decorating contractor relies on subcontracts from two or three main contractors for the bulk of its work and also picks up smaller direct contracts from a number of 'one-off' clients as well as a few 'regulars'. In this case, it is probable that a significant proportion (say, 80%) of the company's turnover is generated by only 20% of its clients.

For this simple example, it would be easy to test the Pareto principle by checking how much work emanated from each client. However, generally speaking, there is no empirical evidence to support the rule which largely stems from observation rather than scientific investigation. Despite there being no strict mathematical foundation to the 80/20 rule, it has some merit, however, because it helps managers to manage by exception and to focus on those aspects of the business that have greatest impact on its success.

With regard to measurement, it has long been generally accepted that 20% of the items in bills of quantities represent 80% of the cost. This can be tested by simply counting the number of items in the bills of quantities and by establishing how many of these are high-value items. The 80/20 rule is more likely to be true in bills of quantities for building work simply because such work is conventionally measured in greater detail, and consequently, there is likely to be a high number of small-value items.

In bills of quantities for civil engineering work, it is probable that the Pareto effect will be less evident; this is because of the likelihood of a higher proportion of 'bill' items with significant quantities or high value or both due to the method of measurement adopted. In civil engineering methods of measurement, the majority of 'minor items' will be deemed included in item coverage rules and therefore not measured separately as they would be in one of the building methods of measurement.

In bidding situations where there is no bill of quantities supplied by the client/employer and the main contractor is consequently responsible for producing its own quantities, it is commonly the case that the contractor's quantity take-off will place most emphasis on the important, high-value items of work. The contractor will be less concerned with measurement precision and accuracy of the quantities than would a PQS working for the employer. The time, effort and cost of measuring 80% of the items that represent 20% of the cost is simply not viable, and in any case, the small-value items will be picked up by the estimator, or by subcontractors, who will make suitable allowances in the rates for the main work items.

4.4 Measurement documentation

Despite the pre-eminence of bills of quantities in construction, measurement information can be conveyed in several different ways according to the nature of the work involved and the type and form of contract used.

JCT contracts offer the choice of bills of quantities of different sorts as well as various schedules as a means of conveying quantity and price information, whereas the choices under the New Engineering Contract family are limited to bills of quantities, activity schedules and quantified and unquantified price lists. In the FIDIC Conditions of Contract for Construction, bills of quantities are, in fact, seen as a subset of 'schedules' which are defined in Clause 1.1.1.7 as documents *completed by the Contractor and submitted with the Letter of Tender* and may include *the Bill of Quantities, data, lists and schedules of rates and/or prices*.

Some construction work lends itself to the use of measured quantities and some to less detailed methods of identifying the work in a project. The choice of documentation will depend on whether a lump sum, measure and value or reimbursement contract is to be used and whether the employer or the contractor is to carry the measurement risk. The range of documents commonly used in the industry may be categorised as follows:

- Formal bills of quantities.
- Formal 'quasi' bills of quantities.
- Formal 'operational' bills of quantities.
- Informal bills of quantities.
- Activity schedules.
- Price lists and contract sum analyses.
- Schedules of actual cost:
  - Daywork schedules.
  - Schedule of cost components.

4.5 Formal bills of quantities

The decline in the use of bills of quantities over recent years has been well documented, the most cited reason being the trend towards non-traditional methods of procurement where bills of quantities are seemingly not required. Whilst there may be some truth in the suggestion that bills of quantities are no longer the *lingua franca* of the UK construction industry, they are, nevertheless, still commonly used for traditionally procured projects and also in non-traditional procurement albeit in a different guise.

Formal bills of quantities are inextricably linked with standardised measurement and with the employer-engaged PQS or 'in-house' specialist in measurement matters. There is good reason for this tradition because tendering contractors are able to rely upon the fact that the bill of quantities is:

- Based on clear and unequivocal rules of measurement.
- The same for each tenderer.
- Accurate in so far as the stage of design development allows.
- A contract document linked with a known method of measurement.
- A sound basis for accurate pricing.
- Subject to clear rules regarding the post-contract adjustment of quantities, errors and omissions and variations that impact on the contractor's original level of pricing.

Formal bills of quantities based on a standard method of measurement may be:

- Firm or
- Approximate.

4.5.1   Firm bills of quantities

The adjective 'firm' is somewhat misleading because it gives the impression that the bill of quantities is of a stable or fixed nature where the quantities are unlikely to change. Most bills of quantities in this category are anything but 'firm'.

Traditionally, firm bills of quantities are used mainly in building work where the design has been developed to a sufficiently detailed stage that accurate quantities can be measured and billed. In practice, it is rarely the case that designs are anywhere near complete at the tender stage, and it is well known that contracts, tendered on the basis of 'firm' quantities, have lots of design issues to be resolved.

Despite tendering a 'firm price', contractors habitually find that the contract bills contain:

- Approximate or provisional quantities.
- Errors and omissions.
- Prime cost sums for work which cannot be quantified with any accuracy.
- Provisional sums for work that has yet to be designed.
- Contingency sums for the unexpected or 'known unknowns and unknown unknowns.'[2]

It is also common in bills of quantities to find that 'firm' measured work items are varied, added to or omitted during the course of the works due to design changes and other circumstances. This can play havoc with the contractor's planning and programming of the works and can have a detrimental effect on profit and overhead recovery. Whilst it is reasonable to suppose that no design can ever be complete in every detail, it is, perhaps, unreasonable that 'firm' quantities based on a supposedly complete design should be subject to the wholesale variations commonly experienced in the UK construction industry. For this reason, it is probably the case that traditional methods of valuing variations inadequately recompense the contractor and that the system of variations based on a contractor's quotation used in some standard contracts (e.g. JCT 2011 SBC/Q) is fairer, at least to contractors!

In any event, all standard forms of contract provide clauses for situations where variations, loss and expense/compensation events and errors in the contract documents occur and standard methods of measurement usually contain protocols that anticipate the need for flexibility in the contract bills. These protocols require, *inter alia*, that provisional quantities are clearly marked as such and the bill of quantities is appropriately structured to accommodate PC and provisional sums.

Risk issue

The so-called 'firm' bills of quantities that are not firm do not constitute a breach of contract because standard forms of contract anticipate a range of circumstances where uncertainty in the quantities and/or in other works required by the contract is expressly provided for in the contract terms. Provided the correct protocols are followed, there can be no breach.

In the same way, bills of quantities prepared on the basis of an incomplete design cannot be construed as not being in accordance with the standard method of measurement if the SMM in question contains protocols for the inclusion of approximate or provisional quantities, PC sums or provisional sums in the bills of quantities. Provided that the SMM rules are followed, there cannot be a breach of contract in such circumstances.

Contractors who sign up to contracts with 'firm' bills of quantities should be fully aware of the consequences of entering into a contractual agreement that may contain a considerable degree of uncertainty in the contract bills.

4.5.2   Approximate bills of quantities

Approximate bills of quantities might be considered to be more 'honest' than the so-called 'firm' bills of quantities because they are what they say they are – approximate. Normally employed on measure and value contracts, such bills of quantities are sometimes referred to as 'quantified schedules of rates'.

The quantities in an approximate BQ are estimated and amount to an approximation of the amount of work that the contractor can expect when the job is started. The final 'firm' quantities are derived by admeasurement as the work proceeds.

There is no need to mark the quantities as 'approximate' in a bill of approximate quantities because the tender documents will make it clear that:

- The tender sum is simply a total of the bill of quantities which enables tenders received to be compared with each other.
- The billed quantities are subject to admeasurement.
- The contract sum will be based on the final quantities derived from the admeasurement process.

This does not necessarily imply that the final quantities can be wildly different from the approximate quantities as some standard forms of contract provide for the adjustment of the contractor's rates when there is a significant increase or decrease in the eventual quantities:

- By providing the contract administrator with the express power to consider the impact of the quantities being greater or less than those stated in the bill of quantities; this could mean revisiting the original BQ rates with a view to deciding whether an increase or decrease is appropriate albeit at the contract administrator's discretion.
- By an express provision in the contract that recognises the contractor's entitlement to additional payment should the difference between the billed and admeasured quantities cause a change in unit cost.
- By an express provision in the contract for additional payment where *the final quantity × the original rate* exceeds *the original quantity × the original rate* by a stated factor or percentage.

In all other respects, approximate bills of quantities follow the same structure and method of measurement as 'firm' bills of quantities.

4.6   Formal 'quasi' bills of quantities

Formal 'quasi' bills of quantities may be taken as *resembling but not actually being*[3] bills of quantities but may nevertheless be issued 'formally' as part of the tender documentation issued by employers to contractors or by contractors to their subcontractors.

The three main categories of formal 'quasi' bills of quantities are:

1. Schedules of rates.
2. Schedules of works.
3. Priced specification.

4.6.1   Schedule of rates

Schedules of rates are used where the nature of the work required is known but not the extent, and they may therefore resemble bills of quantities but do not actually quantify the work required at the tender stage. The quantities are determined when the work is carried out, and the type of contract would thus be a **measure and value contract**.

Schedules of rates are commonly used for repair and maintenance work and on measured term maintenance contracts for social housing, government buildings or highways where the contractor is appointed for a fixed period (say, 3 years) to carry out an unquantified amount of maintenance work. The work would be described in detail in the specification and in the schedule of rates so that tenderers would be aware of the 'scope' of works, but few drawings would be available other than perhaps a site plan, a general arrangement drawing and, perhaps, standard construction details.

At the tender stage, the contractor prices the contract on the basis of a general understanding of the scope and scale of the work as described in the tender documents. The contract prices may be rates per unit for the items of work listed and lump sums where appropriate or, alternatively, a percentage deduction or addition to the rates quoted in the schedule. The eventual total price for the work carried out would be calculated by measuring the quantity of work actually done multiplied by the rate in the schedule.

Types of schedules of rates

There are three types of schedules of rates in common use:

- A bill of quantities-style document, written in SMM format, without any quantities. Such schedules may be standard 'in-house' documents, or they may be produced on an *ad hoc* basis for specific projects. In both cases, the contractor prices the various items of work described, but no tender total may be determined as there are no quantities.
- A bill of quantities-style document, prepared 'in-house' or on an *ad hoc* basis and not necessarily in accordance with a standard method of measurement. Tenderers would be invited to price the scheduled work items by either:
  - Entering unit rates or lump sums (if appropriate) against each scheduled item or
  - Stating a percentage addition to or deduction from rates already entered in the schedule.
- A published schedule, such as the PSA Schedules of Rates, the M3NHF (National Housing Federation) Schedule of Rates or the National Schedules of Rates (NSR) where:
  - Rates are already pre-entered and tenderers quote a percentage addition to or deduction from the scheduled rates or
  - Scheduled items are left blank for tenderers to enter their rates.

Schedules of rates format

There are several ways that schedules of rates may be presented or formatted according to the rates and level of detail required from tenderers.

The National Schedules of Rates (NSR)
The NSR is highly detailed with rates broken down into materials, labour and plant. There is a portfolio of NSR schedules of rates for Building Works, Electrical Services, Highways Maintenance, Housing Maintenance and Mechanical Services.

Table 4.1 shows an extract from the NSR, Part A: *Contractor's General Cost Items*, to illustrate the point.

PSA Schedules of Rates
In Table 4.2, it can be seen that the layout of the PSA schedule has the look of a bill of quantities but with the additional sophistication of a spreadsheet to cater for, in this case, different rates for the removal of different sizes of window. It will be noted that the items are 'written short' in a similar way to bill of quantities format but that the unit of measurement is shown as a heading and not in a 'unit' column.

Table 4.1   National Schedules of Rates.

	A: Contractor's general cost items		Mat. £	Lab. £	Plant £	Total £
A4487 115	5.00–6.00 m from ground level	nr			48.70	48.70
A4490	Lightweight aluminium access units – hire charge (weekly rate)					
200	Chimney scaffold unit to provide working platform to half of centre ridge stack	nr			135.00	135.00
205	Chimney scaffold unit to provide complete working platform around centre ridge stack	nr			270.00	270.00
230	Window access unit, 450 mm wide platform	nr			129,00	129.00
235	Window access unit, 600 mm wide platform	nr			129.00	129.00
240	Staircase access unit, 300–450 mm wide platform	nr			108.70	108.70
245	Staircase access unit, 600–675 mm wide platform	nr			108.70	108.70
A4491	Lightweight aluminium access units – hire charge (daily rate)					
200	Chimney scaffold unit to provide working platform to half of centre ridge stack	nr			27.00	27.00

Source: Reproduced with the kind permission of NSR Management.

A different version of this format is shown in Table 4.3, which illustrates a spreadsheet layout that enables different types of roof covering to be associated with common item descriptions. Once again, the unit of measurement appears as a heading.

The M3NHF Schedule of Rates

The National Housing Federation (M3NHF) Schedule of Rates is more of a 'family' or suite of documents. There are five versions of the schedule which are used by over 600 organisations and 3000 contractors and direct labour organisations.

The M3NHF schedule has an extensive infrastructure of measurement rules. These comprise:

- General rules.
- Measurement rules.
- General *deemed to include* measurement preambles.
- Trade/work section *additionally deemed to include* coverage rules.

The M3NHF measurement rules are illustrated in Table 4.4.

The M3NHF measurement rules underpin the measured items in the various schedules. Additionally, however, the schedules of rates items are subject to price framework rules (see Table 4.5) which, strictly, are contractual but also act as further item coverage for the measured items in the schedules.

Measured items are generally of a 'composite' nature as shown in the example in Table 4.5, but the rules and measured item descriptions, despite being comprehensive, also rely on the common law understanding that items shall include for everything that is *contingently and indispensably necessary* to complete the contract work.

M3NHF Schedules of Rates are revised from time to time, some more frequently than others.

Standard contracts used with schedules of rates

Some schedules of rates, such as the M3NHF schedule, come with a bespoke set of contract conditions which are specifically designed for repair and maintenance work. M3NHF publishes

Table 4.2 PSA Schedules of Rates (1).

C90: Alterations

	Windows			
	1	2	3	4
		Area		
Item C90	Not exceeding 1.00 m²	1.00–2.00 m²	2.00–3.00 m²	ADD for each additional 1.00 m²
Each	£	£	£	£
Taking out single or composite timber casement window and frame: timber surround: accessories: glass				
077 in conjunction with taking down wall..........	6.78	8.91	11.05	2.14
078 preparatory to filling in opening: cutting out fixings..........	31.00	40.26	49.88	9.62
079 preparatory to renewing window and/or frame: cutting out fixings: removing timber surround from frame..........	37.05	48.10	59.86	11.59
Taking out single or multi-light cased frame and sashes: accessories; weights: glass				
080 in conjunction with taking down wall..........	7.12	9.45	11.76	2.33

Source: Reproduced with the kind permission of Carillion.

Table 4.3 PSA Schedules of Rates (2).

				FELTS – continued				
				Bitumen felt – Each layer		Elastomeric bitumen polyester		
							Cap sheet	
	1	2	3	4	5	6	7	8
	Venting base layer	Bottom layer polyester-based bitumen felt	Bottom layer bitumen glass fibre	Bottom or intermediate layer	Sanded finish	Granule finish	Aluminium foil faced	Copper foil faced
Item J41 Square metre	£	£	£	£	£	£	£	£
Roof covering horizontal or to falls, crossfalls or slopes: any width not exceeding								
145 10° pitch	6.19	6.82	6.13	8.00	13.67	15.56	28.59	52.35
146 Sloping 10° to 45° pitch	6.84	7.33	6.73	8.53	14.48	16.51	29.93	53.80
147 Vertical or sloping exceeding 45° pitch	7.35	7.78	7.20	8.96	15.12	17.25	31.00	55.02

Source: Reproduced with the kind permission of Carillion.

Table 4.4 M3NHF measurement rules.

M3NHF measurement rules

GENERAL RULES
Schedule of rates descriptions

001 There are three levels of description for each schedule of rates item. Each of these, in particular, the long description set out in the scope of works envisaged for an order for that item.

002 Each item has a six character numeric code reference and a single-character alpha priority code reference:
Example:

125001 E   Chimney: Ball chimney flue, clear obstruction and clean up including all associated work, and remove waste and debris, – (as an emergency priority (see the following text))   IT   34.39

Items are grouped in the following sections:

MEASUREMENT PREAMBLES
The following are provided as indicative examples only and should be reviewed and adapted as necessary by the client, prior to incorporation into any tender or other contract documentation, to ensure that they are fully compatible with the maintenance service to be provided and the particular schedule of rates with which they are to be used:

Generally
Generally rates deemed to include

A. Rates for all schedule of rates items in all trades generally are deemed to include as appropriate for the following:
1. All work that can reasonably be deemed to be included either as good workmanship, including the provision of materials and plant, or accepted practice whether or not specifically referred to in this document, the client representative's decision on this will be final.

MEASUREMENT RULES

The rules for the measurement of items included in this schedule of rates will be those detailed within the All Trade Preambles and this appendix as follows:

1. For items in the schedule of rates which are measured (indicated in LM/SM/CM), the client will reimburse the service provider for works on the following basis.
The use of an item and the usage rate is less than 1 (one), the charge shall be as for 1 (one) whole. Where, however, more than 1 (one) whole is used reimbursement shall be pro-rata the item schedule rate i.e. 1.27 LM, SM or CM = 1.27 **to two decimal places, (multiplied) by the unit schedule rate.**

2. For items in the schedule of rates where the unit of measure is per no, (number) or IT (item), then the charge shall be as for 1 (one) whole multiplied by the unit schedule rate.

3. For items in the schedule of rates where the unit of measure is per HR (hour), then the

Brickwork and blockwork
Brickwork and blockwork rates deemed to include

A. Rates for brickwork and blockwork are additionally deemed to include as appropriate for the following:

1. All rough and fair cutting.
2. Forming rough and fair grooves, throats, mortices, chases, rebates and holes, stops and mitres and all like labours.
3. Raking out joints and hacking faces to form key for finishings.
4. Labour in eaves filling.
5. Centering to new and rebuilt flat or cambered arches

Table 4.5 Price framework rules and composite rates.

M3NHF schedule of rates

PART 2: Price framework rules

1		Schedule of rates items		
1.1		Service provider to execute works at contract rates		
	951009	R	Bath: install special needs type	NO 729.49
			Bath: renew bath with special needs bath with rim and internal bath seat complete with new taps and waste reconnect to existing supply pipework including any adjustments; provide new service valves if not already installed; complete with new plug and chain, plastic trap; connect to waste pipework; remove existing bath panels and framework and fix new proprietary front and end bath panels including all necessary fixings and supports; make good existing and renew glazed wall tile splashback fixed with adhesive including all rounded edges and labours, silicone sealant between splashback and bath, and crossbond; make good wall and floor finishes; and remove waste and debris.	
	951011	R	Bath: install special needs HIP type	NO 508.53
			Bath: renew bath with special needs hip bath complete with new tans and waste reconnect to existing supply pipework including any	

1.1.1 The service provider's tendered rates include all costs required to undertake the works including the following:

- Labour and all related costs (including travel and other non-productive time);
- Materials supply costs (including delivery and collection costs);
- Equipment (including scaffolding up to two storeys, tools and personal protective equipment);
- All waste, debris and waste disposal costs (including tipping charges, landfill tax and any similar costs) arising from materials;
- All temporary works and reinstatements;
- All payments to utility providers;
- Supervision, transport (including parking and/or congestion charges) depot and storage costs;
- Water supply for the works, including all necessary plumbing and removal of temporary facilities on the completion of void property works;
- Temporary artificial lights and electrical power and/or gas facilities;
- The temporary disconnection and protection of telephone installations including repositioning to maintain services releasing wires and cables before undertaking the

a complete set of tender documents which comprises an invitation to tender, articles of agreement, a price framework, contract details, the contract conditions, preliminaries and a key performance indicator (KPI) framework.

Alternatively, the JCT SBC/XQ 2011 'without quantities' form of contract might typically be employed for repair and maintenance work, but equally, a standard form of term contract could be used such as the JCT Measured Term Contract (MTC) 2011, the NEC3 Term Services Contract (TSC) or the NEC3 Term Services Short Contract (TSSC).

In common with other NEC3 contracts, the NEC3 TSC offers an option structure, thereby facilitating different types of contractual arrangement and various means of allocating price/ financial risk:

- **Option A** is a priced contract with a price list. The risk attached to the agreed prices is largely borne by the contractor.
- **Option C** is a target contract with a price list where the price risk is shared by the employer and contractor in an agreed proportion.
- **Option E** is a cost reimbursable contract where the price risk is largely carried by the employer.

The payment mechanism in the TSSC is based on agreed rates and prices.

For an MTC under the TSC, Option A (priced contract with a price list) would be used in conjunction with secondary Option X19 (Task Order). The price list could be a published or a bespoke schedule of rates. Work is instructed by the service manager on the basis of task orders which should contain a detailed description of the work required and a priced list of items, as well as details of commencement and completion dates and delay damages (if required), and so on.

4.6.2   Schedule of works

Schedules of works are, usually, unquantified lists of work to be undertaken for a building project, usually works involving demolitions, alterations or repairs. They should not be confused with:

- **Work schedules** which are more correctly associated with Gantt or bar charts and other forms of programme or visual displays of work related to time.
- **Activity schedules** which are lists of unquantified construction activities, prepared and priced by the contractor, often, but not necessarily, linked to the contractor's programme.

Schedules of works tend to be prepared by building surveyors rather than quantity surveyors, or they might be drafted by architects as part of the service provided to clients. They will usually have a 'preliminaries' section and will accompany the drawings and specification which, together, will comprise the tender documentation. In its priced form, the schedule of works will be an important contract document and will be used as production information as well as for interim payments and the valuation of variations to the contract.

Schedules of works describe the work required in the form of 'composite' items as opposed to the individual measured items that would appear in a bill of quantities. They do not follow a particular method of measurement, and thus, the style, content and quality of the item descriptions can be very variable. **Composite items** include several work items that would normally be separated in a conventional bill of quantities.

Schedules of works usually consist of unquantified items in which case it is the tendering contractors that must judge the extent of the work required and the amounts of materials needed to complete the contract. In other words, they bear the quantity risk. In this context, it is important to read schedules of works in conjunction with any drawings available, and a site visit is indispensable to understanding what is to be priced.

Different formats are possible with schedules of works, and they can be arranged:

- Elementally (e.g. substructure, superstructure, internal finishes).
- In work packages (e.g. demolition works, groundwork, concrete works, external and internal structural walls).
- In trade or work sections (e.g. concrete work, brickwork and blockwork, drainage).
- On a room-by-room basis (e.g. entrance/reception, consulting room 1, treatment room A).

Where a schedule of works includes details of the specification for materials and the quality of workmanship, it may be referred to as a 'specified schedule of works'.

Schedules of works tend to be used on smaller projects, or for demolition and alteration work, as an alternative to bills of quantities, but it is not inconceivable that the monetary value of such projects could be significant. The main point about schedules of works is simplicity. It is probably not worthwhile measuring the individual items of work involved because the quantities are too small. It is therefore more sensible to combine several individual items in a collective item that the contractor can nonetheless price. It is unlikely that the contractor will attempt to quantify the work involved but will more likely 'spot' price the items using experience of similar work as the price database along with a 'guesstimate' of the materials required (e.g. 1 nr pack of facing bricks, 6 nr bags of multi-finish plaster, 4 nr 10 litre tubs of emulsion paint, etc.).

There are no rules governing the style or content of schedules of works, and the compiler is free to choose whatever form of presentation is preferred. Some guidance in formulating item descriptions may, nevertheless, be found in SMM7 Class C10 (demolishing structures) and Class C30 (spot items). An example of a schedule of works is shown in Table 4.6, but this is not claimed to be 'typical' or 'representative' of industry practice.

Table 4.6   Schedule of works.

Reference	Description	Price
A	Remove skirtings, take up defective flooring and lift existing floor joists and wall plates; remove debris from site; bed new 50 × 100 mm treated softwood wall plates on existing sleeper walls including polymer-based DPC; install new 50 × 150 mm treated softwood floor joists and spike to wall plates; lay new 20 mm softwood tongued and grooved floor boarding; supply and fix new softwood skirtings to match existing. *Approximate areas:* Room A: 6 m × 4.5 m Room B: 4 m × 4 m Room C: 4 m × 3 m Room D: 3 m × 3 m	
B	Remove existing single flight timber staircase, balustrading and strings and remove from site; make good all plasterwork (new staircase and all decorations itemised separately).	
C	Form opening in 250 mm cavity wall faced one side; carefully cut out for and install proprietary galvanised steel boot lintel and pin to existing brickwork; quoin up jambs and point to match existing facework; make good internal plasterwork; supply and install 1200 × 1200 mm softwood casement window complete with brass casement stay and lockable window fastener; supply, cut and fit 25 × 250 mm softwood bullnose window board and plug and screw to brickwork; plugs to be pelleted and sanded flush; remove all debris from site; wall to be temporarily supported with suitable needles and propping whilst the work proceeds (decorations itemised separately). Quantity: 8 nr on two floors	

Risk issue

Schedules of works place more risk on the contractor than a bill of quantities as care has to be taken to ensure that all the necessary costs have been included in the tender price despite the lack of precision in the item descriptions.

It is also much more difficult to value variations with a schedule of works than it is with either bills of quantities or schedules of rates.

Some compilers like to quantify schedules of works, and this might be done by:

- Creating a 'quantity' column similar to that in a bill of quantities.
- Quantifying the component parts of each item within the item itself.

Risk issue

There would seem to be little advantage to be gained from quantifying a schedule of works, the net effect of which is to transfer the quantity risk from the contractor to the employer.

This risk is compounded due to the lack of any supporting measurement rules for schedules of works. Measurement rules provide a degree of protection for the bill compiler/employer as there can be no misrepresentation of the quantities provided that the rules of measurement are followed.

If item descriptions in a schedule of works are incomplete or misleading, the employer may be faced with claims for misrepresentation.

The additional work in quantifying the schedule of works items would also increase the lead time for the preparation of the tender documents, and employers would be faced with additional costs and/or professional fees.

4.7   Formal 'operational' bills of quantities

Operational bills of quantities for production-orientated tendering were first developed in the 1960s at the Building Research Station (now BRE) at Garston, near Watford, United Kingdom, by senior researchers Forbes and Skoyles (1963; Skoyles, 1966, 1967). The late Edward 'Ted' Skoyles was quite a character (I spent the morning with him in 1972) and was someone, on a personal view, ahead of his time.

The operational bill is likened by some to the priced activity schedule used for Options A and C of the NEC Engineering and Construction Contract, but there are a number of significant differences between them:

- The operational bill was a 'formal' tender document prepared the employer's QS (not by the contractor as with the NEC activity schedule).
- It attempted to quantify building work in terms of the activities or operations required and the sequence of construction.
- The operational bill was considerably bulkier even than the equivalent measured bill of quantities and much more so than an activity schedule.
- It required considerable computing power to manipulate all the data (lacking in the 1960s).
- It did not fit conventional estimating methods and databases used at that time, but activity schedules are popular and fit well with modern software.

An operational bill comprised discrete items of construction work (operations) which could be distinguished from other operations. An operation was defined as *a piece of construction work which can be carried out by a gang of operatives without interruption from another gang.* The operational bill was supplemented by a precedence diagram which indicated the sequence and concurrency of operations one to another which could be developed by the contractor using the critical path analysis programming technique that was popular at the time.

From memory, the operational bill had a number of sections, and some of the main features included:

- Operations scheduled with materials quantities in purchasing units (e.g. rolls of damp-proof membrane, bricks per thousand).
- A global labour item with the amount of work described which the contractor would then assess.
- Separate identification of work carried out off-site, such as prefabricated components.
- Management and plant resources at the end of the bill rather than at the front as is the case with the (similar) present-day time and method-related charges.

Jaggar et al. (2002) remember that the operational bill failed for a number of reasons but especially because of design team inertia and because the industry at the time gave it a hostile reception. The researchers clearly came at the problem from the wrong angle – the design team should not be trying to anticipate how the contractor will carry out the work – and they perhaps did not appreciate the impact of the perennial problem of producing bills of quantities from an incomplete design.

NEC activity schedules are clearly related to operational bills but only distantly, so why bring up a subject that resides over 50 years in the past? The answer is BIM!

It is quite clear to this author (who might be considered by some as a heretic!) that there is a great deal of synergy between operational bills and BIM models:

- BIM models comprise objects (e.g. precast concrete wall panel).
- Operational bills comprise operations (e.g. the supply, erection and fixing of precast concrete wall panel).
- BIM objects are 'intelligent' and can carry all sorts of data – structural, thermal, acoustic, **time** and **quantity**.
- BIM data can be quickly manipulated by computer, and the 3D model can be linked to time (4D) and cost (5D).

Using BIM model data in conjunction with a type of 'operational bill' would encourage 'operational estimating' (pricing based on time and resources) rather than the conventional unit price estimating method associated with traditional bills of quantities. This would be a good fit with both activity schedules and linked bar chart software such as Asta Powerproject, Project Commander or Microsoft Project, and the entirety would fit well with a modern family of construction contracts such as the NEC.

Based on a simple method of measurement, suitably aligned with BIM output data, twenty-first century operational bills could be used both formally and informally according to the preferred procurement strategy.

4.8   Informal bills of quantities

An 'informal' document may be defined as being *not of a formal, official… or conventional nature… appropriate to everyday life or use … characterized by idiom, vocabulary, etc., appropriate to everyday conversational language rather than to formal written language.*[4]

This definition is characteristic of the so-called builders' quantities that are widely used by subcontractors, trade contractors and main contractors. Builders' quantities are meant to be 'internal' documents, not seen by outsiders to the company or organisation using them.

The term 'builders' quantities' is something of a colloquialism, however, which has no formal definition and yet is in common use in the construction industry. Everyone understands what it means and yet it can mean different things to different people.

At one end of the spectrum, a plastering contractor can look at a room and work out a price on the basis of one bag of multi-finish and half a day's labour. At the other end, a large contractor pricing a DB tender can produce a bill of quantities fairly quickly without worrying too much about the finer detail.

4.8.1   Builders' quantities

To an extent, there is an understanding that builders' quantities are somewhat 'rough and ready', lacking the precision of a PQS measuring to a standard method of measurement. There is also the understanding that builders' quantities are good enough to produce an estimate but not good enough as a basis for formal competitive tendering.

Whilst there is undoubtedly some truth in this, it is also true that, certainly from the writer's personal experience, builders' quantities have to be good enough for formal competitive tendering because:

- Procurement methods that aim to move the quantity risk to the tendering contractors are on the increase (all available evidence points to this).
- Contractors are increasingly asked to tender for traditional contracts on the basis of partial contractor design; this usually requires the preparation of informal detailed quantities in order to price the 'global' item(s) provided in the formal bill of quantities.
- Design and construct and drawings and specification tenders oblige contractors to produce their own quantities and take the associated risk on board.
- Contractors are comfortable with the Pareto principle that some 20% of the items measured represent 80% of cost and that the remaining 80% of items are not worth the time and effort of measurement.
- The remaining 20% of cost (80% of items) is not an entire risk because estimators habitually make allowances for the low-cost items in their pricing; this is a matter for personal judgement and can be a bit 'hit-and-miss'.
- The risk attached to builders' quantities is 'manageable' within the bidding process, but the judgements made at tender adjudication may have to be made on less precise data than for conventional tendering.

It is true, of course, that some contractors turn to QS practices who offer 'bills of quantities' services to contractors and trade contractors. By providing bills of 'principal quantities', but without detailed descriptions and without measuring the smaller, less cost-significant items, this can be done at a reasonable cost.

Care needs to be taken by estimators to appreciate which items of work have not been measured, and this is why some contractors prefer to prepare the quantities themselves 'in-house'. QS software with the capacity to measure direct from drawings makes this a much easier task than traditional paper-based 'taking-off'.

4.8.2   Uses of builders' quantities

Contractors frequently use builders' quantities 'behind the scenes'. They are used to build up the contractor's price but are not normally seen by the employer. Situations where this can happen include:

- Large contracts where the formal bill of quantities contains 'composite items' that cannot be priced without drilling down into more quantitative detail. Typical examples are where an item is provided in the bill of quantities comprising a dimensioned description of a staircase but no detailed quantities are given or when a manhole or other complex structure is itemised but not measured in detail. In order to price the items in question, detailed quantities are needed but not necessarily in accordance with a standard method of measurement.

- 'Drawings and specification' tenders where no quantities are supplied by the employer. Tenderers are obliged to calculate their own quantities from the drawings supplied with the tender documents, and this enables a lump sum tender figure to be determined. Once the contractor's tender is accepted by the employer or if the tender is of interest before formal acceptance, the contractor may be asked to provide a breakdown of the lump sum. This provides a means for the employer to both scrutinise the prices and check for errors and also value work in progress and variations to the contract once work gets underway.
- Design and build/construct contracts where tenderers not only supply a design but also assume responsibility for quantifying the design to enable a tender sum to be determined. Design and build/construct contracts may be let on either a lump sum or measure and value basis, but in both cases, it is usual for the employer to request a breakdown of the tender figure for pre- and post-contract use. The request for a breakdown may be made in the tender invitations or as a formal requirement of a contract such as the JCT 2011 DB contract.

4.8.3   Preparation of builders' quantities

Perhaps being reflective of the varied structures of the construction industry, there are several ways that builders' quantities can be prepared:

- *Ad hoc* methods:
  - 'Seat of the pants' judgement or the back of a cigarette packet (for smokers!).
  - Handwritten dimensions on drawings, scraps of paper, notepads or notebooks.
  - Handwritten dimensions, perhaps on dimension or estimating paper.
- Direct entry software:
  - Spreadsheets such as Microsoft Excel or OpenOffice.
  - Direct entry into non-SMM-based measurement software from paper-based PDF drawing files and the like.
- On-screen measurement:
  - 2D on-screen measurement from PDF or CAD files.
  - 3D on-screen measurement from CAD files.
- Quantities extracted from BIM models.

For smaller projects, perhaps where the estimator is also producing the quantities, *ad hoc* methods are fine. The estimator understands exactly what has been measured and what has not and can make appropriate allowances in the pricing to cater for smaller, less cost-significant items.

On larger projects, direct entry methods would be the minimum standard, but measurement software is much quicker than spreadsheets unless the work involved is repetitive and a spreadsheet 'template' could be used. The beauty of measurement software, such as Buildsoft Cubit, is that the output can be 'dumped' into Excel spreadsheets, retaining the formatting, which can then be shared and manipulated between members of the estimating team or sent out to subcontractors once the main cells have been locked.

On-screen measurement takes a bit of getting used to but, once mastered, is a great way of producing accurate quantities quickly. Non-SMM-based software, such as Buildsoft Cubit, is probably best because SMM-based software demands too much detail in the item descriptions albeit the less cost-significant items can be overlooked for speed and simplicity.

Irrespective of the method used, however, where there is a team of estimators, some structure to the preparation of the quantities is needed. The obvious answer is to use a familiar standard method of measurement, but this may well be too complex when time is at a premium.

Risk issue

At a time when responsibility for measurement risk is being passed down the supply chain, contractors and subcontractors need to be sure that the quantities they produce follow a structured approach which is essential to avoid costly mistakes and reduce tender risk.

An experienced individual can do this in an *ad hoc* way, but when more than one person is involved, something more formal is needed. A set of 'in-house' procedures would work but takes time and effort to establish, and so, perhaps the easiest solution is to use a recognised standard method of measurement that is not too complex or time-consuming to employ.

POM(I) offers a simple method of measurement with few rules that cover a wide range of building, civil engineering and M&E work. The downside to this method of measurement is that many of the measured items are 'composite' and require further detailed measurement; additionally, the item coverage provided lacks sufficient detail to be clear and unequivocal.

Perhaps a better bet would be to use NRM1. Admittedly, it is designed for early cost estimating and cost planning, but it could equally well be used for producing good-quality, consistent builders' quantities. The big advantage with NRM1 is that the item coverage rules are extensive and clear so that everyone involved should fully understand what is included in the BQ items with the bonus that the time and effort spent to produce the quantities would be reduced.

The advantages of using a simple, yet standard, method of measurement include:

- Subcontract/work package tenders can be invited on the basis of a consistent set of quantities giving all tenderers a 'level playing field'.
- Comparison of subcontract quotes is easier, more consistent and on a 'like-for-like' basis.
- Subcontracts can be awarded with everyone knowing what is included in the rates.
- Remeasurement and the valuation of variations are based on recognised rules that are written into the contract.

4.9 Quantities risk transfer

Where contractors undertake the risk of producing the quantities, a breakdown of the contractor's price is usually required by the employer for pre-contract checking and post-contract valuation of work in progress and variations to the contract.

Risk issue

Entering into a contract or subcontract without an accepted 'list of prices' is a known shortcut to arguments and disputes.

Very few construction projects reach a conclusion without changes being made, and some agreed basis for valuing these changes must be in place to avoid problems. A contract based on a 'fair and reasonable' valuation of changes is not advisable as each party to the contract will have a different idea as to what is 'fair' and what is 'reasonable'.

A breakdown of the contract price is also advisable for the purpose of valuing work in progress. In this regard, employers will wish to avoid the risk of overpaying, and contractors and subcontractors will want to make sure that cash inflows from the project are as positive as possible.

Most standard contracts provide a means for valuing work in progress and variations and other changes. The traditional way, of course, is to use a 'formal' bill of quantities, but where the method of procurement precludes this, another form of analysis is needed. Where the employer chooses not to issue a formal pricing document – DB tenders, drawings and specification tenders, for example – it will be the contractor's responsibility to prepare a breakdown of the contract price or list of prices. However, most contractors would be reluctant to submit their builders' quantities for this purpose because:

- Builders' quantities are a means of pricing a tender and may contain errors, omissions or 'hidden' contingencies that the contractor would prefer to keep private.
- Contractors are generally suspicious of employers' representatives and may fear that their prices may be 'picked apart' to their disadvantage.
- Builders' quantities may be handwritten and may include pricing notes that the contractor would prefer the employer not to see.
- Builders' quantities are 'raw' documents and contractors might prefer to submit a document where the money is in all the right places to ensure the best positive cash flow for the contract.

Popular methods of achieving a breakdown of contractors' tenders appear in standard contracts such as the NEC and JCT families of contracts and include:

- Activity schedules.
- Price lists.
- Contract sum analyses.

4.10   Activity schedules

An activity schedule, or unquantified price list, is a list of activities, determined by the contractor or specified in whole or in part by the employer, wherein the contractor prices lump sums for the various activities listed in order to arrive at the contract sum or target price.

The format and content of an activity schedule will depend upon the form of contract used. In the JCT 2011 SBC/Q, for instance, activity schedules (if supplied by the contractor) are used in conjunction with bills of quantities as a means of simplifying the valuation of work in progress with little prescription as to their content. In the NEC3 ECC form, activity schedules are used 'stand-alone' as a pricing document with prescriptive rules as to how work in progress is to be valued and how activity schedules are to be used in conjunction with the contractor's programme of works.

Activity schedules are invariably prepared by the contractor and are therefore not part of the tender documentation. Once priced and accepted by the employer, however, the activity schedule will be incorporated into the contract along with the other documents supplied at the tender stage such as the drawings and specification.

Activity schedules are often linked to the contractor's programme, and sometimes, there is a contractual requirement to this effect (e.g. ECC Option A). This makes sense from the perspective of ensuring that interim payments are geared to actual progress on-site.

An extract from a simple activity schedule is illustrated in Table 4.7 from which it can be seen that:

- The item descriptions are simple and do not correspond to any specific method of measurement.
- Each activity is referenced A, B, C, etc.
- Each activity is referenced to the contractor's programme.
- A column is provided for the contractor's lump sum price for each activity.

Table 4.8 shows a slightly different presentation where the activity schedule is more detailed:

- Activity descriptions correspond exactly as per the contractor's programme (see Figure 4.1).
- The start and finish dates for each activity and group of activities are shown.
- The duration of each activity is indicated.

Table 4.7   Activity schedule v.1.

	Activity schedule		
Activity reference	**Activity description**	**Programme reference**	**Price, £**
A	Form access and set up contractor's compound and site offices	1	
B	Excavate to remove topsoil and store in temporary spoil heaps on-site	2	
C	Excavate to reduce levels and remove spoil to tips of site	3	
D	Excavate pad foundations, remove spoil from site and prepare formation for concrete	4	
E	Supply and fix steel reinforcement cages to pad foundations	5	
F	Pour ready-mixed concrete to pad foundations against earth faces	6	
etc.			

- A payment preference column is included where the contractor can indicate how payment is required:
  - % completion in accordance with progress.
  - Monthly on completion of the activity.
  - *Pro rata* to time elapsed (e.g. for design activities or preparation of the health and safety file).

Should the programme change during the course of construction, then it makes sense that the activity schedule is changed accordingly so as to correspond to the current programme.

There might be problems under ECC Option A (priced contract with activity schedule) with the activity schedule and programme as illustrated in Table 4.8 and Figure 4.1 respectively, as the programme tasks are grouped under a 'summary' heading and could be regarded as 'grouped activities'. As a consequence:

- The contractor would only be able to apply for payment when all the activities in the group are completed satisfactorily.
- Site establishment could be included in the first valuation, but earthworks, being over 1 month's duration, would have to be deferred until the second payment application.
- Pad foundations are less than 1 month in duration but could not be included in Valuation No. 2 because the concreting would not be completed until after the end of week 8 and would therefore appear in Valuation No. 3.

It can be seen, therefore, even for a simple project, that the contractor's cash flow can quickly be eroded as a result of the provisions of ECC Option A.

'Activity schedule' should not be confused with the term 'work schedule' which is generally taken to be a list of work items related to time in the form of a Gantt or bar chart or other types of programme or visual display of work related to time. However, where a work schedule has a sum of money attached to each 'bar', then this could be an activity schedule.

Risk issue

Except for fairly small projects, it is not sensible to have a 'combined' work schedule and 'activity schedule'. The benefits of each would be compromised, and there is the added danger that the programme could be unwittingly incorporated into the contract.

Legally speaking, this would not be in the employer's interests following the decision in *Yorkshire Water Authority v Sir Alfred McAlpine & Son (Northern) Ltd* (1985).

Table 4.8   Activity schedule v.2.

						Price	
	Activity schedule						
Line	Name	Duration	Start	Finish	Payment preferences	£	p
1	Pre-start procedures	4w	01/06/2015 08:00	26/06/2015 17:00			
2	Design & planning	20w	10/06/2015 08:00	28/10/2015 17:00			
3	Health and safety file	6w	29/06/2015 08:00	07/08/2015 17:00			
4	Gantry crane	19w 3d	09/06/2015 08:00	23/10/2015 17:00			
5	Procurement	12w	09/06/2015 08:00	01/09/2015 17:00			
6	Installation and testing	2w	12/10/2015 08:00	23/10/2015 17:00			
7	Site establishment	3w	29/06/2015 08:00	17/07/2015 17:00			
8	Remedial works	2w	07/07/2015 08:00	20/07/2015 17:00			
	PHASE 1						
9	Bulk excavation	4w	21/07/2015 08:00	17/08/2015 17:00			
10	Piling work	7w 4d	04/08/2015 08:00	28/09/2015 17:00			
11	Access and piling mat	1w	04/08/2015 08:00	10/08/2015 17:00			
12	Piling	3w	10/08/2015 08:00	28/08/2015 17:00			
13	Test piles	4w	01/09/2015 08:00	28/09/2015 17:00			
14	Pad foundations	5w 1d	25/08/2015 08:00	30/09/2015 17:00			
15	Trim piles	2w	25/08/2015 08:00	08/09/2015 17:00			
16	Excavation	2w	04/09/2015 08:00	17/09/2015 17:00			
17	Rebar	3w	08/09/2015 08:00	28/09/2015 17:00			
18	Concrete	2w	17/09/2015 08:00	30/09/2015 17:00			
19	Structural steelwork	5w	01/10/2015 08:00	04/11/2015 17:00			
20	Ground floor slab	7w	23/10/2015 08:00	10/12/2015 17:00			
21	Sub-base	2w	23/10/2015 08:00	05/11/2015 17:00			
22	Formwork	4w	02/11/2015 08:00	27/11/2015 17:00			

Table 4.8   (*Continued*)

			Activity schedule				
Line	Name	Duration	Start	Finish	Payment preferences	Price £	p
23	Rebar	2w	18/11/2015 08:00	01/12/2015 17:00			
24	Concrete	2w	27/11/2015 08:00	10/12/2015 17:00			
25	Envelope	7w 2d	11/12/2015 08:00	15/02/2016 17:00			
26	Brickwork	6w	11/12/2015 08:00	04/02/2016 17:00			
27	Cladding and roofing	5w	12/01/2016 08:00	15/02/2016 17:00			
28	Services	4w 4d	03/02/2016 08:00	07/03/2016 17:00			
29	Plumbing	2w	16/02/2016 08:00	29/02/2016 17:00			
30	Electrical	3w	16/02/2016 08:00	07/03/2016 17:00			
31	HVAC	3w	03/02/2016 08:00	23/02/2016 17:00			
32	Roller shutter door	1w	16/02/2016 08:00	22/02/2016 17:00			
	PHASE 2						
33	Decant existing building	1w	08/03/2016 08:00	14/03/2016 17:00			
34	Asbestos removal	3w	15/03/2016 08:00	06/04/2016 17:00			
35	Internal fit out	6w	07/04/2016 08:00	19/05/2016 17:00			
	PHASE 3						
36	External works	11w 1d	16/02/2016 08:00	06/05/2016 17:00			
37	Drainage	4w	16/02/2016 08:00	14/03/2016 17:00			
38	Roadworks	6w	15/03/2016 08:00	27/04/2016 17:00			
39	Landscape	3w	15/04/2016 08:00	06/05/2016 17:00			
40	Clear site	2w	09/05/2016 08:00	20/05/2016 17:00			
41	Test and commission	2w	23/05/2016 08:00	06/06/2016 17:00			
42	Demobilise site	2w	07/06/2016 08:00	20/06/2016 17:00			

Tender
total £ _____

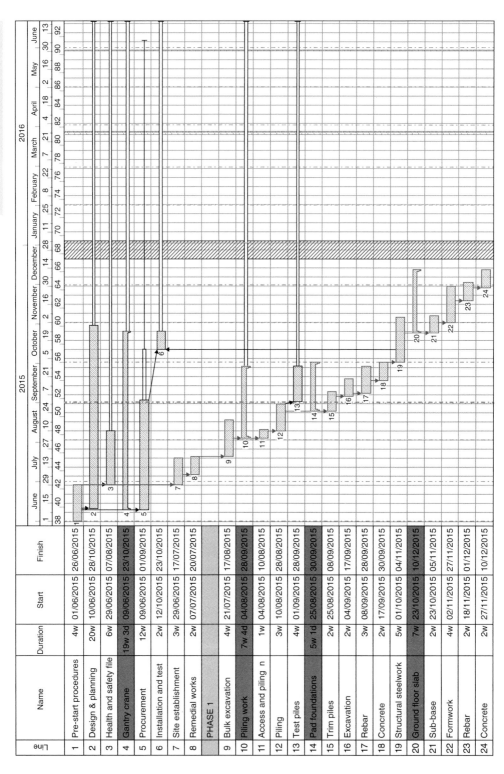

Figure 4.1 Contractor's programme.

4.10.1   Activity schedules generally

The activities appearing on an activity schedule are usually determined by the contractor, but on occasion, the employer may wish to have some input into this. This would normally be done by listing the particular activities to be included in the tender documentation. Overprescription by the employer of what should be in the activity schedule may not be wise particularly where the activity schedule is linked to the contractor's programme.

Activity schedules may equally include administrative activities as work-based construction activities. Therefore, items such as design work and preparation of the health and safety file may appear in the activity schedule alongside the likes of earthworks, drainage or the construction of a complete structure. Temporary work items such as constructing haul roads and steel sheet piling may also be listed as 'activities' as might the testing of completed work.

An activity schedule is essentially a breakdown of a lump sum tender into a series of smaller lump sums with an accompanying description of what each activity is. Activity schedules do not contain any quantities and, as such, need to be underpinned by 'base' quantities so that:

1. The estimator is able to price the work, item by item, in the usual way and can then aggregate items into the 'activities' which will appear in the activity schedule.
2. The contractor can produce a meaningful pretender programme that is based on the quantity of work to be done rather than 'guesswork'.

Just who is to be responsible for preparing the 'base' quantities is a matter for the employer to decide according to the chosen procurement strategy, and this has to be balanced with the advantages and the drawbacks of using activity schedules. Essentially, activity schedules are a means of transferring quantity risk from the employer to the contractor, but Broome (2013) suggests that some employers have found it appropriate to reduce the contractor's risk by using a variety of mechanisms whereby quantities are supplied to tenderers either by or on behalf of employers or by third parties.

4.10.2   Preparation of the 'base' quantities

The foundation to an activity schedule is the underlying or 'base' quantities which enable tenderers to both price the tender and prepare the programme that is usually linked to it. Current industry practice suggests that there are three methods of preparing the 'base' quantities:

a) <u>Each tenderer prepares their own quantities</u>.
   This is the most obvious way to prepare the 'base' quantities and may be used for straightforward projects and for individual 'trades' where there are relatively few items of work to measure.

   For a project of any reasonable size or complexity, however, it may be costly and/or counterproductive to follow this route to quantification, especially in a competitive tendering situation. Not only is this method costly and time-consuming for tenderers, it is also wasteful of scarce industry resources when, statistically, only one contractor in five or six can prepare the 'winning' quantities in any one tender competition.
b) <u>Tendering contractors arrange for a bill of quantities to be prepared and share the cost amongst themselves.</u>

To some extent, this harks back to nineteenth-century contract practice where contractors tendering for lump sum contracts would each appoint their own surveyor to measure the quantities with the contractor making the most mistakes winning the tender. Mainly due to cost and the wide variation in tender prices received by employers, this system developed to the point where tenderers would jointly appoint a surveyor to prepare quantities for all of the contractors with the cost being shared.

This method is being used today, but there are dangers that such practice could contravene collusive tendering legislation.[5]

Risk issue

Contractors who jointly appoint a quantity surveyor to prepare quantities, which they each pay for and use, risk heavy fines if this is done without the knowledge or consent of the employing authority.

In such cases, it is advisable to obtain written permission in advance from the employer.

---

c) **The employer supplies a bill of quantities when choosing to use a priced contract based on an activity schedule.**

At first glance, this seems to be counterproductive as it may be thought that the employer could just as well use a bill of quantities tender arrangement and have done with it. This would, however, ignore both the benefits of using an activity schedule and the disadvantages of bills of quantities.

There appear to be three common methods in current practice:

1. Where the employer has a bill of quantities prepared in-house but the BQ is kept separate from the tender documentation. The BQ is supplied to tenderers on a 'without prejudice' basis but does not become a contract document should the tender be accepted.
2. Where the employer prepares a list of principal quantities to help tenderers with stated confidence limits ± within which variances are at the contractor's risk. Variances of quantity outside the confidence limits would be at the employer's risk.
3. Where the employer provides the documentation necessary for a firm of quantity surveyors to prepare a bill of quantities who then sell the BQ direct to tenderers in order to recoup their costs.

Each of these methodologies has its drawbacks. Quantities prepared for or on behalf of the employer may fall foul of a claim for misrepresentation under the Misrepresentation Act 1967 should the quantities be proved to be wrong. Quantity surveying firms also risk being sued should they make mistakes that cannot be recovered under the contract.

These disadvantages may be outweighed, however, by the reduced cost to the industry of preparing several independent bills of quantities and also by the benefits of careful scrutiny of the design which emanates from the discipline of preparing detailed quantities. This scrutiny may lead to fewer errors, variations and claims as small points of detail would inevitably be unearthed during the measurement process that could be rectified before inviting tenders.

Apart from giving tenderers confidence in their own assessment of the quantities, bills of quantities supplied by the employer, or by a third party, also help to ensure consistency in the tendering process and a more balanced and closer tender competition than would otherwise be the case.

4.10.3   Advantages of activity schedules

Activity schedules avoid many of the drawbacks with bills of quantities including differences of opinion concerning provisional quantities and how 'firm' the 'firm quantities' are. The employer is probably paying a premium for the contractor to take the 'quantity risk' but, conversely, is not paying quantity surveyors to prepare bills of quantities or carrying the risk that the quantities may be wrong.

Activity schedules are popular with users of the Engineering and Construction Contract, including Highways England, and have an attractive list of benefits:

- Brevity and simplicity.
- Linked to the contractor's programme, provides a simple means of payment for work in progress directly related to actual progress.
- With payment for completed activities only, there are no arguments over what is and what is not complete.
- In the absence of unit rates, it is easier for contractors to 'front-load' activities to improve cash flow.
- Easy for the employer to state requirements for the inclusion of provisional sums and the like in the tender documentation or employer's requirements.
- Employers may state what activities are to be included in the activity schedule to suit their own funding and other needs.

4.10.4   Disadvantages of activity schedules

As with all things, activity schedules have drawbacks that need to be weighed in terms of the employer's procurement requirements and the balance of risk desired for a particular contract. With tenderers responsible for preparing them, and the underlying quantities, the use of activity schedules needs to be carefully considered. There is no doubt that the industry carries an additional burden of cost as each tenderer prepares his own submission, but this has to be balanced with the probable reduction of disputes and legal costs associated with bills of quantities and other traditional pricing documents.

The disadvantages of activity schedules include:

- Someone has to prepare the quantities upon which the activities in the activity schedule rely on – the 'base' quantities. This could be the contractor or the employer (see Section 4.10.2).
- Activity schedules require careful drafting, especially when linked to the contractor's programme.
- Employers who state what activities are to be included in the activity schedule risk prejudicing the contractor's ability to link the activity schedule to the master programme.
- Activities with a long duration (exceeding 1 month) may be problematic for payment purposes if they cannot be subdivided or apportioned, unless the contract so provides (c.f. JCT 2011 SBC/Q and NEC3 ECC).
- With no unit rates, variations are more difficult to value, and it is probably best to include a contractual provision requiring contractors to submit quotations for ordered variations.
- The employer has less idea of whether the contractor is being overpaid due to front-loading of the lump sums.
- The employer or employer's agent may decide to 'group' activities for payment purposes which may disadvantage the contractor's cash flow.
- Activities may have to be grouped in order to reduce the length and complexity of the activity schedule.
- Where the contractor decides to 'group' activities, payment may be on the basis of completed groups and not individual activities within the group.
- Grouped activities can lead to considerable capital lock-up in the contract which can adversely affect the contractor's working capital requirements.
- Employer's agents need to be 'on the ball' with regard to checking for defects in 'completed' work prior to certifying payment thereof.

Part 1

4.10.5   Activity schedules under the JCT SBC/Q

The JCT 2011 SBC/Q anticipates that a priced activity schedule may be annexed to the contract should the contractor decide to supply one to the employer. Having said that, there is no reason why the employer cannot require an activity schedule from the contractor, but this would need to be prescribed in the tender documentation.

The main distinction between the use of activity schedules under the JCT contract and ECC Options A and C is that a bill of quantities forms part of the contract under the JCT contract and the activity schedule is optional, whereas under the ECC, there is no 'formal' bill of quantities and the submission of an activity schedule is a contractual requirement.

There is little prescription as to the form or content of activity schedules under the JCT form of contract save to say that:

- Each activity should be priced.
- The sum of the prices should equal the Contract Sum.
- The sum of prices should exclude:
  - Provisional sums.
  - PC sums and any contractor's profit thereon.
  - The value of work for which there is an approximate quantity in the contract bills.

There is no link between the contractor's programme and the activity schedule under the JCT contract, and the list of activities is not required to bear any relation to the operations on the programme.

Where an activity schedule is included in the contract, the JCT 2011 SBC/Q specifies that interim payment is determined by applying the proportion of work completed in an activity to the price stated for that work in the activity schedule. Again, this contrasts with the ECC approach of payment for completed activities only and raises the issue of potential disagreements as to what portion of the work has been completed and whether it has been completed properly.

The major advantage of activity schedules under the JCT contract is that there is a default priced bill of quantities to rely upon for the valuation of variations to the contract. In this sense, the JCT approach is less transparent than that of the ECC, which values compensation events (including variations) on the basis of changes in resource costs, but the upside is that contract administration is less complex under the JCT contract, especially regarding the valuation of variations.

4.10.6   Activity schedules under the ECC

The ECC has six main payment options (A–F) to choose from which determine both how the contractor will be paid under the contract and what the balance of risk will be. Activity schedules are used under Options A (priced contract) and C (target contract).

Option A provides the employer with a lump sum price and ostensibly transfers quantity risk to the contractor. Conversely, the activity schedule under Option C is used as a means of establishing target cost and the eventual pain/gain share with payment being made on the basis of actual cost (i.e. defined cost plus fee). In both cases, however, some means of establishing 'base' quantities is needed as activity schedules rely upon detailed measurement in order that the work can be properly priced.

The use of ECC Options A and C means that a bill of quantities is not supplied by the employer and that tenderers are required to calculate the quantities from the Works Information provided in order to calculate the contract price (Option A) or target price (Option C). Consequently, the quantity risk in the contract lies squarely with the contractor and not the employer, and adjustments to the contract price arise only when variation orders are issued or compensation events occur.

With regard to Option A activity schedules particularly, there are drawbacks which have led the industry to add a layer of complexity to their use. The plain fact of the matter is that, irrespective of how detailed the activity schedule is, quantities are needed in order to establish a price that can be broken down into priced activities as required by the contract.

The NEC3 *Guidance notes* recognise the risk, time and cost involved in preparing quantities and suggest that *employers may wish to calculate quantities before inviting tenders and then issue a copy of the quantities list to all tenderers*. The *Guidance notes* further suggest that some employers instruct tenderers to base their tenders on these quantities and agree any changes with the successful tenderer before entering into a contract.

Where this guidance is not followed, it is possible that tendering contractors may arrange for a bill of quantities to be prepared with the cost shared amongst themselves. Alternatively, the employer may supply, or arrange for a QS firm to supply, a bill of quantities when choosing the activity schedule option.

The issues associated with these arrangements are discussed in Section 4.10.2, but in any event, Broome (2013) considers that disclaimers relating to the accuracy of quantities supplied by the employer are likely to be ineffective and that employers are at risk of claims for misrepresentation when they *get the numbers wrong*.

A further downside to activity schedules under the ECC is that contractors only receive interim payment for completed activities (or groups of activities), and this has led contractors to employ strategies to protect their cash flow position, thereby encouraging the employers to do the same. Broome (2013) reports that:

- Some contractors have resorted to listing hundreds, if not thousands, of items in their activity schedules.
- This results in a programme with a huge number of operations, all of which take less than a month to complete to ensure inclusion in the monthly valuation.
- This then leads to employers imposing restrictions in the tender instructions on the number of activities that can be included in the activity schedule.
- Employers may also place limitations on the minimum value of activities in the activity schedule.

The implication is that common sense should prevail on both sides in order to sustain the principle of the ECC that all parties act *in a spirit of mutual trust and co-operation* (ECC Clause 10.1).

4.10.7   Activity schedules and the contractor's programme

Activity schedules are not necessarily linked to the contractor's programme as the provisions of the JCT 2011 SBC/Q demonstrate, but when there is no other priced document in the contract, it makes sense to synchronise the activity schedule with the programme as there are a number of benefits to be gained:

- Interim payment may be linked directly to physical progress on-site.
- Where payment is for completed activities, there is no dispute as to whether an activity is 40% complete or only 35%.
- The contractor's cash flow can be geared to the programme without resorting to inflating unit rates and front-loading preliminaries as is the case with bills of quantities.
- Post-contract administration is simplified resulting in less costs for contractors and less professional fees for employers.
- Monitoring of physical and financial progress is simplified.

When using an activity schedule in conjunction with the contractor's programme, it is important to consider the vocabulary used, especially where the Engineering and Construction Contract is concerned. Activity schedules naturally contain **activities**, but a glance at any construction management book will reveal that the words 'activity' and 'operation' may equally be used to describe the items of work listed in the programme. Some project management software packages also refer to 'activities' and 'summary activities' (i.e. activities grouped under a heading where more/less detail can be revealed by double-clicking on the summary activity bar). This suggests that common parlance in the industry is that both words are used synonymously.

The importance of the distinction between **activities** and **operations** is made clear by referring to ECC Option A where:

- Clause 31.4 states that the *Contractor* must indicate how each **activity** on the activity schedule relates to each **operation** on the programme.
- Clause 11.2(27) states that payment for priced activities, or groups of activities, is directly linked to the completion of those activities, or groups of activities, and not to the completion of operations.

When used under the ECC contract, activity schedules usually closely mirror the operations on the contractor's programme because of the contractual requirement to show how the activities on the activity schedule relate to the operations on the programme. This impacts on the contractor's administrative workload because:

- Changes in the planned method of working require submission of a revised activity schedule (Clause 54.2).

Under the ECC, there is a considerable distinction to be made between 'activity' and 'operation' because payment for work done to date is on the basis of the total of the prices for:

- Each group of completed **activities**.
- Each completed **activity** which is not in a group.

The prices are the lump sums for each of the **activities** on the activity schedule, and payment is made for activities, or groups of activities, only when they are completed.

Risk issue

Payment is linked to completion of **activities** in the activity schedule and **not** completed **operations** on the contractor's programme.
    This means that 'activities' are not the same as 'operations' as far as the ECC is concerned.

Common industry practice seems to be that the contractor's programme determines the activities on the activity schedule and not vice versa. This means that the contractor must give considerable thought to the programme at the tender stage so as to be able to produce a meaningful activity schedule as well as the quantities to back it up. Consideration must therefore to be given to:

- Choosing a sensible number of operations and grouped operations to appear on the programme.
- Ensuring that the activity schedule is not overly long.
- Making sure that the activity schedule and programme are sensibly referenced to each other even if the number of activities/operations on each is not the same (the programme may be more detailed than the activity schedule or vice versa).
- Choosing of activities/groups of activities bearing in mind that the contractor only receives payment when completion is achieved:
  - Long-duration operations/activities should be divisible into segments of 1 month or less so as to ensure regular monthly payments for work in progress.

- ○ Operations/activities must be clearly identifiable as completed so as not to jeopardise payment and hence cash flow.
- ▪ Ensuring that subcontract operations/activities and terms of payment do not create cash flow problems for the contractor.

An issue with the ECC is that payment is for **completed** activities, or groups of activities, and thus, incomplete activities will not be included in the monthly valuation of work in progress. This is to be contrasted with the JCT contract which gears payment to the proportion of work completed in each valuation period. The JCT approach is more traditional but prone to disagreement as to the extent of 'completeness', whereas the ECC approach is simpler but problematic:

- ▪ Only completed activities will be included in the monthly valuation of work in progress.
- ▪ Activities started but not completed in the valuation period will leave contractors with a cash flow deficit.
- ▪ Long-term activities, including preliminaries, are not included in the valuation of work in progress until completion is reached; this could be at the end of the contract period.
- ▪ Groups of activities will only be included in the monthly valuation when all the activities in the group are complete.

In the case of long-term activities, or groups of activities, there are three options as regards the Engineering and Construction Contract:

- ▪ Subdivide the activities so that they are each capable of being completed within one month (i.e. the valuation period).
- ▪ Invite tenderers to state their '*stage payment requirements*' in the tender documents (as practised by Highways England).
- ▪ Include a suitable 'Z' clause in the contract to permit payment for long-term activities to be included in the valuation on the basis of percentage completion.

Subdividing long-term activities will inevitably lead to a lengthy and overly complicated activity schedule which defeats the object of the exercise.

Broome (2013) identifies other problems with the ECC approach and suggests that:

- ▪ The concept of groups of activities be dropped so that activity schedules contain only 'activities'.
- ▪ Make 'activities' a subset of 'operations' so that operations on the programme, each of which requires statements of methods and resources under ECC Clause 31.2, can be expanded into a detailed 'list' of activities for payment purposes.

These are sensible suggestions that not only simplify the administration of Option A contracts with activity schedules but also provide a good fit with project management software such as Asta Powerproject, Project Commander and Microsoft Project.

4.10.8   Activity schedule size and scope

Bearing in mind the wide variety, size and complexity of building and civil engineering projects conducted by the construction industry, it is difficult to generalise as to how long an activity schedule should be or what it should contain.

A bill of quantities for a typical £5 million building project for a school, leisure centre or office building could well have 200 pages or more of measured items which could amount to 1500 or more items. Based on the Pareto principle (80/20 rule), 300 or more items could be cost significant. If these items are transposed into an activity schedule format, this could result in ±15 pages of activity schedule depending upon the length of the item descriptions.

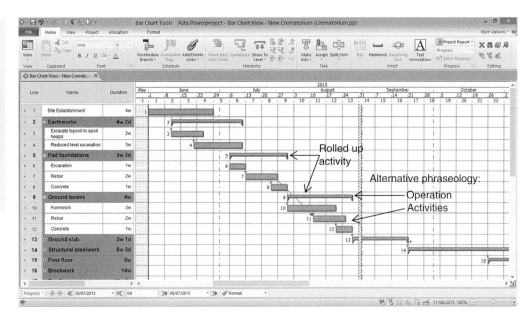

Figure 4.2   Rolled-up activities.

As far as the contractor's programme is concerned, 300 activities/operations may seem a lot but some activities/operations may be grouped under 'summary activities' which can then be scrolled up/down according to the degree of detail required.

Consideration also needs to be given to the number of subcontract packages in the contractor's programme as there could easily be 20 or more different subcontractors on such a project each conducting several different work activities/operations.

Under the Engineering and Construction Contract, a limiting factor on the number of operations on the programme, and consequently on the length of the activity schedule, is that the contractor is obliged to submit details of his working methods and resources for each operation on the programme. This would be a considerable administrative burden were the programme to contain 300 operations, but this would need to be balanced with the contractor's desire to ensure a satisfactory cash flow from the activity schedule.

Taking all these into account, Broome (2013) is most sensible in suggesting that 'operations' on the programme should be used to summarise the 'activities' contained within, as illustrated in Figure 4.2 where the ground slab, structural steelwork and first floor tasks have been summarised.

Rolled-up or summary activities could then be rolled down to reveal further detail within the summarised tasks as illustrated for the earthworks, pad foundations and ground beams tasks in Figure 4.2.

This approach would not only surmount the problem of payment for grouped activities under the ECC but would also reduce the administrative burden on contractors and help avoid the artificial dissecting of long-term activities.

4.11   Price lists

Price lists are common in all walks of life and simply consist of a list of items with an accompanying price and unit. They tend to be simple *ad hoc* documents and may be used for small-scale construction projects and work orders. An example is provided in Table 4.9.

Table 4.9   Price list.

Price list			Leave these columns blank for a lump sum contract		
Item number	**Description**	**Unit**	**Quantity**	**Rate**	**Price**
A	Formwork to sides and soffits of beams	m²	318	45.00	14 310.00
B	Formwork to square columns	m²	144	32.00	4 608.00
C	Formwork to horizontal soffits including propping	m²	432	75.00	32 400.00
D	Rebar as per schedule	tonne	12	1450.00	17 400.00
E	Class C30 concrete in beams	m³	38	265.00	10 070.00
F	Class C30 concrete in square columns	m³	10	275.00	2 750.00
G	Class C30 concrete in horizontal slabs	m³	166	255.00	42 330.00
				Total of the prices	123 868.00

A degree of 'formality' is given to the 'Price List' under NEC3 wherein they are used in the Engineering and Construction Short Contract (ECSC) and in the Engineering and Construction Short Subcontract (ECSSC).

Price lists are typically versatile within the NEC3 family:

- They may be prepared by the employer or the contractor, and this fits with Main Options A–E.
- They are intended for straightforward low-risk projects, but this does not imply any limitation on the value of the contract.
- They may be used for a lump sum contract, where the price paid is not adjusted if the quantity of work changes, by leaving the 'quantity', 'unit' and 'rate' columns blank.
- They may be used for measure and value contracts where the price paid is determined by the quantity of work completed multiplied by the corresponding rate.

In common with other pricing documents in the NEC family of contracts, the price list is not part of the Works Information but is an essential contract document nonetheless. The price list functions as a 'mini-bill of quantities' for a measure and value contract with the actual quantities of work done replacing the original quantities as appropriate. Such changes are assessed as compensation events but only to the extent of the change in quantity.

For a lump sum contract, the contractor takes on the quantity risk, and the price list will contain only lump sums that add up to the total of the prices.

There is no link between the price list and the contractor's programme under the ECC Short Contract, and so applications for payment from the contractor follow conventional lines, that is, an assessment of work completed to date.

4.12   Contract sum analyses

In design and build contracts, it is conventional for the contractor to submit a breakdown of the contract sum in order to check prices and totals and to assist with post-contract administration.

Under the JCT 2011 DB contract, this takes the form of a Contract Sum Analysis. Along with the Employer's Requirements and the Contractor's Proposals, the Contract Sum Analysis is incorporated into the formal contract and becomes a contract document with all that this implies.

There are no similar provisions in the ICC – Design and Construct form of contract and thus the necessary checks and balances needed to administer the contract correctly would need to be incorporated into the tender documents and thence into the contract.

Risk issue

Under the JCT design and build form, there is no prescription as to the form or content of the Contract Sum Analysis. However, it might be prudent to require submission of the Contract Sum Analysis with the Contractor's Proposals and tender bid in order that the employer may be satisfied that the Contract Sum Analysis is suitably transparent before accepting the tender or entering into a formal contract.

The Contract Sum Analysis can take whatever form the contractor likes unless prescribed to the contrary in the Employer's Requirements. A bill of approximate quantities could be used or something similar to an elemental cost plan might be preferred. Unlike ECC Option A with activity schedule, there is no link between the Contract Sum Analysis and the contractor's programme in the JCT 2011 DB contract nor, indeed, is there any connection between the Contract Sum Analysis and the method of payment.

There are two methods of payment under the JCT 2011 DB contract:

- Alternative A: stage payments.
- Alternative B: periodic payments.

One or other of the alternatives should be deleted from the Contract Particulars, but if neither is selected, Alternative B is deemed to apply.

Under Alternative A, the contractor completes the list of stages, with brief descriptions, provided in the Contract Particulars along with a cumulative value for each stage or may provide this information on a separate sheet for incorporation into the contract. Applications for payment are made by the contractor according to the stage of completion reached and the relevant cumulative value stated in the list of stages.

Where Alternative B applies, interim payments are made at the dates stated in the contract, usually monthly. Applications for payment must be accompanied by such details as may be prescribed in the Employer's Requirements. Such requirements may relate to the Contract Sum Analysis but, equally, might follow an ogive curve, formula or an assessment made according to the physical progress of the works on-site relative to the contractor's programme.

4.13   Schedules of actual cost

Traditional construction contracts, such as JCT and ICE/ICC conditions, are drafted in such a way as to confer powers on the contract administrator to issue instructions to the contractor for a variety of reasons:

- Variations to the original contract.
- Postponement of work.
- Expenditure of provisional sums.
- Inspection of completed work.
- Etc.

There are several reasons for this including the need to make changes to the contract without committing a breach and the need to retain a degree of flexibility in how the project is to be completed satisfactorily.

Within the confines of this arrangement, traditional contracts also provide rules for the valuation of such eventualities, and this includes making use of the bill of quantities and the tendered rates as a means of valuing change. However, occasions arise in construction projects where work is required that does not lend itself to measurement in the normally accepted units, that is, m, m², m³, etc., examples of which include:

- The contractor is instructed to search for existing utilities that are not shown on the drawings.
- The contractor discovers an unknown sewer which has to be diverted.
- A subcontract painting and decorating firm is asked back to the site to repair finished work damaged by other trades.

In such cases, the valuation rules provide for the quantity of work undertaken to be determined by measuring the time and resources expended to do the work to which a percentage is added to cover the cost of supervision and other oncosts, overheads and profit.

Payment for work carried out on this basis might be for a complete contract or, more usually, for discrete parts of contracts where unexpected situations arise or where variations are instructed and no other means of valuation is feasible.

This method of payment is a form of cost reimbursement albeit that the costs reimbursed are not exactly the actual costs incurred by the contractor. Labour costs, for instance, are not actual wages paid but are usually determined according to a definition of prime cost to which the contractor adds a percentage. Plant is paid for on a time basis, but here again, the rates paid are not market plant hire rates but rates laid down in a standard schedule to which the contractor adds a percentage to cover his overheads and profit. Materials, on the other hand, are paid for on the basis of invoiced costs plus a percentage addition for overheads and profit.

Under NEC3 conditions, the valuation of change is approached in a completely different way to traditional contracts. Using the novel concept of **compensation events**, ECC Clause 60.1, for example, lists 19 eventualities that qualify as such, including:

- Changes to the Works Information.
- Instructions regarding objects of value or of historical or other interest discovered on-site.
- Physical conditions encountered that an experienced contractor would not have anticipated.
- An event occurs which prevents the contractor from completing the works.

It can be seen that the ambit of compensation events is very wide ranging and includes not only variations to the contract but also eventualities that would lead to a claim for additional payment or loss and expense under traditional contracts.

With this in mind, a further significant difference in approach between the NEC3 family and traditional contracts is that the valuation of compensation events does not rely on the use of bill of quantities rates as a means of valuation (unless only quantities change, and then only within certain limits stated in the ECC 'Black Book').

Part of the reasoning for this is that post-contract administration of NEC3 contracts is very much focussed on using the contractor's programme and method of working as a control tool rather than BQ rates that may have been manipulated at the tender stage for commercial advantage.

In the NEC3 suite, therefore, compensation events are valued according to changes in resources (under the full ECC) or on the basis of actual cost (short ECC), whereas the valuation of work on a 'time and cost' basis in traditional contracts is usually reserved for relatively insignificant amounts of work carried out on daywork.

In any event, express arrangements are needed in any contract for valuation on the basis of resources expended, and it is prudent to do so at the tender stage in order that competitive rates may be obtained for such work. This has an impact on tender documentation preparation and, in particular, drafting of the bill of quantities (if one is used). There are two ways of doing this in common use:

- Daywork schedules – used under JCT and ICC conditions.
- Schedules of cost components – used with NEC3 contracts.

4.13.1　Daywork schedules

The normally accepted meaning of the term 'daywork' is that it is a method for measuring and valuing work on the basis of the resources expended rather than in relation to the quantities of work done. Payment for labour and plant is at an hourly rate, materials are paid for at cost and each resource attracts a percentage addition to cover overheads, profit and oncosts.

Perhaps more correctly, 'daywork' could be defined as a type of provisional sum (included in the bill of quantities) that allows for undefined work which will be paid for on the basis of the time spent and materials used in completing the job (Ross and Williams, 2013). In traditional forms of contract, such work would be treated as a variation to the contract, and the provisional sum would be expended by the contract administrator (if necessary) when no other means of valuing the variation would be appropriate.

Daywork is a normal, but often contentious, feature of most construction projects:

- Contractors and subcontractors are often accused of allocating their most inefficient workers to daywork tasks.
- Daywork record sheets are seen as works of pure fiction.
- Some contract administrators, contractors and, especially, subcontractors do not understand how daywork operates.
- Claims for payment on a daywork basis are made when work should actually be measured conventionally.

The truth is that daywork is a 'necessary evil' because occasions arise on construction projects where there is no fair way to pay for the work other than by daywork. In this regard, perhaps a better way to define 'daywork' is:

- A means of compensating the contractor (or subcontractor) when work is instructed by the contract administrator (or contractor) that wasn't contemplated in the contract (or subcontract) originally and is so small or dissimilar to the measured work that there is no effective way to measure it conventionally.

Provision for daywork is traditionally made in bills of quantities by including provisional items for each class of resource. These are effectively risk allowances because such work is not planned or necessarily expected, and it might not be necessary to spend the money at all. Should the unexpected happen, however, a pot of money is built into the contract sum and is available to the contract administrator without the need to approach the employer for more budget.

The contractual mechanism for determining the daywork rates that tenderers price into the bills of quantities is provided by the **Definition of Prime Cost of Daywork for Building Work 2007** and by the **CECA Schedules of Dayworks Carried Out Incidental to Contract Work 2011** which is the equivalent schedule for civil engineering work.

A typical bill of quantities entry for daywork is illustrated in Table 4.10. This illustrates one of two options for billing daywork recognised by the industry and by the Definition of Prime Cost of Daywork for Building Work 2007 which classes this 'traditional' method as Option A.

Option A allows a series of provisional sums to be adjusted by tenderers according to the percentage addition required to account for any additional costs over and above the Definition of Prime Cost and for the contractor's overheads and profit.

Table 4.10   Billing daywork v.1.

PROVISIONAL SUMS					£
Daywork					
Daywork shall be valued in accordance with the Conditions of Contract and with the Definition of Prime Cost of Daywork Carried out under a Building Contract 2007 published by the BICS.					
A	Allow the Provisional Sum of £5000 for the cost of Labour for work carried out on a Daywork basis.				5000.00
B	Percentage adjustment Add/Deduct		%		
C	Allow the provisional sum of £2000 for the cost of Plant for work carried out on a Daywork basis.				2000.00
D	Percentage adjustment Add/Deduct		%		
E	Allow the provisional sum of £3000 for the cost of Materials for work carried out on a Daywork basis.				3000.00
F	Percentage adjustment Add/Deduct		%		
	TOTAL				
	NB: Rates to be inserted by the Contractor				

In Option B, provisional sums are included for plant and materials, with a place for the required percentage adjustment, whereas labour is provided for by a list of classes of labour against which tenderers quote the 'all-in' hourly rate required with no percentage adjustment, as shown in Table 4.11.

Both Options A and B are recommended good practice as they both achieve the goal of eliciting competitive daywork rates at the tender stage. This is ensured because the sums of money determined by the provisional sums and the contractor's entries will be added into the tender total which can then be compared with competing contractors.

The choice between Options A and B is a personal one, but where NRM2 is the standard method of measurement, Option B is the required choice (refer to NRM2 Paragraph 2.13.3).

Risk issue

To an extent, 'competitive' daywork rates are a fallacy.

In practice, some contractors 'load' the daywork rates and fund this by making savings in other parts of the bill of quantities.

Reducing tenders by making perceived savings on materials, subcontract quotes and errors and omissions in the BQ forms part of the contractor's 'commercial opportunity strategy' as explained by Ross and Williams (2013).

Table 4.11 Billing daywork v.2.

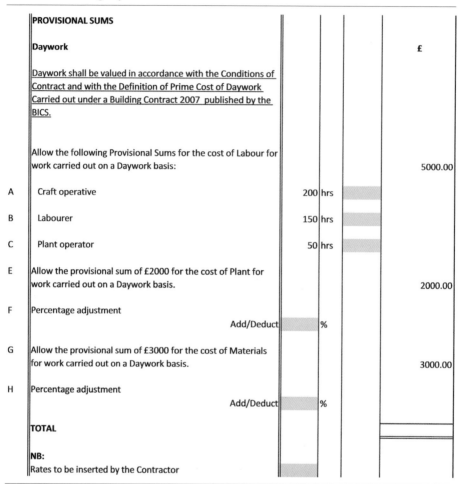

		PROVISIONAL SUMS			£
		Daywork			
		Daywork shall be valued in accordance with the Conditions of Contract and with the Definition of Prime Cost of Daywork Carried out under a Building Contract 2007 published by the BICS.			
		Allow the following Provisional Sums for the cost of Labour for work carried out on a Daywork basis:			5000.00
A		Craft operative	200	hrs	
B		Labourer	150	hrs	
C		Plant operator	50	hrs	
E		Allow the provisional sum of £2000 for the cost of Plant for work carried out on a Daywork basis.			2000.00
F		Percentage adjustment Add/Deduct		%	
G		Allow the provisional sum of £3000 for the cost of Materials for work carried out on a Daywork basis.			3000.00
H		Percentage adjustment Add/Deduct		%	
		TOTAL			
		NB: Rates to be inserted by the Contractor			

Measurement software packages do not support the concept of daywork, and thus, it is necessary to draft the Provisional Sums for Daywork pages in the BQ separately using a word processing package such as Microsoft Word or an Excel spreadsheet.

Standard methods of measurement, such as SMM7, NRM2 and CESMM4, provide sets of rules for including provisional sums for daywork in the bill of quantities, and daywork is referred to in the conditions of contract normally associated with these methods of measurement (e.g. JCT and ICE/ICC forms). The NEC3, however, does not use the term and prefers the concept of 'defined cost' linked to a schedule of cost components.

The various ways of calculating daywork rates and the submission of daywork sheets are beyond the scope of this book, but Ross and Williams (2013) provide a detailed explanation for the interested reader.

4.13.2 Schedule of cost components

Whilst 'daywork' has been used as a means of valuing variations in both the building and civil engineering sectors of the UK construction industry for many years, the approach taken under the NEC3 suite of contracts is conceptually quite different.

Firstly, the NEC3 approach has a broader ambit than 'daywork' which is limited to the valuation of variations when all else fails. The NEC3 suite of contracts, on the other hand, provides a measure of 'change', rather than focussing on the narrower theme of variations, and this is used for:

- Assessing the cost of changes to Works Information (i.e. variations).
- Assessing the cost of other compensation events (i.e. claims or loss and expense in non-NEC-speak).
- Assessing the value of work carried out under cost reimbursement contracts.

The driving force behind the NEC3 concept is the reimbursement of both the time and cost effects of contractual change through its **compensation event** procedure. It is through this mechanism that the contractor is compensated for changes in certain defined costs, to which a fee is added. The 'defined costs' are determined by schedules of cost components, of which there are two:

1. Schedule of Cost Components (SCC) – that is, the full schedule.
2. Shorter Schedule of Cost Components (SSCC) – that is, the shorter, less complex version.

Whilst this system of compensation is meant to be fairer and more rigorous and is based on the contractor's programme and method of working, Broome (2013) recounts that there have been problems. The first and second editions of the ECC, for example, failed to fully recognise how contractors incur additional costs in particular circumstances or how this was reflected in the contractor's compensation. The NEC3 is regarded as an improvement albeit that some anomalies remain.

In order to understand how the NEC3 system of compensation works, it is first necessary to appreciate that:

- ECC Options A and B are for traditional lump sum and measure and value contracts, respectively.
- ECC Options C and D are for target contracts where the contractor is paid on the basis of actual cost plus a fee but within the confines of a target 'pain/gain' arrangement.
- ECC Option E is for pure cost reimbursement contracts.
- Option F is for management contracts where neither of the schedules of cost components applies.
- The ECC Short Contract is a simplified contract for traditional lump sum and measure and value contracts where neither of the schedules of cost components applies.

In all cases, the concept of defined cost plus fee is used but the basis upon which it is used differs according to the contract option used. The concept of defined cost plus fee is also used under the ECC Short Contract but not in conjunction with either of the schedules of cost components. See Chapter 12 for more on this topic.

Notes

1. Wikipedia, 1 December 2012, http://en.wikipedia.org/wiki/Pareto_principle (accessed 2 April 2015).
2. Quote from Donald Rumsfeld, Former US Secretary of State for Defense, February 2002, http://www.youtube.com/watch?v=GiPe1OiKQuk (accessed 2 April 2015).
   - Known-unknowns = things we know we don't know.
   - Unknown-unknowns = things we don't know we don't know.
3. Collins English Dictionary.
4. Paraphrased from Collins English Dictionary.
5. UK Competition Act 1998 and The Treaty of Rome.

References

Broome, J. (2013) *NEC3: A User's Guide*, ICE Publishing, London.

BICS, *Definition of Prime Cost of Daywork for Building Work* (2007), BICS, London.

Davis Langdon (AECOM) (2011) *Contracts in Use, A Survey of Building Contracts in Use During 2010*. RICS, London.

Forbes, W.S. and Skoyles, E.R. (1963) The Operational Bill, *RICS Journal*.

Hibberd, P. (2014) *Building*, 7 August, http://www.building.co.uk/procurement-old-ways-are-the-best/5049677.article (accessed 2 April 2015).

Jaggar, D., Ross, A., Smith, J. and Love, P. (2002) *Building Design Cost Management*, Blackwell Science, Oxford.

Lee, S., Trench, W. and Willis, A. (2014) *Willis's Elements of Quantity Surveying*, 12th Edition, Wiley-Blackwell, Chichester.

M3 Housing Ltd, *The NHF Schedule of Rates*, M3 Housing Ltd, Mitcham, http://www.m3h.co.uk/ (accessed 2 April 2015).

NBS National Construction Contracts and Law Survey 2012, http://www.thenbs.com/topics/ContractsLaw/articles/nbsNationalConstructionContractsLawSurvey2012.asp (accessed 2 April 2015).

NSR Management, *The National Schedule of Rates*, NSR Management, Aylesbury. http://www.nsrm.co.uk (accessed 2 April 2015).

Ostrowski, S. (2013) *Guide to Measurement Using the New Rules of Measurement*, Wiley-Blackwell, Chichester.

Ross, A. and Williams, P. (2013) *Financial Management in Construction Contracting*, Wiley-Blackwell, Chichester.

Royal Institution of Chartered Surveyors (1922) *The Standard Method of Measurement of Building Works*, First Edition.

Skoyles, E. R. (1966) *Examples from Operational Bills*, Building Research Station.

Skoyles, E. R. (1967) *Preparation of an Operational Bill*, Building Research Station, Great Britain.

The Stationery Office, *The PSA Schedules of Rates*, The Stationery Office, London, http://www.tpsconsult.co.uk/psa/ (accessed 2 April 2015).

PART 2

Measurement Risk

Part 2

Managing Measurement Risk in Building and Civil Engineering, First Edition. Peter Williams.
© 2016 John Wiley & Sons, Ltd. Published 2016 by John Wiley & Sons, Ltd.

Chapter 5
New Rules of Measurement: NRM1

NRM1 is part of the suite of documents that make up the RICS New Rules of Measurement. Its full title is *NRM1: Order of cost estimating and cost planning for capital building works*. It was first published in 2009 as *NRM1: Order of cost estimating and elemental cost planning* and is now in its second edition which became operative on 1 January 2013. The NRM suite of documents is also part of the RICS 'Black Book', which is *a suite of guidance notes that define good technical standards for quantity surveying and construction professionals* (RICS).

NRM1 is the 'sister' to two other RICS New Rules of Measurement, namely, *NRM2: Detailed measurement for building works* and *NRM3: Order of cost estimating and cost planning for building maintenance works*. Together, the three documents are intended to provide measurement rules for the effective cost management of construction projects throughout their life cycle.

5.1 New rules: New approach

It is a common theme in the NRM suite of documents that explanations, guidance and 'how-to' examples are included in what is, ostensibly, a set of measurement rules. Historically, methods of measurement have not adopted this approach and have been limited to providing measurement rules that practitioners and other users simply have to learn, understand and apply according to their own interpretation and judgement.

In one sense, the authors of the NRM suite are to be applauded for this novel approach but, in other respects, could be criticised because the documents overall lack precision, contain mistakes and, in some cases, are not very clear.

The authors of NRM1 claim that:

1. The RICS New Rules of Measurement have been written in such a way as to be *understandable by all those involved in a construction project*.
2. The new rules should assist the quantity surveyor/cost manager *in providing effective and accurate cost advice*.

Managing Measurement Risk in Building and Civil Engineering, First Edition. Peter Williams.
© 2016 John Wiley & Sons, Ltd. Published 2016 by John Wiley & Sons, Ltd.

The first of these claims is questionable, to say the least, and the second is debatable. Rules of measurement need to be precise and clear, and NRM1 is neither, but, whatever the case, it is crystal clear that the status of NRM1 as an RICS guidance note should be uppermost in the thoughts of those involved in giving cost advice.

5.2   The status of NRM1

The applicability of the status of NRM1 is more narrowly drawn than that of its sister document, NRM2, as NRM1 refers only to 'member(s)' and 'surveyor', whereas NRM2 also includes the wider term 'user' in the scope of its legal and disciplinary applicability.

NRM1: *Introduction* explains that *these rules* have the status of an **RICS guidance note**. This means that NRM1 is *recommended good practice* because an RICS guidance note is defined as a *document that provides users with recommendations for accepted good practice as followed by competent and conscientious practitioners*.

However, NRM1: *Introduction* also states that *where recommendations are made for specific professional tasks*, such recommendations shall represent *best practice*.

Consequently, 'members' who use NRM1 will be judged by two standards:

1.   **Good practice**, as followed by competent and conscientious practitioners.
2.   **Best practice** when it comes to following recommendations made for *specific professional tasks* within NRM1.

NRM1: *Introduction* additionally explains that **best practice** is defined by the RICS as representing *a high standard of professional competence* and that where recommendations are made in NRM1 for *specific professional tasks*, these are intended to represent best practice.

The meaning of the phrase *specific professional tasks* is not explained in NRM1, nor are any *specific professional tasks* identified in the document. The RICS Rules of Conduct (RICS, 2007a, b) are more explicit and impose upon members, and member firms, two specific obligations with regard to their *professional work*:

- **Competence** – to work with *due skill, care and diligence and with proper regard for the technical standards expected...*
- **Service** – to work *in a timely manner and with proper regard for standards of service and customer care expected...*

Ashworth et al. (2013) shed no light on this issue. They refer to professional 'roles' and 'activities' but not tasks.

It can only be assumed, therefore, that *specific professional tasks* relate to tasks expressly identified in a contract for professional services. If this is the case, then these tasks would be judged to the higher standard of professional competence, that is, best practice.

5.2.1   Professional competence

The status of NRM1 is very important in the context of the increasing tendency towards litigating against construction professionals, such as architects, engineers and surveyors, and there have been a number of cases whereby quantity surveyors have been involved in multimillion £ claims for negligent cost planning and cost advice. A number of landmark cases have arisen from actions brought against quantity surveyors (and others) following significant cost overruns, including *Copthorne Hotel v Arup [1997] 58 Con LR 105* and *George Fischer Holdings v Multi-Design Consultants [1998] 61 Con LR 55*.

Quantity surveyors have obligations and duties of care that originate from a number of sources. They are based in statute law, including the Building Act 1984 and the Building Regulations 2000, regulation, common law, contract and tort. Other regulatory obligations and codes of conduct further define the professional standards expected of a quantity surveyor. In particular, the RICS Rules of Conduct set out standards for its members who are required to carry out work with reasonable care, skill and expertise.

The courts understand the standards expected of professional people and apply the definitive common law test, where the use of some special skill or competence is involved, of *the standard of the ordinary skilled man exercising and professing to have that special skill. A man need not possess expert skill… it is sufficient if he exercises the ordinary skill of the ordinary competent man exercising that particular art.*[1]

This, however, is a standard lower than the *best practice* referred to in NRM1.

Knowles (2012) suggests that where a quantity surveyor provides cost advice to an employer that proves to be incorrect, the employer must demonstrate either that the accuracy of the estimate was warranted or that the reason for its inaccuracy stemmed from a lack of reasonable skill and care on the quantity surveyor's part. Again, this learned interpretation appears to refer to the lesser standard of *good practice* rather than *best practice*.

5.2.2   Negligence

NRM1: *Introduction* makes it clear that conforming to the practices recommended in the document should provide *at least a partial defence to an allegation of negligence* and that non-compliance with recommended practice should only be undertaken with justification as a court, a tribunal or the RICS Disciplinary Panel may ask for an explanation as to why recommended practice was not adopted.

The test applied in cases of professional negligence is, as Patten (2003) suggests, **whether the quantity surveyor acted with the skill and care expected of a reasonably competent member of his profession** and NRM1 confirms that a court or tribunal may take account of the contents of guidance notes as being indicative of 'reasonable competence'.

However, NRM1: *Introduction* also states that *where*, within NRM1, *recommendations are made for specific professional tasks, these are intended to represent 'best practice'* which, in the opinion of the RICS, *meet a high standard of professional competence*, that is, a higher standard than that demanded by the courts.

> ### Risk issue
>
> NRM1: *Introduction* is clearer as to who needs to comply with an RICS guidance note than NRM2, as NRM1 refers only to 'member(s)' and 'surveyor' and not to 'users'.
>
> The most likely litigant in cases of negligent cost advice is, of course, the employer who, having engaged a professional quantity surveyor, feels that he has received inadequate professional advice or services.

5.2.3   Pre-action protocol

The importance of the issue of negligence is not to be understated in that the UK Ministry of Justice has seen fit to publish the *Pre-Action Protocol for the Construction and Engineering Disputes* (Ministry of Justice, 2012) that applies to construction and engineering disputes, including those involving professional negligence claims against architects, engineers and quantity surveyors. The intention of the Protocol is to ensure that:

- The parties are clear as to the issues in dispute.
- Information has been exchanged in a timely and efficient manner.
- The parties have met and have attempted to arrive at a resolution without resort to litigation.
- Should litigation ensue, it can proceed in an efficient manner.

The status of NRM1 as a guidance note is not to be taken lightly!

5.3   Structure of NRM1

NRM1 has four parts and seven appendices:

- **Part 1** contextualises approximate estimating and cost planning in terms of the RIBA and OGC process models and also explains certain symbols and abbreviations. The important aspects of this part are the definitions of terms used in the rules (Paragraph 1.6.3).
- **Part 2** is largely explanatory as regards order of cost estimates – what they are, how they are made up and how they are prepared. It also *sets out the rules of measurement for the preparation of order of cost estimates* using cost per m², cost per functional unit and the elemental method of approximate estimating. Order of cost estimating is beyond the scope of this book, but the measurement rules relating thereto are considered later in the chapter.
- **Part 3** is again largely explanatory as regards the purpose and processes of elemental cost planning along with rules of measurement that apply to putting an elemental cost plan together. Cost planning is also beyond the scope of this book, but the measurement rules relating thereto are considered later in the chapter.
- **Part 4** consists of 5 pages of dialogue explaining how the tabulated rules are to be used and how items are to be codified, as well as 253 pages of measurement rules applicable to the preparation of cost plans.
- **Appendices:**
  - Appendix A – defines 'gross internal floor area' (GIFA) and explains how this is measured and what it includes and excludes. Diagrams C, D and M provide practical (if somewhat limited) examples of dimensions used for the measurement of quantities.
  - Appendix B – lists commonly used functional units.
  - Appendix C – defines 'net internal floor area'.
  - Appendix D – special definitions for shops.
  - Appendix E – sets out the logic and arrangement for an elemental cost plan. Useful as an *aide memoire* for identifying the sub-elements that are included within each element and group element.
  - Appendix F – identifies information requirements for the preparation of formal cost plans. Very useful *aide memoire* to establish a list of tender documents should NRM1 be used to formulate a bill of quantities.
  - Appendix G – template for elemental cost plan.

5.3.1   Measurement rules

At first glance, NRM1 is a daunting document, extending as it does to some 369 pages plus index. However, of this number, 262 pages are taken up by the detailed measurement rules for cost planning, and there are 41 pages for the bibliography and appendices. The remaining 66 pages appear at the 'front end' of NRM1 in Parts 1, 2 and 3.

It is at the 'front end' where most of the problems with NRM1 lie, the main difficulty being finding the 'measurement rules' – they <u>are</u> there, but it takes a lot of reading to find them. The 'measurement rules' within NRM1 Parts 1–3 are interspersed amongst a variety of descriptions, definitions and explanations which, whilst being relevant and useful, mask the actual 'rules'. Consequently, within Paragraph 2.15: *Measurement rules for risk*, for example, there are 2½ pages of text but only one measurement rule.

5.3.2   Measurement rules for order of cost estimating

The measurement rules for order of cost estimating are contained in NRM1 Part 2, Paragraphs 2.5, 2.6, 2.8 and 2.11–2.16.

However, the problem is that these paragraphs also contain explanatory and other general text concerning the constituents of an order of cost estimate and other related general matters. Added to this is that other paragraphs contain what appear to be 'rules', but it is not clear whether or not they are 'rules'.

This may appear to be 'picky', but the expression 'rules is rules' is important in the context of a set of rules of measurement that are meant to be standardised, understood and followed by users of the document and recipients of the output generated by the measurement process.

5.3.3   Measurement rules for cost planning

Part 3 is equally a 'jumble' of descriptions, definitions and explanations, this time concerning cost planning, the constituents of a cost plan and other related general matters.

The actual 'rules' are given in Paragraphs 3.10, 3.11 and 3.14–3.19 albeit that these paragraphs also contain explanatory and other general text.

5.3.4   Distinguishing the rules of measurement

In order to assist the reader to clearly identify the 'rules' of measurement from other text in NRM1, this chapter is structured as follows:

- Measurement rules for order of cost estimates and for cost plans have been collected and presented together in a box under the heading 'NRM1: Measurement rules'.
- Each NRM1 paragraph that contains an obvious measurement rule is listed in the box (e.g. Paragraph 2.6.1 (a) (ii) is a 'measurement rule' which is to be found at NRM1 Paragraph 2.6.1 (a) (ii), and Paragraph 3.11.3 is a 'measurement rule' which is to be found at NRM1 Paragraph 3.11.3).

In all other respects, the headings in this chapter generally follow those of NRM1.

5.3.5   Types of measurement rules

There are two basic types of standardised measurement rules in NRM1 that form the basis of a system of design cost control. One of these is intended for developing order of cost estimates and the other for cost planning purposes.

The measurement rules in NRM1 consist of:

1. Measurement rules relating to techniques of approximate estimating (i.e. for establishing a budget or cost limit):
   - Expressed in m^2 of GIFA.
   - Expressed in functional units (e.g. seats in sports stadium, hospital bed spaces, prison places).

Part 2

2. Measurement rules relating to elemental cost planning:
- Expressed in units of measured quantities (i.e. m, m², m³, nr, tonne).

The point at which the *measurement rules for elemental cost planning* are used instead of the *measurement rules for order of cost estimating* is determined by the design process.

This is likely to be during or towards the end of Stage 3 – Developed Design of the RIBA Plan of Work 2013 when the drawings have become sufficiently developed to make the transition from cost/m² of GIFA, or functional units, to detailed measurement of individual items of building work. This is illustrated in Figure 5.1.

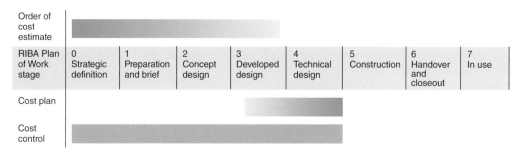

Order of cost estimate								
RIBA Plan of Work stage	0 Strategic definition	1 Preparation and brief	2 Concept design	3 Developed design	4 Technical design	5 Construction	6 Handover and closeout	7 In use
Cost plan								
Cost control								

Figure 5.1   Design cost control.

In this context, the main message from Paragraph 1.4 of NRM1 is that the New Rules of Measurement *represent the essentials of good practice* and provide:

- A *structured basis* for measuring building work.
- A *consistent approach* for dealing with other components of the cost of buildings:
  ○ Contractors' preliminaries.
  ○ Overheads and profit.
  ○ Risk factors.
  ○ Fees.
  ○ Inflation.

5.4   Design cost control: Introduction

Essentially, the purpose of NRM1 is to provide formalised measurement rules for use in the cost control, or cost management, of building design.

This process begins with a 'high-level' estimate, or **order of cost**, which is then developed, alongside the maturing design, into an **elemental cost plan**. The cost plan is then reconciled (or compared) with tenders received from contractors in order to establish any differences between the amount that the employer can afford to pay and what the prevailing market deems to be the right price at the time of tender.

In some respects, NRM1 is stuck in the 'time warp' of traditional procurement, where an employer-engaged professional quantity surveyor (PQS) provides cost advice during the design stages of a project.

Admittedly, NRM1 was rolled out before the RIBA Plan of Work 2013 was published, and whilst written from the employer–PQS perspective, it may be interpreted in different ways to suit different users and different procurement arrangements. In fact, NRM1 Paragraph 3.6.1 refers to the use of the cost plan as a *cost control mechanism* where a project is to be procured using work packages, but this only relates to the *pre-construction* and *construction* stages and not to the design development stages.

Many projects nowadays are procured on a non-traditional basis, and it is not uncommon to find contractors appointed, even for public sector contracts, at very early stages in the maturity of a project, often as early as the concept design stage. This is not recognised in NRM1, but despite its frailties and mistakes, it provides a rich source of guidance and well-structured measurement rules, which should prove attractive to a wider community than the somewhat narrow fraternity of the PQS.

It would be a shame if main contractors, specialists and subcontractors felt that there was little point in looking at NRM1, because it offers the opportunity to take a structured approach to developing cost estimates for design–build projects, early-stage contractor/specialist appointment and target/pain–gain contracts and much more.

Failing to recognise that the cost control of modern construction projects has a much wider purview than the provision of early cost advice and cost planning services by PQSs to employers would be to miss out on a great opportunity.

5.4.1   Symbols, abbreviations and definitions

Paragraphs 1.6.1 and 1.6.2 clarify the measurement symbols and abbreviations used in NRM1, respectively. They are self-explanatory, but the five pages of Paragraph 1.6.3: *Definitions* are more interesting because they contain lots of useful definitions of terms, used in the cost estimating and cost planning process, that help the user to understand what cost estimating and cost planning is about.

NRM1 defines **cost control** as *the process of planning and controlling the cost(s) of buildings…throughout* [the] *complete duration of the construction project*. Ashworth et al. (2013) concur with this definition and point out that cost control is a feature of the construction process that extends from the inception of a project to final account.

NRM1, however, is intended to provide guidance for cost control during design and up to acceptance of the contractor's tender. Conventionally, this is called **design cost control**, but NRM1 offers no definition of this term.

Order of cost estimate is defined as *the determination of possible cost of a building(s) early in the design stage in relation to the employer's functional requirements… and forms the build-up to the cost planning process*. The order of cost estimate forms the basis for deciding on the **cost limit**, which is *the maximum expenditure that the employer is prepared to make* for the building(s) required.

Once the cost limit has been established, the design team can proceed with producing the detail required to satisfy the employer's functional and aesthetic requirements, but this needs to happen within a control framework. This is the function of the cost plan, or more correctly the **elemental cost plan**, which is defined as *the critical breakdown of the cost limit… into cost targets for each element of the building(s)*.

Mention of the 'design team' leads to consideration as to the members of this team and what they do. In NRM1, **design team** *means architects, engineers and technology specialists* who have responsibility for the conceptual design of a building or structure and for its development into drawings, specifications and instructions for construction and associated purposes.

This is the same definition as that used in NRM2, and inexplicably, it excludes the quantity surveyor/cost manager. Considering the importance of the role played by the PQS in the cost planning and cost control of the design and, usually, the construction stages of a project, this is quite simply a bizarre exclusion.

The 'design team' is defined as being part of the 'project team', which, unlike in NRM2, is defined in NRM1. The **project team** is the *employer, project manager, quantity surveyor/cost manager, design team and all other consultants* with the responsibility for the delivery of the project, together with *the main contractor where the main contractor has been engaged by the employer to provide pre-construction* services.

Part 2

This is an even more breathtaking definition than 'design team'. It smacks of the 'old club' which excludes contractors and subcontractors unless they have been involved in the design process. This is not a helpful definition as it perpetuates the traditional division between design and construction and between the employer's 'side' and the contractor's 'side' of the project 'fence' and is hardly in the spirit of 'partnering'.

The definition begs the question as to just which 'team' do contractors and subcontractors belong? Obviously, not the 'exclusive' project team!

5.4.2   NRM1 in context

NRM1 Part 1: *General* places early cost estimating and cost planning in the context of two well-known process models – the RIBA Plan of Work and the OGC Gateway Process. These process models provide a diagrammatic explanation of how the project initiation, design, construction and post-tender reconciliation phases work. The RIBA Plan of Work referred to in NRM1 is the 2007 version which has now been superseded by the RIBA Plan of Work 2013 albeit it is understood that some practitioners do not intend to adopt the 2013 version!

Other than that, NRM1 Part 1 is largely explanatory; it explains the purpose of NRM1 and its intended use and structure and provides details of symbols and abbreviations used in the document.

It is clear that order of cost estimates relate to the early appraisal and briefing stages of the design process and that cost planning comes along later when design work proper has been commenced. As the design develops, NRM1 contemplates three formal cost plan stages, culminating in **Formal Cost Plan 3** at the completion of the technical design. This is the point at which the pre-tender estimate is formulated in anticipation of tenders being received from the competing contractors. The tenders received are then compared (reconciled) with the cost plan as part of the tender adjudication process prior to choosing the successful contractor.

5.4.3   Purpose of NRM1

Paragraph 1.3 identifies the purposes of NRM1, the extent of its application and what it is not intended for:

Paragraph 1.3.1

- *To provide a standard set of measurement rules that are understandable by all those in a construction project, including the employer.*
- The *RICS New Rules of Measurement* should *assist the quantity surveyor/cost manager in providing effective and accurate cost advice.*

Paragraph 1.3.2

- To provide *rules of measurement for the preparation of order of cost estimates and elemental cost plans.*
- To provide *direction on how to describe and deal with costs and allowances forming part of the cost of a building, but which are not reflected in the measurable building work items....*

Paragraph 1.3.3

- The *RICS New Rules of Measurement* <u>do not</u> explain *estimating methods, cost planning techniques, procurement methods or contract strategies.*

Paragraph 1.3.4

- The RICS New Rules of Measurement are based on UK practice but *have worldwide application.*

Paragraph 1.3.3 is, perhaps, the most revealing in the context of the status of NRM1, because 'members' can only be judged, in the context of NRM1, on what the guidance note is concerned with but not on the methods and techniques used when applying NRM1 or on the contractual or procurement advice provided, which are all issues that NRM1 is not concerned with.

Consequently, the *specific professional tasks*, which form the basis of the higher professional standard of **best practice**, must be limited to those with which NRM1 is concerned.

The provision of a standard set of measurement rules for the preparation of order of cost estimates and cost plans (Paragraph 1.3.2) is self-evidently the main purpose of NRM1, but further purposes of the measurement rules in NRM1 are provided elsewhere in the document:

- Paragraph 2.1.4 states that the rules of measurement for element unit quantities (EUQ) can be used as a basis for measuring EUQ for the cost analysis of building projects.
  The phrase *(elemental) cost analysis* is defined in NRM1 Paragraph 1.6.3: *Definitions* as *a product-based cost model* which provides *a full appraisal of costs involved in previously constructed buildings*. Ashworth (2010) casts further light on this by explaining that a cost analysis is based *on data received from an accepted tender* (Ashworth 2010), that is, as opposed to an agreed final account.
- Paragraph 3.1.4 of NRM1 adds that *the measurement rules for elemental cost planning can also be used as a basis for measuring quantities for the application to whole life cycle costing.* This is sensible guidance that should add consistency and reliability to the collection and analysis of such data. If, however, NRM1 can be used for measuring such quantities, it is strange that no mention is made of measuring other sorts of quantities using NRM1 (e.g. bills of quantities/approximate quantities).

5.5   Design cost control: Techniques

Essentially, NRM1 provides standardised measurement rules – where none existed before – which introduce consistency into the preparation of **order of cost estimates** (aka first estimate, budget, cost limit) and **elemental cost plans.**

Order of cost estimates and elemental cost plans are the output of the design cost control process which rely upon the input generated by the estimating methods used.

NRM1 Part 2.1: *Introduction* usefully categorises the three main estimating methods used for preparing order of cost estimates:

- Floor area method.
- Functional unit method.
- Elemental method.

Anyone unused to cost planning will find these estimating methods somewhat alien compared to the more familiar bills of quantities that most people in the industry are used to. All three methods use the familiar **quantity × rate = price** formula, but the items measured and the rates used will not be so familiar.

These methods of approximate estimating are clearly explained in NRM1 itself and by Benge (2014) in a definitive guide to measurement and estimating using NRM1.

Primary sources of cost data for use with approximate methods of estimating include major price books, such as Spons (Langdon, 2014) and Griffiths (Franklin + Andrews, 2014), and the BCIS Cost Information Service.

BCIS, available by subscription from the RICS, supplies rates/m^2 for a wide variety of building types and detailed elemental cost analyses of previous tenders, as well as locational and inflation adjustment indices, building cost and tender price indices and market information. BCIS has produced a new *Standard Form of Cost Analysis* (Fourth Edition) specifically for use with NRM1.

Part 2

5.5.1   Floor area method

The floor area method of approximate estimating relies on the measurement of the GIFA of a building to which a suitable rate per m² is applied to arrive at a price.

The RICS *Code of Measuring Practice* (Sixth Edition) is the 'standard method of measurement' for calculating GIFA which is defined as *the area of a building measured to the internal face of the perimeter walls at each floor level.*

The *Code of Measuring Practice* is reproduced in Appendix A of NRM1 wherein details of what is included and excluded from the GIFA, together with details as to when and how to use the GIFA method of measurement, may be found.

NRM1: Measurement rules

NRM1 Appendix A: Core definition of gross internal floor area (GIFA)

2.0	GIFA
2.1–2.17	Included in the GIFA
2.18–2.22	Excluded from the GIFA
APP 4–8	Applications
GIFA 1–7	Notes
Supplementary definitions 1–4	Definitions adapted from the BCIS *Standard Form of Cost Analysis* (Fourth Edition)

Once the GIFA has been determined, the building works estimate is given by the formula:

GIFA (m²)	×	Cost per m²	=	Building works estimate
Floor area measured inside external walls (see NRM1 Appendix A)		Derived from: ■ BCIS cost analysis ■ Spon's or other builders' price book ■ Internal database		

Example:

New factory unit with internal office

	Quantity	Unit	Rate (£)	Total (£)
Building works estimate	2 000	m²	1 750	3 500 000

Caveat:

It must be pointed out that the unit rate used in this method of estimating is subject to a great deal of adjustment to reflect differences between the data source and its use.

Location of the works, base date of the price data, contract type, prevailing market conditions, the morphology of the building, quality of construction and other factors will influence the quantity surveyor/cost manager's judgement as to what the unit rate should be.

5.5.2   Functional unit method

Certain building types/functions (e.g. schools, hospitals, football stadia) may be given an associated unit of measurement (e.g. per child/student, per bed space, per seat) as an alternative method of approximate estimating, used where cost/m^2 is more difficult to measure or not appropriate.

The basis of this method of measurement is that the functional units are measured (e.g. number of beds in a nursing home) and a rate per functional unit is applied to this quantity which includes ancillary areas of the building (e.g. office, kitchen, public areas)

The various functional units are set out in NRM1 Appendix B which includes the more usual functional units (e.g. per car parking space, per bed space) and the functional unit of per m^2 of net internal area (NIA). Appendix C provides the measurement rules for NIA, reproduced from the RICS *Code of Measuring Practice* (Sixth Edition), and Appendix D provides special measurement rules for shops, again reproduced from the RICS *Code of Measuring Practice* (Sixth Edition).

NRM1: Measurement rules

Appendix B: Commonly used functional units and functional units of measurement

NRM1 Appendix C: Core definition of net internal area (NIA)

3.0	NIA
3.1–3.10	Included in the NIA
3.11–3.21	Excluded from the NIA
APP 9–11	Applications
NIA 1–9	Notes

NRM1 Appendix D: Special use definitions for shops

16.0	Retail area (RA)
17.0	Storage area (StoA)
18.0	Ancillary areas (AA)
19.0	Gross frontage (GF)
20.0	Net frontage (NF)
21.0	Shop width (SW)
22.0	Shop depth (SD)
23.0	Built depth (BD)
APP 19	Applications
RA 1–3; AA 1; GF 1; NF 1; NF 2; SW 1; ShD 1–3	Notes

Once the building function and units of measurement have been determined, the building works estimate is given by the formula:

Functional units (nr)	× Cost per functional unit	= Building works estimate
Number of places, beds, seats, etc. including circulation space (see NRM1 Appendix B)	Derived from: ■ BCIS cost analysis ■ Spon's or other builders' price book ■ Internal database	

Example:

New 20 000-seat football stadium

	Quantity	Unit	Rate (£)	Total (£)
Building works estimate	20 000	nr	2750	55 000 000

Caveat:

It must be pointed out that the unit rate used in this method of estimating is subject to a great deal of adjustment to reflect differences between the data source and its use.

Location of the works, base date of the price data, contract type, prevailing market conditions, the morphology of the building, quality of construction and other factors will influence the quantity surveyor/cost manager's judgement as to what the unit rate should be.

5.5.3   Elemental method

NRM1 Paragraph 2.7 is devoted to an explanation of a further method of approximate estimating – the elemental method. This method of approximate estimating is normally used for elemental cost planning (i.e. later on in the design process) but can also be used for order of cost estimates in a simplified form.

The elemental method of approximate estimating can be used as an extension of the cost/m^2 and functional unit methods:

- The order of cost estimate is calculated using either the cost/m^2 or functional unit methods.
- The figure produced is then analysed into elements (usually using the coarse subdivision of group elements) using either:
  - Predetermined percentages for each element or
  - Percentages calculated from a study of previous similar buildings sourced from the BCIS or an internal database.

Alternatively, if sufficient design information is available, quantities for each group element, and if possible, each element, may be calculated and suitable unit rates applied to derive the order of cost estimate.

The elemental method is explained in NRM1 Paragraph 2.7.1. This lists the elements commonly employed, as shown (in part) in Table 5.1. At the early stages of the design process, it is likely that only the group elements will be used due to the lack of information needed for a more detailed breakdown.

Table 5.1   Elements.

Group element	Elements
0. Facilitating works	
1. Substructure	
2. Superstructure	
3. Internal finishes	1. Wall finishes
4. Fittings, furnishings and equipment	2. Floor finishes
5. Services	3. Ceiling finishes
6. Prefabricated buildings and building units	
7. Work to existing buildings	
8. External works	

Paragraph 2.7.7 further explains that the elemental method can be used at the concept design stage which then provides a frame of reference for developing the first of the formal cost plans – **Formal Cost Plan 1**. It also explains that the initial EUQ and element unit rates (EUR) will eventually be replaced by more detailed measured quantities and unit rates once sufficient design information becomes available. In practice, this would mean using the Part 4: *Tabulated rules of measurement for elemental cost planning* rather than the EUQ.

The elements used for the **elemental method** are taken from the BCIS *Elemental Standard Form of Cost Analysis*, Fourth (NRM) Edition, which is based on nine group elements:

NRM1 Table 2.1: *Rules of measurement for elemental method of estimating* is the place to find these elements, and Paragraph 2.8.2 explains its structure. NRM1 Table 2.1 provides measurement rules and units of measurement for each individual element, within the relevant group element, together with useful notes that explain areas of potential uncertainty.

It shows, for instance, that the **frame** may be expressed in m² by measuring the gross internal floor area (GIFA) of the building, whereas **external walls** may also be expressed in m², but this time by measuring the area of the walls less the area of the windows. The NRM1 units of measurement for the Superstructure element are summarised in Table 5.2.

Table 5.2   Superstructure: units of measurement.

	Element	Unit
1	Frame	m² – the area of the floors related to the frame
2	Upper floors	m² – the total area of the upper floors
3	Roof	m² – the area of the roof on plan
4	Stairs and ramps	Nr – the number of staircases × the number of floors served excluding the lowest floor
5	External walls	m² – the area of the walls less the area of the windows
6	Windows and external doors	m² – the area of the windows and external doors measured over frames
7	Internal walls and partitions	m² – gross area with no deductions for doors and the like.
8	Internal doors	Nr – giving the total number of doors

At order of cost estimating stage, this presumes that a great deal of design detail will be available from which to measure, such as floor plans and elevations at least. Paradoxically, an order of cost estimate is needed at precisely the time when such design information is <u>not</u> available. This doesn't change the fact that the employer needs order of cost information at the earliest possible stage to be able to make a decision on whether to proceed with the project, whether to modify it or whether to not proceed at all.

The choice of measurement unit will depend on the extent of design information available:

NRM1: Measurement rules	
Paragraph 2.8	An order of cost estimate may be expressed in several units of measurement according to the availability of suitable information (refer to Paragraphs 2.8.4 and 2.8.5)
Table 2.1	Various rules for measuring individual elements within group elements 0 – 8 + 9 (main contractor's preliminaries) and 10 (main contractor's overheads and profit)
Paragraph 2.7.3*	■ *If suitable information is available, then the element unit quantities (EUQ) are measured for an element in accordance with the rules* ■ *Where insufficient information is available for a particular element, the EUQ (element unit quantity) for the element is based on the GIFA*
Paragraph 2.7.5	Where measurement is based on the gross internal floor area (GIFA), this shall be determined in accordance with the RICS *Code of Measuring Practice* (reproduced in NRM1 Appendix A)
Paragraph 2.7.6	Where there is more than one building, *each building is to be shown separately*

*This rather clumsy rule means that Table 2.1: *Rules of measurement for elemental method of estimating* shall apply in circumstances where an element has been designed in sufficient detail as to enable it to be measured individually. If not, the GIFA is used.

As mentioned previously, Paragraph 2.7.7 explains that *the initial element unit quantities (EUQ) will eventually be superseded by more detailed measurement of elements, sub-elements and components* once the design has been developed in more detail.

This may sound a little strange but remember that the order of cost estimate is just that – an outline of the likely cost for each group element possibly subdivided into elements. This merely establishes a budget per m² of GIFA, but when the drawings become available, quantities may be measured in more detail in m, m² or m³ and priced accordingly.

The cost target for an element is derived from the formula:

Element unit quantity (EUQ) ×	Element unit rate (EUR)	= Cost target
	Derived from: ■ BCIS elemental cost analysis ■ Spon's or other builders' price book ■ Internal database of cost analyses/ composite rates	

Example:

External wall element for a new house

	Quantity	Unit	Rate (£)	Total (£)
Cost target (based on gross internal floor area)	150	m²	£480*	72 000
or alternatively	**Quantity**	**Unit**	**Rate (£)**	**Total (£)**
Cost target (based on area of external walls)	240	m²	£300**	72 000

*This rate is a proportion of the cost per m² for building a complete house (in this case, 30% of £1600/m²).
**This rate is the rate per m² for a complete wall (e.g. external facing bricks, insulated cavity and internal blockwork).

In the initial stages of a project, the only quantities available will normally be the GIFA, and this will be, therefore, the EUQ for each group element or element in the order of cost estimate.

The EUR will be taken from a previous cost analysis of a similar building and will be a proportion of the total rate per m² of GIFA for such a building. The proportions used will be informed by the previous cost analysis used, but the quantity surveyor will also make a judgement based on experience as to what proportions would be suitable in the circumstances (see Table 5.3).

Table 5.3 Proportion method.

Group element		Cost (£) per m²	%	Allowance (£) per m²	Gross internal floor area (m²)	Totals (£)
1	Substructure	1400	15	210	6000	1260000
2	Superstructure	1400	25	350	6000	2100000
3	Internal finishes	1400	10	140	6000	840000
4	Fittings, furnishings and equipment	1400	5	70	6000	420000
5	Services	1400	35	490	6000	2940000
6	Prefabricated buildings and building units	1400	0	0	6000	0
7	Work to existing buildings	1400	0	0	6000	0
8	External works	1400	10	140	6000	840000
Totals			100	1400		**8400000**

This all sounds a bit arbitrary, but the idea is to provide the employer (and the architect) with an 'order of cost' so that they may make informed decisions about the project and its design.

5.5.4 Unit rates and EUR

Unit rates and EUR are used in approximate estimating and cost planning.

Unit rates are used to price unit measurements of:

- Floor area (e.g. £ × per m² of GIFA).
- Functional units (e.g. £ × per bed space).
- Elements (e.g. £ × per m² of roof).
- Sub-elements (e.g. £ × per m² of roof covering).
- Components (e.g. £ × per linear metre of roof flashing).

EURs are calculated by the formula:

$$\text{Element unit rate} = \frac{\text{Cost of element}}{\text{Element quantity}} \quad \text{For example} \quad \frac{\text{Cost of roof}}{\text{Quantity of roof}} = \frac{£40\,000}{1\,600\,\text{m}^2} = £26 \text{ per m}^2$$

NRM1 Paragraph 2.9 is largely devoted to matters concerning the basis and use of the unit rates used for producing order of cost estimates and should be read in conjunction with Paragraph 1.6.3: *Definitions*.

The majority of Paragraph 2.9 is 'textbook' material covering issues such as the use, interpretation and updating of cost analyses. This is a welcome element of 'added value' in NRM1 – it's just a pity that all such material has not been kept separate from the rules of measurement.

NRM1: Measurement rules

Paragraph 2.9.1	This rule requires that unit rates shall exclude allowances for inflation or deflation which means that separate item must be provided for in accordance with Paragraph 2.16: *Measurement rules for inflation*
Paragraph 2.9.2	This rule requires that the unit rates applied to measured quantities must be applicable to the method of measurement used, that is, per m^2 of GIFA and per functional unit or using element quantities (i.e. count the number of oranges and multiply by the price per orange and not per lemon!)
Paragraph 2.9.3	This rule stipulates that unit rates (whether cost per m^2 of GIFA or cost per functional unit or EUR) shall:

Include:
- Materials, labour and plant
- Subcontractors' or suppliers' design fees
- Subcontractors' preliminaries
- Subcontractors' overheads and profit

Exclude:
- Main contractor's preliminaries
- Main contractor's overheads and profit
- Project/design team fees
- Other development/project costs
- Risk allowances
- Inflation

The essential meaning of Paragraph 2.9 is that separate items need to be provided for the excluded items in the order of cost estimate. Logically, this is a sensible provision as all these items are risk issues that need to be managed.

Practically, however, complying with this rule is another issue. Cost data used for order of cost estimating comes from historic analyses of priced tenders, and BCIS, private QS practices and others have accumulated many years worth of such data that might not be a good fit with the New Rules of Measurement:

- Main contractor's preliminaries have always been shown separately in cost analyses, but contractors often include preliminaries in measured work items as well (e.g. scaffolding could be priced in the prelims or in the measured rates).
- Project/design team fees and other development/project costs are not usually included in cost analyses, but this may not be the case with design and build tenders, for example.
- Main contractor's overheads and profit have always been included in the measured rates, and to some extent in the preliminaries, and are therefore not transparent enough to be extracted in the manner suggested in Rule 2.9.3.
- Design risk allowances (to allow for design changes during construction) have always been catered for in the contingency figure that is usually stated under 'Provisional Sums' in the tender BQ, but they usually include contractor's overheads and profit because provisional sums, if expended, are paid for at prevailing BQ rates (which normally include overheads and profit).
- Price risk allowances (inflation) from tender date to contract completion in firm price tenders are usually included somewhere in the contractor's pricing (in the unit rates, in the preliminaries or both) but are not necessarily shown separately.
- Price risk allowances (inflation) in fluctuating price tenders have always been included in the cost plan by the quantity surveyor, but the contractor also makes provision for under-recovery of inflation on disallowed costs (such as temporary materials, formwork and the like), and such allowances are not necessarily divulged by the contractor.

Consequently, it is not possible to comply completely with Rule 2.9.3 because not all of the information required is transparent from the contractor's tender.

Even if historic cost data could be disaggregated in the fashion required, it would be an enormous task to redo all past analyses to bring them into line with NRM1 Rule 2.9.3 (there are some 18 500 cost analyses in the BCIS database alone). In addition, it is highly unlikely that future cost data may be collected under these headings because contractors habitually price their tenders to suit their own purposes and not to suit future cost planners.

Risk issue

When asked to produce an order of cost estimate in accordance with NRM1, the quantity surveyor/ cost manager may need to draw the employer's attention to the aforementioned limitations as it would be misleading to give the impression that the order of cost estimate is capable of being drafted in the manner required by NRM1.

Allowances for the contractor's overheads and profit and for design and inflation risks may be seen as predictable, but if they cannot be isolated from other costs in the cost analysis data, then they cannot be effectively assessed or controlled.

This may impact on the confidence limits of the cost estimate and on the employer's risk exposure to the project.

5.5.5   Updating unit rates and other costs to current estimate base date

The majority of the data used for order of cost estimating is historic and is, therefore, out of date to some extent. Paragraph 2.10 is concerned with the process of updating such historic cost data to *the current estimate base date*, but there are no measurement rules in this paragraph.

The main focus of Paragraph 2.10, therefore, is on the use of a variety of indices, such as tender price and building cost indices, as a means of converting the base cost data from its original base date to the current base date of the order of cost estimate using the formula:

$$\frac{\text{Current index}}{\text{Base index}} \times 100 = \% \text{ change } (\pm)$$

Paragraph 2.10.2 (Footnote) warns not to update items added to the estimate as a percentage in this way, such as preliminaries and fees. They will be automatically updated when the percentage change is applied.

5.6   Order of cost estimates

NRM1 Part 2 explains the purpose of order of cost estimates and prescribes the information needed from both the employer and the architect to enable order of cost estimates to be prepared. A pro forma layout for an order of cost estimate is also provided which shows how design team fees, risk allowances and inflation allowances may be added to a basic cost estimate.

Also in NRM1 Part 2 are a number of 'measurement rules', some of which relate to the methods and units to be used for measuring quantities when using a variety of order of cost estimating techniques and some of which relate to the manner in which the order of cost estimate is to be laid out or presented.

Whilst NRM1 is intended to provide a consistent approach to the preparation of order of cost estimates, there is, in practice, no compulsion to comply with the NRM1 measurement rules because order of cost estimates are not intended for publication as a tender document or as a document that may lead to a contract coming into existence.

Part 2

Risk issue

An important caveat to this statement is that RICS members are bound by the status of NRM1 as a 'guidance note', and this conditions their freedom to adopt other practices that may not be viewed as equivalent 'best practice'.

However, other users of NRM1 are not bound by the same rules as RICS members.

Notwithstanding this, should the NRM1 rules of measurement for order of cost estimating be intended as a means of producing bills of quantities or other similar tender documents, it might be considered wise to follow the rules implicitly.

This is because such documents will almost certainly need to be interpreted at some stage if there is a 'grey area' or disagreement as to what was intended to be included in the tender price.

In common with NRM2, it is hard work to disassociate measurement rules from guidance and explanatory text in NRM1 Part 2. Even where there is a heading entitled 'Measurement rules…', definitions and other text mask the measurement rules that must be followed.

5.6.1   Purpose of order of cost estimates

Paragraph 2.2.2 explains that order of cost estimates are necessary in order to:

- Establish whether or not a proposed building project is affordable.
- Establish a realistic cost limit for the project.

As this process takes place very early in a project, and it is most likely that several options may be open to the client, Paragraph 2.2.3 goes on to emphasise that alternatives need to be considered at this stage. This would include consideration of alternative types of buildings, whether to refurbish or replace existing buildings or move to a new site. Paragraph 2.2.3 explains that such considerations are called *option costs* or *option costings*.

5.6.2   Information requirements

In order to prepare order of cost estimates, the quantity surveyor relies upon information from various parties involved in the project. NRM1 Paragraph 2.3.1 specifies what information will be required from the employer (Paragraph 2.3.1), the architect (2.3.2), the mechanical and electrical services engineer (2.3.3) and the structural engineer (2.3.4) so that an order of cost estimate may be prepared.

Risk issue

Despite the requirements stated in NRM1:

- There is no guarantee that such information will be forthcoming.
- There is no sanction if it isn't.
- There is no guarantee that the information received will be fully in accordance with NRM1 requirements.

Paragraph 2.3.5 makes the valid point that the accuracy of the order of cost estimate will be *dependent on the quality of information supplied to the quantity surveyor/cost manager* and that *the more information provided, the more reliable the outcome will be.*

Paragraph 2.3.5 adds that *where little or no information is provided, the quantity surveyor/cost manager will need to qualify the order of cost estimate.*

In view of the potential for litigation should the order of cost estimate prove to be inaccurate or misleading, the quantity surveyor/cost manager would be well advised to consider what protocols should be put in place in order to deal with shortcomings in the supply of information from the employer, the architect and others.

5.6.3   Constituents of order of cost estimates

Paragraph 2.4.1 lists the key constituents of an order of cost estimate. These are illustrated in Table 5.4 as an MS Excel spreadsheet, but CATO QuickEst or other methods could equally be used.

Table 5.4   Constituents of an order of cost estimate.

	A	B	C	D	E
1	ORDER OF COST ESTIMATE				
2					
3	Ref	Constituent	Totals	Sub-totals	Total
4			£	£	£
5		a Facilitating works estimate			
6		b Building works estimate			
		c Main contractor's preliminaries estimate			
7					
8		d Sub-total		=SUM(C5:C7)	
		e Main contractor's overheads and profit			
9		estimate			
10		f **Works cost estimate**		=SUM(D8:D9)	
11		g Project/design team fees estimate			
12		Sub-total		=SUM(D10:D11)	
		h Other development/project costs estimate			
13					
14		i **Base cost stimate**		=SUM(D12:D13)	
15		j Risk allowances estimate			
16		k Design development risks estimate			
17		l Construction risks estimate			
18		m Employer change risks estimate			
19		n Employer other risks estimate		=SUM(C16:C19)	
20		o Cost limit (excluding inflation)		=SUM(D14:D19)	
21		p Tender inflation estimate			
		q Cost limit (excluding construction inflation)			
22				=SUM(D20:D21)	
23		r Construction inflation estimate			
24		s Cost limit (including inflation)			=SUM(D22:D23)
25		t VAT assessment			
26					

It will be noted from Table 5.4 that the authors of NRM1 have departed somewhat from convention:

- The main contractor's overheads and profit have been shown as a separate item in the order of cost estimate.
- The usual 'contingency' allowance to allow for design and price (inflation) risk has been replaced by two separate risk estimates:

- ○ A *risk allowance estimate* within which design risks, construction risks, employer change risks and other employer risks have been disaggregated so that separate estimates for each may be calculated.
- ○ Inflation estimates which have been separated into *tender inflation* and *construction inflation* risk estimates.

In many respects these are sensible ideas and the separate inflation estimates reflect normal practice anyway, which is to:

- Update the cost data used for the estimate from the base date (i.e. the tender date of the cost analysis used) to the tender date of the project in hand.
- Project the extent to which construction prices are expected to increase from the tender date to the conclusion of the contract.

However, it is quite fanciful to imagine that estimates of the main contractor's overheads and profit will be any more than educated guesswork because:

- Publically available cost data (e.g. BCIS and the well-known price books) include overheads and profit in the rates per m² or per functional unit because the data is derived from historic tender prices which do not disaggregate these items.
- No self-respecting contractor is ever going to divulge his overheads and profit allowances, and even if he does, the figures will be more fiction than fact.
- Even if future cost data is collected in the new NRM1 format, it is hardly likely to be accurate or reliable as contractors will only divulge the information that they want to divulge.

5.6.4   Facilitating works

NRM1 Paragraph 2.5.1 identifies facilitating works as *specialist works* required prior to the commencement of building works such as demolition work, ground remediation, removal of hazardous material, and soil stabilisation.

Such works should not be confused with 'enabling works' which is commonly taken to mean a package of works that includes facilitating works as well as intrusive site investigations, access roadworks, the provision of main services and so on. Paragraph 2.5.1 also emphasises that facilitating works may be part of an enabling work package but that the two are not synonymous.

Should it be desired that an item for enabling works, including facilitating works, is to be included in the order of cost estimate, presumably the quantity surveyor/cost manager is at liberty to change the heading in the 'pro forma' order of cost estimate accordingly, as there is no indication to the contrary in NRM1 Paragraphs 2.4 and 2.5.

NRM1: Measurement rules

Paragraph 2.5.2   1. *The site area which is defined as either:*
- a. *The total area of the site within the site title boundaries*
  or
- b. *The total area within the site title boundaries defined by the employer as the site for the building or buildings*
  less
- c. *The footprint of any existing buildings, measured on a horizontal plane.*
2. *The area affected measured in* m², *linear metres, enumerated (No) or itemised (item) as deemed appropriate.*

Whilst measurement units are provided for *the area affected*, no units of measurement are stipulated for measuring the site area. The *site area* is defined in Paragraph 2.5.2 and reiterated in Paragraph 1.6.3: *Definitions*, but no measurement units are mentioned.

It is to be assumed that the unit would be m², as this is the unit used for measuring facilitating works under the elemental method of estimating and for measuring like items under Part 4: *Tabulated rules of measurement for elemental cost planning*.

The measurement of the quantities for facilitating works, as detailed in Paragraph 2.5.2, is not exactly crystal clear. From close inspection, it would appear that there are two choices:

1. Measure the site area as one item.
2. Measure individual items, for example, general site clearance (m²), hazardous waste removal (m²), removal of boundary walls (m) and demolition (nr or item).

Risk issue

Although not required by the rules of measurement, where alternative 2 is used to measure the facilitating works, and *the area affected* is measured in units other than m², *the area affected* may need to be defined or indicated on a drawing in order to distinguish the limits of the site of the facilitating works to be undertaken.

Example:

ORDER OF COST ESTIMATE FOR NEW 20 000-SEATER FOOTBALL STADIUM

				Rate	Totals	Subtotals	Total
Ref	Constituent	Quantity	Unit	£	£	£	£
a	Facilitating works estimate	8	ha	250 000	2 000 000		

5.6.5 Building works: Floor area method

The measurement rules used for the building works part of the order of cost estimate depend upon which estimating method is used; they are defined in Paragraph 2.6.1 as:

1. Floor area method – *the total GIFA*.
2. Functional unit method – *by projecting the number of functional units*.

Paragraph 2.6.1 also states that a combination of both methods may be required in certain circumstances.

The measurement rules for the **floor area method** are:

NRM1: Measurement rules

Paragraph 2.6.1 (a) (ii)	The gross internal floor area (GIFA) is measured in accordance with the RICS *Code of Measuring Practice*
Paragraph 2.6.1 (a) (iii)	Where there is more than one building, *each building is to be shown separately*

Part 2

Paragraph 2.6.1 (a) (iv)	Where more than one user function is included within a single building (*e.g. residential, retail, offices*), *the GIFA for each separate function is to be calculated and quantified separately* using *the centre line of the party wall* in order to distinguish the various functions concerned The total GIFA for the building is to equal *the sum total of the GIFA for each separate function*
Paragraph 2.6.1 (a) (v)	Where external works are to be measured separately, *the site area*, as defined in Paragraphs 1.6.3 and 2.5.2, is to be measured **less** *the footprint of the new building(s), measured on a horizontal plane*

5.6.6   Building works: Functional unit method

The measurement rules used for the building works part of the order of cost estimate depend upon which estimating method is used; they are defined in Paragraph 2.6.1 as:

1.   Floor area method – *the total GIFA*.
2.   Functional unit method – *by projecting the number of functional units*.

Paragraph 2.6.1 also states that a combination of both methods may be required in certain circumstances.
   The measurement rules for the **functional unit method** are:

NRM1: Measurement rules

Paragraph 2.6.1 (b) (ii)	*A suitable functional unit of use for the building is to be selected*. This could be a unit of space measured in m^2 or a unit of capacity measured by number (No) of places or seats
Paragraph 2.6.1 (b) (iii)	If the functional unit of measurement is to be the 'net internal area' (NIA), this shall be determined in accordance with the RICS *Code of Measuring Practice* (reproduced in NRM1 Appendix C)
Paragraph 2.6.1 (b) (iv)	If the functional unit of measurement is to be the 'retail area', this shall be determined in accordance with the RICS *Code of Measuring Practice* (reproduced in NRM1 Appendix D)
Paragraph 2.6.1 (b) (v)	A functional unit is deemed to include *all circulation (space) necessary*

Example:

ORDER OF COST ESTIMATE FOR NEW 20 000-SEATER FOOTBALL STADIUM

Ref	Constituent	Quantity	Unit	Rate £	Totals £	Subtotals £	Total £
a	Facilitating works estimate	8	ha	250 000	2 000 000		
b	Building works estimate	20 000	nr	2 750	55 000 000		

Part 2 (side margin)

5.6.7   Main contractor's preliminaries

NRM1 Paragraph 2.11 deals with the measurement of main contractor's preliminaries in the order of cost estimate.

The allowance for subcontract preliminaries, overheads and profit and for associated design fees and risk allowances are to be included in the cost per m² of GIFA used for the estimate.

This also applies where the cost per functional unit or EUR is used.

NRM1: Measurement rules

Paragraph 2.11.1   The main contractor's preliminaries *are to be added* to the order of cost estimate as a percentage of the total cost of the building works

Consequently, any bill of quantities prepared on the basis of the measurement rules for order of cost estimates should include an item for this purpose, but there is no requirement to list separate preliminaries components as would be the case under NRM2.

Example:

ORDER OF COST ESTIMATE FOR NEW 20 000-SEATER FOOTBALL STADIUM

Ref	Constituent	Quantity	Unit	Rate £	Totals £	Subtotals £	Total £
a	Facilitating works estimate	8	ha	250 000	2 000 000		
b	Building works estimate	20 000	nr	2 750	55 000 000		
c	Main contractor's preliminaries estimate	12	%		6 600 000		
d	Subtotal					63 600 000	

5.6.8   Main contractor's overheads and profit

Main contractor's overheads and profit are to be added to the order of cost estimate as a separate item.

Paragraph 2.12.4 refers to a *list of items* to be found within group element 10 in Part 4: *Tabulated rules of measurement for elemental cost planning*. This is meant to provide a guide as to what is typically included within the main contractor's overheads and profit but *is not meant to be definitive or exhaustive*.

The 'list' in group element 10 is painfully short.

NRM1: Measurement rules

Paragraph 2.12.1   The main contractor's overheads and profit are to be based on a percentage addition. The estimated cost of the contractor's overheads and profit *is to be calculated* by applying the percentage additions to the total cost of the building works and the main contractor's preliminaries

Part 2

Example:

ORDER OF COST ESTIMATE FOR NEW 20 000-SEATER FOOTBALL STADIUM

Ref	Constituent	Quantity	Unit	Rate £	Totals £	Subtotals £	Total £
a	Facilitating works estimate	8	ha	250 000	2 000 000		
b	Building works estimate	20 000	nr	2 750	55 000 000		
c	Main contractor's preliminaries estimate	12	%		6 600 000		
d	Subtotal					63 600 000	
e	Main contractor's overheads and profit estimate	8	%			5 088 000	
f	**Works cost estimate**					68 688 000	

5.6.9   Project/design team fees

Project/design team fees are defined in Paragraph 2.13.1 as fees for specialist consultants which may also include the main contractor's pre-construction fees.

A *typical list* of such fees is given *as a guide* in NRM1 Part 4: *Tabulated rules of measurement for elemental cost planning* (group element 11). *This list is not meant to be definitive or exhaustive, but is simply a guide.*

Paragraph 2.13.3 recommends that a single allowance is made for such fees.

NRM1: Measurement rules

Paragraph 2.13.2	*Project/design team fees are to be included in order of cost estimates* unless the employer specifically requests that they be omitted

---

Example:

ORDER OF COST ESTIMATE FOR NEW 20 000-SEATER FOOTBALL STADIUM

Ref	Constituent	Quantity	Unit	Rate £	Totals £	Subtotals £	Total £
a	Facilitating works estimate	8	ha	250 000	2 000 000		
b	Building works estimate	20 000	nr	2 750	55 000 000		
c	Main contractor's preliminaries estimate	12	%		6 600 000		
d	Subtotal					63 600 000	
e	Main contractor's overheads and profit estimate	8	%			5 088 000	
f	**Works cost estimate**					68 688 000	
g	Project/design team fees estimate	10	%			6 868 800	
	Subtotal					75 556 800	

Part 2

5.6.10   Other development/project costs

Paragraph 2.14.1 defines 'other development/project costs' as indirect costs that *form part of the total cost of the building project to the employer,* including insurances, planning fees and fees in connection with party wall awards, and so on.

NRM1: Measurement rules

Paragraph 2.14.1	This rule requires that such costs *are to be included in order of cost estimates unless the employer specifically requests that they be omitted. Other development/project costs are to be added as a lump sum allowance*
Paragraph 2.14.4	*The total estimated cost of other development/project costs is added to the combined total of the works cost estimate and the project/design team fees estimate*

Example:

ORDER OF COST ESTIMATE FOR NEW 20 000-SEATER FOOTBALL STADIUM

Ref	Constituent	Quantity	Unit	Rate £	Totals £	Subtotals £	Total £
a	Facilitating works estimate	8	ha	250 000	2 000 000		
b	Building works estimate	20 000	nr	2 750	55 000 000		
c	Main contractor's preliminaries estimate	12	%		6 600 000		
d	Subtotal					63 600 000	
e	Main contractor's overheads and profit estimate	8	%			5 088 000	
f	**Works cost estimate**					68 688 000	
g	Project/design team fees estimate	10	%			6 868 800	
	Subtotal					75 556 800	
h	Other development/ project costs estimate					2 000 000	
i	**Base cost stimate**					77 556 800	

5.6.11   Risk

Quite bizarrely, in 2½ pages of text under the heading 'Measurement rules for risk', there is only one measurement rule.

The remainder of Paragraph 2.15 concerns itself with an explanation of the risk management process together with recommendations and advice as to how risk may be considered, managed and allowed for in the context of individual projects. Paragraph 2.15.5 recommends that separate allowances be made for the following risks:

- Design development risks.
- Construction risks.
- Employer change risks.
- Employer other risks.

Part 2

NRM1: Measurement rules

Paragraph 2.15.7　*Risk allowances are to be included in order of cost estimates*

Example:

ORDER OF COST ESTIMATE FOR NEW 20 000-SEATER FOOTBALL STADIUM

				Rate	Totals	Subtotals	Total
Ref	**Constituent**	**Quantity**	**Unit**	**£**	**£**	**£**	**£**
a	Facilitating works estimate	8	ha	250 000	2 000 000		
b	Building works estimate	20 000	nr	2 750	55 000 000		
c	Main contractor's preliminaries estimate	12	%		6 600 000		
d	Subtotal					63 600 000	
e	Main contractor's overheads and profit estimate	8	%			5 088 000	
f	**Works cost estimate**					68 688 000	
g	Project/design team fees estimate	10	%			6 868 800	
	Subtotal					75 556 800	
h	Other development/ project costs estimate					2 000 000	
i	**Base cost stimate**					77 556 800	
j	Risk allowances estimate						
k	Design development risks estimate	7	%		5 428 976		
l	Construction risks estimate	5	%		3 877 840		
m	Employer change risks estimate	3	%		2 326 704		
n	Employer other risks estimate	2	%		1 551 136	13 184 656	
o	Cost limit (excluding inflation)					90 741 456	

5.6.12　Inflation

Paragraph 2.16.1 states that *the rules divide inflation over a period of time into two categories, namely:*

a)　*To date of tender (i.e. tender inflation).*
b)　*During the construction period (i.e. construction inflation).*

Tender inflation is defined in Paragraph 1.6.3 as representing cost fluctuations from the estimate base date to the date of tender. *Construction inflation* is an inflation allowance calculated from the date of return of tenders to the midpoint of the construction period.

There appear to be no real rules in Paragraph 2.16, and the remainder of the paragraph largely consists of explanations, calculations and recommendations.

Part 2

NRM1: Measurement rules

Paragraphs 2.16.4 and 2.16.6	The amount of inflation *is ascertained by applying a single percentage rate to the cost limit* which *can be computed using published indices*

Example:

ORDER OF COST ESTIMATE FOR NEW 20 000-SEATER FOOTBALL STADIUM

Ref	Constituent	Quantity	Unit	Rate £	Totals £	Subtotals £	Total £
a	Facilitating works estimate	8	ha	250 000	2 000 000		
b	Building works estimate	20 000	nr	2 750	55 000 000		
c	Main contractor's preliminaries estimate	12	%		6 600 000		
d	Subtotal					63 600 000	
e	Main contractor's overheads and profit estimate	8	%			5 088 000	
f	**Works cost estimate**					68 688 000	
g	Project/design team fees estimate	10	%			6 868 800	
	Subtotal					75 556 800	
h	Other development/project costs estimate					2 000 000	
i	**Base cost stimate**					77 556 800	
j	Risk allowances estimate						
k	Design development risks estimate	7	%		5 428 976		
l	Construction risks estimate	5	%		3 877 840		
m	Employer change risks estimate	3	%		2 326 704		
n	Employer other risks estimate	2	%		1 551 136	13 184 656	
o	Cost limit (excluding inflation)					90 741 456	
p	Tender inflation estimate	2.5	%			2 268 536	
q	Cost limit (excluding construction inflation)					93 009 992	
r	Construction inflation estimate	7.5	%			6 975 749	
s	**Cost limit (including inflation)**						99 985 742
t	VAT assessment						

Part 2

5.6.13   Value added tax assessment

Paragraph 2.17.1 states that value added tax (VAT) *in relation to buildings is a complex area* and is, therefore, best *excluded from order of cost estimates*.

The pro forma order of cost estimate provided in Paragraph 2.4.1 does, in fact, provide a space for the VAT assessment, but following the recommendation of Paragraph 2.17.2, specialist advice should be sought on the matter, and presumably, therefore, it is not the quantity surveyor/ cost manager who is meant to carry out the assessment.

5.6.14   Other considerations

Paragraph 2.18 is concerned with issues such as taxation allowances, taxation relief and grants and, in common with VAT, is the province of specialists whose advice should be sought on the matter.

5.6.15   Reporting of order of cost estimates

Paragraph 2.19 provides useful suggestions for the contents of an order of cost report along with advice as to issues that should be emphasised and brought to the employer's attention.

Paragraph 2.19.1 is categorical in that *costs are to be expressed as 'cost/m² of GIFA'*. However, the following Paragraph 2.19.2 states that functional units may be used as an alternative to, or in conjunction with, cost/m² of GIFA.

Risk issue

In terms of measurement, Paragraph 2.19.4 is perhaps the most significant in that the employer is to be clearly informed as to what is included in, and excluded from, the order of cost estimate.

The quantity surveyor/cost manager would be well advised to ensure that the correct protocols are put in place to ensure that the order of cost estimate is crystal clear to the employer.

Example:

ORDER OF COST ESTIMATE FOR NEW 20 000-SEATER FOOTBALL STADIUM

				Rate	Totals	Subtotals	Total
Ref	**Constituent**	**Quantity**	**Unit**	**£**	**£**	**£**	**£**
s	**Cost limit (including inflation)**						**99 985 742**
t	VAT assessment						
	Cost per functional unit	20 000	Seats				**4 999**

In practice, there is no way that the employer would be given such precise figures. A budget of £99 985 742 and a cost per seat of £4999 are simply the output from the estimating process. More realistically, figures would always be rounded and, in all probability, would be given as a range or 'bracket', within specified confidence limits, especially in the early stages of a project.

5.7   Cost planning

NRM1 Part 3 explains the purpose of cost planning and prescribes the information needed from both the employer and the architect to enable a cost plan to be prepared. A pro forma layout for a cost plan is also provided which shows how design team fees, risk allowances and inflation allowances may be added to a basic works cost estimate.

Part 2

NRM1 Part 3 also provides a number of 'measurement rules', some of which relate to the methods and units to be used for measuring quantities for cost planning and some which relate to the manner in which the cost plan is to be laid out or presented.

Whilst NRM1 is intended to provide a consistent approach to the preparation of cost plans, there is, in practice, no compulsion to comply with the NRM1 measurement rules because cost plans are not intended for publication as a tender document or as a document that may lead to a contract coming into existence.

Risk issue

An important caveat to this statement is that RICS members are bound by the status of NRM1 as a 'guidance note', and this conditions their freedom to adopt other practices that may not be viewed as equivalent 'best practice'.

However, other users of NRM1 are not bound by the same rules as RICS members.

Notwithstanding this, should the NRM1 rules of measurement for cost planning be employed as a means of producing bills of quantities or other similar tender documents, it might be considered wise to follow the rules implicitly.

This is because such documents will almost certainly need to be interpreted at some stage if there is a 'grey area' or disagreement as to what was intended to be included in the tender price.

In common with NRM2, it is hard work to disassociate measurement rules from guidance and explanatory text in NRM1 Part 3. Even where there is a heading entitled 'Measurement rules…', definitions and other text mask the rules that must be followed.

The only paragraph of interest in 3.1 *Introduction* in a measurement context is Paragraph 3.1.4 which states that *the measurement rules for elemental cost planning can also be used as a basis for measuring quantities for the application to whole life costing.*

5.7.1   Purpose of cost planning

Paragraph 3.2 contains no measurement rules but gives a succinct overview of cost planning in the context of the RIBA Plan of Work and the OGC Gateway Process.

The purpose of cost planning is given in Paragraph 3.2.2, and Paragraph 3.2.3 emphasises a key issue – that is, that elemental cost planning is *a budget distribution technique* that breaks down the cost limit into *cost targets for each element of the building.* This is like having a weekly food budget and deciding what to spend it on before going shopping. Anyone who goes shopping without a priced shopping list will risk an overspend!

Paragraph 3.2.4 explains that the cost plan is a statement of how the budget will be spent on each element of the building, and Paragraph 3.2.5 makes the point that this process is repeated several times as more and more design information becomes available.

5.7.2   Constituents of a cost plan

The make-up of a cost plan is given in NRM1 Paragraph 3.3.1 in the same pro forma style as for an order of cost estimate.

The only difference between the order of cost estimate pro forma and that of the cost plan is that the latter shows a detailed breakdown of the **project/design team fee estimate**.

Part 2

Table 5.5　Constituents of a cost plan.

	A	B	C	D	E
1	COST PLAN				
2					
3	Ref	Constituent	Totals	Sub-totals	Total
4			£	£	£
5	a	Facilitating works estimate			
6	b	Building works estimate			
	c	Main contractor's preliminaries estimate			
7					
8	d	Sub-total		=SUM(C5:C7)	
	e	Main contractor's overheads and profit estimate			
9					
10	f	**Works cost estimate**		=SUM(D8:D9)	
11	g	Project/design team fees estimate			
12	g1	Consultants' fees			
13	g2	Main contractor's pre-construction fee estimate (if applicable)			
14	g3	Main contractor's design fees estimate (if applicable)			
15		Sub-total		=SUM(C12:C14)	
	h	Other development/project costs estimate			
16					
17	i	**Base cost stimate**		=SUM(D15:D16)	
18	j	Risk allowances estimate			
19	k	Design development risks estimate			
20	l	Construction risks estimate			
21	m	Employer change risks estimate			
22	n	Employer other risks estimate		=SUM(C19:C22)	
23	o	Cost limit (excluding inflation)		=SUM(D17:D22)	
24	p	Tender inflation estimate			
	q	Cost limit (excluding construction inflation)			
25				=SUM(D23:D24)	
26	r	Construction inflation estimate			
27	s	Cost limit (including inflation)			=SUM(D25:D26)
28	t	VAT assessment			
29					

Table 5.5 shows the pro forma layout with the fee estimate (g) subdivided into g1, g2 and g3. The cost plan is presented as an MS Excel spreadsheet (with formulas), but there are other ways to do this, including the use of software such as CATO Cost Planning.

5.7.3　Formal cost planning stages

Paragraph 3.4 is purely explanatory and positions the formal cost planning stages within the RIBA Plan of Work and OGC Gateway Process Models.

The point is made in Paragraph 3.4.4 that formal cost plans may not be necessary at each and every stage of the design process as this will depend on the procurement arrangements for the project. The example of a design and build contract strategy is given where it may not be necessary to develop the cost plan to the production information stage of the process.

Paragraph 3.4.3 explains the chronology of the cost plan and emphasises that Formal Cost Plans 2 and 3 are reiterations of Formal Cost Plan 1 which is prepared at the point *where the scope of work is fully defined* but where *no detailed design has been commenced.*

5.7.4   Reviewing and approving cost plans

Paragraph 3.5.1 implies that protocols are needed for reviewing the cost plan *which is to be reviewed by the employer and the project team* before proceeding to the next design stage.
   This is to verify that:

- The project is affordable.
- Elemental cost targets are reasonable.
- The cost limit has not been exceeded.

Risk issue

Following the review meeting, NRM1 suggests that the employer:

- Will sign off the cost plan.
- Give any necessary instructions.
- Authorise commencement of the next design stage.

   Whilst the provisions of Paragraph 3.5.2 fall short of qualifying as 'rules', it is clear that protocols are needed to formally record these employer actions in the best interests of all concerned.

5.7.5   Cost control in procurement

Paragraph 3.6.1 relates to the preparation of cost plans where the work is to be procured on the basis of work packages.
   The importance of coding the elemental cost plan is stressed so as to enable *the components allocated to each element and sub-element* to be recoded and allocated to the appropriate work package. This will then create a cost target for each work package, as opposed to each element, which can then be used for the cost management of the project.
   Paragraph 3.6.2 refers the reader to the explanation of the coding system given in NRM1 Paragraph 4.5.

5.7.6   Building projects comprising multiple buildings

Paragraph 3.7.1 is a recommendation that a separate cost plan is prepared for each building when a project comprises more than one building.
   It is also recommended that a 'summary cost plan' is prepared for the entire project.

5.7.7   Information requirements for formal cost plans

Information is the key to developing a cogent cost plan, and this is emphasised in Paragraph 3.8.1 which recognises that more information becomes available as the project develops and as more interaction takes place between the members of the project team.
   Paragraph 3.8.2 points the reader to Appendix G (this is a mistake – it should be Appendix F) wherein lists of information required from the employer, the architect and other consultants at the formal cost planning stages – Formal Cost Plans 1, 2 and 3 – may be found.

Risk issue

Unlike Paragraph 2.3.5 relating to order of cost estimates, there is no suggestion in Paragraph 3.8 that the quantity surveyor/cost manager should qualify the cost plan if the necessary information is not forthcoming.

Presumably, Paragraph 3.5 should be referred to in such circumstances, as it deals with the review and approval of cost plans by the employer and the project team at the various formal cost planning stages. Missing information could then be raised as an issue before the formal cost plan is signed off.

The quantity surveyor/cost manager would do well to remember that:

- There is no guarantee that the required information will be forthcoming.
- There is no sanction if it isn't.
- There is no guarantee that the information received will be fully in accordance with NRM1 Appendix F requirements.
- Protocols may be needed in order to deal with shortcomings in the supply of information from the employer, the architect and others.

5.7.8   Format, structure and content of elemental cost plans

There is no particular standard required by NRM1 for the layout of cost plans, but Paragraph 3.9.1 points to Appendices G and H for exemplar templates.

Appendix G shows a condensed template based on Level 1 codes, whereas Appendix H is an expanded version based on Level 2 codes. Level 1 is the group element level, whereas Level 2 is the element level (refer to NRM1 Paragraph 4.4.2).

A great deal more work goes into preparing a cost plan than is obvious from the examples provided in NRM1, even at the expanded level illustrated in Appendix H.

Taking element 2.7: *Internal walls and partitions* as an example, a cost target for the element has firstly to be established. This will be based on the quantities required for the project in hand and rates taken from a previous project or elemental cost analysis. Once the EUR has been adjusted for time, location and quality of specification, the elemental unit quantities can be priced and a cost target established. Table 5.6 illustrates this process.

At a later stage, when more design information becomes available, cost targets need to be checked to see whether they are valid or not, and this requires a different type of measurement as illustrated in Table 5.7.

Table 5.6   Cost target.

Element		Cost target			
Internal walls and partitions					
Rates taken from analysis No. 16443	**EUQ**	**Unit**	**EUR, £**	**Total, £**	
EUR Price level $\underline{219} \times$ 172 Quality +15% **Total**	$\underline{83.65}$ 106.51 $\underline{15.98}$ $\underline{122.49}$				
	9 000	m²	122.49	**£1 102 410**	

Table 5.7   Cost check.

Cost check				
Rates taken from BQ 1679 (current)				
Item	**Quant**	**Unit**	**Rate, £**	**Total, £**
100 mm block walls	8 000	m²	36.00	288 000
Stud partitions dry lined both sides	10 000	m²	45.00	450 000
Hardwood glazed screens	1 500	m²	375.00	562 500
				£1 300 500

In the cost check, quantities are measured from the drawings, and unit rates are applied to arrive at a total for the element. The unit rates may come from a variety of sources – previous priced bills of quantities, an internal database or a builders' price book – or they may be built up from first principles based on current rates for labour, materials, plant and so on.

Once the cost check has been established, this needs to be compared to the initial cost target to see whether or not there is any variance (see Table 5.8). If the difference is significant, this could result in several implications, such as the need to redesign the element, to redesign other elements to save money or to spend contingency.

Table 5.8   Variance.

Element		Cost target			
Internal walls and partitions					
Rates taken from analysis No. 16443		**EUQ**	**Unit**	**EUR (£)**	**Total (£)**
EUR	83.65				
Price level					
<u>219</u> ×	106.51				
172					
Quality +15%	<u>15.98</u>				
Total	**122.49**				
		9 000	m²	122.49	**£1 102 410**
Cost check					
Rates taken from BQ 1679 (current)					
Item		**Quant**	**Unit**	**Rate (£)**	**Total (£)**
100 mm block walls		8 000	m²	36.00	288 000
Stud partitions dry lined both sides		10 000	m²	45.00	450 000
Hardwood glazed screens		1 500	m²	375.00	562 500
					£1 300 500
Underspend					
Overspend					**£198 090**

The measurement rules in Part 4: *Tabulated rules of measurement for elemental cost planning* need to be scrupulously applied, as it must be demonstrable that they have been followed, because the cost planning process will almost certainly be subject to internal audit through the office quality assurance régime.

5.7.9   Facilitating works

NRM1 Paragraph 3.10 is a little unclear, but it appears to mean that there are two sets of measurement rules that apply to facilitating works (which is group element 0) – 'general rules' and 'specific rules'. The 'general rules' are those contained within Paragraph 3.11: *Measurement rules for building works*, and the 'specific rules', relating to the measurement of sub-elements and components, are in NRM1 Part 4, the part that makes up the bulk of NRM1.

Without wishing to appear pedantic or obtuse, NRM1 seems to have been written in such a way as to distinguish between 'measurement rule' and 'rule of measurement' in that the former appears to refer to general rules and the latter to rules for measuring specific elements or components of work. Whether or not this distinction is intentional is unclear.

NRM1: Measurement rules	
'General rules'	*Shall be the same as those for the measurement of building works* as detailed in Paragraphs 3.11.2–3.11.5 inclusive (refer to Paragraph 3.10.2). These rules explain *the method of measuring quantities* for Formal Cost Plans 1–3
'Specific rules'	Relating to the measurement of building work as detailed in Part 4: *Tabulated rules of measurement for elemental cost planning*, which determine how work shall be described for group elements 0–14 inclusive and what units it shall be measured in (e.g. m, m², m³, nr)

5.7.10   Building works

NRM1 Paragraph 3.11.1 states that the rules of measurement for building works (i.e. group elements 1–8) are to be found in Part 4: *Tabulated rules of measurement for elemental cost planning*. Group element 0 (facilitating works) is dealt with separately in Paragraph 3.10.

There is no clear guidance in Paragraph 3.11 regarding the development of the various formal cost plans save to say that:

- Formal Cost Plan 1 begins with a condensed list of elements which is then *developed into a full list of elements, sub-elements and components* as more design information becomes available (refer to Paragraph 3.11.4 (iv)).
- *Cost-significant items are to be measured by means of approximate quantities* (Paragraph 3.11.2).

Judging by the units of measurement used in NRM1 Part 4: *Tabulated rules of measurement for elemental cost planning*, it would be logical to assume that, at some point, the formal cost plan will eventually resemble a bill of approximate quantities albeit that elemental cost plans are to be reported to the employer on the basis of cost/m² of GIFA. This is logical as the cost plan report is likely to be presented in summary form.

Paragraph 3.11 contains a number of 'general rules' for the measurement of building works for the purposes of Formal Cost Plans 1–3 and for contractor-designed work. In the inimitable style of NRM1, these 'general rules' are somewhat 'lost' within the explanations and other guidance provided.

NRM1: Measurement rules

Paragraph 3.11.2
- Cost-significant items *are to be measured* by approximate quantities where possible
- Non-cost-significant items *are to be ignored*, but unit rates are to be increased appropriately (e.g. by adding a suitable percentage)
- Composite items consisting of several work items that individually may have different units of measurement are to be measured using *common forms of measurement*:
  - Example – basement
    - Excavation and disposal – m³
    - Earthwork support – m²
    - Concrete slab – m³
    - Brick retaining walls – m²
    - Common unit of measurement – m² of GIFA

NB:
A composite rate is derived by pricing the individual components using appropriate unit rates and then dividing the resultant total amount of money by the GIFA to give a rate expressed in £/m² of GIFA

Paragraph 3.11.3
- Quantities *shall be given* to nearest whole unit (or unity if less than one)
- Quantities for rebar, steelwork and the like *shall be given* to two decimal places

Paragraph 3.11.4 **Formal Cost Plans 1–3**

- Quantities *shall be determined* in accordance with Part 4: *Tabulated rules of measurement for elemental cost planning* (i.e. m, m², m³, nr, etc.)
- Where this is not possible, GIFA *is to be* the unit of measurement

Paragraph 3.11.5 **Contractor-designed works**

- Such work *shall be identified and described separately* in the cost plan as 'contractor designed works'

NB:

1. There are no particular rules of measurement specified, and so it is assumed that Paragraph 3.11.4 will apply in this case
2. Where the contractor is to design the entire project, the main contractor's design fees are to be determined in accordance with Paragraph 3.16.7

Part 2

5.7.11   Unit rates used to estimate the cost of building works

The unit rates, EUR and composite rates used in cost planning are usually derived from a number of sources:

- Internal 'in-house' database of historical costs (QS consultants, contractors or subcontractors).
- BCIS elemental cost analyses.
- BCIS price information (formerly Wessex price books).
- Builders' price books such as Spon's (prepared by Davis Langdon) and Griffiths' (prepared by Franklin + Andrews, 2014).
- First principles (i.e. rates built up from basic costs of materials, plant and labour).

NRM1: Measurement rules

Paragraph 3.12.1	Unit rates are to **include:** ▪ Materials, labour and plant ▪ Subcontractors' or suppliers' design fees ▪ Subcontractors' preliminaries ▪ Subcontractors' risk allowances ▪ Subcontractors' overheads and profit
Paragraph 3.12.2	Unit rates are to **exclude:** ▪ Main contractor's preliminaries ▪ Main contractor's overheads and profit ▪ Project/design team fees ▪ Other development/project costs ▪ Risk allowances ▪ Inflation

The list of inclusions and exclusions in Paragraphs 3.12.1 and 3.12.2 is the same list as Paragraph 2.9.3 relating to order of cost estimates, except that subcontractors' risk allowances are to be included in the unit rates used for cost planning. Whether there is some reason for this difference or whether it is a mistake in the drafting is not clear.

Paragraph 3.12.3 warns that rates should be appropriately adjusted to remove built-in allowances for construction inflation.

5.7.12   Updating unit rates and other costs to current estimate base date

Paragraph 3.13 is simply concerned with making sure that the unit rates and EUR used for cost planning are always current at the time the estimate is made.

Therefore, the unit rates used in the order of cost estimate have to be updated for inflation to the base date of Formal Cost Plan 1, and the unit rates used in Formal Cost Plan 1 have to be updated to the base date of Formal Cost Plan 2 and so on.

This sounds a bit complicated, but the idea is that the rates used in each cost plan are current when that cost plan is compiled and that there is an allowance for inflation in the cost plan summary which allows for inflation from the date of the particular formal cost plan to the date of tender for the project. This allowance reduces over time until the final version of the cost plan is completed.

5.7.13   Main contractor's preliminaries

Main contractors' preliminaries represent a significant cost target in the cost plan which therefore has to be cost checked for each formal cost plan stage.

The extent of measurement of preliminaries items is limited as it is usual to allow a percentage on the building works estimate to establish their projected cost.

However, when more information becomes available, it is feasible that the contractor's preliminaries may be cost checked in detail using the extensive (but not *exhaustive*) list of items included in group element 9 of NRM1 Part 4: *Tabulated rules of measurement for elemental cost planning*.

Despite its title, *Measurement rules for main contractor's preliminaries*, Paragraph 3.14 contains very little by way of rules.

NRM1: Measurement rules

Paragraph 3.14.4	If this can be called a 'rule', the sole requirement is that where the main contractor's preliminaries, or any part thereof, is to be based on a percentage addition, the percentage *is to be* applied to the building works estimate. The alternative, given by Paragraph 3.14.5, is to add a lump sum, but this is not a 'rule'
Paragraph 3.14.7	This states that any cost plan allowance for subcontractors' preliminaries *is to be made* in the unit rates that are applied to the measured quantities

5.7.14   Main contractor's overheads and profit

The requirement to create a separate provision for main contractor's overheads and profit in the order of cost estimate, and in the cost plan, is one of the most inexplicable features of NRM1.

The database of elemental cost analyses held by BCIS contains some 18 500 analyses of tender prices for a wide range of projects of different types, sizes, values and locations over many years. Whilst there is no data available to indicate how many of these analyses reveal the contractor's overheads and profit addition, it is a 'racing certainty' that the vast majority, if not all, do not show this information.

This means that since the dawn of the cost planning era in the early 1950s, the rates used for order of cost estimating and cost planning have been inclusive of main contractor's overheads and profit. Consequently, the rates per m² of GIFA, EUR and the rates used for approximate quantities and cost checking include overheads and profit.

The reason for this is that these rates are derived from bills of quantities submitted by contractors at the tender stage, and the rates and prices contained in the bills of quantities also invariably include overheads and profit. Hence, the vast database of information used by the industry for order of cost estimating and cost planning, whether in-house data or data publically available through BCIS or the usual builders' price books, is inclusive of main contractor's overheads and profit.

Being largely based on contractors' tenders, this database is incapable of being interrogated to reveal the true amount included for overheads and profit because contractors habitually do not volunteer this information in their tenders. Admittedly, some contractors do like to show their OH and P allowance as a lump sum addition to their tenders for a variety of reasons – but these instances are relatively few and far between and are certainly not 'the norm'. It is only, therefore, the tendering contractor who knows how much has been allowed in the tender to cover overheads and profit for each specific project.

How then can the quantity surveyor/cost manager hope to produce an order of cost estimate or cost plan showing a separate amount for main contractor's overheads and profit? The answer is simple – it isn't possible, at least with any degree of accuracy.

Example

Take a rate – any rate – let's say a rate of £69.12/m² inclusive of overheads and profit.
Without knowing the contractor's percentage addition for overheads and profit, it is impossible to derive the net rate, that is, the rate <u>exclusive</u> of overheads and profit. The only way is to estimate the percentage and thereby calculate the net rate

Say, the estimated overheads and profit allowance is 8%

The net rate is therefore £69.12/1.08 = **£64.00**

So where does the 8% come from? It could be:

- An intelligent guess.
- From the quantity surveyor's assessment of the construction market at the time that the tender was submitted.
- Derived from an in-depth analysis of the contractor's profit and loss account from the annual financial report (turnover less cost of turnover).
- Derived from statistics of contractor's overheads and profit percentages which might have a mean of, say, 10% and a standard deviation of 2%, giving a maximum of 12% and a minimum of 8%.

None of these methods will give an accurate answer, and even the contractor's annual account figures will only be an average across the company and not the specific figure for a particular tender.

So, if the quantity surveyor/cost manager has to 'guess' the overheads and profit figure, what is the point of having two sums (net cost and overheads and profit) that are both inaccurate?

- It might be that the cost planner wants to add different OH and P figures to different parts of the cost plan due to the nature of the work involved.
- It might be that a cost target is required because the overheads and profit figure/percentage is the basis of the contractor's appointment (e.g. stage one of a two-stage tender).
- It might be that OH and P is to be 'ring-fenced' or protected due to the procurement strategy adopted for the project (e.g. partnering).
- It might be that the employer wishes to make overheads and profit a negotiable part of the tender competition (e.g. a framework agreement).

We could go on surmising forever, but the point is that the only clue given in NRM1 as to why main contractor's overheads and profit should be shown separately is that an *agreed level of overheads and profit* may have been established *as part of a two-stage tendering process* (refer to Paragraph 3.15.8).

Whilst NRM1 explains in great detail how overheads and profit is to be added into the cost plan (Paragraphs 3.15.2–3.15.5 and 3.15.7), it does not suggest any means by which the percentage may be arrived at other than *from a properly considered assessment of main contractor's overheads and profit found on previous building projects* (refer to Paragraph 3.15.6). NRM1 is silent on how this is to be done!

Risk issue

Admittedly:

- An order of cost estimate is in effect only a budget.
- A cost plan is only an expression of how this budget is to be spent.
- The cost control of building design is not a precise science.

However, it would seem to be an (almost) pointless exercise (in the majority of cases) to separate overheads and profit in the order of cost estimate or cost plan, and there would seem to be no tangible justification for spending the time (and professional fees) for doing so. If this is to be done *from a properly considered assessment of main contractor's overheads and profit found on previous building projects*, then the best that could be achieved from available data is <u>a guess</u> at the level of overheads and profit and <u>a guess</u> at the net cost of the work involved in a project (i.e. the material, plant and labour costs).

The question therefore arises as to the quantity surveyor's legal position should it be decided to ignore this provision in NRM1.

The introduction to NRM1 (page 1) states that:

- *It is for each surveyor to decide on the appropriate procedure to follow in any professional task.*
- *However, where members do not comply with the practice recommended in (NRM1) they should do so only for a good reason.*
- *In the event of a legal dispute, a court or tribunal may require them to explain why they decided not to adopt the recommended practice.*
- *If members' actions are questioned in an RICS disciplinary case, they will be asked to explain the actions they did take and this may be taken into account by the Panel.*

Estimates of the main contractor's overheads and profit can be added in to the cost plan either as one item (or *cost centre* in NRM1) or as separate cost centres, one for overheads and the other for profit (refer to NRM1 Paragraph 3.15.1).

In view of the likelihood that little or no historic information will be available as a basis for the estimate, it is surprising that Paragraph 3.15.6 states that such information *is to be derived* from previous building projects.

NRM1: Measurement rules

Paragraph 3.15.1	*Main contractor's overheads and profit <u>are to be based</u> on a percentage addition*
Paragraph 3.15.1: one separate cost centre	*The estimated cost of any main contractor's overheads and profit <u>is to be calculated</u> by applying the selected percentage addition for overheads and profit to the combined total cost of the building works estimate and the main contractor's preliminaries estimate*
Paragraph 3.15.6: two cost centres	*The percentages addition to be applied for main contractor's overheads and main contractor's profit <u>are to be derived</u> from a properly considered assessment of main contractor's overheads and profit found on previous building projects*
Paragraph 3.15.8	*Where the main contractor has been appointed early (e.g. as part of a two-stage tendering process), the actual agreed level of overheads and profit <u>is to be included</u> in the cost plan*

Paragraph 3.15.8 is the only workable measurement rule for main contractor's overheads and profit, and therefore, where an allowance has already been made in the cost plan, Paragraph 3.15.8 requires that the cost plan shall be adjusted according to the agreement reached.

5.7.15 Project/design team fees

Paragraph 3.16.1 indicates that consultants' fees are a necessary part of a building project, and Paragraph 3.16.2 refers to the *list of typical project/design team fees* in Part 4: *Tabulated rules of measurement for elemental cost planning*, group element 11.

These fees include consultants' fees and fees payable to the main contractor for pre-construction services and design fees, where applicable – see Paragraph 3.16.3 which recommends that *separate allowances* should be made for each.

The remainder of Paragraph 3.16 is given over to an explanation of how such fees are to be calculated and what considerations should be made where the main contractor is to accept either full or partial design liability. The total project/design team fee estimate is given by the total of consultants' fees, main contractor's pre-construction fee and main contractor's design fees, each of which *is to be derived from a properly considered assessment of fees charged on other similar previous building projects*.

Part 2

Paragraph 3.16.6(d) warns that care must be exercised to ensure *that sufficient allowance has been made for main contractor's overheads and profit on the pre-construction fee.*

NRM1: Measurement rules

Paragraph 3.16.4	*Project/design team fees* are to be included in cost plans

5.7.16   Other development/project costs

Other development costs form part of the building project but are not directly associated with the cost of building works (Paragraph 3.17.1). Examples include insurances, planning, party wall fees, etc.

NRM1: Measurement rules

Paragraph 3.17.2	*Other development/project costs are to be included in cost plans,* and they are to be added as a lump sum allowance

Other development/project costs form part of the base cost estimate of the order of cost estimate and of the cost plan to which risk allowances and inflation estimates are added in order to arrive at the project cost limit.

They *are to be added* to the order of cost estimate/cost plan *as a lump sum allowance* priced on the basis of fixed cost or time-related items. Part 4: *Tabulated rules of measurement for elemental cost planning* (group element 12) provides examples of such items which include land acquisition, archaeological fieldwork and decanting and relocation costs (NRM1 Paragraph 3.17).

5.7.17   Risk allowances

Paragraph 3.18.1 states that:

- *Risk allowances <u>are to be included</u> in each formal cost plan.*
- They are to be *based on the results of a formal risk analysis.*
- They are not to be standard percentages.

Strangely, Paragraph 3.18.7 suggests that risk allowances may in fact *be based on a percentage addition.*

Also strangely, Paragraph 3.18.5 recommends that *risk allowances be treated as <u>three</u> separate cost targets* but does, in fact, list <u>four</u> recommended (not mandatory) cost targets:

a) Design development risks.
b) Construction risks.
c) Employer change risks.
d) Employer other risks.

Perhaps (c) and (d) are intended to be combined?

The lack of clarity in Paragraph 3.18 is further exemplified by a failure to distinguish between **risk** and **uncertainty** and to emphasise that the purpose of risk assessment is to:

- Identify aspects of a project (threats) that could be detrimental to the cost limit.
- Provide a measure of the risk of a particular threat occurring determined by the likelihood of it happening and the consequence if it does.

- Eliminate risk where possible.
- Manage remaining (residual) risk.

In the context of cost planning, the presence or absence of design and other information is a determining factor in any assessment of risk, and this is why it is important to distinguish between risk and uncertainty. Ross and Williams (2013) cite Winch (2010) who suggests that there _is_ a distinction to be made:

- **Uncertainty** relates to the absence of information required for decision-making.
- **Risk** is where a probability distribution can be assigned to an occurrence.

The need to reassess risk registers and risk estimates at regular intervals throughout the design process is emphasised in Paragraph 3.18.3, and Paragraph 3.18.4 explains that _successive assessments are to show decreasing risk_ as a consequence of the more certainty attached to the project as time goes by. Paradoxically, the same paragraph notes _that risk does not always decrease_ but offers no explanation for this statement.

Paragraph 3.18.10 refers to guidance offered in Part 4: _Tabulated rules of measurement for elemental cost planning_ (group element 13) as to typical risks to be considered; this is a useful resource to help the quantity surveyor/cost manager to arrive at a balanced assessment of risk so that _considered risk allowances_ (Paragraph 3.18.2) may be included in the cost plan.

NRM1: Measurement rules	
Paragraph 3.18.1	▪ Risk allowances _are to be included_ in each formal cost plan ▪ Based on the results of a formal risk analysis ▪ Are not to be standard percentages
Paragraph 3.18.7	_Where any aspects of risk allowances… are to be based on a percentage addition,_ such allowances _are to be_ calculated by multiplying the base cost estimate by the percentage additions

5.7.18   Inflation

NRM1 Paragraph 3.19.1 makes the statement that _elemental cost plans are to be prepared using rates and prices current at the time the cost plan is prepared._ This statement has a number of implications that are not explained in NRM1 but are illustrated in Figure 5.2:

- The cost data upon which the cost plan is based will be historic information and may date back some time before the project inception.
- Whatever the source of cost data, it will have to be updated from its base date to the current date.
- The current date will be the base date of:
  - The order of cost estimate
  - The Formal Cost Plans 1, 2 and 3
- The base date of the cost data will vary:
  - Cost data used for the order of cost estimate (e.g. cost per m² GIFA, functional units) will probably date back some months, maybe years.
  - Cost data from elemental cost analyses of similar buildings may also date back some time.
  - Cost data for cost checking (e.g. unit rates from previous tenders) may be more up to date.
- This means that each formal cost plan will have to be updated from the previous one or, for Formal Cost Plan 1, from the base date of the order of cost estimate.
- Formal Cost Plan 3 will have to reflect the extent of inflation up to the tender base date.
- The total allowance for inflation up to the tender base date will not necessarily reflect the degree of inflation from the origin in Figure 5.2 because different data drawn from different base dates will have been used throughout the cost planning process.

Part 2

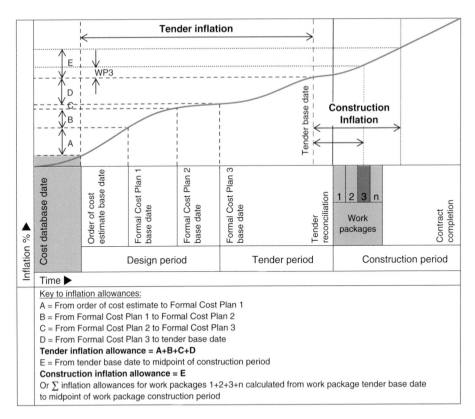

Figure 5.2   Inflation.

As far as the measurement rules are concerned, NRM1 Paragraphs 3.19.4 and 3.19.6 distinguish respectively between **tender inflation** and **construction inflation**:

- **Tender inflation**
  This is defined as *the period from the estimate base date to the date of tender return* (refer to Paragraph 3.19.2 (a)).

- **Construction inflation**
  This is defined as *the period from the date of tender return to the mid-point of the construction period* (refer to Paragraph 3.19.2 (b)).

Paragraphs 3.19.4 and 3.19.6 also issue the warning that the measurement rules for tender and construction inflation may be oversimplistic where the project is to be procured with separate work packages.

This is because:

- The work package procurement programme will be different for each work package.
- Therefore, both the tender and construction inflation allowances should be calculated for each work package according to the procurement programme for the specific package concerned.

The guidance notes in NRM1 Paragraphs 3.19.4 and 3.19.6 are not quite clear, but it should be emphasised that:

- The tender inflation allowance for each work package should be related to the tender base date of the package concerned.
- The construction inflation allowance for each work package should be related to the mid-point of the construction period of the package concerned.

This is illustrated in Figure 5.2.

NRM1: Measurement rules

Paragraph 3.19.1	*Elemental cost plans are to be prepared using rates and prices current at the time the cost plan is prepared*
Paragraph 3.19.6(a)	*Where procurement is to be on the basis of separate work packages, a separate allowance for construction inflation for each work package* will be required

Part 2

5.7.19   VAT assessment

Paragraph 3.20.1 recommends *that VAT is excluded from cost plans* due to problems of complexity, and Paragraph 3.20.2 recommends *that specialist advice is sought on VAT matters to ensure that the correct rates are applied.*

5.7.20   Other considerations

Paragraphs 3.21.1–3.21.3 are concerned with capital allowances, land remediation allowances and grants, respectively, and in each case, it is recommended that *specialist advice is sought* on appropriate allowances and that such matters are *excluded from cost plans.*

5.7.21   Reporting of elemental cost plans

Paragraph 3.22.4 provides a helpful list of typical items to be included in a cost plan report, and Paragraph 3.22.3 requires that the employer is clearly informed as to the items included in and excluded from the cost plan.

The essential part of a cost plan report is, of course, the cost plan. Depending upon how advanced the cost plan is in its development (Formal Cost Plan 1, 2 or 3), the cost plan will be presented as an elemental breakdown expressed in:

- Group elements (e.g. superstructure).
- Elements (e.g. frame, upper floors).
- Sub-elements (e.g. steel frames, concrete frames, timber frames).

The unit of measurement for each constituent part of a cost plan is 'cost/m^2 of GIFA' (Paragraph 3.22.1), but functional units or even 'cost/ft^2 of GIFA' may be used where appropriate (Paragraph 3.22.2).

Example:

COST PLAN FOR NEW 20 000-SEATER FOOTBALL STADIUM

Group element	Element	Sub-element	Constituent	Subtotals £	Totals £	Subtotals per seat £	Totals per seat £
3			**Internal Finishes**				
	3.1		Wall finishes	720 000		36.00	
	3.2		Floor finishes	932 000		46.60	
	3.3		Ceiling finishes	764 000		38.20	
					2 416 000		120.80

Whilst not stated in Paragraph 3.22, it would seem sensible to also include absolute costs in the cost plan report, totalling to the agreed cost limit, as such values are likely to mean more to the majority of employers than cost/m².

5.8 Part 4: Tabulated rules of measurement for elemental cost planning

The basic idea of cost planning is to break down the order of cost estimate so as to represent the evolving design in more detail and thereby aid the designer to make informed design choices.

As the design develops, and more detail becomes available, group elements may be subdivided into elements, and elements into sub-elements, and the rates used for estimating change from the 'approximate' **cost per functional unit**, or **cost per m² of GIFA**, to the more precise EUR. The order of cost estimate is thus developed into the outline cost plan and, subsequently, into the detailed cost plan.

Later on, at some point in the development of the cost plan, the 'component' level of detail is reached which requires a more detailed method of estimating. This is the point when 'quantities' can be produced that can be priced using 'unit rates'.

The purpose of Part 4 of NRM1 is to provide the 'tabulated' rules for measuring the component parts of elements and sub-elements for cost planning (and other) purposes. The rules are tabulated in the irritating landscape presentation redolent of SMM7 using, in the main, 'normal' units of measurement – m, m², m³, nr, t – albeit that there is the occasional use of % and the choice between *item* and *nr* for certain items.

The measurement rules for cost planning are contained in NRM1 Parts 3 and 4. The Part 3 'rules' might be more accurately referred to as 'general rules', whereas the rules for the detailed measurement of building work are contained in Part 4: *Tabulated rules of measurement for elemental cost planning*.

The *Tabulated rules of measurement for elemental cost planning* are likely to be used at some point during or towards the end of Stage 3 – Developed Design of the RIBA Plan of Work 2013 – and this point marks the transition from the measurement of group elements and elements in cost/m² of GIFA to the measurement of building components in m, m², m³, etc.

In this regard, it is unhelpful to think of order of cost estimates and cost plans as being separate entities measured in different units.

Formal Cost Plan 1 is a development of the *Order of cost estimate* and, as such, will be measured in cost/m² of GIFA until such time as more detailed measurements can be taken – probably for the development of *Formal Cost Plan 2*. Nonetheless, all iterations of the cost plan must still be expressed in cost/m² of GIFA, in accordance with NRM1 Paragraph 3.22.1, accompanied by absolute values for clarity.

The other important function of the *Tabulated rules of measurement for elemental cost planning* is the cost checking of elements, sub-element and components. This process is akin to measuring approximate quantities in order to see whether the allowance for a particular item stands up to scrutiny when the detailed design of that item is available.

Risk issue

In terms of measurement, Paragraph 2.19.4 is perhaps the most significant in that the employer is to be clearly informed as to what is included in, and excluded from, the order of cost estimate.

The quantity surveyor/cost manager would be well advised to ensure that the correct protocols are put in place to ensure that the order of cost estimate is crystal clear to the employer.

5.8.1   Introduction

One of the potential further uses for the tabulated rules of measurement is signalled in NRM1 Paragraph 4.1.2 – whole life cycle costing – but bills of quantities for building work could also be prepared with NRM1 rules where the detail required by NRM2 rules is not practical or desired.

Paragraph 4.1.1 also explains that Part 4 of NRM1 contains advice on how the reallocation of costs from elements/sub-elements to work packages may be achieved and also on how elemental cost plans may be coded to achieve this, and other, objectives.

5.8.2   Use of tabulated rules of measurement for elemental cost planning

The rules of measurement for cost planning are organised into 15 tables; the first 9 deal with building works and the remainder, the 'softer' issues such as preliminaries, project/design team fees and risks.

NRM1 Paragraphs 4.2.2 and 4.2.3 explain the hierarchy of the tabulated measurement rules for facilitating works (group element 0) and building works (group elements 1–8):

Level	Hierarchy	Example	
1	Group element	2	Superstructure
2	Element	2.1	Frame
3	Sub-element	2.1.1	Steel frames
4	Component	2.1.1.1	Structural steel frame
		2.1.1.1.2	Fire protection
		2.1.1.1.3	Factory applied paint systems

There may be more than one element in a group element, one or more sub-elements in an element and one or more components in a sub-element.

It can be seen that the tabulated rules are based on a work breakdown structure and that the four levels in the hierarchy provide the basis for a coding system which acts as a frame of reference

Part 2

for individual parts of the cost plan and as a means of redistributing discrete items into a work package structure if desired.

However, whilst all the levels are used to uniquely identify a component, not all parts of the hierarchy are measured:

- Level 1 is the group element to which a Level 2 element belongs but neither is measured.
- Level 3 is the sub-element to which a component belongs, but sub-elements are not measured.
- Level 4 is the component level; it is the components that are measured, and it is the components to which the measurement rules and the 'included' and 'excluded' items apply.

Table 5.9 illustrates how the NRM1 tables work.

It is important to observe the horizontal lines in the tabulated structure of the measurement rules as they denote the end of a sub-element (e.g. sub-element A) and the start of the next one (e.g. sub-element B), and they also help to divide sub-elements into one or more components (e.g. A1, A2 and A3).

Group elements 9–14 have a slightly different tabular structure, but the principles are the same. The tabular structure for each group element is described in NRM1 Paragraphs 4.2.5–4.2.10 respectively.

5.8.3   Work not covered by the rules of measurement for elemental cost planning

Paragraph 4.3.1 recognises that the NRM1 rules may not be exhaustive enough to cover every possible circumstance of components to be measured and that, in such situations, NRM1 rules are to be followed where possible.

Where not possible, the rules adopted shall be stated in the cost plan. This is sensible guidance especially where there is more than one cost planner or where personnel change for some reason or other.

5.8.4   Method of coding elemental cost plans

The NRM1 coding system for elemental cost plans is relatively straightforward (if you like brain teasers, try NRM2!), and guidance can be found in Paragraph 4.4 which provides a worked example.

The NRM1 example refers to two concrete beams which are part of a reinforced concrete frame:

- Concrete beams 1200×800 mm.
- Concrete beams 2000×800 mm.

The first place to begin the search for a code number for these items is Appendix E: *Logic and arrangement of Levels 1–3 for elemental cost planning* which is identified in Paragraph 4.4.1.

Appendix E identifies three levels:

- Level 1 – Group element.
- Level 2 – Element.
- Level 3 – Sub-element.

However, Paragraph 4.4.2 says that further code levels may be added as necessary.

Within each level is a classification system and each item listed in each level classification is given a number. This is illustrated in Table 5.10.

Table 5.9 Tabulated rules.

Group element
Element

Sub-element		Component		Unit	Measurement rules for components	Included	Excluded
A A sub-element is part of an element	**A1**	Components are part(s) of a sub-element For example, component A2 is part of sub-element A	**m**	Different units may apply to different components. For example, component A2 is measured in m^2 and A3 in m^3	Rules that apply to each component within a sub-element, that is: • What to measure • How to measure it • What must be measured and described separately	The work items deemed to be **included** in each component of the sub-element	The work items deemed to be **excluded** from each component of the sub-element
A definition of the sub-element is to be found in this column	**A2**		**m^2**				
	A3		**m^3**				
B	**B1**		**m^2**				
	B2		**m^3**				

Part 2

Table 5.10 Classification system.

Level 1 Group element		Level 2 Element		Level 3 Sub-element		Reference
1	Substructure	Substructure		1 2 3 4 5	Standard foundations Specialist foundations Lowest floor construction Basement excavation Basement retaining walls	
2	**Superstructure**	1	**Frame**	1 2 3 **4** 5 6	Steel frames Space frames/decks Concrete casings to steel frames **Concrete frames** Timber frames Specialist frames	 **2.1.4**
		2	Upper floors	1 2 3	Floors Balconies Drainage to balconies	
		3 Etc.	Roof Etc.	1 Etc.	Roof structure Etc.	

Risk issue

There is a mistake in Paragraph 4.4.2 following the first set of bullet points.

The codes given in this paragraph for the concrete beams example are stated as 3.1.4.3 and 3.1.4.4 which is not correct – the Level 1 reference should be 2 (superstructure) and not 3 (which is internal finishes).

It is worth reiterating at this stage that an elemental cost plan develops over time as the design progresses, and the starting point for the elemental cost plan is the order of cost estimate.

The order of cost estimate is not likely to be at a more detailed level than Level 1 – group element – and the first formal cost plan (Formal Cost Plan 1) *will use a condensed list of elements* which will then become more detailed as more information becomes available (refer to NRM1 Paragraph 3.11.4 (a)(iv)).

At some point in the cost planning process, the Level 3 (sub-element) level of detail will be too crude, and therefore, sub-elements will need to be broken down into **components** and **sub-components** for more accurate estimating.

Referring back to the concrete beam example earlier, beams are clearly a subset of **frames** and may be further subdivided into **concrete, formwork** and **rebar**. Consequently, further coding levels are needed to identify these specific components and sub-components in the cost plan.

NRM1 Paragraph 1.6.3: *Definitions* defines 'component' as *a measured item that forms part of an element or a sub-element*, and therefore, a concrete beam is a 'component' because it forms part of a concrete frame, which is a sub-element at Level 3.

Referring to Appendix E, the numbered reference for a concrete frame is 2.1.4 which is given by:

- Level 1 – Group element **2: Superstructure**.
- Level 2 – Element **1: Frame**.
- Level 3 – Sub-element group element **4: Concrete frames**.

Referring now to NRM1 Paragraph 4.4.2, this states that:

- *Codes for levels 1–3 are provided by the measurement rules* and that *codes for level 4 (i.e. components) will be <u>user defined</u>* due to the large variety of components that could be found in any particular sub-element.

However, looking at NRM1 Part 4: *Tabulated rules of measurement for elemental cost planning* (specifically group element 2: *Superstructure* on page 103 of NRM1), it can be seen that the items listed under the heading 'Component' also have a reference number notwithstanding that Paragraph 4.4.2 says that the component level reference (i.e. Level 4) *will be user defined*.

The logical conclusion to this statement is that the numbering under 'Component' in the rules of measurement is to be ignored in favour of a number to be made up by the person compiling the cost plan.

However, the question arises as to why components should be numbered in the measurement rules when the numbers are to be ignored? This doesn't make any sense. Admittedly, the items listed in the *Measurement rules for components, Included* and *Excluded* columns are also referenced, but these are not measured items and would not feature in the referencing of a component or sub-component.

Looking back at Paragraph 4.4.2, it is clear that the idea of *user-defined* codes at Level 4 is so that components may be added that are not listed in NRM1. This is fair enough, as no method of measurement can anticipate every mortal thing, but it might have been a better idea to encourage users to <u>continue</u> the numbering at Level 4 rather abandoning the codes already provided.

A case in point in element 2.5: *External walls* is Rule C10 which states that *Contractor designed work is to be described and identified separately*. Therefore, an item at the component level for external walls <u>to be designed</u> by the contractor would require a *user-defined* code to distinguish it from 2.5.1.1 for external walls <u>not to be designed</u> by the contractor.

Finally, Paragraph 4.4.2 goes on to say that a user-defined Level 5 reference could be created, if desired, so that 'sub-components' such as formwork and reinforcement could also be given unique code numbers.

A simple example may help to illustrate the issues and, hopefully, provide a sensible solution.

Consider two below ground drainage items:

- Drain run A: depth to invert 0.9 m, pipe diameter 100 mm.
- Drain run B: depth to invert 1.2 m, pipe diameter 150 mm.

These items would be measured under NRM1 group element 8: *External works*, and their codes would be derived as shown in Table 5.11.

It can be seen that, using the NRM1 Component (Level 4) reference number results in two problems:

- Each drain run is coded with the same number.
- If there were, as is likely in practice, several drain runs with different pipe diameters and depths to invert, all the items would have the same code.

Clearly, this is not practicable as the two drain runs, A and B, are different items of work with different pipe diameters, different depths to invert and different costs, and each item merits its own code.

Ideally, each item would be coded uniquely, but also be traceable back to its root in NRM1, as this would be beneficial for auditability of the cost plan. It would be tempting to add a decimal point and further code number after the '2', but this would create a Level 5 code and not a unique Level 4 code.

Table 5.11   Measurement codes.

Item	Code			
	Level 1	Level 2	Level 3	Level 4
External works External drainage Surface water and foul water drainage Drainage runs; below ground	Group element **8**	Element 8.6	Sub-element 8.6.1	Component
				The NRM1 Level 4 reference number for drainage runs is 2. Using this reference, the item codes would be:
• Drain run A				State trench depth (m) and pipe diameter (mm) · 8.6.1.2
• Drain run B				Ditto · 8.6.1.2

Therefore, it would be logical if all below ground drain runs could be referenced under the umbrella of the 8.6.1.2 Level 4 code because this would trace the items back to the NRM1 Component level (Level 4) *Drainage runs; below ground* but with something added to make each item unique. This would mean changing the Level 4 code from 2 to some other number but preferably retaining the '2' so as to maintain the link with *Drainage runs; below ground*.

QSPro for Windows provides a clue as to how this might be achieved, and this is illustrated in Figure 5.3 which shows the respective descriptions and codes that might be employed.

It can be seen that QSPro gives a primary code of 08.060.010.020 for both measured items, which is the equivalent to 8.6.1.2, but with 'zeros' added for greater flexibility. This provides the opportunity to change the Level 4 code (020), thereby giving drain runs A and B unique codes if desired whilst at the same time retaining the '2' link to the NRM1 Level 4 descriptor for *Drainage runs; below ground*.

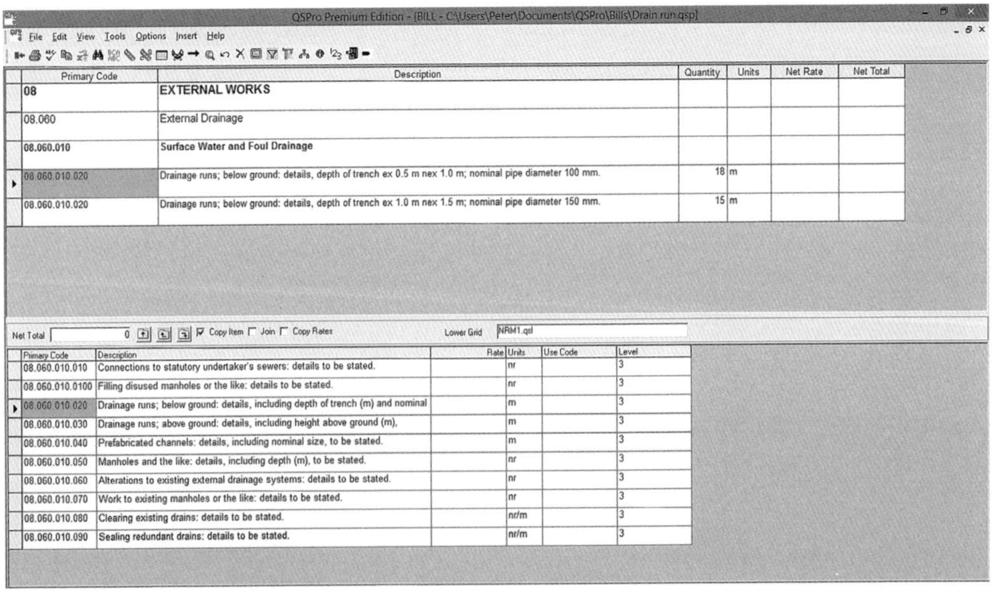

Figure 5.3   Coding items – 1.

However, bearing in mind that there could be several 100 mm diameter drain runs with different depths to invert, a Level 4 code needs to be chosen that will accommodate a reasonable number of items.

For example, 100 mm diameter drain runs could be coded 210 to 219 depending on depth:

- 08.060.010.210 – Depth not exceeding 0.5 m.
- 08.060.010.211 – Depth exceeding 0.5 m but not exceeding 1.0 m.
- 08.060.010.212 – Depth exceeding 1.0 m but not exceeding 1.5 m.
- Etc.

Accordingly, 150 mm diameter drain runs could be coded from 220 to 229, and 225 mm diameter drain runs could be coded 230 to 239 and so on.

The revised codes for our example are shown in Figure 5.4 with the 100 mm diameter drain run coded 08.060.010.211 and the 150 mm diameter drain run coded 08.060.010.223.

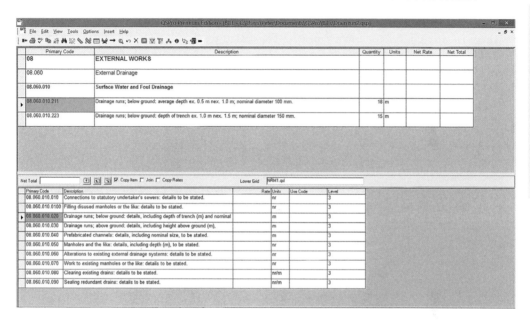

Part 2

Figure 5.4   Coding items – 2.

The user of NRM1 is now faced with a dilemma:

- Ignore the 'component' numbering in Part 4: *Tabulated rules of measurement for elemental cost planning* and employ a completely different numeric reference.
- Adopt the NRM1 'component' reference as part of a more logical system of referencing that provides flexibility and auditability.

Example

- Use the NRM1 referencing system down to Level 3.
- If suitable, use the Level 4 component reference number.
- If the list of Level 4 component reference numbers in the measurement rules is not sufficiently exhaustive, introduce *user-defined* Level 4 reference numbers as required.
- Use the 'component' reference number listed in the measurement rules as part of the *user-defined* Level 4 reference.
- If the Level 4 reference is not sufficiently detailed (e.g. where there is more than one type of pipe material), create a *user-defined* reference (Level 5).
- Reference sub-components (e.g. different trench backfill details) with a unique *user-defined* Level 6 reference.

In the context of NRM1 Paragraph 4.4, there is nothing wrong with this approach as Level 4 codes (and beyond) are meant to be *user defined* in any event.

Whichever approach the quantity surveyor/cost manager takes to referencing the cost plan, it should always be remembered that:

- Each component should *be continuously and sequentially numbered under the sub-element* (refer to NRM1 Paragraph 4.4.3) which infers that only numeric codes can be used.
- The very purpose of a work/cost breakdown structure is to create a cascade of items each with a unique code, and therefore the use of alpha codes (letters) would not work.

5.8.5   Method of coding elemental cost plans for work packages

For some projects, the quantity surveyor/cost manager may need to present the cost plan in a work package rather than an elemental format. This could happen, for instance, where:

- There is early contractor involvement in the design process (e.g. stage one of a two-stage tender), and it makes both commercial sense and practical sense to divide up the project into work packages.
- The contractor is engaged early on in the project on the basis of an agreed fee (e.g. target cost contract) who then works with the employer's quantity surveyor to develop work package-based target costs.
- A project is to be procured on a traditional basis, but the work is to be tendered on the basis of bills of quantities for individual work packages to be managed by an in-house project management team (e.g. multistorey, high-value–high-quality speculative housing/commercial development).

The best time to determine the work packages would be when the contractor is on board.

In each case, a decision will have to be made as to the format of the cost plan and, in particular, the make-up of the individual work packages. To do this, and to avoid unnecessary duplication of effort, it would seem sensible to the allocate elements to appropriate work packages at the earliest possible stage, but this idea is not mentioned in NRM1.

Notwithstanding this, it should be understood that recoding the items in the cost plan does not imply any significant degree of extra work. It simply means that the cost plan will be sub-divided into work packages instead of group elements, but the building work measured in each work package will still be measured and described in the NRM1 elemental format. This type of reordering can be easily done in a cost planning software package such as CATO or QSPro.

This situation is contemplated in NRM1 Paragraph 4.5 which suggests a numeric suffix codi-fication framework such as that provided in NRM1 Figure 4.1. A selection of these can be seen in Table 5.12.

Table 5.12   NRM1 work package suffixes.

Work package	Suffix
Main contractor's preliminaries	/001
Substructure and groundworks	/002
Piling	/003
Concrete works (including precast concrete components)	/004
Carpentry	/006
Masonry	/007
Curtain walling	/011
Etc.	

There is no compulsion to use the NRM1 codes, but whatever the choice, Paragraph 4.5 suggests two methods of providing the additional coding needed to reallocate elements and sub-elements to individual work packages:

1. By *simply introducing one or more numeric suffix to each item in the cost plan* (Paragraph 4.5.1).
2. By using *one or more character(s)* as a suffix in order *to identify a work package* (Paragraph 4.5.2).

The first of these suggestions seems straightforward enough, but it is not clear why Paragraph 4.5.1 only refers to the reallocation of *elements and sub-elements* to work packages and fails to mention 'components' as it is quite possible that the cost plan may be developed to this level of detail at some point.

The first method of recoding may be explained by taking the simple example of a strip foundation shown in Figure 5.5.

The NRM1 code for this item (0$\underline{1}$.0$\underline{1}$0.0$\underline{1}$0.0$\underline{1}$0) is given by:

Group element	1	Substructure	1
Element	1.1	Substructure	1.1
Sub-element	1	Standard foundations	1.1.1
Component	1	Strip foundations	1.1.1.1

In order to reallocate this item, from group element 1 to the 'substructure and groundworks' work package, a suffix (002) should be added to the elemental code. The eventual code number, however, will depend upon the level of detail reached in the development of the cost plan. This could be done at the elemental level (group element level is insufficiently detailed), in which case the code would be 1.1/002 or 1.1.002. At the finest level of detail, 'component' level, the code would be either **1.1.1.1/002 or 1.1.1.1.002.**

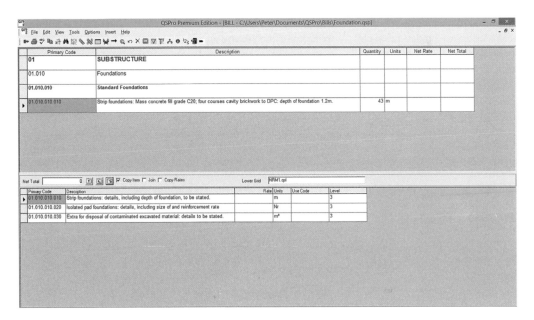

Figure 5.5 Recoding.

Part 2

Further items in the cost plan would be similarly dealt with in order to populate the 'substructure and groundworks' work package and also to create other work packages as appropriate. Paragraph 4.5.3 emphasises that elements may be broken down into further detail by introducing *additional levels of code* as needed.

The second method of recoding elements to work packages is not at all clear, and it is less than obvious what the words *one or more character(s) can be used as a suffix to identify a work package* actually mean:

- A 'character' is defined in the Oxford Dictionary as:

    A letter, number, or other mark or sign used in writing or printing, or the space one of these takes.

- *On the face of it, it would seem that Paragraph 4.5.2 simply means using some other* code or reference – alpha or numeric – than those listed in NRM1 Figure 4.1
- This character(s) would then be used as a suffix for items to be reallocated to the 'substructure and groundworks' work package (e.g. 'B')
- The suffix 'B', therefore, could be allocated to the foundation item giving a component level code of **1.1.1.1.B**, for instance.

5.8.6   Group elements 0–8

Group elements 0–8 of NRM1 Part 4: *Tabulated rules of measurement for elemental cost planning* provide the detailed rules for measuring building work items for inclusion in elemental cost plans.

The rules are structured in a tabular format as shown in Table 5.13.

Table 5.13   Measurement rules.

Sub-element	Component	Unit	Measurement rules for components	Included	Excluded
Sub-element title + definition of what the sub-element is	A named measurable part of a sub-element + details of what should be included in the description of the component	The unit of measurement applicable to the component	Rules that apply to the way in which each component is to be measured	Definitions of the work that is included in each component	Definitions of the work that is excluded from each component
Each sub-element has a unique numeric reference	Each component has a unique numeric reference	m, m², m³, t, nr, item, %	Each rule has a unique reference prefixed with the letter C	Each definition has a unique numeric reference	Each definition has a unique numeric reference

Some 'general rules' that apply to the measurement and description of group elements 0–8 can be found in NRM1 Part 3: *Measurement rules for cost planning*. However, Part 4: *Tabulated rules of measurement for elemental cost planning* also contains several rules that might be considered as 'general rules', but instead of being included in Part 3, they are repeated throughout Part 4. Examples of such rules include:

- *Other cost-significant components are to be described and identified separately.*

- *Contractor designed work is to be described and identified separately (**note:** where the contractor is only responsible for designing specific elements and/or components of the building project (i.e. not the entire building project)).*
- *Curved work is to be described and identified separately.*
- *Work to existing buildings is to be described and identified separately.*
- *Work within existing buildings is to be described and identified separately.*

5.8.7   Group elements 9–14

Group elements 9–14 of NRM1 Part 4: *Tabulated rules of measurement for elemental cost planning* provide the detailed rules for measuring non-building work items for inclusion in elemental cost plans. Such items include main contractor's preliminaries, risks and inflation allowances.

The rules are structured in a tabular format, but the layout of the tables is slightly different to that for group elements 0–8. For group elements 9, 11 and 12, the tables are laid out as shown in Table 5.14.

The layout for the group element 10 table is different as Table 5.15 illustrates.

The layout of the table for group element 14 is different again (see Table 5.16).

Table 5.14   Measurement rules: group elements 9, 11 and 12.

Component	Included	Unit	Excluded
A named measurable part of an element	Definitions of what is included in each component	The unit of measurement applicable to the component	Definitions of what is excluded from each component
Each component has a unique numeric reference	Each definition has a unique numeric reference	nr, item, per week, %	Each definition has a unique numeric reference

Table 5.15   Measurement rules: group element 10.

Element	Included	Excluded
A named measurable part of a group element	Definitions of what is included in each element	Definitions of what is excluded from each element
Each element has a unique numeric reference	Each definition has a unique numeric reference	Each definition has a unique numeric reference

Table 5.16   Measurement rules: group element 14.

Element	Included	Unit	Excluded
A named measurable part of a group element	Definitions of what is included in each element	The unit of measurement applicable to the element	Definitions of what is excluded from each element
Each element has a unique numeric reference	Each definition has a unique numeric reference	%	Each definition has a unique numeric reference

Part 2

Finally, group element 13: *Risks* has no tabulated structure at all but consists of four elements, each of which is defined by a 'boxed' numeric list of items appropriate to the particular element concerned. The four elements are:

Element	Risk
13.1	Design development risks
13.2	Construction risks
13.3	Employer change risks and
13.4	Employer other risks

The 'risks' listed in group element 13 are *not meant to be exhaustive* but should be considered *merely a guide*.

Strictly speaking, these 'risks' are not risks at all but are items that might be considered by the cost planner when deciding what *risk allowances… are to be included in each formal cost plan* (Paragraph 3.18.1). Element 13 provides a good *aide memoire* of possible hazards or threats to the robustness of the cost plan, but like all such threats, the extent of the 'risk' should be a considered judgement based on the likelihood of the event happening and its impact or severity.

NRM1 Paragraph 3.18.1 calls for a *formal risk analysis* as the basis for risk allowances in the cost plan and guidance for this may be found in Paragraph 2.15.

The phrase 'risk analysis' is clearly used loosely as the guidance provides no indication as to how an analysis may be conducted nor is it likely in practice that this would be done with any degree of sophistication.

A risk register, and *a properly considered estimate of risk* (Paragraph 2.15.4), is about the best that might be expected in most cases.

Note

1. *Bolam v Friern Hospital Management Committee* [1957] 1 WLR 582.

References

Ashworth, A. (2010) *Cost Studies of Buildings*, Fifth Edition, Prentice Hall, Harlow.

Ashworth, A., Hogg, K. and Higgs, C. (2013) *Willis's Practice and Procedure for the Quantity Surveyor*, Wiley-Blackwell, Chichester.

Benge, D.P. (2014) *NRM1 Cost Management Handbook*, Routledge, London.

Franklin + Andrews (2014) *Griffiths Building Price Book 2015*, 60th Edition, Franklin + Andrews.

Knowles, R. (2012) *200 Contractual Problems and Their Solutions*, Wiley-Blackwell, Chichester.

Langdon, D. (2014) *Spon's Architects' and Builders' Price Book 2015*, 140th Edition, CRC Press, Boca Raton.

Ministry of Justice (2012) *Pre-Action Protocol for Construction and Engineering Disputes*, Ministry of Justice, London.

Patten, B. (2003) *Negligence in Construction*, Routledge, London.

RICS (2007a) *Rules of Conduct for Members*, Version 6 – effective 1 January 2013, RICS, London.

RICS (2007b) *Rules of Conduct for Firms*, Version 5 – effective 1 January 2012, RICS, London.

Ross, A. and Williams, P. (2013) *Financial Management in Construction Contracting*, Wiley-Blackwell, Chichester.

Part 2

Chapter 6
New Rules of Measurement: NRM2

6.1 Introduction

NRM2 is part of the suite of documents that make up the RICS New Rules of Measurement. Its full title is *NRM2: Detailed measurement for building works*, and it became operative on 1 January 2013. NRM2 replaces the *Standard Method of Measurement of Building Works (SMM)* that has been used in the United Kingdom since the first edition was published in 1922. Consequently, NRM2 replaces the existing Standard Method of Measurement of Building Works 7th Edition (SMM7) that has been in use since its publication in 1988.

NRM2 has been developed to provide a modern, detailed set of measurement rules for the production of bills of quantities (BQ) and priced schedules of rates for capital building works or maintenance works. It sits between NRM1: *Order of cost estimating and cost planning for capital building works* and NRM3: *Order of cost estimating and cost planning for capital building maintenance works*. Together, the three documents are intended to provide measurement rules for the effective cost management of construction projects throughout their life cycle.

6.1.1 Standard methods of measurement

Standard methods of measurement in the building sector of the construction industry have a long history. The first SMM was published in 1922 and there have been seven editions all told. Some of these are remembered with fondness by users, whilst others have disappeared into the mists of time and are long forgotten. The original purpose of SMMs was to facilitate the production of BQ based on an agreed method of measurement that everyone understood. The creation of standardised BQ benefitted both the industry and construction clients to the extent that tendering costs were lower, tenders received from contractors were more consistent than hitherto, a clear definition of what was included in the tender price could be obtained, and work executed on-site could be remeasured (if desired) using the same agreed principles of measurement.

Managing Measurement Risk in Building and Civil Engineering, First Edition. Peter Williams.
© 2016 John Wiley & Sons, Ltd. Published 2016 by John Wiley & Sons, Ltd.

Subsequent editions have introduced ideas, 'modern' at the time, of using standard phraseology for the framing of bill descriptions, and SMM7 was devised with the use of IT in mind and was arranged in accordance with a Common Arrangement of Work Sections (CAWS) rather than the traditional work sections or trades.

Few will mourn the passing of SMM7 (the author included) which, conventional wisdom suggests, has become out of step with the industry and with modern methods of procurement. It remains to be seen, however, whether NRM2 will achieve the popularity of previous editions of the SMM such as the well-regarded (and still used) SMM6. A book consisting of more than 300 pages, variously presented in portrait and (annoyingly) in the landscape style of SMM7 and costing £45 (or subscription to isurv for the interactive version), is not a step in the right direction, despite the 'free' PDF download available to RICS members. Nonetheless, the shortcomings of SMM7 have at least been recognised by RICS Quantity Surveying and Construction Professional Group, and a Measurement Initiative Steering Group concluded *that significant improvements were required*.

With any new method of measurement, comparison with the previous edition is the traditional means of identifying differences in general methodology and measurement detail. NRM2, however, is an entirely new set of measurement rules, and thus, the approach taken in this book is to identify 'risk issues' in order to draw attention to key factors that could influence BQ preparation, tender pricing or potential claims that might arise during the ensuing contract. Notwithstanding, it is hoped that the reader will forgive the occasional backward glance at SMM7 for old time's sake!

6.1.2   International appeal

Many countries around the world, including the Republic of Ireland, the Republic of South Africa, Australia and New Zealand, have developed their own methods of measurement. Invariably, these have been based on UK SMMs but bespoked to local construction methods and practice.

NRM2 Paragraph 1.3.4 states that the *coordinated set of rules and underlying philosophy behind each section* of the RICS New Rules of Measurement *have worldwide application* despite being based on UK practice. No doubt this is true, but no set of measurement rules can be viewed in isolation from the law of the country in which it is used or the method of procurement and conditions of contract that will be used for an individual project.

In view of the wide variety of procurement methods and standard conditions of contract available in various countries, this chapter will focus on the JCT 2011 Standard Building Contract with Quantities and Approximate Quantities used, respectively, for lump sum and measure and value contracts in the United Kingdom.

Prior to NRM2 coming into force, the standard method of measurement referred to in these contracts was the SMM7.

6.1.3   Rules of measurement

The advent of NRM2 marks the passing into history of the phrase 'standard method of measurement', at least for building works in the United Kingdom, and the industry now has a set of new 'rules of measurement' to familiarise itself with. Just what the distinction between the two is remains to be seen as no clarification can be found in the Foreword to NRM2 which claims that it *provides a uniform basis for measuring and describing building works and embodies the essentials of good practice* just as General rules Paragraph 1.1 of SMM7 did.

Speaking of SMM7, it may be useful to compare the 'bulk' of the two documents:

Main section	Pages in SMM7	Pages in NRM2	Variance
General	5	43	+38
Main contract preliminaries	6	50	+44
Subcontract preliminaries	0	20	+20
Measured Work Sections	154	141	−13
Appendices	6	17	+11
Totals	**171**	**271**	**+100**

The 'General' section of NRM2 is much more verbose than its SMM7 counterpart and it is a shame that this material was not kept separate (say, in an appendix). This would have made the general rules much clearer and also provide a useful reference resource for students and practitioners alike.

Notwithstanding the new work section dealing with subcontract preliminaries, a more than sevenfold increase in the number of pages for the measurement of main contract preliminaries is quite beyond reason, whereas the 9% reduction in the measured work sections is to be applauded albeit insufficient. However, the striking similarity with SMM7 in much of the work section content is disappointing.

6.1.4   Amendments to JCT contracts

According to the JCT NRM Update issued in August 2012, all relevant JCT contracts and subcontracts entered into on or after 1 January 2013 should be using NRM2 instead of SMM7. This has necessitated the issue of amendments to the following contracts:

- Standard Building Contract – SBC/Q, SBC/AQ, SBC/XQ.
- Intermediate Building Contract – IC and ICD versions.
- Construction Management Trade Contract.
- Standard Building Sub-Contract – SBCSub/D/C and SBCSub/C.
- Design and Build Sub-Contract – DBSub/C.
- Intermediate Sub-Contract – ICSub/D/C and ICSub/C.
- Intermediate Named Sub-Contract – ICSub/NAM/C.
- Management Works Contract – MCWC/C.

The implications of these amendments for the Standard Building Contract (SBC/Q) 2011 are to remove reference to the phrase 'Standard Method of Measurement' and replace it with 'Measurement Rules'. In particular:

- Clause 1.1 – the definition of provisional sum:
  - **Delete** 'General Rule 10 of the Standard Method of Measurement'.
  - **Insert** 'Paragraph 2.9.1 of the Measurement Rules'.
- Clause 1.1 – definition of Standard Method of Measurement:
  - **Delete** the definition of 'Standard Method of Measurement'.
  - **Insert**:

  'Measurement rules: the RICS New Rules of Measurement – Detailed Measurement for Building Works (NRM2), in the form published at the Base Date, unless otherwise stated in the Contract [Bills/Documents]*.'

  * Delete as applicable

- Clause 1.4, etc.
  - ○ <u>Either</u> insert the following in clause 1.4.6:
    - 'references to the Standard Method of Measurement shall be read as reference to the Measurement Rules'
  - ○ <u>Or</u> in each of the following clauses, delete 'Standard Method of Measurement' and insert 'Measurement Rules':
    - 2.13.1
    - 2.14.1
    - 5.6.3.3.

Similar amendments apply to the other JCT contracts and subcontracts albeit that the Construction Management Trade Contract requires:

- A choice to be made between the RICS New Rules of Measurement, the Civil Engineering Standard Method of Measurement (CESMM) or another method of measurement as the basis for preparing the Bills of quantities.
- Revisions to the definitions of 'Defined Provisional Sum' and 'Provisional Sum' contained in the Trade Contract.

It is interesting to note that, where BQ are to be used, the Construction Management Trade Contract relies on the NRM2 definitions of 'Defined Provisional Sum' and 'Provisional Sum' as contained within NRM2 Paragraphs 2.9.1.2 and 2.9.1, respectively, <u>and not</u> Paragraph 1.6.3: *Definitions* <u>and</u> that these definitions shall apply <u>irrespective of the method of measurement used</u>.

The JCT NRM Update suggests that the modifications proposed may be incorporated in the various contracts and subcontracts <u>either</u>:

- By amending the contract document, initialling each amendment and executing the contract accordingly

  or
- By attaching the Update to the Contract and inserting an Article to the effect that the Contract Agreement and Conditions are modified as set out in the New Rules of Measurement Update.

6.2   What is NRM2?

The Foreword to NRM2 explains that NRM2 grew out of a perception that there were *problems associated with the measurement of building works at all stages of the design and construction process* and that *significant improvements were required*.

From this research grew a suite of three documents that are intended to cover *all aspects of the measurement and description of a building project from – 'cradle to grave'*. These documents are:

- NRM1: Order of cost estimating and cost planning for capital building works.
- NRM2: Detailed measurement for building works.
- NRM3: Order of cost estimating and cost planning for building maintenance works.

NRM2 is intended to:

- Provide fundamental guidance on the quantification and description of building works for the purpose of preparing:
  - ○ BQ.
  - ○ Quantified schedules of works.

- Provide a sound basis for designing and developing standard of bespoke schedules of rates.
- Provide direction on how to deal with items that are unquantifiable, such as:
  - Preliminaries.
  - Overheads and profit.
  - Contractor-designed works.
  - Risk transfer.
  - Fluctuations.
- Provide a uniform basis for measuring and describing building works.
- Embody the essentials of good practice.

NRM2 is part of the RICS 'Black Book' suite of guidance notes that define good technical standards for quantity surveying and construction professionals.

6.3 Status of NRM2

Unlike previous editions of the SMM, which were silent on this issue, the status of NRM2 is expressly covered in Part 1.1: *Introduction*.

This explains that the RICS New Rules of Measurement in general, and NRM2 in particular, have been given the status of an **RICS guidance note** which is defined as a *document that provides users with recommendations for accepted good practice as followed by competent and conscientious practitioners*. The status of NRM2 is therefore *recommended good practice*.

From this inclusion, it is clear that those responsible for drafting NRM2 were mindful of the increasing tendency towards litigating against construction professionals, such as architects, engineers and surveyors, and that there have been a number of cases where quantity surveyors have been involved in multimillion £ claims for negligent cost planning, cost advice, certification and measurement.

The status of NRM documents, and the standards expected of their users, is discussed, in detail, in Chapter 5.

6.3.1 Negligence

The *Introduction* to NRM2 makes it clear that conforming to the practices recommended in the document should provide *at least a partial defence to an allegation of negligence* and that non-compliance with recommended practice should only be undertaken with justification as a court, tribunal or RICS Disciplinary Panel may ask for an explanation as to why recommended practice was not adopted.

The test applied in cases of professional negligence is, as Patten (2003) suggests, **whether the quantity surveyor acted with the skill and care expected of a reasonably competent member of his profession**, and NRM2 confirms that a court or tribunal may take account of the contents of guidance notes as being indicative of 'reasonable competence'.

However, NRM2 – *Introduction* also states that *where*, within NRM2, *recommendations are made for specific professional tasks, these are intended to represent 'best practice'* which, in the opinion of the RICS, *meet a high standard of professional competence*, that is, a higher standard than that demanded by the courts. This issue is developed in detail in Chapter 5.

Part 2

Risk Issue

NRM2 – *Introduction* is unclear as to whom the need to comply with an RICS guidance note refers and variously refers to 'members', 'surveyor' and 'user'. Conventionally, the most likely litigant in cases of professional negligence is the employer who, having engaged a professional quantity surveyor, feels that he has received inadequate professional advice or services.

Patten (2003) suggests that quantity surveyors engaged by the employer have less to fear from contractors as, unlike architects and engineers, they do not normally have a certification role. Thus, where the employer is forced into insolvency due to over-certification, based on a quantity surveyor's valuation of the work in progress, the contractor is unlikely to have a claim following the judgement in *Pacific Associates Inc. v Baxter* which held that there was no duty of care owed to the contractor to prevent economic loss in such circumstances.

A more likely possibility is **contribution proceedings** taken against a quantity surveyor by an architect or engineer who has been sued by the employer for negligent certification. In such cases, a 'contribution' may be sought should it be established that the quantity surveyor's measurement or valuation was negligent.

Notwithstanding the foregoing, the popularity of design and build and drawings and specification procurement, and the decline in the use of formal PQS-produced bills of quantities, means that it is subcontractors who may be potential litigants where NRM2-based documents are poorly, inaccurately or misrepresentatively drafted. Consequently, a main contractor or, perhaps, an employer-engaged construction manager may have something to fear should NRM2-based work packages or composite items prove to be misleading.

6.3.2   Pre-action protocol for construction and engineering disputes

The importance of this issue of negligence is not to be understated in that the UK Ministry of Justice has seen fit to publish a *Pre-Action Protocol for Construction and Engineering Disputes* (Ministry of Justice, 2012) that applies to construction and engineering disputes including those involving professional negligence claims against architects, engineers and quantity surveyors. The intention of the protocol is to ensure that:

- The parties are clear as to the issues in dispute.
- Information has been exchanged in a timely and efficient manner.
- The parties have met and have attempted to arrive at a resolution without re-sort to litigation.
- Should litigation ensue, it can proceed in an efficient manner.

The status of NRM2 as a guidance note is not to be taken lightly!

6.4   NRM2 structure

In addition to its Foreword, Acknowledgements and Introduction, NRM2 comprises:

- Part 1: General.
- Part 2: Rules for detailed measurement of building works.
- Part 3: Tabulated rules of measurement for building works.
- Appendices.

NRM2 Part 1, in common with Part 2, contains a considerable amount of dialogue that is more in keeping with a textbook than a standard method of measurement. As a consequence, it is tempting to 'skip' some of the text in order to get into the 'meat' of the rules of measurement themselves. This would be a mistake as there are a number of 'gems' in Parts 1 and 2 that deserve careful consideration, especially in the context of what might be loosely termed 'risk issues'.

Navigating NRM2 is easy. Each of the three main parts and seven appendices has a hierarchy of numbered paragraphs. For example, Part 2: *Rules for detailed measurement of building works* has 17 paragraphs, most of which are subdivided into sub-paragraphs. The paragraph numbering structure is akin to a work breakdown structure (WBS) with a series of levels. Some paragraphs have a title or heading and some do not. Table 6.1 (Section 6.8.1) illustrates how the paragraph numbering system works.

6.5   Part 1: general

This part of NRM2 is both contextual and explanatory:

- Measurement is contextualised in terms of the design process relative to two common process models – the RIBA Plan of Work and the OGC Gateway Process (see Chapter 4).
- The purpose, use and structure of NRM2 are explained.
- Symbols and abbreviations used in NRM2 are explained.
- Some 33 words and phrases used later on in the document are defined.

Risk Issue

The status of Part 1: *General* is unclear.

Parts 2 and 3 are clearly entitled 'rules', but Part 1 does not have 'rules' in its title (unlike SMM7: General rules). Paragraph 1.1.1, however, does state that Part 1 *explains the symbols, abbreviations and definitions used in the rules*, and, as such, it may be the case that Part 1 carries the status of 'rules'.

This is an important distinction because, where there is doubt in interpreting the 'rules' in Parts 2 and 3, the definitions in Part 1 may well assume a key clarification role, especially if there is a dispute, as to the meaning of the rules of measurement.

6.5.1   *Measurement in context with the RIBA Plan of Work and OGC Gateway Process*

NRM2 Paragraph 1.2 briefly explains the methodology of the RIBA Plan of Work and the OGC Gateway Process and identifies the point at which measurement is carried out for the purposes of producing BQ, quantified schedules of works or quantified work schedules. This point is defined as RIBA Work Stage G (Tender Documentation). This is the traditional point when such documentation would be prepared – when the design is sufficiently well developed – but this fails to acknowledge that pricing documents are prepared at much earlier stages in non-traditional procurement methodologies.

In this context, it is disappointing to note that Part 3: *Tabulated rules of measurement for building works*, bears striking similarity to SMM7, and it would seem that a golden opportunity has been overlooked to design a much simpler and less detailed standard method of measurement that could be used by a variety of 'users' in a number of procurement situations.

Risk Issue

Notwithstanding the above, NRM2 Rule 3.3.3.12 provides for the creation of composite items where separate components or sub-components may be combined to form a single item. Such items would be 'user defined' and, as such, may potentially lead to a dispute or litigation should the description of a composite item lack clarity as to item coverage (see also Section 6.10.1).

Part 2

6.5.2   Purpose of NRM2

The stated purpose of NRM2 is explained in Paragraph 1.3 which is to:

- Provide a standard set of measurement rules for the procurement of building works that:
  - Are *understandable by all those involved in a construction project, including the employer*.
  - Facilitate *communication between the design/project team and the employer*.
  - Set out information requirements *from the employer* and *consultants to enable a BQ to be prepared*.
  - Deal *with the quantification of non-measurable work items, contractor designed works and risks*.
  - Create a *coordinated set of rules and underlying philosophy* that will *have worldwide application*.

6.5.3   Use of NRM2

Paragraph 1.4 explains that the RICS New Rules of Measurement:

- Provide the means to measure building work in a structured and consistent way.
- Represent *the essentials of good practice*.
- Address the production of BQ for entire projects and for discrete work packages.
- Can also be used to prepare:
  - Quantified schedules of works.
  - Quantified work schedules.
  - Standard and bespoke schedules of rates for:
    - Discrete contracts.
    - Term contracts.
    - Framework agreements.

Nowhere within NRM2 are the terms 'quantified schedules of works' and 'quantified work schedules' defined. In fact, Part 2: *Rules for detailed measurement of building works* is completely given over to BQ production without reference to other forms of documentation.

The unit of measurement adopted in NRM2 is the metric system with a point (.) as a decimal marker and a comma (,) as a thousands spacer.

6.5.4   Structure of NRM2

Paragraph 1.5 expands on the structure of NRM2 with particular reference to Parts 1, 2 and 3 and the Appendices. There is no dialogue of importance here, and there seems to be little point in this inclusion which is no more than an expanded list of contents.

6.5.5   Symbols, abbreviations and definitions

Paragraph 1.6 consists of:

- 1.6.1: Symbols used for measurement.
- 1.6.2: Abbreviations.
- 1.6.3: Definitions.

Paragraphs 1.6.1 and 1.6.2 are self-explanatory, but Paragraph 1.6.3 contains a number of important definitions, some of which deserve particular mention.

6.6   Definitions

NRM2 Paragraph 1.6.3: *Definitions* provides a list of 33 words or phrases used in the document which should be understood. Some of these definitions are self-explanatory, whilst others are significant either because they are:

- important in terms of the rules of measurement

or because they are:

- ill defined or could lead to confusion or even dispute.

A number of the definitions raise risk issues that ought to be carefully considered by users of NRM2, and thus, this section concentrates on 19 definitions that are considered to be important.

6.6.1   Bill of quantities

This is defined as *a list of items*, with *detailed identifying descriptions* and *firm quantities*.

Part 2

> **Risk Issue**
>
> This is a misleading definition because it excludes the very likely possibility that the bills of quantities might contain provisional quantities or defined/undefined provisional sums. These terms are defined later on in Paragraph 1.6.3.
>   There is no definition of bills of approximate quantities although these are specifically mentioned in Paragraph 2.4.3 in the context of types of bills of quantities.

6.6.2   Daywork

This is defined as a *means of valuing work on the basis of time spent by the contractor's workpeople, the materials used and the plant employed.*

> **Risk Issue**
>
> This definition is at odds with a further definition given in Paragraph 2.13.3.1 which refers to contractor's <u>employees</u>. Employees are engaged on a contract of service or employment contract, and this excludes subcontractors.

There is no mention of the RICS *Definition of Prime Cost of Daywork for Building Work*, or any other definition for that matter, nor is reference made to a definition of prime cost (PC) of daywork in Paragraph 2.13.3 which expands on the subject of daywork. This may be a reflection of the 'neutrality' of NRM2 which is intended to be used in other jurisdictions beyond the United Kingdom.

Paragraph 2.13.3.2 states that if a schedule of dayworks is incorporated into the BQ, then a statement as to how the contractor will be paid <u>is to be given</u> in the preliminaries bill or schedule of dayworks.

> **Risk Issue**
>
> Paragraph 2.13.3.2 sounds like a 'rule'.
>   However, payment terms are the normal province of the conditions of contract, and care needs to be taken in complying with Paragraph 2.13.3.2 in order to avoid a conflict between the conditions and the contract bills.

Paragraph 2.13.3.3 requires that *the method of calculating labour time charge rates…for work carried out outside of normal hours (i.e. non-productive time) shall be defined* in the schedule of dayworks and that the definition of 'normal working hours' *shall be given* in either the preliminaries bill or schedule of dayworks. This is a sensible inclusion as it is often a 'bone of contention' as to what constitutes 'premium time' (e.g. weekend working).

6.6.3   Defined provisional sum

This definition is the same as that given in General Rule 10.3 of SMM7 and relates to the expenditure of sums for work that is envisaged but cannot be accurately quantified due to incomplete design. The nature, location, scope and extent of the work, together with indicative quantities and any specific limitations, *shall be* provided.

This definition should be read in conjunction with Paragraph 2.9.1: *Provisional sums* and, in particular, with Paragraph 2.9.1.2 which provides an amplified definition of the term 'defined provisional sum'.

There seems little point having a definition of a term that has to be redefined elsewhere in the document, however.

6.6.4   Design team

In NRM2, this *means architects, engineers and technology specialists* who have responsibility for the conceptual design of a building or structure and for its development into drawings, specifications and instructions for construction and associated purposes.

The 'design team' is defined as being part of the 'project team', but the term 'project team' is not defined in NRM2. Unless quantity surveyors are 'technology specialists', they would appear not to be members of the design team nor, would it seem, are contractors or subcontractors who may nevertheless contribute to the design concept and its development under some methods of procurement.

The reasoning for this bizarre definition is obscure as the term 'design team' is not used anywhere in NRM2. Paradoxically, the terms 'project team' and 'employer's project team' are used in several places, but neither term merits a definition. Even more strange is the lack of a definition for the 'quantity surveyor/cost manager' who has a number of defined duties under NRM2. The quantity surveyor is identified in Article 4 of JCT 2011.

Paragraph 2.7: *Preliminaries* of NRM2 explains how the 'preliminaries' section of a BQ might be set out and quantified. Paragraph 2.7.3.1: *Information and requirements* is stated as being the descriptive part of the main contract preliminaries which sets out, inter alia, the names and contact details of the 'employer' and the 'employer's project team'.

Risk Issue

The 'employer's project team' may well include the architect and other designers, the (CDM) principal designer, the quantity surveyor, maybe a clerk of works or inspector and, possibly, personnel within the employer's organisation that deal with accounts and payments and other administrative matters.

Care needs to be exercised in defining just who is included in the 'employer's project team' in order to make sure that the list includes those named in the Articles of Agreement in the contract who have particular duties or authority under the contract (see, e.g. the Articles of Agreement in JCT 2011 and in particular Articles 3–5).

Being a statutory appointment that must be made by the employer under CDM Regulation 5(1)(b) and named in JCT 2011 Article 6, the question arises as to whether or not the principal contractor is a member of the 'employer's project team'.

6.6.5   Director's adjustment

This refers to *a reduction or addition to the tender price* compiled by the contractor's estimating team *offered by the director(s)*.

The use of the word 'offered' implies something capable of being accepted or refused, but the true meaning is undoubtedly that of a final adjustment to the tender figure which constitutes the contractor's 'offer' which the employer may subsequently accept or reject. The positioning of the apostrophe in the heading *Director's Adjustment* implies the action of a single director, whereas the definition contemplates more than one.

Paragraph 2.13.2 expands on the reasons for a director's adjustment where it is also stated that *separate provision is to be incorporated* in the BQ *for the contractor to insert a 'director's adjustment'*; Appendices D and E show how this might be achieved in a pricing summary template.

NRM2 does not explain whether the director's adjustment is:

- To be regarded as an adjustment to the BQ rates.
- To be added to or deducted from interim payments in proportion to the value of work carried out and the contract sum.
- A lump sum to be added or deducted at final account stage.

Risk Issue

Nowhere in NRM2 is there a rule stating how the director's adjustment is to be applied when it comes to interim payments, the valuation of variations, the expenditure of provisional sums or calculation of the final account.

This is an important omission as, without such a rule, the contract administrator (as certifier) and/ or the quantity surveyor (as valuer) could be liable to:

- The employer for negligent certification should the contractor be overpaid.
- The contractor for breach of duty of trust should the contractor be underpaid.

A comparison of the implications of the various methods for dealing with the director's adjustment post-contract is provided in Chapter 14, Section 14.2.

6.6.6   Employer

This term is taken to mean 'the employer' in the normally accepted sense in construction contracts, or it could mean 'end user' or someone with delegated powers in central government, such as 'senior responsible owner' or 'project sponsor'.

This definition would seem to be of most importance in relation to the provision of information for measurement purposes (NRM2: Paragraph 2.14 refers).

6.6.7   Fixed charge

This relates to work where the cost is unrelated to duration.

Fixed charges appear as a 'pricing method' in Part 3: *Tabulated rules of measurement* in the Part B: *Pricing schedule* of Preliminaries (main contract) and Preliminaries (work package contract). See also 6.6.16.

6.6.8   Main contractor

This is another curious definition because, whilst the term 'main contractor' is used in NRM2, the predominant reference is to 'the contractor'.

> ## Risk Issue
>
> The distinction is important because 'the contractor' has specific duties in Part 3 of NRM2 including the provision of documents relating to the design, production information, as-built drawings and documents required before practical completion (Tabulated Work Section 1.6.3 refers).

The definition makes no distinction between the main contractor (under the civil contract) and the principal contractor (who carries the statutory duty under the CDM Regulations). Very often, the main contractor and the principal contractor are one and the same, but there are occasions when the statutory duty is undertaken by another party (the employer or a contractor appointed as a construction manager, for instance).

The inclusion within the definition of the term 'prime contractor' reflects current practice in central government procurement.

6.6.9   Main contract preliminaries

This is defined as *items that cannot be allocated to a specific element, sub-element or component*.

This definition is taken from NRM1 and fails to refer to the NRM2 phrase *item or work to be measured*.

In terms of NRM2, preliminaries would consist, *inter alia*, of items that cannot be allocated to a specific measured item in the BQ, irrespective of the format of the BQ (i.e. elemental, work section, work package). Therefore, even in an elemental BQ, which would be subdivided into elements and sub-elements, the BQ would contain items described in accordance with NRM2 (i.e. items or work to be measured), and the NRM1 'component' level of detail would not apply.

Notwithstanding, items/work and components are measurable items, and preliminaries are non-measurable except in units of time or money.

The definition specifically excludes *costs associated with subcontractors' or work package contractors' preliminaries* from the main contract preliminaries. This statement merits some consideration because the contractor needs to be sure where to price his subcontractors' preliminaries (if there are any) because:

- Subcontractor preliminaries need to be included somewhere in the contractor's tender.
- The contractor may prefer to separately identify subcontractor preliminaries as they may be material to the valuation of a variation or a claim.
- The contractor may wish to include subcontractor preliminaries in the method-related charges.

Domestic subcontractors who carry out traditional 'trades', such as joinery, plastering and painting, do not normally price any preliminaries items in their quotations, but others do. Piling subcontractors can have significant fixed and time-related charges to consider as can those responsible for formwork and structural concrete packages, structural steelwork erection, installation of heating and air-conditioning plant, etc.

Risk Issue

Normal practice is that main contractors price their subcontractors' preliminaries in either the measured rates or in the main contract preliminaries. However, NRM2 poses some challenges to convention because these two places are specifically excluded by NRM2 for pricing subcontractor/work package contractor preliminaries, namely,

- The *costs associated with subcontractors' or work package contractors' preliminaries* are <u>excluded</u> from the main contract preliminaries in the definition.
- The coverage rule for measured items of work <u>does not include</u> subcontractors' preliminaries items *unless specifically stated otherwise in the BQ* (NRM2 Paragraph 3.3.3.13 refers).

For more on work package contract preliminaries, see Section 6.6.19.

6.6.10   Overheads and profit

This is defined to mean:

- *The contractor's costs associated with head office administration apportioned to each building contract* plus
- *The main contractor's return on capital investment.*

This definition is entirely misleading.

To begin with, the definitions of 'overheads' and 'profit' are well understood in the industry:

- Overheads

The Chartered Institute of Building (CIOB, 2009) *Code of Estimating Practice* defines 'head office overheads' as *the incidental costs of running a business as a whole*. This implies much more than the costs of administration as stated in the NRM2 definition and includes directors' salaries, company cars, company pension contributions, auditing fees and the cost of financing working capital. In the annual accounts, head office overheads are normally classed as 'administration expenses' which, together with any interest repayments on borrowings, makes up the 'overhead'. This is entirely consistent with the CIOB definition.

- Profit

Contractors' profit is normally expressed as a percentage of annual turnover and not as return on capital employed. Return on capital employed is often used as a financial ratio in accounting circles and, by investors, as a measure of the how efficiently working capital has been used; it is not a consideration at tender stage nor is it normally used as a means of adding profit to a contractor's tender.

Secondly, contractors do not necessarily apportion overheads to *building* contracts – the contractor may undertake all sorts of other work as well which will attract an overhead allowance.

Thirdly, profit, when added as a 'markup' on BQ rates, is invariably added as a percentage related to turnover and not to the return on capital invested in the business.

Paragraph 2.11.1 states that:

- Provision shall be made in the BQ to enable the contractor to apply a percentage for overheads and profit on:
  - Preliminaries.
  - Measured work.
  - Risk allowances.
  - Work relating to the expenditure of provisional sum.

Paragraph 2.11.2 says that 'overheads' and 'profit' can be treated as two separate items.

What Paragraph 2.11 does not say is:

- Where the percentage addition shall be inserted in the BQ; Appendices D and E are templates provided in NRM2 which show this as being on the final summary, but there is no rule to this effect, however.
- That provision could be made at the end of each work section, if desired, without breaching the rules of measurement.
- That provision for a percentage shall be a separate percentage for each listed item (i.e. preliminaries, measured work, etc.).
- Whether or not the contractor is obliged to insert a percentage.

Adding to the confusion, Paragraph 3.3.3.13 says that each component/item in the BQ shall *be deemed to include*:

(7)   *Establishment charges* – normally a synonym for 'overheads'

> ### Risk Issue
>
> The definition of overheads and profit is likely to lead to disagreement at best and disputes at worst.
>   It is misleading, wrong in parts and conflicts with other parts of NRM2, particularly when read in conjunction with Paragraphs 2.11 and 3.3.3.13.
>   The issue of 'overheads and profit' is further discussed in Section 6.7.11 of this chapter.

6.6.11   Prime cost sum

This is *a sum of money included in a unit rate to be expended on materials or goods from suppliers*. This is a 'supply-only' rate for materials or goods whose precise quality is unknown.

Presumably, whilst the precise quality is not known, the precise quantity is. The PC sum excludes the cost of fixing, fees, preliminaries, overheads and profit, etc.

In the specific paragraph (3.3.7) that deals with the NRM2 procedure where the exact type of product or component is not specified, the term 'PC sum' is not used. In this paragraph, the term 'PC item' is used, and it seems that this is the term that should be used in the relevant item description.

The idea of the contractor pricing work using a PC sum for the cost of materials is not new. The process is that the contractor prices the item using the PC sum as the basis for the cost of materials/goods (e.g. PC sum of £700/thousand for facing bricks), and he then adds any ancillary costs (such as mortar), labour, plant (cement mixer) and margin. When the exact quality (and cost) of the materials in question is known, the unit rate is adjusted by omitting the PC sum (say, £700) and adding back the actual cost (say, £590/thousand).

This is pretty straightforward contract administration, but it is a process that worked in the days when the expenditure of PC sums was catered for in the JCT standard contract. This is now not the case and, consequently, there may be problems with administering such items in the context of the interface between NRM2 and the conditions of this particular form of contract.

Risk Issue

Clause 3.10 of JCT 2011 SBC/Q requires the contractor to comply with instructions which the architect/contract administrator is <u>expressly empowered</u> to give.

The JCT 2011 SBC/Q contains no provision for PC sums, whether included in a unit rate or otherwise, nor is there an express power for the architect/contract administrator to issue instructions regarding the expenditure of PC sums.

Clause 3.16 deals with architect/contract administrator instructions regarding the expenditure of <u>provisional sums</u> only.

On the face of it, the obvious way round the problem is to adjust the PC sum by issuing a variation instruction under Clause 5.1 of JCT 2011 SBC/Q. However, the definition of 'variation' in Clauses 5.1.1 and 5.1.2 does not provide for such circumstances.

Clause 5.1.1 refers to *the alteration or modification of the design, quality or quantity of the Works*, and Clause 5.1.2 deals with *the imposition by the Employer of any obligations or restrictions* regarding issues such as access to the site, limitations of working space or working hours or the execution or completion of the work in any specific order.

This definition needs to be read in conjunction with Rules 3.3.7 and 3.3.3.13 of the Tabulated rules of measurement.

6.6.12   Provisional quantity

This is defined to mean *a quantity which cannot be accurately determined (i.e. an estimate of the quantity)*.

The definition is clear enough, but NRM2 is not at all clear on the general use and application of provisional quantities which practitioners will remember are referred to as 'approximate quantities' in SMM7.

The definition of 'provisional quantity' needs to be read in conjunction with Paragraph 2.4: *Types of bill of quantities* and Paragraph 3.3.8 of NRM2 which are discussed in Section 6.7.4 of this chapter.

6.6.13   Provisional sum

This is defined as a sum of money that is set aside to carry out work that cannot be described and quantified in accordance with Part 3: *Tabulated rules of measurement for building works*.

The definition says that a provisional sum <u>*will be*</u> identified as 'defined' or 'undefined' not <u>shall be</u> as is required by General Rule 10.2 of SMM7 and by NRM2 Rule 2.9.1.1.

It is disappointing that the definition is so imprecise and that users of NRM2 are obliged to search for a further definition, in Paragraph 2.6.7, only to find out that the rules relating to provisional sums are to be found in yet another paragraph of the document. Paragraph 2.6.7.3 stipulates that the rules are given in Paragraph 2.9.1.

NRM2 Paragraph 2.9.1.1 specifies that work items that cannot be measured in accordance with the tabulated rules of measurement _shall be_ given as a 'provisional sum' and that the work shall be identified as 'defined work' or 'undefined work' as appropriate.

See also Section 6.7.9 on provisional sums.

6.6.14   Residual risk (or retained risk)

This _means the risks retained by the employer._

Just why there is a need for this definition is unclear as, nowhere in NRM2, are such risks defined or otherwise identified. Additionally, it would seem that there is no mechanism within the measurement rules for the contractor to be notified as to exactly what these risks are.

Paragraph 2.10.4.1 gives a clue such that residual risks retained by the employer are represented by risk allowances in the cost plan. Conventionally, such risks are translated into the BQ by means of provisional quantities, provisional sums or contingency allowances, but there is no mechanism in NRM2 for this to happen.

Other clues are to be found in the Appendices:

- Appendix D
  The template for a condensed 'pricing summary' provides for 'risks' under cost centre 11.0, but these are not defined.
- Appendix E
  Similarly, the expanded 'pricing summary' template includes 'risks' under cost centre 10.0, but no detail is given.
- Appendix F
  A template for a 'schedule of construction risks' is provided, but these would appear to be risks borne by the contractor and not those retained by the employer.

Paragraph 2.10.4.2 indicates that the contractor may be approached to undertake such risks at a premium, but there is no suggestion within NRM2 as to how this might be done or how the contractor might be asked to price such risks at tender stage. Under JCT 2011 SBC/Q, this might be taken to indicate that the contractor may be asked to provide a 'variation quotation' for undertaking employer's retained risks; the contractor, however, has the right to object under the contract conditions and, with proper notice, is not required to provide a quotation.

Paragraph 2.10: _Risks_ should be consulted for additional information albeit that it casts very little light on the foregoing.

6.6.15   Subcontractor

This _means a contractor employed by the main contractor_ to carry out _specific work_ and is a synonym for _specialist, works, trade, work package and labour-only contractors._

The question as to whether a 'subcontractor' is a 'work package contractor' in respect of preliminaries is discussed in Sections 6.6.18 and 6.6.19.

6.6.16   Time-related charge

This relates to work where the cost is dependent on duration.

Fixed and time-related costs in construction are normally regarded as 'preliminaries' or site 'oncosts' which are items largely concerned with the management of a project and the way that the contractor intends to carry out the works. In the United States, this is called 'means and methods' which rather graphically conjures up the idea of the resources and practicalities of how the job will be done.

The significance of the definition of fixed and time-related costs is to be found via the somewhat tortuous route of:

- Part 3: *Tabulated rules of measurement for building works*
  - Tabulated work sections
    1. Preliminaries (main contract)
       Part B: Pricing schedule
    1.1 Employer's requirements
    1.2 Main contractor's cost items.

This 'pricing schedule' provides a list of preliminaries items where the 'pricing method' is stated as 'fixed charge', 'time-related charge' or sometimes both.

6.6.17   Undefined provisional sum

This is a sum of money provided for work where the nature, location, scope and other information requirements of a defined provisional sum cannot be stated. Unlike General Rule 10.5 of SMM7, the NRM2 definition clarifies that undefined provisional sums are *for work that is not completely designed.*

Part 2: *Rules for detailed measurement of building works* also provides information about provisional sums (Paragraph 2.6.7 refers). However, no actual rules are stated here, and thus, it is necessary to look at Paragraph 2.9.1: *Provisional sums* to discover how provisional sums are to be dealt with in the BQ. As if there was not enough unnecessary dialogue in NRM2, Paragraph 2.9.1 goes on to repeat all of the definitions relating to provisional sums that can be found in Paragraph 1.6.3: *Definitions.*

The reader has to search for the important bit about provisional sums which is to be found in Paragraph 2.9.1.3 where it is stated that the work to which defined provisional sums apply *shall be deemed* to have been allowed for in the contractor's *programming, planning and pricing* [of] *preliminaries.*

Political correctness having now arrived in a standard method of measurement for the first (and hopefully the last) time means that 'his' or 'her' is used in Paragraph 2.9.1.3 when referring to the contractor!

Risk Issue

This is not the sort of imprecise language usually found in a standard method of measurement, and perhaps a sharp barrister could make capital out of the fact that most standard contracts refer to the contractor as 'he' and that many contractors are limited companies and therefore merit the sobriquet of 'it' (i.e. a legal entity, not a person)!

6.6.18   Work package contractor

Inexplicably, a distinction is drawn between 'subcontractor' and 'work package contractor' who is defined as *a specialist contractor who undertakes particular identifiable aspects of work.*

Just why this should be so is somewhat confusing as the definition of 'subcontractor' includes work package contractors (Paragraph 1.6.3 refers).

The definition concludes by explaining that *works contractors* (i.e. not work package contractors) *may be employed directly by the employer or by the main contractor* according to the *contract strategy* employed. This would appear to fit with management contracting and construction management procurement in particular.

6.6.19   Work package contract preliminaries

These are defined as *preliminaries that relate specifically to the work that is to be carried out by a work package contractor*. Part 3: *Tabulated Work Section – Preliminaries* makes specific provision for the separate pricing of work package contract preliminaries.

NRM2 Paragraph 2.7.4 explains that work package contract preliminaries consist of two components:

1. *Information*
   The descriptive part of work package contract preliminaries where general information is to be found such as the project particulars and description of the works.
2. *Pricing schedule*
   Where the work package contractor prices the preliminaries items relating to the particular work package in question.

NRM2 is not exactly crystal clear as to how, or in what circumstances, work package contract preliminaries are employed. Paragraph 2.7.4.2 seems to suggest that work package contract preliminaries represent part of a work package tender submission together with measured items of work that relate to the particular work package in question which, if the tender is successful, will form part of the subsequent work package contract.

It is difficult to see this happening for a traditionally tendered contract as it is the employer's quantity surveyor/cost manager who must determine the number and content of the work packages as Paragraph 2.15.3.1(3) of NRM2 suggests. Consequently, work package contract preliminaries are most likely to arise for contracts procured on a management contracting or construction management basis where the management contractor or construction manager is appointed on a fee-only basis.

This issue is discussed further in 6.7.7.

6.7   Part 2: Rules for detailed measurement of building works

6.7.1   Introduction

Notwithstanding Part 1: *General* Paragraph 1.1.1, Part 2 of NRM2 is devoted exclusively to the preparation of BQ of various sorts. Little mention is made of schedules of works or work schedules despite the claim in Paragraph 1.3.2 that the purpose of NRM2 is also to provide rules for the preparation of schedules of works (quantified) and to develop bespoke and standard schedules of rates.

Paragraph 2.1.1 explains that Part 2 of the rules:

- Describes the purposes and uses of NRM2.
- Describes the types of BQ.
- Gives guidance on the composition and preparation of BQ.
- Defines the information required to enable a BQ to be prepared.
- Sets out the rules of measurement of building items.
- Sets out the rules for dealing with:
  - Preliminaries.
  - Non-measurable works (provisional sums – see Paragraph 2.6.7).

- ○ Contractor-designed works.
- ○ Risks.
- ○ Overheads and profit.
- ○ Credits.

Paragraph 2.1.2 states that Part 2 of the rules:

- Deals with other aspects of BQ production including:
  - ○ Price fluctuations.
  - ○ Director's adjustments.
  - ○ Daywork.
  - ○ VAT.
- Provides guidance on:
  - ○ The codification of BQ.
  - ○ The use of BQ for cost control and cost management.
  - ○ The analysis of a BQ to provide cost data.

Part 2 is, therefore, a combination of descriptions, guidance, definitions and rules.

Whilst much of this information is no doubt well intentioned, most practitioners will need to quickly, easily and unequivocally understand the rules of measurement so that:

- BQ can be structured correctly.
- The work involved in a project can be described clearly and consistently and measured accurately.

To do this, it is necessary to do a lot of reading in order to distinguish 'optional' descriptive text and guidance (bearing in mind the status claimed for NRM2) from the actual rules of measurement that must be followed. Without clear rules of measurement that must be followed and, above all, are clear to all those concerned, there is no point in having a standard method of measurement.

A starting point to discover where the rules of measurement are to be found is in Paragraph 2.6: *Composition of a bill of quantities* albeit that a considerable amount of reading is needed to find them:

- 2.6.5: *Measured work* – The rules are given in Paragraph 2.8.
- 2.6.6: *Risks* – The rules are given in Paragraph 2.10.
- 2.6.7: *Provisional sums* – The rules are given in Paragraph 2.9.1.
- 2.6.8: *Credits* (for materials arising from the works) – The rules are given in Paragraph 2.12.
- 2.6.9: *Dayworks (Provisional)* – The rules are given in Paragraph 2.13.3.

Inexplicably, there is no reference to 'rules' in Paragraph 2.6.4: *Preliminaries*, and precious little of the content of this paragraph even approaches the status of a rule. Paragraph 2.6.4.3 does say that *the quantification of preliminaries is dealt with in paragraph 2.7*, but this is hardly a rule. Even the lengthy Paragraph 2.7: *Preliminaries* contains only two items that vaguely resemble rules.

However, the penultimate paragraph of Paragraph 2.7.3.2 imposes the following duties on the quantity surveyor/cost manager:

- Instruct the main contractor to return a full and detailed breakdown of the total price for preliminaries with the tender.
- Request that this information is appended to the priced BQ.
- Ensure that the supporting calculations are presented in an easy-to-read and logical format.
- Instruct the main contractor to ascertain the price for preliminaries in accordance with the rules of measurement for main contractor's preliminaries – Part B (pricing schedule) of Table 1 (preliminaries (main contract) at Part 3: *Tabulated rules for measurement* of these rules).

- <u>Make it clear</u> to the main contractor *that costs relating to items that are not specifically identified in* [the] *full and detailed breakdown will be deemed to have no cost implications or have been included elsewhere within (the) rates and prices.*

As far as Paragraph 2.7.4: *Preliminaries* (works package contract) is concerned, Paragraph 2.7.4.2 emphasises that *it is essential* for the work package contractor to be *instructed to provide a full and detailed breakdown* that clearly shows how the *price for each* [preliminaries] *item has been calculated* and *how the total price for preliminaries has been arrived at.* The paragraph stops short of imposing this duty on the quantity surveyor/cost manager.

Consequently, the Part 2 obligatory rules of measurement (or reference thereto) are to be found in:

- 2.7 Preliminaries.
- 2.8 Measurement rules for building works.
- 2.9 Non-measurable works.
  - 2.9.2 Contractor-designed works.
  - 2.9.4 Works to be carried out by statutory undertakers.
- 2.10 Risks.
  - 2.10.1 Risks generally.
  - 2.10.2 Risk transfer to the contractor.
  - 2.10.3 Risk sharing by both employer and contractor.
- 2.11 Overheads and profit.
- 2.13 Other considerations.
  - 2.13.1 Price fluctuations.
  - 2.13.2 Director's adjustment.
  - 2.13.3 Dayworks (provisional).
- 2.14 Information requirements for measurement.
  - 2.14.3 Specification.
  - 2.14.4 Drawn information.
  - 2.14.5 Schedules.
  - 2.14.6 Reports and other information.

To some, the foregoing observations may appear 'picky' albeit that they are genuinely meant to be constructive. Had Part 2 not been entitled 'Rules for detailed measurement of building works', there would have been no room for criticism because much of the text provides sensible and logical guidance and information. However, Part 2 does have 'Rules' in its title, and it is a shame that these rules have not been more clearly identified.

Consequently, there is little point concentrating on the remainder of Part 2 here as this would simply be a reiteration of the contents of NRM2 Part 2 save to say that Paragraphs 2.7 and 2.8 are worth reading because they explain good practice in the composition of the various components of a BQ.

What follows, therefore, is purely concerned with issues arising from the <u>rules</u> of measurement as stated under Paragraphs 2.7–2.14 inclusive. The one exception is Paragraph 2.4: *Types of bills of quantities* which is material to both Paragraph 1.6.3: *Definitions* and Paragraph 3.3.8: *Procedure where quantity of work cannot be accurately determined* despite containing no actual rules.

6.7.2   Purpose of bills of quantities

Paragraph 2.2.1 explains that the primary purposes of BQ are:

- As a tender document.
- As a contract document which is used for:
  - The valuation of work carried out so that interim payments can be made to the contractor.
  - The valuation of varied work.

It further explains that a BQ is a coordinated list of items with identifying descriptions and quantities. There are no 'rules' in this paragraph.

Whilst BQ (and the like) have a number of other important purposes, NRM2 classifies these as 'benefits'.

6.7.3   Benefits of BQ

The benefits accruing from BQ are explained in Paragraph 2.3.1 which also points out the need for one party or the other to quantify the extent of the works at some stage irrespective of the chosen procurement strategy. This could be:

- The employer's quantity surveyor/cost manager.
- The main contractor.
- Work package contractors.

Again, there are no 'rules' in this paragraph.

6.7.4   Types of BQ

Paragraph 2.4 distinguishes between two types of BQ:

- Firm BQ (Paragraph 2.4.2).
- Approximate BQ (Paragraph 2.4.3).

Firm BQ

Paragraph 2.4.2.1 explains that the reliability of tender prices *will increase in relation to the accuracy of the quantities provided*. The paragraph also explains that the final cost of a project will be equal to the tender price should there be no design changes, but that such changes will happen in practice, and therefore, the BQ will provide a good basis for the cost control of the project.

Following on from this, Paragraph 2.4.2.2 says that *the firmer the bill of quantities the better it will be a means of financial control*. This tends to beg the question as to whether there may be degrees of firmness in firm BQ.

Risk Issue

Nowhere under Paragraph 2.4.2 is the issue of provisional quantities mentioned. There is no suggestion that firm bills of quantities should be anything other than 'firm' albeit that there may be a question relating to the degree of firmness from BQ to BQ.

Rule 10.1 of SMM7 requires that quantities that cannot be accurately determined <u>shall be</u> identified as an approximate quantity and that an estimate of the quantity <u>shall be</u> given.

In NRM2, we have to look to Paragraph 3.3.8: *Procedure where quantity of work cannot be accurately determined* to discover an equivalent provision.

Approximate BQ

Paragraph 2.4.3.1 explains that approximate BQ are used in circumstances where there is *insufficient detail* to enable a firm BQ to be prepared or where the time and cost of doing so are not warranted. The paragraph further explains that approximate BQ lead to contracts that are subject to remeasurement on completion and warns of the greater variability of outturn cost compared to a firm BQ.

Paragraph 2.4.3.3 stresses that, whilst the quantities in an approximate BQ may be approximate, *the descriptions of work items should be correct.*

Paragraph 3.3.8

Paragraph 3.3.8 of NRM2 relates to the *procedure where the quantity of work cannot be accurately determined.*

It is not at all clear whether this may be viewed as a reference to both provisional quantities in the 'firm BQ' and the 'approximate' quantities in the 'approximate BQ' described in Paragraph 2.4 of NRM2.

Paragraph 3.3.8.1 states that, where the quantity of work cannot be accurately determined, *an estimate of the quantity shall be given and identified as a 'provisional quantity'.*

The quantities in an 'approximate BQ' cannot, by definition, be accurate, and thus, Paragraph 3.3.8.1 could be interpreted as referring to both 'firm' and 'approximate' BQ. If this is the case, then the provisions of Paragraph 3.3.8.2 may equally apply to 'approximate' BQ; this states that the *'approximate quantity' shall be substituted by the 'firm quantity' measured* for work items identified as a 'provisional quantity', once the work has been remeasured.

Paragraph 3.3.8.2 also requires that differences between the *provisional quantity* and the *firm quantity* of less than 20% shall not warrant a review of the BQ rate but variances in excess of this will.

However, NRM2 Paragraph 3.3.8.2 is unclear as to whether or not the '20% rule' shall also apply to the review of the tender rates in approximate BQ as well as those contained in firm BQ.

Risk Issue

The NRM2 20% rule is at odds with both the valuation rules for variations in JCT contracts and the compensation event clause in NEC3 ECC Option B Clause 60.4.

JCT 2011 SBC/Q Clause 5.1.1, for instance, defines changes in quantity as a variation, and, as such, they are subject to the hierarchy of valuation rules in the contract, that is:

- BQ rates if appropriate.
- Based on BQ rates if not.
- Fair rates.
- Daywork.

It is for the contract administrator, in discussion with the contractor, who decides what is fair and reasonable.

In the ECC, there is a clear rule determining whether or not a change in quantity is a compensation event:

- The change in quantity must not be as a result of a change to the works information (i.e. not a variation).
- There must be a change in the defined cost per unit quantity (i.e. the contractor must show that unit costs changed as a result of the quantity change) and
- The final quantity × BQ rate must be more than 0.5% of the total of the prices at the contract date (i.e. the tender total).

Risk Issue

Without some amendment to the contract, the NRM2 '20% rule' would be ineffective.

6.7.5   Preparation of BQ

Paragraph 2.5 explains the stages in a project at which BQ are prepared:

- Using the RIBA Plan of Work process model, NRM2 states that BQ will be prepared at Work Stage G (Tender Documentation). Since publication of NRM2, the new RIBA Plan of Work 2013 has been rolled out and thus, whilst BQ preparation will depend upon the chosen procurement method, it is likely to take place either at Stage 3: *Developed design* or Stage 4: *Technical design.*
- If the OGC Gateway process model is used, NRM2 states that BQ production will be *an intrinsic part of Gateway 3C (Investment Decision).*

Whatever the case, Paragraph 2.5.2 emphasises that BQ production is reliant on the availability of technical designs and specifications in sufficient detail, and, to this end, Paragraph 2.5.4 refers to the information requirements for measurement listed in Paragraph 2.14 which comprises:

- Specification.
- Drawn information.
- Schedules.
- Reports and other information.

Whilst there are no 'rules' in Paragraph 2.5, a useful explanation as to who is likely to quantify the building works for a variety of procurement methods is given at Paragraph 2.5.5 and Figure 2.1 of NRM2.

6.7.6   Composition of BQ

Paragraph 2.6, whilst containing no 'rules', is informative because it identifies and explains the constituent parts of a BQ:

- Form of tender.
- Summary/main summary.
- Preliminaries.
- Measured work.
- Risks.
- Provisional sums.
- Credits.
- Dayworks.
- Annexes.

Paragraph 2.6.2.1 explains that the form of tender may be a separate document which would be necessary where a priced BQ is not submitted at tender stage.

Paragraph 2.6.3.1 provides a comprehensive list of items that should make up a final summary, but, strangely for a method of measurement based on work sections, the 'measured works' given in the example are shown in elemental BQ format (presumably for brevity).

Part 2

In this regard, Paragraph 2.6.5.1 draws attention to the main part of the BQ (the measured works) and states that *the quantities and descriptions of items should be determined in accordance with the tabulated rules in measurement in Part 3*. The term 'should be' is somewhat less emphatic than the 'shall be' found in most other methods of measurement and implies that the bill compiler has some choice in the matter. This could lead to problems.

Risk Issue

The bill of quantities is normally a contract document, and the method of measurement used for its preparation is stated in the form of contract (e.g. JCT 2011 Clauses 1.1 and 2.13.1).

Where NRM2 is stated as being the 'Measurement Rules' in the contract, the parties are entitled to expect that the bill of quantities has been prepared in accordance with these rules in order to give certainty to the contract.

It may be the case, provided the item description is clear and unambiguous, that the legal axiom that the particular (i.e. the item description) should override the general (i.e. the method of measurement) may prevail. Alternatively, it may be that a departure from the method of measurement could be construed as a discrepancy between documents (e.g. between the BQ and the drawings) meriting an appropriate architect's instruction.

Depending upon the seriousness of the departure from the measurement rules, non-compliant item descriptions and quantities could be construed as a misrepresentation that could lead to a claim for damages.

In any event, contractors will always find ways to start an argument, and so it is inadvisable to gift them an opportunity by not following the measurement rules closely.

Paragraph 2.6.5.2 directs the reader to Paragraph 2.8 where we are informed that *the rules relating to the quantification and description of measured work are given*. Paragraph 2.8 contains no rules of any sort, however, and merely points the way to Part 3: *Tabulated rules of measurement for building works*. The point of Paragraph 2.8 is a mystery known only to the authors of NRM2!

For information on how item descriptions shall be presented in BQ and how quantities shall be determined, refer to NRM2 Paragraph 3.3 and to Section 6.9.3 of this chapter.

6.7.7   Preliminaries

Paragraph 2.7 represents the starting point for understanding how 'preliminaries' are provided for in NRM2, and this needs to be read in conjunction with Part 3: *Tabulated Work Sections – Preliminaries*.

Paragraph 2.7.1 firstly explains that preliminaries for the most part represent *the cost of administering a project* and the provision of *items that are not included in the rates for measured work*.

Paragraph 2.7.2 then identifies two types of preliminaries:

1. Preliminaries (main contract).
2. Preliminaries (work package contract).

The content of main contract preliminaries is detailed in Paragraphs 2.7.3.1 and 2.7.3.2, and the content of work package contract preliminaries is identified in Paragraphs 2.7.4.1 and 2.7.4.2. In both cases, preliminaries consist of:

- *Information.*
- *Pricing schedule.*

There are no 'rules' in Paragraph 2.7 except, perhaps, for:

- Paragraph 2.7.3.2: *The preliminaries bill is therefore to include a pricing schedule....*
- Paragraph 2.7.4.2: *The preliminaries bill for a works package shall comprise a pricing schedule....*

There is no explanation in NRM2 as to when a BQ shall include a preliminaries (main contract) bill and when a preliminaries (work package contract) bill shall be included.

It can only be presumed that:

- For **traditional** procurement:
  - There will only be a preliminaries (main contract) bill irrespective of whether the BQ is structured in work sections or elementally.
- For **management contracting** procurement:
  - The management contractor will be appointed on the basis of a preliminaries (main contract) bill and management fee.
  - The measured work will be based on work packages, and therefore, a preliminaries (work package contract) bill will precede the measured work section for each of the work packages.
- For **construction management** procurement:
  - The construction manager will be appointed on the basis of an agreed fee.
  - Each work package will have a preliminaries (work package contract) bill preceding the measured items.
  - The site infrastructure preliminaries will be provided by:
    - The employer or
    - The construction manager on the basis of a priced preliminaries (main contract) bill or
    - One of the work package contractors, who will be asked to price both a preliminaries (main contract) bill and a preliminaries (work package contract) bill.
- For **framework** procurement:
  - The Tier 1 contractor will price a preliminaries (main contract) bill.
  - The Tier 2/3 contractors will price a preliminaries (work package contract) bill.

The pricing schedule for preliminaries (main contract) is in two parts:

- Employer's requirements comprising items listed and defined in Part A of Table 1 of Part 3: *Tabulated rules of measurement.*
- Contractor's main cost items comprising items listed and defined in Part B of Table 1 of Part 3: *Tabulated rules of measurement.*

The pricing schedule for Preliminaries (work package contract) is similarly subdivided.

For the contractor's main cost items, the quantity surveyor/cost manager is required by Paragraph 2.7.3.2 to:

- *Obtain a full and detailed breakdown* of the preliminaries that:
  - *clearly identifies the items,*
  - *shows how the price for each item has been calculated and*
  - *how the total price for preliminaries has been calculated.*

In order to do this, the quantity surveyor/cost manager is required to:

1. *instruct the main contractor to return along with his or her tender a full and detailed breakdown* of the main contractor's *total price* for preliminaries in order to show how this has been calculated.
2. *request that the main contractor append this information to his* [but not her!] *priced bill of quantities.*
3. *ensure that the main contractor's detailed supporting calculations* are presented *in an easy-to-read and logical format by*

4. _further instructing_ the main contractor to _ascertain the price for preliminaries_ according to Part B of Table 1 of Part 3: _Tabulated rules of measurement._
5. _make it clear_ to the main contractor that:
   a) _costs relating to items that are not specifically identified by the main contractor_ in the detailed breakdown _will be deemed to have no cost implications_ or
   b) _have been included elsewhere_ in the rates and prices.

The quantity surveyor/cost manager is to achieve requirements 1–4 _as part of the conditions of tender_ and requirement 5 _in the preliminaries bill and/or preliminaries pricing schedule._

Risk Issue

The requirement for a full and detailed breakdown refers to the contractor's total price for preliminaries, but it would be misleading to assume that this means the total of the preliminaries bill.

On careful reading of the penultimate three paragraphs of Paragraph 2.7.3.2, it is clear that the quantity surveyor/cost manager's duty is to require a much more detailed breakdown of the contractor's preliminaries than a series of lump sums attached to the relevant preliminaries items listed in Part B of Table 1 of Part 3: _Tabulated rules of measurement._

Why? Because:

a) If the 'Part B' list included in the preliminaries (main contract) bill was a sufficiently full and detailed breakdown, then why ask the contractor to append the breakdown to his priced bill of quantities? It would already be in the bill of quantities
b) The quantity surveyor/cost manager's duty is to:
   ▪ Obtain a full and detailed breakdown of the preliminaries that:
     ○ clearly identifies the items,
     ○ shows how the price for each item has been calculated and
     ○ how the total price for preliminaries has been calculated.

It would seem, therefore, that the intention of Paragraph 2.7.3.2 may be to obtain a breakdown of each of the lump sums priced by the contractor in the preliminaries (main contract) bill and that it is the quantity surveyor/cost manager's duty to obtain this information.

For preliminaries (work package contract), Paragraph 2.7.4.2 explains that the 'information' and 'pricing schedule' are also to be drafted in accordance with Parts A and B of Table 1 of Part 3: _Tabulated rules of measurement_, respectively. Additionally, the final paragraph of Paragraph 2.7.4.2 says that:

▪ _It is essential that the work package contractor is instructed to provide a full and detailed breakdown_ of the preliminaries that:
  ○ _clearly identifies the items,_
  ○ _shows how the price_ for each item _has been calculated and_
  ○ _how_ the total price _for preliminaries has been calculated._

Risk Issue

In the case of preliminaries (work package contract), the duty to obtain _a full and detailed breakdown_ falls on no one.

Work package tenders will normally be invited by the main contractor, and it could, therefore, be his duty to obtain the necessary pricing information. On the other hand, the quantity surveyor/cost manager will normally be responsible for vetting work package tenders, and so it could equally be argued to be his/her duty.

Another agenda item for the pre-contract meeting!

6.7.8   Measurement rules for building works

Paragraph 2.8 is brief. It merely says that the rules for measuring building work *are set out in the tabulated rules of measurement...at Part 3...of these rules.*

This paragraph is more of a pointer than a rule and indicates where to find the detailed rules of measurement which apply to the quantification of measured items of work.

6.7.9   Non-measurable works

Non-measurable works consist of:

- Provisional sums.
- Contractor-designed works.
- Risks.
- Works to be carried out by statutory undertakers.

Provisional Sums

Empirical evidence suggests that the vast majority of construction projects are not fully designed before tenders are invited. This may be the result of a conscious decision to choose a particular procurement strategy or, more normally, may be a function of the desire to start work on-site as early as possible for commercial, practical or other reasons. In any event, unforeseen work, or work envisaged but not designed, may need to be catered for in the contract documentation.

The usual way to do this is to include provisional sums in the BQ (or other pricing document) which serve the purpose of:

- Making a provision in the tender price or tender total.
- Alerting tenderers to the likelihood of additional work during the contract period.
- Giving the contract administrator a sum of money to expend pursuant to the particular terms of the contract which provides the authority to issue instructions as to how the money is to be spent.

NRM2 provides for this eventuality by providing a set of rules which explain the meaning of the phrase 'provisional sum' and by stipulating how such sums shall be defined in the contract bills. This idea was first introduced in SMM7 in order to add certainty to tender submissions, to reduce the disruptive effect of unforeseen work during the contract and to avoid costly disputes later on when the claims start flowing. The NRM2 rules concerning provisional sums are to be found in:

Paragraph 2.9.1.1	*Where building components/items cannot be measured and described in accordance with the tabulated rules of measurement <u>they shall be given</u> as a 'provisional sum' and identified as either 'defined work' or 'undefined work' as appropriate.*
Paragraph 2.9.1.2	*for defined work...the following <u>shall be</u> provided:* ■ *the nature and construction of the work* ■ *a statement of how and where the work is fixed to the building and what other work is to be fixed thereto* ■ *a quantity or quantities which indicate the scope and extent of the work* ■ *any specific limitations identified.*
Paragraph 2.9.1.4	*Where <u>any aspect</u> of the information required by Paragraph 2.9.1.2 cannot be given, work <u>shall be</u> described as an 'undefined' provisional sum....*

| Paragraph 2.9.1.5 | Where a provisional sum *does not comprise the information required under 2.9.1.2* it *shall be* construed as a provisional sum for undefined work irrespective that it was described as defined work in the BQ. |
| Paragraph 2.9.1.6 | *Provisional sums shall be exclusive of overheads and profit. Separate provision is to be made in the BQ for overheads and profit.* |

The impact of these rules is to be found in:

| Paragraph 2.9.1.3 | Where the contractor *shall be deemed to have made due allowance* in the programming and planning (of the works) and in the pricing of the preliminaries items for provisional sums for 'defined work'. |
| Paragraph 2.9.1.4 | Where the contractor *will be deemed not to have made* such allowances in respect of provisional sums for 'undefined work'. |

This means that 'defined work' is to be included in the contractor's master programme and sequenced with the other items of work that have been measured according to the *Tabulated rules of measurement for building works*.

Risk Issue

Under the JCT SBC/Q 2011 contract, the expenditure of provisional sums for defined work does not rank as:

- A relevant event for an extension of time (Clause 2.29.2.1).
- A relevant matter for loss and expense (Clause 4.2.4.2.1).

Notwithstanding the above, where a provisional sum is described in the BQ as being for 'defined work', but the information required by Paragraph 2.9.1.2 is not provided, the item *shall be construed as a provisional sum for 'undefined work'*.

As far as the valuation of the work carried out under a provisional sum is concerned, normal practice is that the quantity surveyor will value the work according to the valuation rules in the contract (e.g. JCT SBC/Q 2011 Clause 5.2.1). It is also normal practice to use applicable BQ rates as the basis for such valuation, or as the basis for establishing 'fair rates', pursuant to the Valuation Rules (JCT SBC/Q 2011 5.6 1).

However, it should be carefully noted that Paragraph 2.9.1.6 states that provisional sums *shall be exclusive of overheads and profit* and that *separate provision* for overheads and profit on provisional sums *is to be made* in the BQ. This is an unusual feature of NRM2 as provisional sums are normally valued using BQ rates which traditionally include for overheads and profit.

Paragraph 2.9.1.6 also makes reference to Paragraph 2.11 which states provision *shall be made* in the BQ for the contractor to price overheads and profit on:

- Defined provisional sums.
- Undefined provisional sums.

Whilst this provision is a departure from normally accepted practice in the industry, Paragraphs 2.9.1.6 and 2.11 are quite clear that provisional sums exclude overheads and profit. Paragraph 2.11 is also quite clear that overheads and profit may be treated as separate items when required.

This provision should be read in conjunction with the definition of 'overheads and profit' in Paragraph 1.6.3 and also Paragraph 2.6.3.1 which provides an example of a BQ summary (for an elemental BQ); this clearly shows provision of a separate item for overheads and profit.

Part 2

Notably, however, overheads and profit is not mentioned in Paragraph 2.6: *Composition of a bill of quantities* nor do they warrant an explanatory paragraph in Paragraph 2.6.3: *Summary (or main summary)*.

Appendices D and E do, however, provide templates for the pricing summary in elemental BQ (condensed and expanded versions) where the main contractor's overheads and profit is given as a separate item to be priced as a percentage. Notably, the percentage is to be applied *after* the provision for provisional sums and risks has been added to the total of building works and main contractor's preliminaries. Consequently, this is contrary to Paragraph 2.9.1.6 which states that *separate provision* for overheads and profit on provisional sums *is to be made* in the BQ.

As far as BQ based on either work section or work package layout formats are concerned, there is no template provided. However, Appendix A does give guidance on the preparation of BQ for a variety of breakdown structures for BQ albeit that no mention is made of the main contractor's overheads and profit. Paragraph A.12: *Price summary* of Appendix A does, nonetheless, suggest that the *structure of pricing summaries* for BQ other than those in elemental format *should follow the same principles* as those provided in Appendices D and E.

A further template is provided at Appendix F which includes a *schedule of provisional sums*, but no mention is made of main contractor's overheads and profit. This leaves the BQ compiler with the dilemma of deciding whether to provide an 'overheads and profit' item with the list of provisional sums or to include it in the main summary.

Added to this, a further issue arises in that Paragraph 3.3.3.13 states that, *unless specifically otherwise stated in the BQ or in these rules* – presumably the Part 3 rules – *each building component/item shall be deemed to include*:

> (7)   *establishment charges*

The term 'establishment charges' is normally taken to be a synonym for 'overheads', but the term 'establishment charges, overheads and profit' is also used. If 'establishment charges' are taken to mean the cost of running the head office establishment, then 'overheads' must refer to the other indirect costs of running a contracting business. It is, therefore, less than clear as to the meaning of Paragraph 3.3.3.13(7) or whether the BQ rates for components/items include overheads or not. One thing is clear – rates do not include for profit.

The lack of clarity as to how provisional items are to be billed leaves BQ compilers and contractors with a number of issues to resolve.

Part 2

Risk Issue

As provisional sums are clearly not components/items within the meaning of NRM2 Part 3, care must be taken both when compiling bills of quantities, and when pricing provisional sums, at tender stage and also when valuing defined or undefined work arising from the expenditure of provisional sums during the contract:

- To ensure that overheads are not priced into measured items AND provisional sums as this will inflate the tender price.
- To ensure that work carried out under defined or undefined provisional sums is not overvalued by including overheads twice in the calculations.

BQ compilers, therefore, have a number of choices:

- Perhaps the simplest choice is include an item for overheads and profit in the main summary but insert it <u>before</u> the provisional sums total and then qualify the BQ to the extent that Paragraph 3.3.3.13 (measured items) is deemed to <u>exclude</u> establishment charges; rates and prices in the BQ will then be priced net, and a separate item for overheads and profit on provisional sums can then be inserted in the provisional sums section of the BQ in accordance with Paragraph 2.9.1.6.

- Include an item for overheads and profit in the main summary and trust to luck that the main contractor will exclude overheads (and profit) on measured items (i.e. the BQ rates and prices will be net).
- Include an item for overheads and profit in the main summary and qualify the BQ to the extent that Paragraph 3.3.3.13 (measured items) is deemed to <u>exclude</u> establishment charges; rates and prices in the BQ will then be priced net.
- Omit the overheads and profit item from the main summary, provide an item for overheads and profit in the 'preliminaries' section, make sure that provisional sums have separate items for overheads and profit (in accordance with Paragraph 2.9.1.6) and then qualify the BQ to the extent that Paragraph 3.3.3.13 (measured items) is deemed to include establishment charges *and* profit; rates and prices in the BQ will then be priced gross.
- Include an item for overheads and profit in the main summary and leave it to chance that the contractor will do all the necessary calculations to arrive at a competitive price.

Contractors will, no doubt, price the BQ exactly how they wish to, but they also have choices:

- Price overheads (and profit) with the defined/undefined provisional sums, if an item (or items) is (are) provided for this purpose, price the measured work items and preliminaries gross leaving the main summary overheads and profit item marked 'included' (rates and prices in the BQ will therefore be gross).
- If there is no overheads and profit item with the defined/undefined provisional sums, price the main summary overheads and profit item as a percentage and price the measured work items and preliminaries net (rates and prices in the BQ will therefore be net).
- Price the main summary overheads and profit item as a percentage and exclude overheads from the measured work items (rates and prices in the BQ will therefore be net).

Contractors will need to choose whether to:

- Reveal their overheads and profit percentage by pricing the main summary overheads and profit item as a percentage.
- Price the measured work items and preliminaries gross and mark the main summary item as 'included'.
- Allocate their overheads and profit to selected BQ items/preliminaries in order to maximise cash flow and/or commercial opportunity.
- If provisional sums are billed with a separate item for overheads and profit, price the required percentage for overheads and profit and then either price the main summary overheads and profit item, or mark it included, and price the measured work items and preliminaries gross.

Contractor-Designed Works

Partial contractor design is a 'halfway house' between a traditional lump sum and measure and value contract, with architect/engineer design, and full contractor design and build. It is a popular method for procuring a specialist design for a part or parts of a project, and there are a number of advantages for the employer with this approach:

- The overall design integrity is kept under the control of an employer-engaged architect or engineer.
- A competitive tender can be obtained which fulfils the employer's requirements.
- The choice of the successful tenderer can be informed by the various responses to the employer's contractor-design requirements along with the price/time/quality bid.
- The design risk for such part(s) is shifted to the contractor who, whether subcontracting the work or not, becomes responsible to the employer for the sufficiency of the design as well as the construction of that part of the project.

- The employer obtains a specialist design (e.g. a structural frame or roof) without the complexities of nomination; nomination is not now available under the JCT SBC/Q 2011 contract.
- A detailed price breakdown can be obtained by careful structuring of the tender BQ which is contained in the contractors' responses.

A number of standard contracts facilitate this method of procuring the design including JCT SBC/Q 2011 and the ICC – Measurement Version. However, whilst NRM2 is not the only method of measurement that specifically provides for partial contractor design of the works (MMHW does as well), the authors are to be applauded for this inclusion, despite some unfortunate shortcomings.

Whilst NRM2 is not contract specific, it is pretty clear that the JCT SBC/Q 2011 contract was in mind when the rules of measurement were drafted. There is clear reference to the phrase 'contractor designed portion' (CDP) in NRM2 albeit that JCT SBC/Q 2011 refers to contractor's designed portion. Whether this is intentional or a drafting or proofreading error is not clear.

NRM2 Paragraph 2.9.2 deals specifically where the contractor (or, more likely, a subcontractor) is to design part of the works:

Paragraph 2.9.2.1	Where any work is *not clearly identified as contractor designed works*, the employer *shall be deemed responsible* for such works.
Paragraph 2.9.2.3	Where discrete parts of a building are to be designed by the contractor, *the work items shall be identified as 'contractor designed works'*.
Paragraph 2.9.2.5	Should contractor-designed works be capable of being *measured and described in accordance with the tabulated rules of measurement*, a preamble is to be given such that performance objectives or criteria are *clearly defined*.
Paragraph 2.9.2.6	Where contractor-designed works are a *complete element or works package*, … the works *are to be* measured and described as *one or more item*. The number of items is *at the discretion of* the quantity surveyor/cost manager but *must be* sufficient to provide an analysis of the price....
	The rule emphasises that it is essential that the quantity surveyor/cost manager obtains a detailed breakdown of the contractor's price.
Paragraph 2.9.2.7	This rule states that *the quantity surveyor/cost manager is to obtain* details of *performance objectives and/or criteria from the relevant design consultant*.
Paragraph 2.9.2.8	Contractor-designed works *shall be deemed to include*: - all costs included in Rule 3.3.3.13 and - *all costs in connection with* the design and design management - design and construction risks - due allowance in the programme and planning *for all design works*.

In any event, the NRM2 rules dealing with such works need to be read in conjunction with the terms of the specific form of contract to be used for the procurement of the project; for the purposes of this section, the JCT SBC/Q 2011 contract will be referred to.

JCT SBC/Q 2011 is a lump sum contract with the option to have the contractor, or his preferred subcontractor, design a specific part or parts of the work if desired. This is achieved via the traditional procurement process where, after successfully tendering for the job, a contract will be entered into by the parties, that is, the employer and the contractor.

Partial contractor design is provided for in various places within the JCT SBC/Q 2011 contract as indicated below:

- Articles of Agreement:
  - Recitals.
  - Articles.
  - Contract Particulars Part 1.
- Conditions:
  - Sections 1–9 (of 9).
  - Schedules 1 and 5 (of 7).

NRM2 Paragraph 2.9.2.1 states that the employer *shall be deemed responsible* for works that are *not clearly identified as contractor designed works*. What this means is unclear. It may simply mean that the employer is responsible for the design of works that are not to be designed by the contractor, or it could be construed to have a wider meaning. What the difference is between 'identified' and 'clearly identified' is equally unclear as Paragraph 2.9.2.3 says that *where the contractor is required to take responsibility for the design of discrete parts of the building* such work items *shall be identified* (NB: not 'clearly' identified) *as 'contractor designed works'*.

As far as the measurement of contractor-designed work, and the provision of a means of pricing such work, is concerned, Paragraph 2.9.2.4 states that *the method of quantifying contractor designed work* depends upon *the nature of the work*. This statement is not especially helpful, and the reader is left to discover for himself/herself the two methods of quantification tucked away in Paragraphs 2.9.2.5 and 2.9.2.6:

- NRM2 Paragraph 2.9.2.5 deals with work that can be measured according to the tabulated rules of measurement. Two examples of such work are given – windows and precast concrete components – the list is no more exhaustive than that. In addition to the detailed quantities, a preamble to the work items is required that states the performance objectives or criteria that are to be met by the contractor; these are *required … to be clearly defined*. Additionally, detailed documents defining the performance objectives or criteria to be met are to be included in an annex to the BQ and *clearly cross-referenced in the preamble*. In short, the required structure is:
  1. A preamble stating the performance objectives or criteria.
  2. Work items measured in detail in accordance with the tabulated rules of measurement.
  3. An annex containing detailed documents defining the performance objectives or criteria.
- Paragraph 2.9.2.6 concerns contractor-designed work when it comprises *a complete element or works package*; the entire electrical and mechanical engineering service for a building is given as an example of this. In this case, the works are to be measured and described as <u>one or more item</u> with the number of items being *at the discretion of the quantity surveyor/cost manager*. In any event, the number of items must be sufficient to provide an analysis of the contractor's price for such work. Where the BQ is in elemental format, the elements will be those defined in *NRM1: Order of cost estimating and cost planning for capital building works*.

  The paragraph further requires that, *irrespective of the structure of the analysis*, it is essential to obtain *a <u>full and detailed breakdown</u>* that clearly shows how the contractor has calculated his price <u>*for each item*</u> in the analysis. This duty falls upon the quantity surveyor/cost manager, but how this can be achieved in one item is left to the imagination! Clearly, it is impossible to have one item <u>and</u> a detailed breakdown, and thus, the choices are either:
  - Take a 'punt' on what the contractor will design and provide a list of suitable items to be priced or
  - Ask tenderers for a detailed breakdown of the lump sum price in a Preamble to the Bill of Quantities.

There is no mention of performance objectives or criteria in this paragraph which is a somewhat startling omission unless the intention is to employ a prescriptive specification.

Risks

Paragraph 2.9.3 simply points to Paragraph 2.10 where the *method for dealing with the employer's residual risks* at the BQ production stage is to be found.

Works to be carried out by statutory undertakers

Paragraph 2.9.4.1	Such works <u>are to be given</u> as a 'provisional sum'.
Paragraph 2.9.4.2	The contractor *is to be deemed to have made due allowance* in his *programming, planning and pricing of preliminaries for all general attendance on statutory undertakers.*

> **Risk Issue**
>
> The term 'general attendance' is not defined in NRM2 Paragraph 1.6.3: *Definitions* or elsewhere.

Paragraph 2.9.4.3	Provisional sums for statutory undertakers' work <u>are to be exclusive of overheads and profit</u> which is to be provided for separately under Rule 2.11.

6.7.10   Risks

Paragraph 2.10 is subdivided into:

- Risks generally.
- Risk transfer to the contractor.
- Risk sharing by both employer and contractor.
- Risk retention by the employer.

Whilst generally comprising dialogue, there are some important rules in this paragraph.

2.10.1	**Risks generally**
Paragraph 2.10.1.2	Where there are remaining risks present at the time that the works are to be quantified, *a risk response* will be needed which <u>will take the form</u> of one or more of the following: • Risk transfer to the contractor. • Risk sharing by both employer and contractor. • Risk retention by the employer.

This paragraph is acknowledgement that NRM2 does not recognise the concept of 'contingency allowance' and that a more considered approach is needed in order to provide for the unexpected.

2.10.2	**Risk transfer to the contractor**
Paragraph 2.10.2.3	Where the contractor is to manage specific risks, they <u>are to be fully described</u> and <u>are to be listed in the BQ</u> under the heading of 'schedule of construction risks'.

Part 2

It should be noted that the risks are _to be fully described_ so that it is clear:

- What risks the contractor is required to manage.
- What services and/or works the employer is paying for.

The distinction between such risks and defined/undefined provisional sums is not clear.

Appendix F contains a template for including risks in the BQ, but no examples of the sort of things that might be covered are given. In view of the 'textbook' style of NRM2, this is somewhat of a disappointment.

Paragraph 2.10.2.4	The contractor _will be deemed to have made due allowance_ for programming, planning and pricing preliminaries _in his risk allowances_.

This paragraph puts 'risks' between undefined and defined provisional sums in the hierarchy of non-measurable work items.

Paragraph 2.10.2.5	Risk allowances _shall be_ exclusive of overheads and profit for which a separate provision _should be made_ in accordance with Rule 2.11.

2.10.3   Risk sharing by both employer and contractor

Paragraph 2.10.3.2	Shared risks _will normally be dealt with using 'provisional quantities'_.

Risk Issue

Where provisional quantities are used, the **pricing risk** is to be taken by the contractor and the **quantity risk** by the employer.

On the face of it, this seems equitable, but careful consideration needs to be paid to the contractor's right to a re-rate should the provisional quantities prove to be inaccurate.

2.10.4   Risk retention by the employer

Paragraph 2.10.4.1 merely states the obvious that the employer (or the project team) may retain and manage certain risks and that these will have been (or should have been) included in the cost plan. Paragraph 2.10.4.3 points out that 'retained risks' are not necessarily controllable.

Interestingly, Paragraph 2.10.4.2 suggests that the employer may like to find out what premium the contractor will charge for resolving a 'retained risk'. This will then give the employer a choice:

- Pass on the risk to the contractor at a price.
- Retain the risk.

If it is decided to pass on the risk, Paragraph 2.10.4.2 requires this to be _dealt with as a risk transfer in accordance with Paragraph 2.10.2_. Risks that the contractor is required to manage are to be:

- Fully described and transparent.
- Listed in the BQ under the heading of 'schedule of construction risks'.

- Deemed to include due allowance for programming, planning and pricing preliminaries in his risk allowances and
- <u>Shall be</u> exclusive of overheads and profit for which separate provision <u>should be</u> made.

6.7.11   Overheads and Profit

Paragraph 2.11.1	Provision <u>shall be made</u> in the BQ for the contractor to apply <u>a percentage</u> for overheads and profit on the following:
	■ Preliminaries.
	■ Measured work, including contractor-designed work.
	■ Risk allowances.
	■ Work resulting from the expenditure of provisional sums:
	  ○ Defined provisional sums.
	  ○ Undefined provisional sums.
	  ○ Works to be undertaken by statutory undertakers.

Paragraph 2.11.2	<u>When required</u>, 'overheads' and 'profit' <u>can be</u> treated as two separate cost items. This provision is unique to NRM2, and contractors will need to be extremely cautious when pricing a tender document containing this provision. The following 'risk issues' should serve to indicate why.

Part 2

Risk Issue

There are no rules within NRM2 stating how this provision will be administered or how the contractor will be paid, particularly regarding variations and the expenditure of provisional sums.

Risk Issue

Bearing in mind the shortcomings of the definition of overheads and profit in Paragraph 1.6.3, there remains a great deal of uncertainty as to the precise meaning of 'overheads' and 'profit' and, therefore, how such shortcomings will play out in terms of the financial administration of the contract.

Risk Issue

Paragraph 3.3.3.13 of NRM2 states that the component parts of BQ items shall include:

- Establishment charges.
- Cost of compliance with legislation, including health and safety legislation and disposal of waste.

Read in conjunction with Paragraph 3.3.3.13, Paragraph 2.11 is unclear because:

- 'Establishment charges' is an acknowledged synonym for 'overheads' or may be taken to mean 'head office overheads' when the phrase 'establishment charges, overheads and profit' is used.
- Compliance with health and safety and waste disposal legislation is normally considered a 'preliminaries' item.

Just how a contractor is supposed to price the unit rates and preliminaries items is therefore less than certain as is the question of how variations will be valued based on BQ rates.

Risk Issue

Many contractors apply a different 'margin' to their BQ rates according to whether the work items in question are the contractor's own work or the work of subcontractors or whether they are preliminaries items.

It is not clear from Paragraph 2.11.1 whether the 'overheads and profit' percentage would be applied individually to <u>each</u> of the totals for preliminaries, measured work, risk allowances and provisional sums or whether the percentage would be applied as a 'global' percentage. Appendices D and E give the impression that a place would be provided in the final summary for the percentage, and therefore, it would be a 'global' percentage.

Risk Issue

Many contractors like to distribute their 'margin' disproportionately throughout the BQ. This is done in order to take advantage of under-/over-measured BQ items or mistakes (commercial opportunity) or to front-load items in order to improve cash flow. Disaggregating overheads and profit from unit rates and other BQ items will mean that the contractor will be unable to do this.

Risk Issue

The reasoning behind disaggregating overheads and profit might work in a partnering context where the overheads and profit allowances could be protected (or 'ring-fenced') in order to encourage proactive value and risk management. Exactly how this might be achieved is not clear.

For a traditional contract, NRM2 is not clear as to whether or not overheads and profit would be subject to remeasurement. On the face of it, the presumption is that if work were to be omitted from the contract by way of a variation, then the contractor's overheads and profit recovery would automatically be less than expected as a result of the application of the percentage to the remaining work.

However, many contractors (and indeed PQSs) work on the basis that variations are valued to reflect loss of overheads and profit. This is probably justified because contractors budget for overheads and profit recovery on contracts and variations of omission can lead to serious under-recovery. In any event, it is sensible to reflect loss of overheads and profit in the valuation of variations of omission as a 'least cost' way of avoiding expensive disputes.

Notwithstanding the foregoing, Paragraph 2.11.1 merely states that *provision shall be made in the bill of quantities for the contractor to apply their percentage addition for 'overheads and profit'*. There is no rule as to where this provision shall be made in the BQ because the pricing summaries given in Appendices D and E are not mandatory. Consequently, there would appear to be nothing to prevent:

- Making a provision for overheads and profit at the end of each relevant section of the BQ, that is, preliminaries, measured work, risk allowances and provisional sums.
- Tenderers from pricing the BQ rates and prices 'gross' and ignoring the BQ provision for overheads and profit.
- Tenderers pricing the rates and prices 'gross' whilst also stating their OH&P percentage in the pricing summary and writing 'included' in the extension column.

6.7.12   Credits

NRM2 Paragraph 2.12.1 acknowledges that the issue of 'credits' only normally arises on projects where there is an element of refurbishment, rehabilitation or demolition of existing buildings or structures. This is not to say that such work is uncommon – quite the contrary – but that the term 'credits' applies where materials, components, equipment and the like are generated for disposal or recycling and the owner/client/employer is content to relinquish ownership to the contractor.

Some projects may generate large quantities of valuable materials, and, in some cases, the entire project may be 'credit driven' due to the intrinsic value of such materials. Whatever the case, the topic of 'credits' should not be passed over lightly as there are several strategies available to the property owner which should be considered when measuring and drafting a BQ:

- Surplus materials become the property of the contractor who may dispose of, recycle or sell the items.
- Surplus materials (or parts thereof) remain the property of the employer, and the contractor is required to reuse them in the works.
- Surplus materials (or parts thereof) remain the property of the employer and are set aside for reuse elsewhere (e.g. on another project).
- Surplus materials are disposed of by the contractor who may then recycle or sell the items, but the employer is credited for their value.

In NRM2, Paragraph 2.12.2 perceives two ways of dealing with 'credit items':

- A list of items is inserted into the BQ, and this is priced by the contractor on the basis of the value that may be attached to the items in question.
- Tenderers may be invited to submit a list of items with their tenders indicating the value of each of the items on the list.

There is no suggestion in NRM2 that a special work section should be created for credit items and, apart from providing a simple template for 'credits' in Appendix F, NRM2 takes the subject no further. Appendices D and E (for elemental BQ) each show a template for a pricing summary with an item called 'credit (for retained arisings)' shown as having a negative value (i.e. a credit), but, otherwise, the BQ compiler is left to decide just how to deal with the alternatives proposed by Paragraph 2.12.2.

Traditionally, demolitions and alterations were included in BQ by providing a work section called 'spot items', 'demolitions and alterations' or 'works to be priced on-site' (where the tenderers walk around the site and make an 'on-the-spot' assessment of cost/value). SMM7 and now NRM2 provide specific sections for such work, but neither Work Section 3: *Demolitions* nor Work Section 4: *Alterations, repairs and conversion* makes any provision for 'credits'. Each of these work sections refers to 'recycling' and to 'retained materials' and to 'materials to remain the property of the employer' but not to 'credits'.

As far as drafting the BQ is concerned, conventional practice is to provide two cash columns in the appropriate work section so that tenderers may price the cost of the work entailed, and any credit value, for specific items of work. This might be a good 'fit' with NRM2 because both the total of the work section and the total of credits could be carried to the pricing summary and separately identified as illustrated in NRM2 Appendices D and E. The other possibility would be to repeat appropriate work section items in a 'credits' section (NRM2 Appendix F refers) which tenderers could then price.

Risk Issue

Note 3 of the first table of Work Section 4: *Alterations, repairs and conversion* states that *all materials arising from these works become the property of the contractor unless otherwise stated*.

There is no similar note in Work Section 3: *Demolitions* which is surprising as items such as structural steelwork, bar reinforcement, copper and the like can have considerable 'scrap' value.

BQ compilers would be well advised to make sure that the issue of 'ownership' of materials arising from the site is well defined somewhere in the contract documents and that suitable 'notwithstanding' caveats are included where any of the NRM2 rules are not to apply.

6.7.13   Other considerations

Paragraph 2.13.1   **Price fluctuations**

This rule distinguishes between 'fixed price contracts' and 'fluctuating price contracts' where, on the one hand, the contractor takes the risk for price increases during the contract and, on the other, the employer undertakes to pay for such increases under some sort of price fluctuation agreement. Paragraph 2.13.1 recognises that most standard forms of contract contain provisions for either alternative by means of including or deleting specific relevant clauses.

Where the conditions of contract do not incorporate price fluctuations provisions, Paragraph 2.13.1.2 (1) states that *separate provision is to be incorporated in the bills of quantities* so that the contractor can *tender his fixed price adjustment*. This *is to be* referred to as the *'main contractor's fixed price adjustment' or the 'work package contractor's fixed price adjustment'* as appropriate.

Risk Issue

The quantity surveyor/cost manager, when preparing the bills of quantities, shall ensure that the conditions of contract do not contain any provision relating to the recovery of price fluctuations.

2.13.2   Director's adjustment

Paragraph 2.13.2.1   This rule requires that *a separate provision is to be incorporated* in the bills *of quantities* so that the contractor can insert a director's adjustment.

Risk Issue

Where the director's adjustment results in a reduction in the tender price, the BQ rates would normally be reduced in proportion to the amount of the reduction and the total value of the BQ items. Where the adjustment is an addition to the tender price, the converse would be the case. Such an adjustment is important to ensure that:

- The contractor is neither underpaid nor overpaid when it comes to interim payments.
- The final account is corrected accordingly when the contract is complete.

There is no such rule in NRM2 dealing with the adjustment of the BQ rates, or other items, to take account of the director's adjustment in which case the director's adjustment presumably would have to be treated as a lump sum adjustment that bears no relationship to the contract sum or the eventual final account figure.

2.13.3   Dayworks (provisional)

Paragraph 2.13.3.2   This rule provides the option that *a schedule of dayworks is to be incorporated* in the bills of quantities but only if required.

Paragraph 2.13.3.3   The method of calculating the labour time charge for work carried out in normal working hours *shall be defined* in the schedule of dayworks and *the definition of normal working hours shall be given* in either the preliminaries bill or the schedule of dayworks.

Risk Issue

The employer may be exposed to risk if a daywork schedule is not included in the bills of quantities.

Most standard conditions of contract provide for varied work to be valued on a daywork basis where it cannot be valued by measurement or revised rates.

It is quite probable that very few contracts are completed without some daywork being carried out, and so it would be prudent to include a daywork schedule to ensure that competitive rates are obtained for such work.

Risk Issue

Paradoxically, as far as subcontracts or work package contracts are concerned, it is not unknown for the main contractor to refuse to accept the concept of daywork, especially where the main contractor is unable to pass on such costs to the employer or, as a contra-charge, to another subcontractor.

In such circumstances, subcontractors would be unwise to enter into a subcontract without an agreed daywork schedule.

Disagreements over daywork are common in subcontracting, and very often the only solution for the subcontractor is to agree a 'horse deal' or take the matter to adjudication.

Paragraph 2.13.3.4   This rule states that *the total amount included for daywork by the contractor shall be omitted* from the contract sum and that the *rates and percentage additions included in the BQ shall be used* to value works *authorised to be valued on a daywork basis*.

This is a contract administration issue for which there is usually a provision in the form of contract.

Paragraph 2.13.3.4   **Note:**
The note to Rule 2.13.3.4 provides guidance such that the monetary total of a schedule of dayworks *can be* included *or* excluded from the contract sum.

When included in the contract sum, the daywork *is to be* treated as a provisional sum, and when excluded, *it shall be* clearly *stated* that the daywork rates/percentages tendered *are included in the contract*.

The wording of Rule 2.13.3.4 is curious.

Presumably, *the total amount included for daywork by the contractor* refers to a provisional sum or provisional number of hours included in the BQ that have been priced by the contractor using the tendered daywork rates.

Seemingly, reference to the valuation of authorised daywork is in the context of interim payments and, more particularly, settlement of the final account. If so, the conditions of contract should also be referred to as there are normally protocols for each written into the contract.

The note to Rule 2.13.3.4 is more instructive although it is again curious that a note should contain rules as to the treatment of daywork in relation to the contract.

Paragraph 2.13.4

Rule 2.13.4 is unequivocal.

VAT *shall be* excluded from the BQ, but, where required by the employer, a VAT assessment *can be* incorporated in the form of tender.

6.7.14   Information requirements for measurement

For a section with 'requirements' in the title, there are few 'requirements' and no rules at all.

Paragraph 2.14.2

This paragraph explains that certain information *will be required* by the quantity surveyor/cost manager when preparing BQ. Four main classes of information are listed:

- Specification.
- Drawings.
- Schedules.
- Reports and other information.

Just who is to supply this information is not defined, but Paragraph 2.14.6 does refer to some employer's requirements and policy documents and to details of *any planning conditions and informatives that the contractor is required to comply* with.

2.14.3   Specification

This lengthy paragraph is largely given over to explaining the two main types of specification that are used – **prescriptive** and **performance** specifications.

Paragraph 2.14.3.2 (1)

Where prescriptive specifications are used and any materials are not named, *reference will be made* to published materials such as British or other country-specific standards.

Paragraph 2.14.3.2 (2)

Where performance specifications are used, *the benefit to the employer is that design will not need to be advanced...before inviting tenders from contractors.*

2.14.4   Drawn information

Paragraph 2.14.4.1

Drawn information *is required* to **describe** the assembly of the building, as well as any temporary works.

Risk Issue

This is a very misleading statement that concerns the complex issues of 'design intent' and 'design liability'.

According to a reference given by Crotty (2012), Pittman (2003) maintains that ambiguity in design is necessary so that the architect may express his/her broad design intent in sufficiently clear terms so as to enable a contractor to construct the building whilst at the same time avoiding giving explicit instructions for how to do so. Pittman (2003) goes on to say that ambiguity is necessary in order to minimise the architect's liability should something go wrong during the construction process.

This view resonates with the general law which provides a clear understanding of the obligations of employers, architects and engineers for the 'buildability' of a design.

For instance, in *Oldschool v Gleeson (Construction) (1976)*,* his Honour Judge Stabb QC stated that *It seems abundantly plain that the duty of care of an architect or of a consulting engineer in no way extends into the area of how the work is carried out.*

Furthermore, the general law position is that an employer under a construction contract does not impliedly warrant the fitness of the site to enable the contractor to complete the work† nor does he warrant the feasibility of the design set out in the contract documents.‡

In *Clayton v Woodman & Sons Ltd [1962]*,§ a personal injury case, Pearson L.J. said that t*he architect is engaged as the agent of the owner* and that *his function is to make sure that…the owner will have a building properly constructed in accordance with the contract* and that the *architect does not undertake to advise the builder…as to how he should carry out his building operations.* In the same case, the judgement also stated that *inter alia, it might be suggested that the fault of the architect was in not advising the builder…as to how the work required by the specification should be executed. If he had done so, the architect would have been stepping out of his own province and into that of the builder.*

---

* *Oldschool v Gleeson (Construction)* (1976) 4 B.L.R. 103 D.C.
† *Appleby v Myers* (1867) L.R. 2 C.P. 651.
‡ *Thorn v London Corporation* (1876) 1 App. Cas. 120.
§ *Clayton v Woodman & Son (Builders)* Ltd [1962] 1 W.L.R. 585 at 593 (CA).

Paragraph 2.14.4.1	*Drawings shall be to a suitable scale.*

Self-evidently but, nonetheless, a rule is warranted.

2.14.5   Schedules

Paragraph 2.14.5.1   Schedules *shall be deemed to be* drawings where they *provide the information required by the tabulated rules.*

This rule is necessary because the first two rows of each work section of Part 3: *Tabulated rules of measurement for building work* concern the drawn information ('drawings') required for measurement purposes.

The list of schedules given is stated as *not definitive or exhaustive, but simply a guide.*

2.14.6	**Reports and other information**

A long list of information that may possibly be required for the preparation of BQ is given. There are no rules save to say that such reports and other information _may_ include *details of planning conditions or informatives that the contractor is required to comply with.*

The list is *not meant to be definitive or exhaustive, but merely a guide* intended to be used by the quantity surveyor/cost manager.

6.7.15   Codification of BQ

NRM2 Paragraph 2.15 is devoted to an explanation of:

- The different types of BQ breakdown structures.
- How to code BQ.

On the face of it, these are relatively straightforward aspects of BQ production, but the six pages of NRM2 that deal with these topics are, to say the least, challenging. So much so that a special section (Section 6.8) has been provided in this chapter to try to unravel and explain how it is all intended to work.

Suffice to say that, if the authors of NRM2 had followed the advice of Albert Einstein, the world would be a much simpler place:

> Most of the fundamental ideas of science are essentially simple, and may, as a rule, be expressed in a language comprehensible to everyone.

6.7.16   Cost management/control

The main purpose of this part of NRM2 is to explain the cost planning and cost control process and the important part that BQ play in it. There are no rules as such, but there are one or two issues that merit consideration.

Paragraph 2.16.3	Post-tender estimate
	The timing and role of the post-tender estimate are explained, but, more particularly, it is emphasised that the quantity surveyor/cost manager _should_ include a summary of the post-tender estimate(s) when reporting the outcome of the tendering process to the employer.
Paragraph 2.16.5	Pricing variations
	This paragraph states that *the rates in a priced bill of quantities provide a basis for the valuation of varied work.*

Risk Issue

Whilst this statement may be to some extent true, it is not entirely correct in a legal context.

It is the **contract conditions** that determine how varied work is to be valued and different contracts do this in different ways.

The provisions for the valuation of variations in the JCT 2011 SBC/Q contrast sharply with the ECC compensation event arrangements, for example.

Risk Issue

As far as **approximate quantities** are concerned, JCT2011SBC/Q states that the rates stated in the contract bills shall be the basis of valuation provided that the quantities stated represent a reasonably accurate forecast. Where this is not the case, a fair allowance shall be made to reflect the difference between the estimated quantity and the actual quantity.

NRM2 Rule 3.3.8.2 provides a mechanism for this 'fair allowance', but it should be noted that NRM2 refers to 'provisional quantities', whereas JCT2011SBC/Q refers to 'approximate quantities'.

NRM2 also states that 'pro rata' and 'analogous' rates can be derived from the priced BQ in order to price components not specifically measured in the BQ. In all practical senses, this can be likened to the 'fair rates and prices' valuation process stipulated in the JCT conditions.

6.7.17   Analysis, collection and storage of cost data

Paragraph 2.17.1 emphasises the value of the *real-time cost data* that priced BQ provide, and 2.17.2 explains that this information can be used in a variety of ways.

This is little more than informative and there are no measurement rules in 2.17, or any other rules for that matter, despite the implication to the contrary in the title of Part 2: *Rules for detailed measurement of building works.*

6.8   Codification of bills of quantities

This section of Chapter 6 is somewhat lengthy for the simple reason that the codification system and classification tables in NRM2 are complicated and poorly explained and, frankly, do not work terribly well.

The easy fix is to allow a measurement software package to do the work.

The problem with this approach is that some software packages do code the items in the take-off, but they do not code the BQ as suggested by NRM2 Paragraph 2.15.3.1. It also has to be said that there are 'glitches' in some of the measurement software packages, perhaps because NRM2 is new and the coding is complex, but, in any event, a competent professional should understand what these packages are doing, and this requires an understanding of the NRM2 system and its frailties.

6.8.1   Work breakdown structure

The basis of both the NRM1 and NRM2 coding systems is the work breakdown structure (WBS), and consequently, it is imperative to understand how a WBS works in order to make the most of the flexibility offered by the New Rules of Measurement.

WBS is defined in NRM2 Paragraph 1.6.3: *Definitions*:

Work breakdown structure (WBS) – *is used to sub-divide a building project into meaningful elements or work packages.*

A WBS is *a tree structure that starts with the end objective* and is then successively sub-divided *into the main components and sub-components that make up the entire building project* (Paragraph 2.15.1.1). This provides a *hierarchical breakdown* of the project similar to a filing system as illustrated in Table 6.1

Table 6.1   Hierarchical structure.

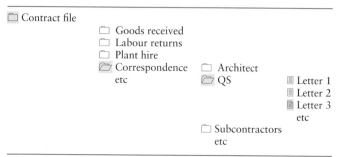

In NRM2, WBS is used in the context of BQ and therefore refers to the way that the BQ is subdivided. The advantage of a WBS is that, appropriately coded, measured items can be sorted into the desired BQ format using suitable software with a multiple sort facility, rather like shuffling a pack of playing cards. Additionally, the priced BQ items can be related back to the original cost plan in order to reconcile tenders received with the intended spend profile for the project.

This also means that a BQ that has been structured in a certain way can be restructured into any other BQ format desired by virtue of the WBS coding system. This facility is not only useful for the quantity surveyor/cost manager but also for the contractor who, furnished with the coded BQ and suitable software, could convert a BQ based on traditional work sections into one based on work packages in order to obtain subcontract prices.

6.8.2   Cost breakdown structure

The usefulness of the WBS becomes evident when the contractor has priced the BQ. The existence of competitive market rates enables the actual cost of elements or work sections or, alternatively, work packages, to be compared with the 'theoretical' cost plan allowances developed during the design stage. This is the point at which a WBS becomes a cost breakdown structure (CBS) which is defined in NRM2 Paragraph 1.6.3:

> **Cost breakdown structure (CBS)** – *is the financial breakdown of a building project into cost targets for elements or work packages.*

A CBS is basically a WBS with money attached to the items in the breakdown structure. Therefore, a cost plan WBS becomes a CBS when it has been priced by the quantity surveyor/cost manager, and a BQ WBS becomes a CBS when the BQ has been priced by the contractor.

6.8.3   BQ Structures

A BQ is essentially an output from the measurement process. This may be carried out by a quantity surveyor/cost manager working for the employer or may be undertaken by a contractor or subcontractor working from drawings provided during the tendering period. Consequently, a BQ could be for an entire project or for a number of projects within an overarching project, or it could be a small part of a project such as a trade (e.g. painting and decorating) or work package (e.g. groundworks).

The starting point for coding any BQ is the codification of the measured items included in the take-off. Lee et al. (2014) warn that coding measured items on paper for later entry into measurement packages is prone to error but that building descriptions from standard libraries, and direct entry of dimensions using software packages, provide a reliable record of the logic used for the quantity take-off.

Once the measured items have been coded, the BQ can be prepared. Software packages that offer standard libraries of descriptions will code measured items automatically as item descriptions are chosen from the library menus, and this will then enable a BQ to be produced based on the chosen method of measurement library. Some software packages allow BQ to be sorted in a variety of ways. In NRM2, there are three different ways that BQ can be structured as shown in Table 6.2.

Table 6.2   Bill of quantities (BQ) structures.

NRM2 BQ structures								
Elemental			**Work Section**			**Work package**		
Ref	**Element**		**Ref**	**Work Section**		**Ref**	**Work package**	
1	Preliminaries (main contract)		1	Preliminaries (main contract)		1	Preliminaries (main contract)	
2	Facilitating works		3	Demolitions		3	Demolition works	
3	Substructure		5	Excavation and filling		4	Groundworks	
4	Superstructure		7	Piling		5	Piling	
5	Internal finishes		11	In situ concrete works		6	Concrete works	
	Etc.			Etc.			Etc.	
NB								
a	This is the NRM1 default structure		a	This is the NRM2 default structure		a	This structure is defined by the user	
b	The reference numbers are taken from NRM2 Appendix A Figure A.1		b	The reference numbers are taken from NRM2 Appendix A Figure A.2		b	The reference numbers are taken from NRM2 Appendix A Figures A.3	
c	Each element may be subdivided into finer levels of detail		c	Each Work Section may be subdivided into finer levels of detail		c	Each work package may be subdivided into finer levels of detail	
						d	Any other breakdown structure and/or reference numbering may be chosen	

According to NRM2 Appendix A.1.3, the elemental breakdown structure *makes it easier for the quantity surveyor/cost manager to analyse a contractor's tender price and collect real-time cost data,* whereas the work sectional breakdown structure *is often preferred by contractors for the purpose of pricing as all like products and components are grouped together.*

The work package breakdown is seemingly *used by contractors to procure packages of work from their supply chain.*

6.8.4   NRM2 Part 3: Tabulated rules of measurement for building works

NRM2 *Part 3: Tabulated rules of measurement for building works* consists of 41 work sections which includes preliminaries (Work Section 1). The structure of each of the 40 measured work sections comprises:

- A main title (numbered).
- In most work sections, there is one or more subheading indicating the categories of work that are included in the work section (no numbering).

- Tables specifying the drawings and other information that must be provided, a list of deemed included items, etc. (no numbering).
- A greyed-out heading under which there is:
  - A column of items to be measured (numbered) followed by
  - Three columns (Levels 1, 2 and 3) of descriptive features (numbered) and
  - A final column consisting of notes, comments and glossary (numbered).

This is illustrated in Table 6.3 which is extracted from Work Section 5: *Excavating and filling*. The list of measured items is effectively a 'spreadsheet' which is read both vertically and horizontally.

Table 6.3 Work Section structure.

5. Excavating and filling
   Site clearance/preparation
   Excavations
   Disposal
   Fillings
   Membranes					

Omitted for clarity					
Omitted for clarity					
Item or work to be measured	**Unit**	**Level 1**	**Level 2**	**Level 3**	**Level 4**
6. Excavation	m³	1. Bulk excavation	1. Not exceeding 2 m deep	1. Details of obstructions in ground to be stated	*Omitted for clarity*
	m³	2. Foundation excavation	2. Over 2 m not exceeding 4 m deep 3. And thereafter in stages of 2 m		

Looking at Table 6.3, it is tempting to think that the NRM2 classification structure (i.e. *Tabulated rules*) is designed to enable each BQ item to be given a discrete reference or code when combined with the relevant descriptive features listed. Such a code would distinguish measured items one from another and facilitate re-sorting should a different BQ layout be required. Thus, any individual measured item could have a maximum of five digits in its unique code.

For example, an item of **excavation for foundations 1.5 m deep** would theoretically create a code of **5.6.2.1**. If the same item was to be measured where there were **boulders or old foundations** present in the ground, the code could be **5.6.2.1.1** using the additional Level 3 code.

However, NRM2 states that:

- *for the quantity surveyor/cost manager to manage the cost plan during the procurement and construction phases of the building project…, the codification framework used for **cost planning** <u>must be used</u> as the basis for the codification of building components/items in the bill of quantities (BQ).*

This is an unusual and confusing feature of NRM2, a consequence of which is that:

- NRM2 requires BQ to be coded using the codes in the NRM1 classification tables and not those of NRM2 <u>but</u> only when there is a cost plan in place and it is desired to relate the priced BQ back to the cost plan <u>or</u> when a BQ is required with an elemental breakdown structure.

The net effect of this is that NRM2 has a third objective as regards BQ coding to add to the two discussed in Chapter 4:

- *Objective 1*
  So that each item can be distinguished from other items in the BQ.
- *Objective 2*
  In order that the distinguishing characteristics of each item may be traced back to their origin, that is, the method of measurement used.
- *Objective 3*
  To ensure that the BQ is coded in such a way as to relate it back to the design stage cost plan.

As a result, where there is a cost plan in place, but a BQ with a work section or work package breakdown structure is required, the BQ must also be coded with NRM1 codes. However, pursuant to NRM2 Paragraphs 2.15.3.1(2) and (3), a distinguishing suffix must be added to the primary (NRM1) code. The idea here is to be able to re-sort the work section or work package items into an elemental format so that the priced tender can be reconciled with the cost plan.

Consequently,

- The work sections and the measured items therein will be those defined in NRM2 Part 3: *Tabulated rules*, but they will have an NRM1 code and a suffix.
- The work packages and the measured items will be those defined by the quantity surveyor/cost manager (or the contractor), but they will have an NRM1 code and a suffix.

In circumstances where there is no cost plan, or it is not desired to reconcile priced tenders with a cost plan:

- An elemental BQ will be coded with NRM1 codes because NRM1 is organised by elements and NRM2 is based on work sections.
- A work section BQ can be coded in any way that the bill compiler desires because there is no system of coding in NRM2 other than that described in Paragraph 2.15.
- A work package BQ can be coded as desired because there is no rule in NRM2 to determine what the coding should be.

It is quite understandable that NRM2 should seek to satisfy Objective 3 as, amongst other benefits, the full functionality of software such as CATO can be used to re-sort BQ and analyse tenders.

What is difficult to appreciate, however, is why NRM2 should be exclusively concerned with Objective 3 when it is clear that contractors, subcontractors and others may also wish to create BQ but without a cost plan being in place. Such users would also need a suitable coding system in order to reference and audit measured items and to restructure BQ from work section to work package format as required.

The codification of BQ under NRM2 is not straightforward therefore, and three issues need to be considered in order to understand the system:

1. The codification of the quantity take-off (i.e. the measured items on the dimension sheets).
2. The practicalities of coding measured items.
3. The codification of the finished BQ in order to achieve Objective 3.

This is easier said than done as NRM2 is most unclear and, to a certain extent, misleading due to mistakes in the worked examples provided in NRM2 Part 2.

6.8.5   Coding the quantity take-off

The obvious place to look for guidance on coding measured items in NRM2 is Paragraph 2.15: *Codification of bill of quantities*, but despite being six pages long, this does not actually explain how to code BQ using the codes in NRM2 *Part 3: Tabulated rules of measurement for*

Part 2

building works, nor does it offer an explanation of how to code measured items in the quantity take-off.

Ostrowski (2013) clearly believes that a quantity take-off should be referenced using NRM2 codes and both CATO and QSPro use the Part 3 classification tables as a hierarchy for building item descriptions. Without criticising Ostrowski, his worked examples clearly demonstrate the weaknesses of the NRM2 classification tables, and both CATO and QSPro adopt different approaches to formulating item descriptions more akin to SMM7.

The fault here lies with the authors of NRM2 because, not only are the classification tables badly organised, there is no explanation of how to code NRM2 measured items in Paragraph 2.15 or anywhere else in NRM2 for that matter. This is left to the user to work out for himself/herself, unlike SMM7 which provides a simple and understandable explanation of how to use the classification tables (in General Rule 2).

Another place to look for guidance is Appendix A: *Guidance on the preparation of bill of quantities*. This explains, *inter alia*, that BQ may have an elemental breakdown structure or, alternatively, a work section or work package structure. It also provides examples of typical BQ formats for each of the three breakdown structures. Looking at Appendix A.4, it becomes immediately obvious that the items shown are not coded in accordance with either NRM1 or NRM2 and that the coding used is a simple cascade or WBS which, when checked against NRM1 and NRM2, has little relevance to either document.

Finally, guidance may be sought from NRM2 Part 3: *Tabulated rules of measurement for building works* and Paragraph 3.2: *Use of tabulated rules of measurement for building works* in particular, but this provides no help either.

Under Paragraph 3.3: *Measurement rules for building works*, there is some limited help in the form of Paragraph 3.3.3.3 which informs us that:

Descriptions <u>shall</u> state:

- *the building components/items being measured (taken from the first column of the tabulated rules) and*
- *include <u>all</u> Level 1, 2 and 3 information (taken from the third, fourth and fifth columns respectively) applicable to that item.*
- *Where applicable, the relevant information from column five <u>shall be</u> included in the description.*

The conclusion to be drawn from Paragraph 3.3.3.3 is that:

- There is, in fact, no requirement to take only one descriptive feature from each column (Levels 1, 2 and 3).
- Item descriptions may be compiled on a 'pick-and-mix' basis, provided that all relevant descriptive features are included.

Consequently, there is seemingly, at best, only a partial but inconsistent NRM2 coding system for quantity take-off or for the production of BQ. In fact, Paragraph 2.15.3.1 makes it clear that BQ should be coded for the primary purpose of relating the priced BQ back to the cost plan by using NRM1 identification numbers as the primary code for an elemental BQ and the addition of suffixes as a secondary code should a work section or work package BQ be preferred.

In practical terms, this means purchasing both NRM1 and NRM2 (total cost circa £90), having both documents open at the same time in order to code the BQ and a great deal of extra time and effort. This is quite bizarre.

Having said all that, there is no reason why the NRM2 code numbering cannot be used in whatever fashion the user wishes, ignoring NRM1 altogether, and the leading QS software providers seem to have taken this approach. However, if it is desired to create measured items with accompanying NRM2 codes, there are further problems in that a number of the work sections in Part 3: *Tabulated rules of measurement for building works* are arranged in a way that prevents the creation of unique codes for certain items of work. This issue is discussed in Section 6.8.6.

6.8.6   NRM2 coding: Practicalities

Coding measured items of work using the NRM2 classification tables is, *prima facie*, straightforward, but, in several work sections, there are instances where it is impossible to assign a unique code to measured items. This arises when more than one of the descriptive features under the Level 1, 2 and 3 columns are needed to describe items fully. The following example will illustrate the point:

> *A 150mm diameter uPVC push-fit pipe is to be laid in a trench with an average depth of 1.25m on a granular bed and surround. The trench backfill is to be MOT Type 1 sub-base. The drain is to be laid in water bearing ground adjacent to an existing building.*

Table 6.4 shows part of NRM2 *Work Section 34: Drainage below ground*, and this indicates that drain runs are measured in linear metres. The item coverage is extensive and includes excavation, disposal, earthwork support, levelling trench bottoms, backfilling, pipes and pipe bedding and so on.

Part 2

Table 6.4   Work Section 34 (part).

34 Drainage below ground				
Storm water drain systems				
Foul drain systems				
Pumped drain systems				
Land drainage				
Omitted for clarity				
Omitted for clarity				
Item or work to be measured	**Unit**	**Level 1**	**Level 2**	**Level 3**	**Level 4**
1. Drain runs	m	1. Average trench depth in 500mm increments 2. Type and nominal diameter of pipe 3. Multiple pipes stating number and nominal diameter of pipes	1. Method of jointing pipes 2. Pipe bedding and or surround details stated 3. Type of backfill if not obtained from the excavations	1. Vertical 2. Curved 3. Below groundwater level 4. Next to existing roadway or path 5. Next to existing building 6. Specified multiple handling details stated 7. Disposal of excavated material where not at the discretion of the contractor: details stated	*Omitted for clarity*

The item to be measured, that is, 'drain runs', can be seen in the table of measured items, and this has the code number 34.1. As far as the addition of further descriptive features is concerned, NRM2 Paragraph 3.3.3.3 stipulates that *all Level 1, 2 and 3 information... applicable* shall be included in the items descriptions.

This is clear enough but Table 6.5 illustrates that, at Level 1, two of the three choices available would be needed to describe the item of drainage, that is, trench depth and pipe diameter. At Level 2, all of the choices are required, and at Level 3, several of the choices are needed to describe the item. This means that a discrete code cannot be assigned to the item at Levels 1, 2 and 3.

Table 6.5 Level 1, 2 and 3 information.

Level 1		**Level 2**		**Level 3**	
1	Average trench depth exceeding 1.00 m not exceeding 1.50 m	1	Push-fit	3	Below groundwater level
2	150 mm diameter uPVC pipe	2	Granular bed and surround	5	Next to existing building
		3	MOT Type 1 sub-base trench backfill	7	Location of disposal point if not at the contractor's discretion

The resulting measured item is demonstrated in Figure 6.1.

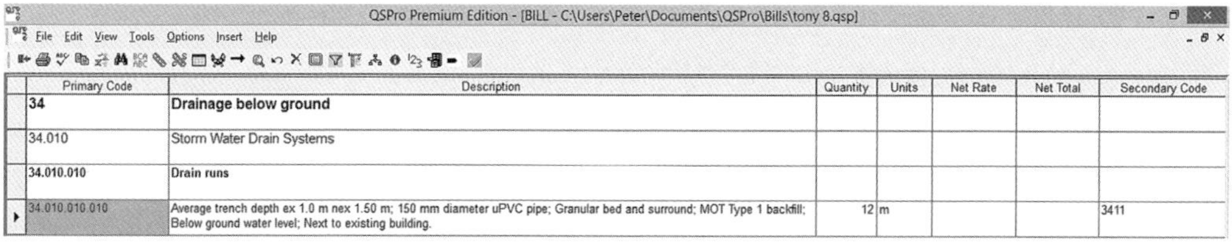

Primary Code	Description	Quantity	Units	Net Rate	Net Total	Secondary Code
34	Drainage below ground					
34.010	Storm Water Drain Systems					
34.010.010	Drain runs					
34.010.010.010	Average trench depth ex 1.0 m nex 1.50 m; 150 mm diameter uPVC pipe; Granular bed and surround; MOT Type 1 backfill; Below ground water level; Next to existing building.	12	m			3411

Figure 6.1 Measured item.

Risk Issue

There is no simple solution to this problem as NRM2 does not have a system for dealing with such situations. There is not even an official 'asterisk (*) system', such as in SMM7, which is used when a measured item description requires all of the items listed in a column to be chosen.

Consequently, it is impossible to give certain measured items a unique code without devising a supplementary coding system. This would seem to defeat the object of NRM2 coding as the BQ compiler would be spending more time coding that taking-off quantities!

Another solution would be to amend the method of measurement in a similar way to CESMM Class I: Pipework – Pipes. This would mean that the Level 1, 2 and 3 descriptors would be:

- Type of pipe.
- Diameter.
- Depth of trench.

Other descriptors would be included in column five (notes, comments and glossary).

This anomaly makes nonsense of the work breakdown structure principle underpinning NRM2.

This problem arises in several work sections including (3) Demolitions, (6) Ground remediation and soil stabilisation, (14) Masonry and (23) Windows, screens and lights as well as (34) Drainage below ground.

6.8.7   Coding the BQ

BQ can potentially be produced at any stage in a construction project by any participant using any breakdown structure, for instance:

- By the employer's quantity surveyor/cost manager at the usual stage in the traditional procurement process.
- By a professional quantity surveyor (PQS) or contractor or both, early in the design process, in order to develop a target cost.
- By a contractor, early in the design process, as part of a two-stage design and build tender.
- For a design and build project in order for the main contractor to obtain quotations from subcontractors.
- By subcontractors or work package contractors when invited to submit tenders on a 'drawings and specification' basis.
- By a main contractor or a subcontractor/work package contractor, at any time during or after the design stage, where partial contractor design is envisaged.

Consequently, it is not necessarily the PQS who produces the BQ – it depends on the procurement method used.

NRM2 Paragraph 2.15 provides the rule set for coding BQ, but it doesn't work in the way that might be imagined.

It may be helpful to understand that Paragraph 2.15.3.1(1) of NRM2 is worded in such a way as to give the impression that the procurement route will be a traditional one or, in other words, that there will be a cost plan prepared by the employer's quantity surveyor/cost manager who will then assume responsibility for preparation of the BQ, and that this is the starting point of the NRM2 coding system.

Paradoxically, however, the authors of NRM2 specifically designed the rules of measurement to be flexible enough to be used for a variety of purposes and by a variety of users at various stages of the design and construction process employing various methods of procurement. They haven't made it easy to do so.

All the problems with the NRM2 coding system stem from the fact that it is largely directed at the quantity surveyor/cost manager who, having produced a cost plan, wishes to code the BQ in such a way that the contractor's pricing can be related back to the cost plan for cost management purposes. The importance of the collection of cost data is emphasised in several places in NRM2, and in Paragraph 2.3.1 and Appendix A.1.3(a) in particular, which both state that the priced BQ is *one of the best sources of real-time cost data*.

In point of fact, there are several instances within the text of NRM2 where it is obvious that the collection of contractors' pricing data is the main driving force behind NRM2 rather than providing *a standard set of measurement rules for the procurement of building works that are understandable by all those involved in a construction project* (NRM2 Paragraph 1.3.1) with the flexibility to suit a variety of procurement circumstances.

It must be said that, for many years, contractors have provided 'free' cost information to PQSs in the form of priced tenders.

This data is used as the basis for the BCIS cost information database which the RICS sells back to the industry on a subscription basis, admittedly with some added value. Builders' price books also use this 'free' data, and the authors of NRM2 confirm how valuable this *real-time cost data* is (Paragraph 2.17.1 refers). For the many contractors and subcontractors who will be supplying this 'free' information, this is a bit of a 'kick in the teeth', especially as NRM1 does not even define them as members of the 'project team'. However, we digress!

In order to understand the NRM2 coding system, it is vital to appreciate that:

1. The Paragraph 2.15 explanation of the coding system in NRM2 exclusively relates to the coding of BQ using the <u>NRM1</u> coding system.
2. This is because there is a presumption in NRM2 that an elemental cost plan has been prepared with codes based on *NRM1: Order of cost estimating and cost planning for capital building works* when coding items for NRM2-based BQ.
3. When measuring work using NRM2, and when there is a cost plan in place, the eventual BQ codes are driven by the NRM1 coding system and not by those of NRM2.

Consequently, anyone wishing to code a BQ that has been measured in accordance with the NRM2 *Tabulated rules* will be obliged to code the items using NRM1 codes and not those in NRM2 if it is desired to electronically rearrange the priced BQ from work section/work package format to elemental cost plan format. This strange anomaly arises because:

- NRM2 Paragraph 2.15.2.2 says that:
   - *the codification framework used for cost planning <u>must be used</u> as the basis for the codification of building components/items in the bill of quantities (BQ).*
- NRM2 Paragraph 2.15.3.1(2) says that:
   - where a work section bill of quantities is required *it is essential that the work sectional breakdown structure can be easily reconciled with the original cost plan breakdown structure.*
- NRM2 Paragraph 2.15.3.1(2) says that:
   - where a work package bill of quantities is required, the primary code is *that used for BQ based on an elemental breakdown structure.*

Albeit that NRM2 Paragraph 2.15 is purely concerned with the task of producing BQ codes that will enable an elemental, work section or work package BQ to be re-sorted into an elemental cost plan (NRM1) format:

- The measured items of work will be measured in accordance with the NRM2 rules of measurement.
- All item descriptions and units of measurement will be as per NRM2.
- Whichever BQ structure is chosen, it will contain exactly the same items, descriptions and quantities based on the NRM2 *Tabulated rules of measurement.*

Therefore, in order to re-sort the priced BQ into elemental cost plan format, each item of work needs to be coded with an <u>NRM1</u> unique code.

NRM2 Paragraph 2.15.3.1(1) states that *the resultant (NRM1) codes can be inserted in the right-hand column of the bill paper or in brackets after the bill description.* This is not the normal place for the item code, which usually appears in the left-hand column of the bill paper, and the NRM1 code should not be confused with the code generated by the measurement process. It is this code which is used to create the bill pages and to distinguish the BQ items one from another. This additional code will be needed to help the computer package to allocate the items to the appropriate headings so that a BQ can be produced for tender purposes. It will probably be an alpha/numeric code following custom and practice.

The distinction between elemental, work section and work package BQ is simply that the measured items will be allocated to different headings and subheadings in the relevant BQ format and will therefore be in a different order. Preliminaries are an exception and this section of the BQ will look the same irrespective of the BQ structure chosen.

As previously discussed, where there is no cost plan or where the user simply wishes to produce a BQ (e.g. for a single work package or to establish a target cost), there is no alternative coding system in NRM2 or any explanation of how to:

- Code measured items in accordance with NRM2 during the taking-off process.
- Code the BQ items using a WBS.
- Amend/restructure the coded items so as to produce elemental, work section or work package BQ as required.

6.8.8 Coding an elemental BQ

The WBS for an elemental BQ is based on the group elements used for cost planning which are to be found in NRM1: *Order of cost estimating and cost planning for capital building works*. Once again, however, confusion reigns!

NRM1 Paragraph 4.4.2 explains the coding system, and this is illustrated in NRM1 Appendix E. This is underpinned by NRM1 Paragraph 4.2.3 which explains that the NRM1 Part 4: *Tabulated rules of measurement* is based on four principal levels. Paragraph 4.4.2 informs, however, that *further code levels can be added to suit user requirements*.

The NRM1 coding system is summarised in Table 6.6.

Table 6.6 NRM1 coding system.

Level	Description	Function	Code defined by	Example Ref	Example
1	Group element	The primary headings in a cost plan. Group elements 0–8 represent building works, and 9–14 are for preliminaries, fees, risk allowances, etc.	Identification numbers in NRM1 tabulated measurement rules	2	Superstructure
2	Element	Part of a group element. There may be several elements that make up a group element	Identification numbers in NRM1 tabulated measurement rules	1	Frame
3	Sub-element	Part of an element. There may be several sub-elements that make up an element	Identification numbers in NRM1 tabulated measurement rules	4	Concrete frames
4	Component	Building work items that are part of a sub-element. There may be several components that make up a sub-element	User: • NRM1 does not pretend to list all possible components within a sub-element • User-defined codes should be sequential within the sub-element	1 2 3	Columns Beams Walls
5*	Sub-component	Where a component needs to be subdivided in more detail	User: • Columns, beams and walls may need to be subdivided into their constituent parts	1 2 3	Concrete Formwork Reinforcement
Example		**Formwork to reinforced concrete beams**		**Code**	**2.1.4.2.2**

*Further levels may be introduced as desired in order to provide each item with a unique code.

This simple and logical WBS starts to get complicated and confusing when we turn to NRM2 in order to find out how to code an elemental BQ with codes from the cost plan (in order to be able to re-sort the priced BQ into cost plan format once tenders have been received).

NRM2 Paragraph 2.15.3.1 suggests that *five to six levels of code are considered sufficient in cost planning*, and it lists *the main identification numbers* as illustrated in Table 6.7.

Table 6.7   The main identification numbers.

Level	Description	Identification number
0	Project number	User defined
1	Cost plan number	User defined but not required for a single cost plan
2	Group element	Predefined by NRM1
3	Element	Predefined by NRM1
4	Sub-element	Predefined by NRM1
5	Component	User defined

However, it can be seen that the identification numbers for the various levels are now different to those used in NRM1. This is because NRM2 introduces the idea of a project code (Level 0) and a cost plan code/reference number (Level 1), neither of which is mentioned in NRM1. Consequently, the group element, element and sub-element identification numbers now become Levels 2, 3 and 4, respectively, instead of 1, 2 and 3!

Moving swiftly on, NRM2 Paragraph 2.15.3.1 adds that a further level (Level 6) *will need to be introduced for each sub-component of a component that is to be measured in accordance with NRM2*. NRM2 Figure 2.3 illustrates how this can be done with reference to the example of a pile cap. There are, disappointingly, several errors in NRM2 Figure 2.3, and these are identified in Table 6.8.

The net effect of all this is that:

- The **group elements** provide the section headings for the BQ.
- The **elements** and **sub-elements** provide the various headings and subheadings within each section.
- The NRM1 components and sub-components are the items that will be measured under NRM2.
- The NRM2 classification tables refer to measured items as *items or work to be measured*.
- The 'pile caps' component is not a measured item under NRM2.
- Therefore, the items to be measured under NRM2 are the sub-components:
  - Work Section 5: *Excavating and filling*
    - **Excavation.**
    - **Disposal.**
  - Work Section 11: *In situ concrete works*
    - **Concrete.**
    - **Formwork.**
    - **Reinforcement.**
- Each of these items will be coded with the relevant NRM1 code.

The NRM2 quantity take-off for the pile cap items is illustrated in Figure 6.2.

These items would be billed under the **Substructure** group element as shown in Figure 6.3 which also shows that all items are allocated the same (NRM1) code, that is, 1.1.2.12.

6.8.9   Coding a work section BQ

NRM2 Part 3: *Tabulated rules of measurement for building works* is structured using a work sectional breakdown structure. There are 41 work sections with the first being devoted to preliminaries and the remaining 40 to the creation of measurement rules for a variety of building

Table 6.8   Analysis of NRM2 Figure 2.3.

Level	Description	Item	ID No.	Resultant codes in NRM2	Notes
0	Project no.		DPB27		The project and bill ID references
1	Bill no.	Bill no. 3	3		have been omitted from resultant NRM2 codes for clarity
2	Group element/ BQ no.	Substructure	1		Correct
3	Element	Foundations	1		The element name in NRM1 is Substructure
4	Sub-element	Piled foundations	2		a. Incorrect b. 'Piled foundations' is a component heading within the sub-element of 'Specialist foundations' in NRM1 c. Therefore, the sub-element is Specialist foundations
5	Component	Pile cap	1		a. The component description in NRM1 is Piled foundations: Pile caps b. The component reference in NRM1 is 2 not 1

6	Sub-component	Excavation	1	1.1.2.1.1	a. The codes given in NRM2 Figure 2.3 are correct but with some reservation (see b iii)
					b. The fourth digit of the codes is shown shaded because:
					  i. They are user defined
					  ii. The reference used is (1), whereas the component reference in NRM1 is 12
					  iii. Although this code is user defined, it makes little sense to use (1) as a reference because there are several components listed under 'Piled foundations' that will also be included in the cost plan
					c. The final digits of the codes are 'user defined' and could equally be alpha or numeric

6	Sub-component	Disposal	2	1.1.2.1.2	
6	Sub-component	Concrete	3	1.1.2.1.3	
6	Sub-component	Formwork	4	1.1.2.1.4	
6	Sub-component	Reinforcement	5	1.1.2.1.5	

components/items. It should be noted that the preliminaries work section is divided into two parts – main contract preliminaries and work package contract preliminaries.

Each work section is referenced with a serial number (e.g. 14 Masonry), and, within each work section, the *item or work to be measured* and the Level 1, 2 and 3 descriptors also have a reference number. The *item or work to be measured* is the equivalent of *component* in NRM1.

Part 2

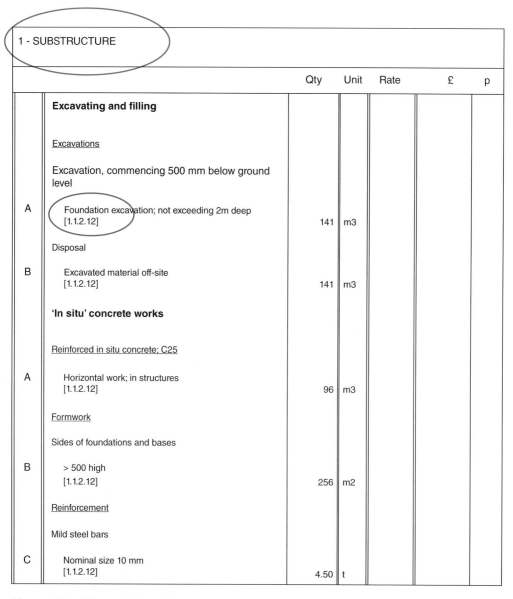

Figure 6.2 Quantity take-off for pile cap excavation.

Figure 6.3 Billing of pile cap items.

NRM2 Paragraph 2.15.3.1(2) deals with the coding of BQ based on work sections, but, once again, the recommended way to code items assumes that the work section BQ is to be reconciled with the cost plan once priced tenders have been received.

As a result, the method recommended by the NRM2 rules is to:

- Use a primary code equivalent to that used for a BQ based on an elemental WBS.
- Add a secondary code which acts as a suffix to the primary code.

The primary code would be derived from NRM1, and the secondary code would be the work section serial number taken from NRM2. Figure 6.4 shows a masonry item taken from Group element 2: *Superstructure* in the cost plan. It can be seen that this is a 'composite' item comprising brickwork, blockwork and cavity insulation and that the item has an NRM1 code of 2.5.1.1.

The ensuing code for the billed items in the work section BQ would be the NRM1 code (2.5.1.1) with the addition of the work section serial number (14), that is, 2.5.1.1/14. This code would be applied to all the items in Work Section 14 included within the external walls component of sub-element 2.5.1 in the cost plan because:

- The item for **external walls** in the cost plan is a 'composite' item.
- The work included in the **external walls** component is measured in detail in NRM2 Work Section 14.

This is illustrated in Figure 6.5.

Should there be no cost plan in place, or if it is not desired to relate the priced BQ to a cost plan, then the codes for the work section BQ would be any code that the bill compiler wishes to use. In all practicality, the NRM2 codes would normally be used (despite their frailties), and the BQ items would be coded as shown in the left-hand column of Table 6.9.

6.8.10 Coding a work package BQ

The third and final BQ WBS in NRM2 is the work package breakdown structure. This is where all the NRM2 measured items of work that would otherwise be included in several work sections

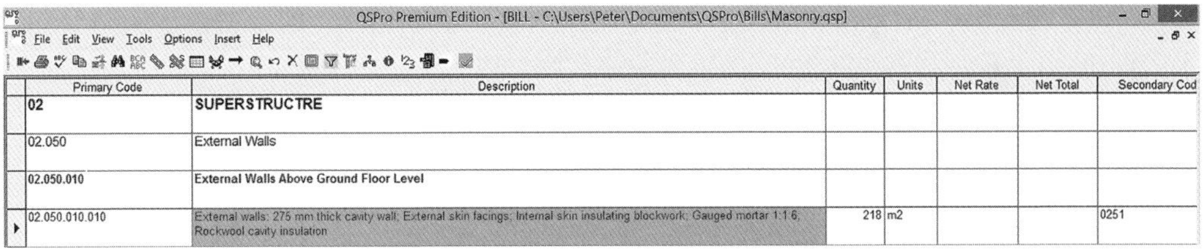

Primary Code	Description	Quantity	Units	Net Rate	Net Total	Secondary Cod
02	**SUPERSTRUCTRE**					
02.050	External Walls					
02.050.010	**External Walls Above Ground Floor Level**					
02.050.010.010	External walls; 275 mm thick cavity wall; External skin facings; Internal skin insulating blockwork; Gauged mortar 1:1:6; Rockwool cavity insulation	218	m2			0251

Figure 6.4 Coding – 1.

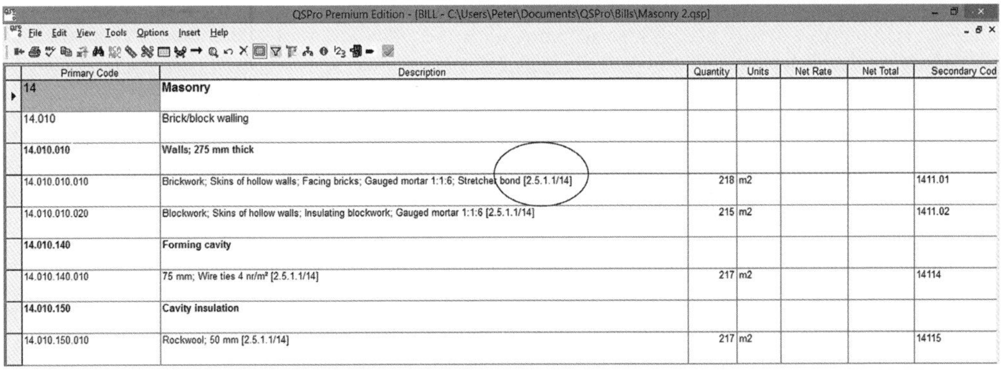

Primary Code	Description	Quantity	Units	Net Rate	Net Total	Secondary Cod
14	**Masonry**					
14.010	Brick/block walling					
14.010.010	**Walls; 275 mm thick**					
14.010.010.010	Brickwork; Skins of hollow walls; Facing bricks; Gauged mortar 1:1:6; Stretcher bond [2.5.1.1/14]	218	m2			1411.01
14.010.010.020	Blockwork; Skins of hollow walls; Insulating blockwork; Gauged mortar 1:1:6 [2.5.1.1/14]	215	m2			1411.02
14.010.140	**Forming cavity**					
14.010.140.010	75 mm; Wire ties 4 nr/m² [2.5.1.1/14]	217	m2			14114
14.010.150	**Cavity insulation**					
14.010.150.010	Rockwool; 50 mm [2.5.1.1/14]	217	m2			14115

Figure 6.5 Coding – 2.

Table 6.9 Coding – 3.

Item	Code	Ref	NRM2 source	Item descriptor
Brickwork	**14.1.1.1.1**	14	Work Section 14	**Masonry**
		1	Subheading (unnumbered)	**Brick/block walling**
		1	Measured item	**Walls**
		1	Level 1	**Brickwork**
		1	Level 2	**Skins of hollow wall**
Blockwork	**14.1.1.2.1**	14	Work Section 14	**Masonry**
		1	Subheading (unnumbered)	**Brick/block walling**
		1	Measured item	**Walls**
		2	Level 1	**Blockwork**
		1	Level 2	**Skins of hollow wall**
Form cavity	**14.1.14.1**	14	Work Section 14	**Masonry**
		1	Subheading (unnumbered)	**Brick/block walling**
		14	Measured item	**Forming cavity**
		1	Level 1	**Width and method of forming**
Cavity insulation	**14.1.15.1**	14	Work Section 14	**Masonry**
		1	Subheading (unnumbered)	**Brick/block walling**
		15	Measured item	**Cavity insulation**
		1	Level 1	**Type and thickness**

are collected together in one work package according to the perception of risk and in order to facilitate procurement of the work in question.

A simple example of this might be where a large site is to be developed with a hypermarket, access roads and car parking. The first job will be to reduce the levels on the site and to install the main surface water and foul sewers prior to commencing the building and external works.

In a work section BQ, the excavation work and drainage would be measured in Work Sections 5: *Excavating and filling* and 34: D*rainage below ground*. Assuming, however, that a work package is to be compiled for this work, all the measured items from Work Sections 5 and 34 would be collected in a work package that might be called 'Site works'. Consequently, one of the sections within the BQ would have the title of **Site works,** and this might be given the reference number 02, for example.

In trying to follow what the authors of NRM2 intended, it is interesting to note that the emphasis in NRM2 Paragraph 2.15.3.1(3) is different to that in Paragraphs 2.15.3.1(1) and (2). In the latter two paragraphs, the guidance on coding focuses on coding the BQ so that the priced BQ can later be related back to the elemental cost plan.

In Paragraph 2.15.3.1(3), however, the emphasis is on the restructuring of <u>cost plans</u> from elements to work packages which seems to indicate that the priced work package BQ would be compared to a work package-based cost plan. This resonates with NRM1 Paragraph 4.5.1 which suggests that cost plans can be coded in such a way that *the works allocated to elements and sub-elements can be reallocated to the applicable work package.*

This is somewhat confusing as NRM2 Paragraph 2.15.3.1(3) later goes on to say that the number and content of work packages will have to be carefully considered by the quantity surveyor/cost manager *before commencing the preparation of the <u>bill of quantities</u>* as opposed to before the cost plan is commenced.

In any event, Paragraph 2.15.3.1(3) points out that the primary code used for the measured items in the work package BQ should be the code that would have been used for an elemental BQ with the addition of a suffix to indicate to which work package the relevant items belong.

Assuming that a work package **02 Site works** is to be compiled for the bulk excavation and main drainage for our major hypermarket development, the contents of the work package may be expected to contain the following items of work:

- Removal of topsoil over the site and disposal to stock piles for reuse.
- Reduced level excavation and disposal off-site.
- Connections to existing sewers.
- Main foul and surface water drainage and manholes.

In order to code the elemental cost plan, the first job is to find the above items in NRM1. Then, they will have to be coded with appropriate NRM1 codes. Table 6.10 illustrates where to find the various items in NRM1.

It should be noted that some of the items of work are not specifically measured in NRM1 and, therefore, cannot be given a precise code. Removal of topsoil and disposal to spoil heaps on-site, for instance, are included in the *Lowest floor construction* sub-element of the *Substructure* group element (along with lots of other work items). These items, therefore, cannot be separately identified and, as such, cannot be coded for transfer into the work package.

Table 6.10   Coding – 4.

Item	Group element		Sub-element		Code
Removal of topsoil over the site area	1	Substructure	1.1.3	Lowest floor construction	No specific item
	8	External works	8.1.2	Preparatory groundworks	No specific item
			8.2.1	Roads, paths, pavings and surfacings	No specific item
			8.2.2	Special surfacings and pavings	No specific item
Stockpiling the topsoil for future use (e.g. landscaping)	1	Substructure	1.1.3	Lowest floor construction	No specific item
	8	External works	8.1.2	Preparatory groundworks	No specific item
			8.2.1	Roads, paths, pavings and surfacings	No specific item
			8.2.2	Special surfacings and pavings	No specific item
Bulk excavation to reduce levels over the site	1	Substructure	1.1.3	Lowest floor construction	No specific item
	8	External works	8.1.2	Preparatory groundworks	No specific item
			8.2.1	Roads, paths, pavings and surfacings	No specific item
			8.2.2	Special surfacings and pavings	No specific item
Disposal of excavated material off-site	1	Substructure	1.1.3	Lowest floor construction	No specific item
	8	External works	8.1.2	Preparatory groundworks	No specific item
			8.2.1	Roads, paths, pavings and surfacings	No specific item
			8.2.2	Special surfacings and pavings	No specific item
Connections to existing sewers	8	External works	8.6.1	Surface water and foul drainage	8.6.1.1
Main surface water drainage and manholes					8.6.1.2
Main foul water drainage and manholes					8.6.1.2
Manholes					8.6.1.5

Part 2

This problem illustrates that the work package items (i.e. the NRM2 items) must be identified first and then coded to the cost plan and not vice versa. Thus, the structure of the work package must be determined, and then the items can be coded with the NRM1 codes. These codes will be the elemental codes derived from NRM1 which means, therefore, that the BQ items, when re-sorted, will re-sort into an elemental cost plan and not a work package cost plan. If a work package cost plan is required, the NRM1 codes will have to be supplemented with a suffix relating to the work package (in this case 02) following the guidance in NRM1 Paragraph 4.5.1. Consequently, the primary code will identify the group element/sub-element/component and the suffix will direct the items into the correct work package (i.e. 02).

In order to code the BQ items with the appropriate NRM1 codes and a suffix, however, the NRM2 measured items will have to be taken off in such a way that they can be identified with a specific group element/sub-element/component.

This adds a further layer of complication in that the topsoil excavation and disposal items, for instance, would have to be split between the *Substructure* and *External works* group elements and further split into the appropriate sub-elements in order to allocate the correct NRM1 code.

A way round this would be to allocate all of the quantities measured to one item in the cost plan (e.g. Group element 8: *External works* could include Site preparation works, Preparatory groundworks, Forming new site contours and adjusting existing site levels). All of the items could then be coded 8.1.2.1/02, and the bulk excavation could be coded 8.1.2.2/02 (see Table 6.11).

Table 6.11 Work package coding – 1.

02 SITE WORKS			Excavating and filling			
		Qty	Unit	Rate	£	p
	Excavating and filling					
	Excavations					
	Site preparation					
A	Remove topsoil: Average 150 mm deep [8.1.2.1/02]	135888	m2			
	Excavation; Commencing level 150 mm below existing ground level					
B	Bulk excavation; Over 2m not exceeding 4m deep [8.1.2.2/02]	298954	m3			
	Disposals					
	Disposal					
C	Excavated material off site [8.1.2.2/02]	298954	m3			
	Retaining excavated material on site					
D	Top soil; To temporary spoil heaps; Average distance 250 m [8.1.2.1/02]	20383	m3			

This problem doesn't arise, thankfully, with some of the other measured items.

Drainage runs, manholes and *connections to existing sewers*, for instance, are specifically measured items under NRM1 Element 8.6: *External drainage* as they are under NRM2 Work Section 34: *Drainage below ground*. This is illustrated in Table 6.12.

Table 6.12   Work package coding – 2.

02 SITE WORKS				Drainage below ground		
		Qty	Unit	Rate	£	p
	Drainage below ground					
	Storm Water Drain Systems					
	Drain runs					
A	Average trench depth 2.0 m; Precast concrete pipe 450 mm diameter; Granular pipe bedding and surround [8.6.1.2/02]	78	m			
B	Average trench depth 2.5 m; Precast concrete pipe 450 mm diameter; Granular pipe bedding and surround [8.6.1.2/02]	46	m			
	Connections					
C	Local authority sewer; Depth 3.0 m [8.6.1.1/02]	1	it			
	Foul Drain Systems					
	Drain runs					
D	Average trench depth 2.50 m; Precast concrete pipe 300 mm diameter; Granular pipe bedding and surround [8.6.1.2/02]	61	m			
E	Average trench depth 3.0 m; Precast concrete pipe					

As if the coding of elemental and work section BQ were not complicated enough, the coding of work package BQ is even more tortuous.

Risk Issue

Unless this author is sadly mistaken, the BQ coding arrangements in NRM2 are horribly complicated.

If the NRM family of documents is to be adopted as recommended best practice in line with their status as RICS guidance notes, QS professionals may find themselves having to justify to clients the amount of time and effort needed to code items back to the cost plan and the level of fee charged for this work.

It may well be concluded that the cost is not justifiable, and thus, the PQS will then have to wrestle with the conundrum 'do I follow the NRM or do I take the risk of having to justify an alternative approach before a court or other tribunal?'

6.9   Part 3: Tabulated rules of measurement for building works

6.9.1   Introduction

NRM2 Paragraph 3.1.1 explains that Part 3 of the rules is made up of:

- The information <u>and</u> requirements for main contractor and work package contractor preliminaries.
- The <u>rules</u> for preparing the preliminaries pricing schedule.
- The <u>rules</u> of measurement applying to building components and items.

Paragraph 3.1.2 states that Part 3 explains how the tabulated rules are used, and Paragraph 3.1.3 makes the two-part statement that:

- *Bill of quantities (BQ) <u>are to</u> fully describe and accurately represent the quantity and quality of the works to be carried out.*
- *More detail than is required by these rules <u>should be given</u> where necessary to define the precise nature and extent of the required work.*

These statements stop short of their SMM7 General Rule 1.1 equivalents which both use the word 'shall' in order to emphasise that they are, in fact , 'rules' to be followed rather than discretionary options.

However, Paragraph 3.1.3 is, in fact, made redundant by the later Paragraph 3.3.1 which does use the word 'shall', and it can only be concluded, therefore, that the entire Paragraph 3.1 is to be ignored as consisting of purely 'throwaway remarks' that have no importance or meaning within the tabulated rules.

6.9.2   Use of tabulated rules of measurement for building works

Paragraph 3.2 explains how the rules of measurement are structured and that they are set out in tables. In particular, Paragraph 3.2.1.1 states that the tables are divided into two categories:

- Preliminaries.
- Measurement of building components/items.

This is, in fact, misleading.

There is no categorisation under Part 3: *Tabulated Work Sections* which merely comprises 41 work sections of which 'preliminaries' is Work Section 1. There <u>is</u> categorisation <u>within</u> Work Section 1, however, which comprises preliminaries (main contract) and preliminaries (work package contract), both of which are divided into Parts A (information and requirements) and B (pricing schedule).

This is explained in NRM2 Paragraphs 3.2.2.1 and 3.2.2.2 and illustrated in Table 6.13.

Table 6.13   Classification of preliminaries.

Work Section	Subsection		Part	Purpose
1	Preliminaries (main contract)	A	Information and requirements	The descriptive part of preliminaries
		B	Pricing schedule	The part where preliminaries prices are inserted by the contractor
	Preliminaries (work package contract)	A	Information and requirements	The descriptive part of preliminaries
		B	Pricing schedule	The part where preliminaries prices are inserted by the work package contractor

The distinction between Work Section 1 and the remaining 40 work sections is that:

- The tables relating to Preliminaries have different layouts.
- There are two types of table in Work Section 1.

Paragraphs 3.2.1.2–6 are self-explanatory, but careful note should be taken of:

- Paragraph 3.2.1.4
  Horizontal lines divide the tables and rules into zones to which different rules apply.
- Paragraph 3.2.1.4
  A broken line (-----) between units of measurement or measurement rules *denotes a choice of units or choice of ways of measuring the work.* The best method to suit the situation *shall be* chosen.

Tables: Preliminaries

Paragraphs 3.2.2.3 and 3.2.2.4 explain the function of the various columns in the tables for *information and requirements* and for the *pricing schedule.* These explanations apply to both preliminaries (main contract) and preliminaries (work package contract).

Both *Part A: Information and requirements* and *Part B: Pricing schedule* comprise a number of tables, each of which has a heading that indicates the subject matter of the table in question. This is illustrated in Table 6.14.

Table 6.14   Tables: preliminaries.

1   Preliminaries (main contract)
   Part A: Information requirements
1.7   Employer's requirements: management of the works

Subheading 1	Subheading 2	Information requirements	Supplementary information/notes
Preliminaries items to be considered	Sub-items to be considered	Information which shall be included	Information that might be needed

1   Preliminaries (main contract)
   Part B: Pricing schedule
1.1   Employer's requirements
1.1.1   Site accommodation

Component	Included/notes on pricing	Unit	Pricing method	Excluded
Preliminaries items to be considered	The sub-item of preliminaries items to be considered		Stipulates if the component is a: • Fixed charge • Time-related charge • Combination of both	Describes items excluded from a component

In *Part A: Information and requirements*, it should be noted that:

- Subheadings 1 and 2 list items to be considered; consequently, these items are optional and will be chosen, or not used, according to the requirements of the project.

- The 'Information requirements' column lists information that <u>shall be included</u> should the particular item or sub-item be included in the schedule of preliminaries.
- The 'Supplementary information/notes' column is an *aide-memoire* that provides additional information that might be needed in preliminaries descriptions and guidance on the drafting of preliminaries statements.

In *Part B: Pricing schedule*, it should be noted that:

- The items listed under the headings 'Component' and 'Included/notes on pricing' are optional and will be chosen, or not used, according to the requirements of the project.
- The heading 'Pricing method' provides for items to be priced as 'fixed' or 'time-related' charges or both.
- The 'Excluded' column identifies items that are not included in a component.

Tables: Building Components/Items

Paragraph 3.2.3 is an important part of NRM2 because it states that the rules of measurement are laid out in tables and also explains how the tables work. It does not, however, explain how to formulate item descriptions from the classification tables which is dealt with in Paragraph 3.3.3 and in Paragraph 3.3.3.3 in particular.

Paragraph 3.2.3.1 lists the 40 Work Sections, other than preliminaries, that appear in Part 3 of the document, whilst Paragraph 3.2.3.2 explains the structure of the tables that are used for items of measured work.

Each work section (e.g. Carpentry) has:

- A reference number (e.g. 16).
- A title (e.g. Carpentry).
- A list of the types of work included in the Work Section, for example:
  - Timber framing.
  - Timber first fixings.
  - Timber, metal and plastic boarding, sheeting, decking, casings and linings.
  - Metal and plastic accessories.
- A two-part table:
  - Information requirements and general rules.
  - Items or work to be measured together with descriptive features (*Levels 1, 2 and 3*) and *notes, comments and glossary.*

The NRM2 tabular layout for Work Sections 2–41 is illustrated in Table 6.15.

Each work section has a reference number and title, and it is this title that forms the *heading* needed for each work section as required by NRM2 Paragraph 3.3.3.1.

The list of items below the title is interesting. There is no reference to this list in the text of NRM2, and, at first glance, it would appear that its function is little more than informative. Its importance is explored in Section 6.9.3 (Descriptions) of this chapter.

The next part of the table layout consists of two rows which are read from right to left. They set out:

- Drawings:
  - required for measurement purposes.
  - that <u>shall</u> *accompany the bill of quantities when issued.*
- Mandatory information:
  - *that is to be provided* in each Work Section.
- Minimum information *that <u>shall be shown</u>:*
  - *on the drawings or*
  - *any other document that accompany each Work Section.*

Table 6.15 Tables: building components/items.

Ref no	Title	List of types of work included in the Work Section			Notes, commentary and glossary
		Level 1	**Level 2**	**Level 3**	
Drawings that must accompany this section of measurement	NOT USED	For example: General arrangement drawings Site survey Plans Sections		**Mandatory information to be provided** For example: Specification information	**Notes, commentary and glossary** For example: Description rules Explanatory rules
Minimum information that must be shown on the drawings that accompany this section of measurement	NOT USED	For example: Extent, position or location of the work		**Works and materials deemed included** For example: Items not measured but understood to be included in the measured item	Additional measurement rules
Item or work to be measured	Unit	**Level 1**	**Level 2**	**Level 3**	**Notes, commentary and glossary**
Components/items	m, m^2, m^3, etc.	↑	Item descriptors that shall be included where applicable	↑	Relevant information that shall be included in item descriptions
Horizontal lines that denote units of measurement and/or descriptive features that apply to the particular component/item residing between the lines	Broken line denotes a choice of unit				

- Works and materials:
  - *that are not measured* but are
  - *deemed to be included* in the components/items measured.

Following the first two rows are six columns that are to be read vertically:

- Building component/items to be measured are in the first column followed by:
- Units of measurement followed by:
- three columns (Levels 1–3) listing:
  1. Information.
  2. Supporting information.
  3. Further supporting information:
     - That <u>*shall be included*</u> in the item descriptions.
- Levels 2 and 3 also contain any additional dimension requirements <u>*which shall be included*</u> in the item descriptions.
- The final column:
  - <u>*Explains*</u> what is *deemed to be included* in specific items.
  - <u>*Clarifies*</u> *the approach to quantification and description* of items.
  - <u>Defines</u> *specific terms and phrases used* in any particular component/item.

There are three work sections which have a slight, but important, variation to this layout: Work Sections 11, 22 and 41. The difference is easily missed.

Work Sections 11, 22 and 41 introduce intermediate rows in the table which contain headings. They are not highlighted but nevertheless represent 'subheadings' in the classification that must be observed. Referring to Table 6.16, it can be seen that there is a row in between the preceding two rows and the following six columns described earlier. In Work Section 11: *In situ concrete*, for instance, there are four headings that must be used to distinguish between various types of concrete work – Plain in situ concrete, Reinforced in situ concrete, Fibre-reinforced in situ concrete and Sprayed in situ concrete.

Paragraph 3.2.3.2 culminates with a statement which says that the list of building components/items in the tables represents those commonly encountered in building work but that they *are not intended to be exhaustive*. Curiously, the reader is not directed as to what to do when items not listed are to be measured. In such circumstances, it is necessary to trawl through two more pages to discover Paragraph 3.3.5 which provides appropriate guidance (see also later in Section 6.9.3: *Measurable work not covered by the tabulated rules*).

Further 'trawling' is needed to discover that descriptions for components/items to be measured are to be compiled using <u>all</u> Level 1, 2 and 3 information and also, where, applicable, information contained in the *Notes, commentary and glossary* column (column five) (Paragraph 3.3.3.3 refers).

6.9.3   Measurement rules for building works

Paragraph 3.3.1 corrects the imprecise Paragraph 3.1.3 and emphasises that the BQ <u>*shall*</u> *fully describe and accurately represent the quantity and quality of works to be carried out* and that, where necessary, additional detail <u>*shall be given*</u> in order to ensure that *the precise nature and extent* of the works is conveyed.

Bill compilers should carefully heed these rules as the implications of not doing so could be serious and costly.

Table 6.16   Intermediate headings.

Ref no		Title			
Drawings that must accompany this section of measurement	NOT USED	For example: General arrangement drawings	**Mandatory information to be provided**	For example: Specification information	**Notes, commentary and glossary** For example: Description rules Explanatory rules
Minimum information that must be shown on the drawings that accompany this section of measurement	NOT USED	For example: Extent, position or location of the work	**Works and materials deemed included**	For example: Items not measured but understood to be included in the measured item	Additional measurement rules
Item or work to be measured	**Unit**	**Level 1**	**Level 2**	**Level 3**	**Notes, commentary and glossary**
Headings that distinguish between various types of work to be measured For example: **Work Section 11: In situ concrete** Plain in situ concrete Reinforced in situ concrete Fibre-reinforced in situ concrete Sprayed in situ concrete					
Work to be measured such as mass concrete	m²	Concrete thickness	Use of concrete such as filling voids or trench filling	Further description such as whether the concrete is poured against earth or unblinded hardcore	Additional description or deemed to be included rules or whether different types of concrete may be aggregated

Risk Issue

It is not inconceivable that a bill of quantities either:

- Does not fully describe or accurately represent the work to be carried out

or

- Lacks detail that could or should have been given

or

- Both

It is arguable that such shortcomings are not covered by the term 'variation' (e.g. JCT 2011 Clause 5.1.1) and that they could, in fact, constitute a misleading statement that could lead to a claim for damages either:

- Under the Misrepresentation Act 1967 or
- In tort on the principle of *Hedley Byrne & Co Ltd v Heller & Partners* (1963).

It is also possible that a breach of contract (e.g. under Clause 2.13.1 of JCT 2011) may be claimed as such shortcomings would not be in accordance with the Rules of measurement.

An important issue arises in Paragraph 3.3.1 is that there is no statement that *the rules apply to measurement of proposed work and executed work* such as may be found, for example, in General Rule 1.2 of SMM7.

This may be an oversight, or it may have been a conscious decision by the authors of NRM2 that the JCT 2011 SBC/Q (a lump sum contract) would be the default contract to use with NRM2. As such, JCT 2011 Clause 5.6.1.3 states that the measurement of variations *shall be in accordance with the same principles as those governing the preparation of the Contract Bills* referred in Clause 2.13.

Clause 2.13 concerns preparation of the contract bills, and Clause 2.13.1 refers expressly to the *Measurement Rules* (i.e. NRM2) pursuant to the JCT August 2012 NRM Update.

Risk Issue

Without a clause stating that *the rules apply to measurement of proposed work and executed work*, there is no express or implied undertaking that executed work (i.e. work actually carried out) shall be measured in accordance with the New Rules of Measurement (see Chapter 12).

There is no problem with the JCT lump sum contract, because remeasured work is dealt with as a variation, but care needs to be exercised in deciding how the omission of such a rule may play out should non-JCT forms of contract be used in conjunction with NRM2.

Quantities

Paragraph 3.3.2 provides 'rules' for the measurement and billing of items and for dealing with voids. These rules concern issues such as the following:

- Work shall be measured net as fixed in position.
- Laps, joints, seams and the like shall be deemed included in the net quantity (thereby avoiding the need for repetitive 'coverage' rules in the work sections).
- Quantities to be given to the nearest whole number (or unity where less than 1) with the exception that items measured in tonnes are to be given to two places of decimals.
- The treatment of deductions for voids and the like.

Descriptions

The writing of item descriptions is, arguably, the most important part of any standard method of measurement as this is where the ability of the bill compiler to formulate descriptions that *fully describe and accurately represent the quantity and quality of the works to be carried out* (Paragraph 3.1.3) is tested.

In order to satisfy this requirement, the BQ needs to be laid out in the manner prescribed by the method of measurement, and the item descriptions need to include all the descriptive features delineated in the classification tables.

To begin with, NRM2 Paragraph 3.3.3.1 requires that *each Work Section shall have*:

- A heading.
- A description stating the *nature and location of the work*.

This is a significant departure from SMM7 General Rule 4.5 which states that a description of the nature and location of the work is to be provided *unless evident from the drawn or other information required to be provided by these rules*. In other words, where the nature and location of the work is clear from the drawings or other document(s) provided (e.g. specification), a description does not have to be provided in the BQ under SMM7.

Next, it is conventional to subdivide BQ work sections into distinct parts where related items may be arranged together. This is usually done by using subheadings. NRM2 is silent on this issue except in Appendix A.3(2)(b) which refers to *Subdivisions*, but even here the guidance is most unclear.

Risk Issue

BQ compilers will be faced with an additional administrative burden in dealing with this requirement which seems unnecessarily onerous and, in some work sections, is a duplication of effort.

In Work Sections **5: Excavation** and **33: Drainage above ground**, for instance, locational information is <u>expressly required to be given</u> in the **Mandatory information to be provided** part of the Work Section table. In Work Section 5, Paragraph 3.3.3.1 is specifically referred to. This is not the case, however, in Work Section 33.

An additional consideration is that the Paragraph 3.3.3.1 requirement may also offer contractors the opportunity to make a claim for additional payment where the description is:

- Not given.
- Unclear.
- Inaccurate or misleading.

The simplest way round the problem would appear to be to:

- Include a 'notwithstanding' clause in the BQ.
- Add a caveat to the Paragraph 3.3.3.1 requirement.
- In NEC3 ECC Options B and D, add an amendment to the method of measurement in Part 1 of the Contract Data.

Appendix A.3(2) refers to the order of items in work section BQ:

- Appendix A.3(2)(a) requires that the BQ is firstly divided into Work Sections.
- Appendix A.3(2)(b) requires that there shall be subdivisions:
  - (i)   Of work sections as contained in NRM2: *Detailed measurement for building works*[7].
  - (ii) As required by NRM2: *Detailed measurement for building works*,[2] such as *external paintwork*.

Appendix A.3(2), therefore, infers that the order of items in BQ should be hierarchically subdivided into headings and subheadings, and this resonates to some extent with NRM2 Paragraph 3.3.3.2 which refers to *headings for groups of components/items*. It could thus be argued – but NRM2 is not at all clear on the matter – that within each work section, there will be groupings of items under subheadings.

If this is the case, then Appendix A.3(2)(b)(i) might be suggesting that the source of these subheadings is the list of items that appears under the main title of each work section. The only problem with this idea is that the list is not referenced at all and therefore would seem to be outside of the WBS propounded by NRM2.

The other type of subdivision referred to in A.3(2)(b)(ii) is where it is required by the *Tabulated rules* – the example of *external painting* is given. This makes less sense because *external painting* is a descriptive feature of Work Section 29: *Decoration* at Level 2 and, as such, makes up part of an item description. For example, *external painting to general surfaces over 300 mm girth* would have a reference of 29.1.2.2.

The authors of QSPro seem to have been persuaded by the first idea of subdivision as not only have they chosen to use the list of items under the main work section headings as work section subdivisions but have also given this subheading a code or reference number as illustrated in Figure 6.6.

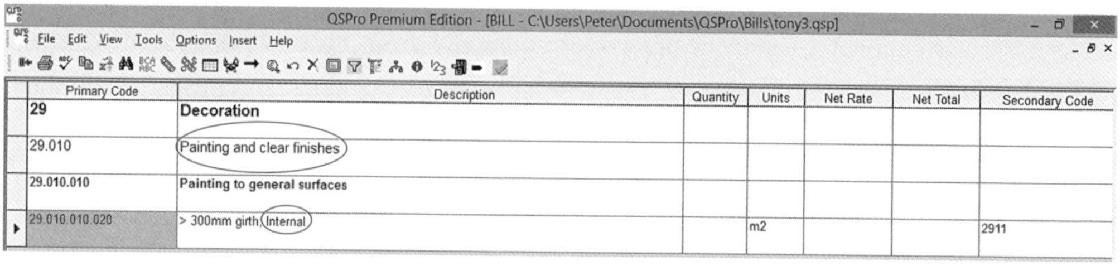

Figure 6.6   Approaches to subdivision – 1.

It should be noted that:

- The use of the NRM2 subheadings in this example is similar to the subheadings used in SMM7.
- The SMM7 subheadings are coded.
- The code used in the QSPro example is consistent with the idea of a WBS proposed in NRM2.
- The introduction of a code for these subheadings is not contrary to the rules of NRM2 because there are no rules that apply to the coding of measured work items.
- NRM2 only has a coding system for BQ items where it is desired to link them to an elemental cost plan.

CATO, on the other hand, employs the idea of a blank level under the main work section heading to enable the user to insert any subheading desired as illustrated in Figure 6.7. Again, this is consistent with the WBS of NRM2.

Figure 6.7   Approaches to subdivision – 2.

In passing, it will be noted that:

- The QSPro and CATO codes are slightly different which is consistent with the 'no rule' arrangement within NRM2 as regards the coding of measured items (as opposed to the coding of BQ items).
- Neither QSPro nor CATO use a subheading to distinguish *internal and external paintwork*. This is contrary to NRM2 Appendix A.3(2)(b)(ii), but as this appendix is entitled 'Guidance', it is not a rule and can thus be ignored. The main thing is that the BQ item is properly described.

Moving swiftly on!

Turning to the writing of item descriptions, this is covered in three paragraphs:

- Paragraph 3.3.3.3:
  - the component/item to be measured is to be *taken from the first column of the tabulated rules.*
  - *all of the level 1, 2 and 3 information (taken from the third, fourth and fifth columns respectively)* applicable to the item shall be included.
  - *relevant information from column five[3] shall be included* where applicable.
- Paragraph 3.3.3.4:
  - Item descriptions shall include:
    - *the type and quality of the material.*
    - *critical dimensions* of the materials concerned.
    - the method of fixing or installation *where not at the discretion of the contractor.*
    - the character of the background that the material is to be fixed to.
- Paragraph 3.3.3.5:
  - The character of background material stated is to be chosen from a list of six types:
    1. *timber* (including boards).
    2. *plastics.*
    3. *masonry* (including brick, concrete, block, etc.).
    4. *metal* (any).
    5. *metal-faced timber or plastics* and
    6. *vulnerable materials* (such as glass, marble, tiles etc).

Therefore, when fixing plasterboard to a concrete background, this is to be stated as 'masonry' because the term 'masonry' includes concrete.

Paragraph 3.3.3.10 is an interesting one as it has its origins in SMM7 General Rule 4.2 but is slightly different in meaning. The SMM7 rule is clear in that item descriptions in the BQ have to be drawn from the three columns of the classification tables, together with supplementary information from the Supplementary information column of the tables. Where necessary, however, the same information may be given on a drawing or in a specification so long as *a precise and unique cross reference is given* in the BQ item description.

In NRM2, however, *information required by these rules may be given* in another document (e.g. specification or catalogue – no mention of drawings) provided that *a precise and unique cross reference* is given. This sounds much the same as the SMM7 rule, but the underlined words *by these rules* mean that the derogation applies to all the rules in Part 3 and not simply the rules in Paragraphs 3.3.3.3 and 3.1.3 which are the SMM7 equivalents.

Risk Issue

Paragraph 3.3.3.10 could involve the tenderer in all sorts of additional research to discover what a BQ item means.

This might require a search of the NBS or other specifications as well as searching out manufacturers' catalogues and the like as there is no provision in NRM2 that requires such information to be given.

Part 2

Paragraphs 3.3.3.11 and 3.3.3.12 are important from the perspective of what to measure where a composite item could be construed. This could arise where, for instance, 12.5 mm plasterboard is to be fixed to walls on timber battens. In such circumstances, there are two ways to proceed:

1. Paragraph 3.3.3.11 requires that:
   ○ The timber battens <u>and</u> the plasterboard shall be measured and described separately.
2. However, Paragraph 3.3.3.12 states that:
   ○ A single composite item could be billed providing that the item description is unequivocal as to what is included in the item.

Assuming the first method is adopted, i.e. where the component items are to be measured separately, the item descriptions would be made up as follows:

- The timber battens are measured in Work Section 16: *Carpentry*:
  ○ The item reference is, therefore [16.3.1.2]:
    • 3 – Backing timber.
    • 1 – Nominal size 25 mm × 38 mm.
    • 2 – Battens.
- The plasterboard is measured in Work Section 28: *Floor, wall, ceiling and roof finishes*:
  ○ The item reference is, therefore [28.7.2]:
    • 7 – Walls; 12.5 mm plasterboard.
    • 2 – exceeding 600 mm wide.

As to what is deemed to be included in a component or item, reference must be made to Paragraph 3.3.3.13 which is broadly similar to SMM7 General Rule 4.6 but with one notable exception. SMM7 deems that the contractor's *establishment charges, overheads and profit* is included, whereas NRM2 states that only *establishment charges* are included in any building component/item. The reasoning for this is that the *Main contractor's overheads and profit* is to be stated as a separate item (%) in the final summary of the BQ under NRM2.

There may well be justifiable reasons why overheads and profit should be disaggregated from BQ rates and prices where, perhaps, there is a partnering agreement and the main contractor's margin is to be protected or 'ring-fenced'. In normal circumstances, however, no contractor in their right mind will wish to divulge such confidential information, and most contractors, it is suggested, will price the BQ in exactly the way that they wish to price it.

Risk Issue

Neither SMM7 nor NRM2 is clear on the distinction between establishment charges and overheads, but, in SMM7 at least, it is quite clear to all concerned that the BQ items are priced 'gross'.

NRM2 Paragraph 3.3.3.13, however, leaves considerable doubt in the mind both as to what the distinction is and what is deemed included in any particular BQ rate.

The question arises, therefore, as to how variations to the contract should be valued when overheads and profit have been priced separately but the BQ item is nevertheless deemed to include establishment charges (i.e. overheads).

Work of special types

Paragraph 3.3.4.1 requires that work of a particular type *shall be separately identified*, and Paragraph 3.3.4.2 stipulates that *specific details* of such work shall be provided *at the start of each applicable* Work Section. The rules detailed in Paragraph 3.3.4.1 must be read in conjunction with the rules of measurement given in the various work sections (Paragraph 3.3.4.3). Work of special types includes:

- Work to existing buildings.
- Work to be carried out and subsequently removed.
- Work outside the curtilage of the site.
- Work carried out in extraordinary conditions, such as work in or under water, tidal work and compressed air working.

Risk Issue

Where work of special types is to be included in the BQ, Paragraph 3.3.4.4 requires that details of additional preliminaries *shall be given in the description.*
  It is not clear, however, whether:

- The *description* referred to is an overall description or a preamble that precedes the measured items of work if, indeed, they are to be separately billed.
- This is meant to be a pointer to additional items that have been included in Work Section 1: *Preliminaries.*

Measurable work not covered by the tabulated rules

Occasions arise where the tabulated rules of measurement do not cover the items to be measured.
  A pedestrian underpass giving access to a new shopping centre development may have to be tunnelled under an existing highway, for instance. Such circumstances are provided for in NRM2 under Paragraph 3.3.5 which suggests two possible solutions:

- Paragraph 3.3.5.1:
  - Use the rules for similar work.
  - Clearly state and fully define the rules adopted in either:
    - The preliminaries or
    - With the relevant BQ items.
  - Ensure, *as far as possible*, that the rules adopted conform to the tabulated rules of measurement for similar items.
- Paragraph 3.3.5.2:
  - Use 'bespoke' rules of measurement.
  - The rules adopted shall be *reiterated in full* in either:
    - The preliminaries or
    - In the BQ above the relevant items.

The use of the word *reiterated* is puzzling because this means 'repeated' or 'restated', implying, perhaps, that the rules may not actually be 'bespoke' (tailor-made, adapted, customised) but may already exist in another document. This might suggest, for instance, that such 'bespoke' rules may be derived from another standard method of measurement (such as CESMM4).
  There would seem little point in reiterating the rules from another standard method of measurement when a simple cross reference to the relevant section(s) of the document could be used instead.

Procedure where work cannot be quantified

Paragraph 3.3.6 refers the reader to Paragraph 2.9 of the rules where detailed guidance for such circumstances is provided. Section 6.7.9 of this book discusses the issues arising from Paragraph 2.9.

Part 2

Procedure where exact type of product or component is not specified

It is quite common to find that a BQ is produced at a point in time when the technical design is insufficiently developed to enable certain materials or products to be precisely identified so that the contractor can price a competitive rate. This situation is normally overcome by providing the contractor with a 'price' for the material in question which can then be used for building up the unit rate. This approach has the dual advantage that:

- The BQ compiler can complete the BQ to a reasonable level of accuracy.
- A competitive rate can be obtained from the contractor which has to be adjusted for the material element only once the precise specification is established.

Under NRM2 Paragraph 3.3.7.1, an estimated price for the product or component in question *shall be given* in the item description and this shall be stated as a *PC price*. An example is provided in Figure 6.8 which shows a PC price for facing bricks.

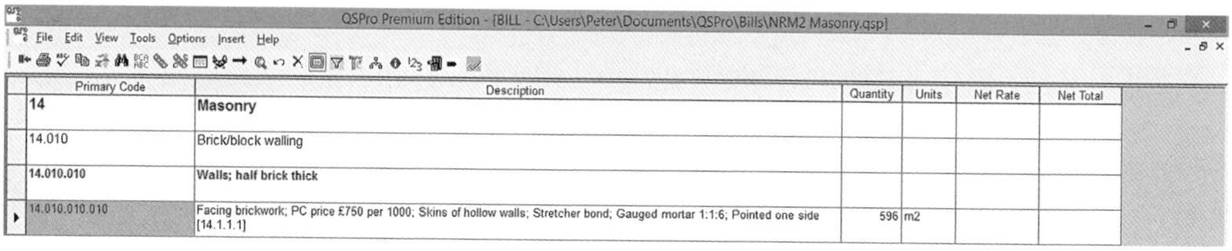

	Masonry					
	Brick/block walling					
	Walls; half brick thick					
A	Facing brickwork; PC price £750 per 1000; Skins of hollow walls; Stretcher bond; Gauged mortar 1:1:6; Pointed one side [14.1.1.1]	596	m2			

Figure 6.8 Prime cost item.

Paragraph 3.3.7.3 states that *PC prices shall exclude any allowance for the contractor's overheads and profit;* this is because *provision shall be made* in the BQ for the pricing of overheads and profit in accordance with the Paragraph 2.11 item in the final summary.

Where a PC price is to be incorporated into a BQ rate, the contractor *shall be deemed* by Paragraph 3.3.7.2 to have allowed in the rate for all the items listed in Paragraph 3.3.3.13; this list includes (7) *establishment charges*.

Notwithstanding the above, the PC price arrangement in NRM2 is problematic as discussed earlier in Section 6.6.11.

Procedure where quantity of work cannot be accurately determined

Paragraph 3.3.8 deals with situations where items are to be inserted into the measured work sections of a BQ which can be described in accordance with the tabulated rules of measurement *but* the quantity cannot be measured precisely.

Paragraph 3.3.8.1 requires that *an estimate of the quantity* of such work *shall be given <u>and</u> identified as a 'provisional quantity'*. There is no guidance as to how this should be done, but custom and practice indicates that the word 'provisional' is given in brackets at the end of the item description.

Risk Issue

Paragraph 3.3.8.2 goes on to say that provisional quantities *shall be* subject to *remeasurement when they have been completed.*
  This could be taken to mean that:

- Incomplete work items shall not be remeasured.
- The contractor's progress payments shall be based on the BQ and not actual quantities.
- Progress payments may not accurately reflect the quantity of work carried out.
- Actual cost may be greater than actual value until such time as the work is completed and remeasured.
- There may be a delay of several weeks, if not months, until the contractor receives the correct payment for work in progress.
- The employer's quantity surveyor/cost manager is obliged to undertake a remeasure as soon as the work has been completed.

Paragraph 3.3.8.2 also says that where there is a 'provisional quantity', *the 'approximate quantity' shall be substituted by the 'firm quantity' measured.* There is no definition of either of these terms in NRM2, and their use is both superfluous and confusing as, in any event, the paragraph concludes with a procedure for dealing with differences between the 'provisional quantity' and the 'firm quantity'.

Risk Issue

The distinction between 'provisional' and 'firm' quantities is important for contractors.
  Where the conditions of contract state that NRM2 is the applicable standard method of measurement, NRM2 Paragraph 3.3.8.2 introduces the notion of a 'rule' (the '20% rule') for dealing with variations between provisional quantities and the eventual remeasured quantity.
  This provision in NRM2 may lead to a conflict with the prevailing conditions of contract for the project, even where JCT conditions are used.

Paragraph 3.3.8 is far more complex than its counterpart General Rule: 10.1 in SMM7 and is discussed further in Chapter 12 in the context of final accounts.

6.10   Tabulated work sections

The *Tabulated work sections* part of NRM2 is a huge disappointment.
  When the CESMM was first published in 1990, the industry was faced with something novel and new, and the old standard method was soon confined to history. With NRM2, the same impact could have been achieved, and, at the same time, a building standard method of measurement aligned to the exiting world of BIM could have been produced.
  What has been achieved is arguably an improvement on SMM7, but the overall effect leaves the feeling that SMM7 is still with us but wearing a different 'coat'. This is, of course, a personal view, and other commentators may well have a different opinion, but reading the tabulated rules, there are a lot of similarities with SMM7 that do not bear repeating in this book.
  The *Tabulated work sections* in NRM2 provide the specific rules for measuring excavation, concrete work, brickwork, plastering and the like which are known as components/items in NRM2. The work sections appear in Part 3: *Tabulated rules of measurement for building works*

of NRM2 and, more particularly, may be found under Paragraph 3.3: *Measurement rules for building works – Tabulated Work Sections.*

There are 41 tabulated work sections, including preliminaries, which is subdivided into:

- Preliminaries (main contract).
- Preliminaries (work package contract).

The tabulated work sections must be read in conjunction with the preceding text in NRM2 Parts 1, 2 and 3 which set the scene for how building work is to be measured and billed.

6.10.1   Changing from SMM7 to NRM2

It is not the function of this book to provide a line-by-line comparison between NRM2 and SMM7, but comparisons are, nonetheless, inevitable when changing from one method of measurement to another. Some of the differences are, therefore, highlighted in this part of the book, especially where the differences raise risk issues.

Although there are some significant differences between NRM2 and SMM7, it will no doubt be comforting for SMM7 users to find that much of the content of the NRM2 *Tabulated Work Sections* has been 'cut and pasted' from its predecessor and that the NRM2 rules of measurement are fairly easy to follow. However, the implications of how the rules work and how they differ from SMM7, despite the striking similarities, may not be so evident.

Lee et al. (2014) provide a useful overview of the differences between SMM7 and NRM2, presented in a tabular format, which is also available at the free *Designing Buildings Wiki* knowledge base.[4] This comparison identifies the SMM7 Work Section reference and the NRM2 equivalent as illustrated in Table 6.17.

Table 6.17   SMM7 v NRM2.

Ref	SMM7	Ref	NRM2
D	Groundwork	5	Excavation and filling
D20.2	Excavating	5.6	Rules now simplified into either bulk excavation or foundation excavation with depth ranges now only in 2 m stages
		5.7	Earthwork support has been simplified in that it is now only measured where specifically called for in the contract documents (specification)

Adapted from Lee et al. (2014).

Of particular concern with NRM2 are the alarming deficiencies in item coverage that appear in several work sections which leave the bill compiler with a problem as to how to formulate item descriptions that *fully describe and accurately represent the quantity* (NRM2 Paragraph 3.3.1) of the work concerned.

Risk Issue

The crucial difference between NRM2 and SMM7 is the way that item descriptions are compiled.

Lee et al. (2014) suggest that *the framing of descriptions is…not as simple following these new rules* and counsel that *care should be taken to ensure that each description adequately reflects the work to be priced.*

This is good advice as it is easy to be seduced into thinking that the NRM2 classification tables work the same way as those in SMM7. They don't, and incomplete, or even misleading, item descriptions could result from such a misunderstanding along with all that this implies.

It is important to remember, therefore, that the NRM2 classification table structure is quite different from that of SMM7.

The tables are not hierarchical and item descriptions are framed differently to those of SMM7, albeit that the NRM2 structure and numbering system may persuade otherwise. This important difference is illustrated in Table 6.18.

Table 6.18   Framing descriptions.

NRM2		SMM7	
Paragraph 3.3.3.3	Descriptions shall **state the building components/items being measured** (taken **from the first column** of the tabulated rules) **and** include <u>**all**</u> **Level 1, 2 and 3 information (taken from the third, fourth and fifth columns**, respectively) **applicable** to that item. **Where applicable**, the relevant **information from column five** shall be included in the description	General Rule 2.6	Each item description shall identify the work with respect to **one descriptive feature** drawn **from each of the first three columns** in the classification table **and as many of the descriptive features in the fourth column as are applicable** to the item

Each NRM2 Work Section lists the item or work to be measured and the attendant unit of measurement, but the Level 1, 2 and 3 descriptors provide 'information' to describe the item and not a hierarchy from which item descriptions are formulated. It is worth reiterating that NRM2 Paragraph 3.3.3.3 is instrumental to formulating item descriptions and that:

- the component/item to be measured is to be *taken from the first column of the tabulated rules.*
- *all of the level 1, 2 and 3 information* applicable to the item shall be included.
- *relevant information from column five shall be included* where applicable.

Therefore, whilst there is no rule to prevent the NRM2 tables from being used hierarchically, and it can be in some instances, the system is essentially a 'pick-and-mix' arrangement with the emphasis on creating items descriptions that fully describe the work concerned.

This creates some problems with library-based measurement software, which are essentially hierarchical in the way they work. With NRM2, a great deal more editing of item descriptions is needed in some instances, and this might been viewed by some practitioners as a backward step.

6.10.2   NRM2 and measurement

The measurement process using NRM2 is no different to using any other method of measurement, irrespective of whether traditional 'dim' sheets, cut and shuffle, on-screen measurement or measurement from a BIM model is used.

What must be carefully considered, however, is the final output required, that is, the BQ. NRM2 Paragraph 2.15.1.2 envisages three principal breakdown structures for BQ:

1. Elemental breakdown structure – where each section of the BQ is represented by a group element as defined in *NRM1: Order of cost estimating and cost planning for capital building works* (e.g. substructure, superstructure, internal finishes, etc.).
2. Work sectional breakdown structure – where the BQ is divided into work sections as defined in NRM2 (e.g. excavating and filling, piling, in situ concrete works, etc.).

3. Work package breakdown structure – where the BQ is divided into work packages which might be a specific 'trade' (e.g. plastering, painting and decorating) or a single package made up of a number of 'trades' (e.g. concrete works comprising formwork, rebar and in situ concrete trades). The work packages may be defined either by the employer, the quantity surveyor or by the contractor according to the procurement method chosen.

Further detail on these BQ breakdown structures is given in Paragraph A.1 of Appendix A.

Consequently, a means of re-sorting the take-off into one of these BQ formats must be anticipated. This could be done by using the traditional 'take-off-abstract-bill' system, by the 'cut and shuffle' method or by the bill sort facility within a measurement software package.

However, a coding system must also be thought about, and this would not be the same as the codes generated by using the tabulated work section references. Some sort of additional prefixing and suffixing, as suggested in NRM2 Paragraph 2.15, would be needed to arrive at the desired BQ format.

6.10.3   Phraseology

There are some words and phrases that appear in this part of Chapter 6 that do not appear in NRM2:

Bill compiler – This means the person responsible for preparing the BQ who could be the *quantity surveyor/cost manager* or *surveyor* referred to in NRM2 but could equally be a main contractor's or subcontractor's quantity surveyor.

Descriptor – This refers to the Level 1, 2 and 3 choices for creating item descriptions which are referred to in NRM2 Paragraph 3.3.3.3 as *information*.

General notes – This refers to the information given in the first two rows of each Work Section where the *mandatory information to be provided* and *works and materials deemed included* lists reside. These notes have general application to the whole of the relevant work section.

Notes – This refers to the *Notes, comments and glossary* given in the sixth column of the classification tables. These notes have particular application to the items, units and levels included within the boundaries of the solid horizontal lines drawn across the classification table.

6.10.4   Preliminaries

Most standard methods of measurement, including SMM7, CESMM4, MMHW and POM(I), take a simple and pragmatic approach to the 'preliminaries bill', four pages in CESMM4 and six pages in SMM7 being sufficient to:

- Briefly describe the works and the contract particulars.
- Specify the employers' requirements relating to the management, supervision and control of the works.
- Provide a place for the contractor to price general cost items such as site supervision, temporary accommodation and method-related charges.

Traditionally, the 'preliminaries bill' is the first 'bill' in the BQ for several logical reasons:

- Preliminaries is usually the first work section in the standard method of measurement.
- Part of the preliminaries bill provides essential 'general' information about the project which the contractor needs to read first before starting any pricing:
  ○ The identity of the employer and his consultants.
  ○ The conditions of contract and any amendments thereto.

- ○ The information about start and completion dates.
- ○ The rate of liquidated and ascertained damages.
- ■ The 'general' part of the preliminaries bill will alert tenderers to any specific conditions, hazards, restrictions or employer-imposed limitations on how the contractor may go about planning and organising the works.
- ■ The priced part of the preliminaries bill 'overarches' the measured work as it is largely linked to the contractor's intended programme and proposed method of working.
- ■ The priced method-related charges 'overarch' specific bill items as they represent the fixed costs associated with particular measured work items.

NRM2 Preliminaries structure

The approach to 'preliminaries' in NRM2 is more complex:

- ■ Work Section 1: *Preliminaries* spans 70 pages of text with a further 3½ pages of explanation, guidance and rules in NRM2 Paragraphs 2.7 and 3.2.2.
- ■ There are two categories in Work Section 1:
  - ○ Main contract preliminaries.
  - ○ Work package preliminaries.
- ■ Each category has two distinct parts:
  - ○ *Part A: Information requirements.*
  - ○ *Part B: Pricing schedule.*
- ■ NRM2 Appendices B and C provide the bill compiler with a choice of formats for the layout of the preliminaries section in the BQ:
  - ○ Condensed version.
  - ○ Expanded version.
- ■ According to NRM2 Paragraph 2.7.3.2, the bill compiler *should*, as part of the conditions of tender, *instruct the main contractor to return* with his tender *a full and detailed breakdown* of how the price for preliminaries has been calculated which should be appended to the priced BQ.

Part of the reasoning for this complexity is revealed in Paragraph 3.2.2.4(1) which explains that the 'pricing schedule' tables for preliminaries are structured so that *the contractor's pricing of preliminaries are captured under a number of headings*. This resonates with much of the *raison d'etre* of NRM2 which is the collection of contractors' pricing data for cost planning purposes as discussed earlier in Section 6.8.7.

Preliminaries (Main Contract)

The NRM2 preliminaries (main contract) work section is, at first glance, daunting because it is extremely long and detailed. However, it can be broken down into 'bite-sized' chunks:

- ■ **Part A: Information and requirements:**
  Part A is straightforward with nothing to measure. Its purpose is to describe the project, identify the conditions of contract and the names the participants, provide information about the site and existing buildings and list the drawings and pre-construction information.
- ■ **Part B: Pricing schedule:**
  Part B is in two parts:
  - ○ Employer's requirements (B 1.1)
    - • Part B 1.1 identifies the employer's site accommodation requirements, should separate facilities from those of the contractor be needed.
    - • It also identifies the site records to be provided by the contractor and certain completion and post-completion issues such as handover requirements and operational and maintenance matters.

○ Main contractor's cost items (B 1.2)
- Part B 1.2 lists the main contractor's preliminaries items.
  * The main contractor's pricing schedule can be:
    - Condensed (NRM2: Appendix B) (see Table 6.19).
    - Expanded (NRM2: Appendix C).

It should be noted that there is seemingly no compulsion to use any or all of the expanded NRM2 preliminaries headings as Paragraph 3.2.2.4 explains that the 'Component' column of the pricing schedule *lists the preliminaries items to be considered under each main heading*.

The billing of Part A is a straightforward 'word-processing' exercise to be added to the BQ outside the measurement package. Notwithstanding, QSPro does provide the NRM2 Part A 'menu' as an *aide-memoire*.

Table 6.19   Pricing schedule – main contract preliminaries (condensed).

Ref	Component (condensed list)	Time-related charges	Fixed charges	Total charges	Component (expanded list) example
	Preliminaries – Main contractor's cost items				
A	Management and staff				
B	Site establishment				
C	Temporary services				
D	Safety and environmental protection				
E	Control and protection				
F	**Mechanical plant**				• Generally
G	Temporary works				• Tower cranes
H	Site records				• Mobile cranes
I	Completion and post-completion requirements				• Hoists
					• Access plant
J	Cleaning				• Concrete plant
K	Fees and charges				• Other plant
L	Insurances, bonds, guarantees and warranties				
	Totals				
	Carried to pricing summary				

Preliminaries (Work Package Contract)

The preliminaries section for work package contracts follows the same pattern as for the main contract preliminaries, but Parts A and B are much briefer in content.

There is no provision within the Part B: *Pricing schedule* for 'employer's requirements', and this would seem to indicate that work package contracts are intended to come under the supervision of either a main contractor, a Tier 1 contractor or a 'supervising' work package contractor who will price a preliminaries (main contract) bill.

Measurement Units

In both the main contract and work package contract sections of Tabulated Work Section 1: *Preliminaries*, the Part B: *Pricing schedule* is measurable.

Common units of measurement include *week, item, nr, m²* and *m*. Alongside the 'Unit' is a column headed 'Pricing method' which indicates whether specific preliminaries items are to be priced by a *fixed charge* or a *time-related charge* or both. So far, so good!

Turning to NRM2 Appendices B and C, templates suggesting the manner in which the pricing schedule is to be laid out are supplied. These are in a 'spreadsheet' format as indicated in Table 6.20.

Table 6.20   Template for preliminaries.

Cost centre	Component	Time-related charges	Fixed charges	Total charges
		£ p	£ p	£ p
1.2	Main contractor's cost items			
1.2.1	Management and staff			
1.2.1.1	Project specific management and staff	A	B	A+B
1.2.2	Site establishment			
1.2.2.1	Site accommodation	C	D	C+D
1.2.3	Temporary services			
1.2.3.1	Temporary water supply	E	F	E+F
1.2.3.3	Temporary electricity supply	G	H	G+H
1.2.3.5	Temporary drainage	I	J	I+J
1.2.4	Security			
1.2.4.1	Security staff	K	L	K+L
1.2.4.3	Hoardings, fences and gates Etc.	M	N	M+N

The 'spreadsheets' provide columns for inserting prices for time-related and fixed charges and total charges, as indicated in Table 6.20. There are, however, no quantity, unit and rate columns which is a departure from convention.

Whether it should be assumed that lump sums are to be priced in these columns, or whether tenderers are meant to contrive their own columns, is not clear.

Both QSPro and CATO present the pricing schedule in a 'traditional' BQ format. In CATO, each item is accorded a unit, as illustrated in Table 6.21, the default quantity being '1'. Presumably, it is intended that each tenderer completes the quantities for each preliminaries item according to their intended programme and method of working.

An alternative to this approach would be to follow the NRM2 Appendix B and C layout at tender stage. The units of measurement would then come into play at a later stage when the contractor's *full and detailed breakdown* of the preliminaries is prepared (as required by NRM2 Paragraph 2.7.3.2).

Part 2

Table 6.21   Pricing schedule (part) – main contract preliminaries.

					£	p
	1 PRELIMINARIES (MAIN CONTRACT) : PRICING SCHEDULE					
	1.2 MAIN CONTRACTOR'S COST ITEMS					
	1.2.1 Management and staff					
	Project-specific management and staff					
A	time-related charge; construction manager	1	week			
	1.2.2 Site establishment					
	Site accommodation; offices					
B	fixed charge; purchase charges	ITEM				
C	time-related charge; hire charges	1	week			
	1.2.3 Temporary services					
	Temporary water supply					
D	fixed charge; temporary connections	1	nr			
E	time-related charge; temporary connections	1	nr			

This would seem logical, as only the contractor would be able to measure the quantities of items such as site accommodation, temporary works or hoardings, etc., to the level of detail indicated in NRM2.

Risk Issue

Contractors must be prepared to compile a *full and detailed breakdown* of their preliminaries at tender stage and *append this information to* [the] *priced bill of quantities*. Under NRM2, the quantity surveyor/ cost manager (PQS) is under a duty to ask for this information and, undoubtedly, will (NRM2 Paragraph 2.7.3.2 refers).

If this is the case, the breakdown will have to comply with NRM2 Work Section 1 and be measured in the units stipulated.

This will require a great deal more time and effort than pricing preliminaries under SMM7, or any other method of measurement for that matter.

Temporary Works

The *pricing schedule* of the main contract preliminaries of Work Section 1 provides for the contractor's pricing of temporary works, including:

- **Site accommodation:** *Temporary works in connection with site establishment* (e.g. bases and foundations).
- **Mechanical plant:** *Tower cranes* (e.g. temporary bases).
- **Site services:** *Temporary works* (e.g. temporary screens and façade retention works).

Similar, but not the same, provisions are included in the work package contract *pricing schedule*.

Temporary works are also provided for at 1.2.8: *Temporary works* of the main contract *pricing schedule*. They are categorised as:

1. Access scaffolding.
2. Temporary works.

Access scaffolding (1) includes *Common user access scaffolding*, but scaffolding specific to works packages and scaffold inspections are excluded (see 1.2.5). Temporary works (2) includes *Common user temporary works* such as:

- Support scaffolding and propping.
- Crash decks.
- Temporary protection to trees and vegetation.
- Floodlights.

Expressly <u>excluded</u> from this classification are:

- Temporary earthwork support (EWS) to basement excavations – presumably because basement excavation is measured under 5: *Excavation and filling* and EWS is deemed included.
- Temporary props and walings to support contiguous bored piled walls of basement excavations – inexplicably because such temporary works are <u>not</u> deemed included in either 5: *Excavation and filling* or 7: *Piling*.

However, there is no provision anywhere in the NRM2: *Preliminaries* pricing schedules for the contractor to price method-related charges, especially those in connection with excavation, geotechnical works and drainage:

- Earthwork support and working space are not measured under NRM2, and excavation and filling for temporary works is *deemed included* (see Work Section 5: *Excavation and filling*):
  - **NB:** Earthwork support **is measured** where <u>not</u> at the contractor's discretion (e.g. where designed by the employer's engineer and shown on the drawings).
- Earthwork support is also *deemed included* in Work Sections 9: *Diaphragm walls and embedded retaining walls*, 10: *Crib walls, gabions and reinforced earth*, 8: *Underpinning* and 34: *Drainage below ground*.
- Temporary works in connection with 3: *Demolitions*, 4: *Alterations, repairs and conservation* and 8: *Underpinning* are also *deemed included*.

Risk Issue

Under SMM7, where site conditions warranted the use of steel sheet piling, this was a measurable item by default, under D32.2, whether or not its use was at the contractor's discretion. The consequences of this were:

Advantages	Disadvantages
- Reduced risk for the contractor	- Increased risk for the employer
- The avoidance of a loss and expense claim	- A legitimate claim for an extension of time (with costs) as a variation is a relevant event
- A tender price not unnecessarily inflated by a contractor risk allowance for sheet piling that might not be required	- The problem of agreeing suitable rates for the piling work

Considering the detailed scrutiny that NRM2 places on the pricing of the contractor's preliminaries, it is unlikely that the contractor will 'get away with' pricing method-related charges in the temporary works parts of the preliminaries bill, and so the question arises as to just where they <u>can</u> be priced:

Part 2

Under NRM2, the situation would be different as steel sheet piling is not measurable unless specified in the contract:

Advantages

- Reduced risk for the employer

Disadvantages

- Increased risk for the contractor
- A tender price inflated by a contractor risk allowance for sheet piling that might not be required
- An inevitable loss and expense claim if the bore-hole data is inaccurate or misrepresented or if an experienced contractor could not have foreseen the need for steel sheet piling
- A legitimate claim for an extension of time (with costs) as part of the entitlement

- The quantity surveyor/cost manager could create a subset to 1.2.8: *Temporary works*, and this would appear to be in accordance with NRM2 (NRM2 Paragraph 3.2.2.4 4 refers).
- They could be included in the *schedule of construction risks* (NRM2: Appendix F) prepared by the quantity surveyor/cost manager and would then appear in the BQ *pricing summary*.
- The *schedule of construction risks* could be partially or fully left blank for the contractor to complete.
- The contractor could price the items in the rates.

Risk Issue

In view of the cost of temporary works to deep excavations, basements and the like and to the support of existing buildings and façades, contractors need to be alert as to how the bill of quantities has been prepared and where these items may be priced.

It is likely that such items will be scrutinised much more than under SMM7, and there is also the attendant risk that method-related charges apportioned to BQ rates may be under-recovered should the measured items be admeasured or varied.

6.10.5   Off-Site Manufactured Materials, Components or Buildings

Work Section 2 largely deals with the measurement of proprietary building components, units or structures that are manufactured off-site and then delivered to site for final incorporation into the building.

In providing this work section, NRM2 has embraced the trend towards modular system building for providing pre-finished fully serviced hotel bedroom and bathroom pods and the like which are designed and factory-built using BIM design and coordination methods. The advantages of defect-free construction, just-in-time delivery and fast track erection sequences are obvious.

The NRM2 Work Section 2 preamble makes it clear that the measured items are fully inclusive and include transport to site, fixing in position and connection to services. The work section *Notes* also make it clear that other work ancillary to the items measured in Work Section 2 is to be measured elsewhere. This would include such work as the provision of plumbing and electrical installations up to but excluding the final connection.

Measurement and description of the work items are straightforward. For an item of 256 nr bathroom pods for a new hotel, the NRM2 item code is 2.3.1.1, and the item might be billed as illustrated in Table 6.22.

Table 6.22 Off-site manufactured component.

					£	p
2 OFF-SITE MANUFACTURED MATERIALS, COMPONENTS OR BUILDINGS						
Component; Messrs Acme Bathroom pod; Metal studding; Bolted to structure						
Prefabricated building units; toilet/bathroom units						
A 1.76 x 2.15 x 2.27 overall dimensions	256	nr				

6.10.6   Demolitions

Demolition of all existing structures, individual structures or parts of structures are itemised (3.1.*) as is decontamination of the site, such as the removal of hazardous materials (3.4.*).

The contractor is advised to visit the site and *the surveyor* is required to provide additional information concerning temporary works *if not readily ascertained from the drawings*. The temporary works concerned are measured items (3.3.*.*.*) and could well be extensive.

Part 2

> **Risk Issue**
>
> There is no definition of who *the surveyor* is in NRM2 Paragraph 1.6.3, despite there being a clear duty to provide supplementary information to tenderers.
>
> As the supplementary information may be required to convey *the full extent and scope* of the work, it might be assumed that the quantity surveyor/cost manager will undertake the duty.
>
> In view of what might have to be measured, the duty could well be onerous.

Temporary works other than these items are *deemed included*. This might include temporary propping between floors, propping openings and refuse chutes.

Disposal of debris is *deemed included*.

6.10.7   Alterations, Repairs and Conservation

'Spot items' – measured in previous methods of measurement, including SMM7 – do not appear in NRM2. Instead, provision is made for the measurement of a variety of works of adaption, alteration, repair and renovation, all of which are to be measured in detail.

A choice of units of measurement is available, and, in a number of instances, the notes dictate that this shall be at *the discretion of the surveyor*.

> **Risk Issue**
>
> In common with Work Section 3: *Demolitions, the surveyor* has a duty to provide supplementary information to tenderers without there being a definition of who *the surveyor* is.
>
> This duty applies to Temporary works (4.24.*.*.*) where there appears to be a typographical error with the word 'Roads'. This should be a Level 1 descriptor (4.24.4) and not a work item.

6.10.8   Excavating and Filling

Excavation

The measurement rules dealing with excavation have changed, both in terms of the types of excavation and the depth categories.

Excavation is now classed as *bulk excavation* and *foundation excavation* only with stage depths to be given in 2 m steps and any obstructions in the ground to be stated. Obstructions are nothing to do with ground conditions but relate to piles and manholes that must remain undisturbed during the excavations. Strangely, Note 5.6.1.*.2 only relates to bulk excavation.

Bulk excavation includes excavating to form basements, pools, ponds and the like, and each may be described separately if desired (5.6.1.* Note 1). Foundation excavation includes strip and pad foundations, pile caps and *all other types of foundations* (5.6.2.* Note 1) which may equally be measured separately (5.6.2.* Note 2).

This is all somewhat prosaic but, more important, is the issue of excavation and filling to working space or earthwork support.

Filling

Filling, with both excavated material and imported fill, is classed as less than or exceeding 500 mm, with the exception of imported blinding not exceeding 50 mm thick.

Excavation and Filling to Working Space and Earthwork Support

Working space is not measured in NRM2, and neither is earthwork support, unless specified or instructed (see 5.8.*). However, where the type of backfill to *the extra space taken up by working space or earthwork support* is <u>not</u> left to the discretion of the contractor, then an allowance shall be made in the quantities for excavation and filling (see Note 3 in Work Section 5 'General rules'). The same rule applies to the 'extra over' items measured under 5.7.*.

Consequently, if a drawing note or specification clause specifies that the contractor shall backfill working space, and/or the space taken up by earthwork support, with a particular type of fill, then both the additional excavation and additional fill would be measurable. There is no stated limit to the type of fill that might be specified, and therefore, a provision for *backfilling with selected excavated material* would seem to be caught by this measurement rule.

This would be the case even if the choice of working space or earthwork support provision was at the contractor's discretion.

Risk Issue

This is an impossible item for the quantity surveyor/cost manager to measure unless a provisional quantity is to be given in the BQ.

As there is no provision within NRM2 to measure a provisional item exclusively for the excavation and filling of working space/earthwork support, the additional quantities would have to be added to the general excavation and filling items. This would make the entire excavation and filling quantity 'provisional'.

If so, the provisional quantity would be subject to the provisions of NRM2 Paragraph 3.3.8.2 which would also mean subjecting the provisional quantity to the '20% rule' test for a possible re-rate to the item.

Hard Materials

The breaking up of rock, and other hard materials, is measured extra over all types of excavation, irrespective of depth, in m³ under 5.7.2.* The SMM7 categories of breaking out 'existing materials' and 'existing hard pavings' have disappeared, with the latter having moved to the site preparation item (5.5.3.*)

This is a straightforward measurement item except that the usual 'chestnut' of the definition of 'rock' must be considered.

In NRM2, 'rock' is slightly different to the SMM7 definition, being:

*any hard material which is of such size or **location** that it can only be removed by the use of wedges, **rock hammers**, special plant or explosives*; the differences are highlighted in bold. (Note 1 of item 5.7.2.1)

- This is not a geological definition and therefore, even if the strata to be removed comes within a geological definition of rock, such material is not to be measured as 'rock' if its size or location enables it to be removed with normal excavating plant not deemed to be *special*.

Risk Issue

Note 3 of item 5.7.2.1 is less precise, because whilst this states that *degraded or friable rock that can be scraped out by the excavator bucket does not constitute rock*, it also implies that, if the excavator being used on-site is not 'man enough' to dig the rock, it will be measured as rock should a larger machine have to be brought to site, as, presumably, this would qualify as 'special plant'.

Note 2 of item 5.7.2.1 limits rock to:

1. *A boulder ≤5 m³* or
2. *one that can be lifted out in the bucket of an excavator.*

There are problems with this provision of NRM2.

Risk Issue

1. The first part of the provision says that a boulder <u>less than</u>, but more importantly equal to 5 m3, is <u>not</u> classed as rock. The cube root of 5 is 1.71, and therefore, a boulder measuring up to 1.71 m × 1.71 m × 1.71 m is not classed as rock according to Note 2. This must be a mistake (suggest 0.5 m³).
2. As for the second part, the ability of an excavator to pick up a boulder in its bucket is a matter of opinion, and not a matter of fact, and is a test that requires the presence of an employer's representative on-site at a precise moment in time. Methods of measurement depend on rules that are clear and definitive, and this provision is neither.

Contractors should be wary of this provision and would be wise to look for a 'notwithstanding NRM2' preamble that at least deals with the first part of Note 2 more sensibly.

Support to Faces of Excavation

Earthwork support, or more correctly *support to faces of excavation*, is not measurable under Work Section 5: *Excavating and filling* unless:

- Support to the faces of excavation is not at the contractor's discretion (General note) <u>and</u> is specified in the contract documents (Note 5.8.1) <u>or</u> is instructed by the contract administrator during the course of the works (Note 5.8.1).

> ## Risk Issue
>
> It would be unwise for contractors to assume that the existence of a provision in Work Section 5, to measure support to faces of excavation, is similar to the SMM7 D20 *Definition Rule* D5. This provides for interlocking steel piling to be measured if this means of earthwork support should become necessary when on-site.
>
> The D5 rule was a relief granted to the contractor who knew, when pricing the earthwork support item measured in the contract bills, that the item coverage did not include interlocking steel piling.

It is submitted that the NRM2 provision is not the same as SMM7 D20 *Definition Rule* D5 and would be used only where:

Pre-contract
1. The engineer wishes to employ an earth support system compatible with or integral with the foundation design (e.g. a contiguous piled wall or diaphragm wall).
2. The employer wishes to relieve the contractor of the below-ground risk for earthwork support.
3. The employer's insurers are not prepared to risk the possibility of relieving adjacent buildings of their support on the basis of a contractor's design.
4. The cost of the earth support system and its design would disproportionate to the value of the remainder of the works.
5. Contractor design is either unwanted or impractical in view of the time allowed for tendering.

Post-contract
1. Ground conditions were misrepresented in the tender documents.
2. Ground conditions on-site were such that an experienced contractor could not have anticipated them.

The NRM2 measured item is for the *Support to the face(s) of excavation where not at the discretion of the contractor*, measured in m², with the location and method of forming the support stated.

It is strange, therefore, at the Level 3 descriptor, to find that the method of forming the support is to be stated in the item description *only where not left to the discretion of the contractor*.

This could imply that:

- The support system has been designed by the employer's engineer, and there is a drawing and specification for the work (i.e. no contractor discretion).
- The engineer has prepared more than one design (i.e. not at the contractor's discretion) with the choice being left to the contractor (i.e. at the contractor's discretion).
- There could be a performance specification with the tender documents (i.e. not at the contractor's discretion), and the method of forming the support is to be left to the contractor (i.e. at the contractor's discretion).
- The engineer decides that site conditions are such that the contractor's choice of earth support system (i.e. contractor discretion) was forced upon him by unforeseen ground conditions (i.e. not really at the contractor's discretion).

Water and Water-Bearing Ground

Excavation in water-bearing ground poses a significant risk for contractors both at tender stage and during construction. At tender stage, tenderers need to be sure that they are competing on a 'level playing field', and during construction, contractors need to know where their tender risk allowance stops and grounds for a legitimate entitlement claim begin.

Some standard conditions of contract make the issue clear, and both the ECC and ICC forms place the risk firmly with the contractor up to the point where an experienced contractor could not have judged the prevailing ground conditions. The JCT forms are silent on the issue, and this places the risk for dealing with below groundwater squarely with the contractor according to case law. Traditionally, however, JCT contracts are used in conjunction with SMM7 which grants contractors relief from certain ground conditions via the rules of measurement.

The rules of measurement dealing with water-bearing ground are more complex under NRM2, however, as the comparison with SMM7 in Table 6.23 indicates.

Table 6.23   Water-bearing ground – SMM7 v NRM2.

	SMM7			NRM2	
Item ref.	**Measured item**		**Item ref.**	**Measured item**	
D20.3.1	Extra over excavation for excavating below groundwater level	m³	5.7.1.3	Extra over for excavating below groundwater level	m³
			5.7.1.5	Extra over for excavating in unstable ground[5]	m³
D20.7.*.*.2	Earthwork support belowground water level[1]	m²			
D20.7.*.*.3	Earthwork support; unstable ground[1,2]	m²			
			5.8.*	Support to faces of excavation where not at the discretion of the contractor	m²
D20.8.2	Disposal of groundwater[3]	Item	5.9.1.*	Disposal of groundwater	item
D32.2.*	Interlocking steel piling is measured if needed[4]	m²			
			6.1.*	Site dewatering	Item

Notes

[1] Measured full depth of excavation

[2] Unstable ground is running silt, running sand, loose **gravel** and the like

[3] Only measured where there is a corresponding D20.3.1 item

[4] D20 Definition Rule D5

[5] Unstable ground is running silt, running sand, loose **ground** and the like

Under NRM2, the measurement of items for excavating and filling are deemed to include for disposal of surface water as per the 'general rules' of Work Section 5. This does not include dealing with **groundwater** which is a measurable item *Extra over all types of excavation irrespective of depth* (5.7.1.*).

Groundwater is defined in Work Section 5.7.1 Note 2 as *any water encountered below the established water table level*, excluding water arising from streams, broken drains, culverts or surface flooding, and also excludes running water from springs, streams or rivers. The water table is required to be re-established at the time each excavation is carried out, and this becomes the *post-contract ground water level* as per Work Section 5 'general rules'.

Part 2

Also measurable in Work Section 5 is *Excavating in unstable ground*, which is measured extra over all types of excavation (5.7.1.5). Unstable ground is defined as *running silt, running sand, loose ground and the like* (Note 4 of 5.7.1.5 refers).

In water-bearing ground, running silt and running sand, groundwater is naturally present, and, as such, tenderers could rightly expect to see a measured item in the BQ according to the rules of Work Section 5. The same could be expected for unstable ground conditions, which condition is also measurable under Work Section 5.

However, the quantity surveyor/cost manager is faced with a potential problem when measuring work of this nature because:

- water-bearing ground is not necessarily unstable and
- unstable ground is not necessarily water bearing, but the two conditions could be present in the same excavation.

A further issue with NRM2 is that Work Section 6: *Ground remediation and soil stabilisation* provides a measurable item for site dewatering, whether or not the choice of dewatering method is at the contractor's discretion.

Dealing with water and water-bearing ground is a tricky measurement issue in NRM2, and it is felt that a case study might help to explain the complex issues and provide a worked example of how this problem could be dealt with. Consequently, Chapter 14 is devoted to measuring the excavation items relating to a deep basement to be excavated in difficult ground.

Disposal

Work Section 5.9 contains two measurable items:

5.9.1   Groundwater:

- Unlike SMM7, the disposal of surface water is not measurable in NRM2 but is *deemed included* in Work Section 5 'general rules'. The disposal of groundwater, on the other hand, is measurable, stating the depth below original ground level and any known polluted water (5.9.1.*).

- This item needs to be read carefully in conjunction with measurable items for extra over for excavating in water-bearing ground (5.7.1.3–5) and with 6.1.*.1.* – site dewatering (Section 6.10.8.2 refers).

- Unlike SMM7 (as well used phrase in this chapter!), there is no measurement rule in NRM2 that limits the measurement of a disposal of groundwater item to the measurement of a corresponding item for extra over for excavating below groundwater level (SMM7 D20.8 M12 refers).

5.9.2   Excavated material off-site:

- Excavated material to be disposed of off-site is measured in m³, with a stated destination if not left to the contractor's discretion.

- Material for disposal that is a 'controlled waste' also has to be classified as hazardous or non-hazardous, but there is no classification of 'inert' waste.

- The only real issue with the 5.9.2.* disposal item in NRM2 is the lack of a unit of measurement. Ostrowski (2013) believes that disposal off-site should be measured as an 'item' because this is the unit of measurement for the other 'disposal' item – 5.9.1.* disposal of groundwater. However, as there is a solid line between the two items, the unit of measurement is not shared (see NRM2 Paragraph 3.2.1.4), and there can be no presumed common unit of measurement.

- Common sense would indicate that the unit of measurement should be m³ as, not only is m³ the unit of measurement for excavation, but it is also custom and practice in the industry across all common methods of measurement. In Work Section 7: *Piling*, disposal of excavated material is measured in m³, and the common library-based software packages adopt the same pragmatic approach.

Risk Issue

The lack of a unit of measurement in a standard method of measurement is not so much a major drama as an inconvenience.

Clearly, if the quantity surveyor/cost manager adopts a unit of measurement that is contrary to the rules, this would be an issue, unless highlighted in a preamble to the bill of quantities.

Should there be no unit of measurement stated in the method of measurement, no measurement unit adopted could be contrary to the rules, by definition, but it might be prudent to clarify the chosen unit in a preamble.

6.10.9   Ground Remediation and Soil Stabilisation

To SMM7 aficionados, NRM2 Work Section 6 is 'new' as there were no specific rules for ground investigation, soil stabilisation or site dewatering in the outgoing method of measurement. By contrast, the NRM2 provisions reflect modern geotechnics by providing measurement rules for ground freezing, gas venting, soil nailing, ground anchors, pressure grouting and the like.

Dynamic deep compaction (DDC), used for compacting weak soils on marginal sites, is also included (*Compacting*: 6.9.*.1). This technique can improve the subsoil to the extent that shallow foundations can replace the need for piling and deep excavation, following treatment. As with other geotechnical process, DDC is designed to a performance specification, and the rules of measurement reflect the need to state the method and extent of treatment required.

There is little technical detail provided in the NRM2 item descriptor tables, and library-based measurement packages leave the bill compiler to 'fill in the blanks' with such detailed design and specification information as may be available. Table 6.24 illustrates how an item description may be 'fleshed out' according to the rules of measurement.

Table 6.24   Ground remediation.

					£	p
	6 GROUND REMEDIATION AND SOIL STABILISATION					
	Compacting					
	Generally					
A	Dynamic deep compaction; Site areas A and B; Drawing No BD137/D; 10 tonne compactor; Fall height 15 m; Five drops per pass to grid pattern; Three passes	11808	m2			

Part 2

Site Dewatering

Site dewatering is now a measurable item.

The method of disposal of water may be at the contractor's discretion or, alternatively, may be stated in the item description, as shown in Table 6.25, together with both the pre-contract water level and the level to which groundwater must be lowered and maintained.

Table 6.25 Site dewatering.

						£	p
6 GROUND REMEDIATION AND SOIL STABILISATION							
Site dewatering							
Well point dewatering							
A	area of site to be dewatered 950 m2; maximum depth of boreholes 4.5 m; pre-contract water level 109.75 m; water level to be lowered to 105.25 m	ITEM					

The coding for such an item would be 6.1.*.1.* as both Level 1 and 3 item descriptors are required to complete the description. It is a frailty of the NRM2 classification system that, if there are unwanted choices within the item descriptor 'levels', coding is impossible beyond the point where the unwanted choices reside.

More to the point, however, is the relationship of 6.1: *Site dewatering* to the items for extra over excavation for excavating in water-bearing and unstable ground in 5.7.1.4/5: *Excavating and filling*.

6.10.10   Piling

The measurement rules for piling have been simplified in NRM2 (Work Section 7) with the different pile types collected sensibly in one place. Steel sheet piling has also been added to this new 'family', and vibro-compacted piles and trench fill make a welcome appearance.

Risk Issue

There is no item coverage in Work Section 7 for props and walings to sheet piled or contiguous piled walls.

This is no different to the provisions of SMM7, but it is surprising that this deficiency has not been redressed in NRM2, particularly as the *Preliminaries* Work Section 1.2.8.2 (Note 6), expressly excludes *props and walings to support contiguous bored pile walls of basement excavations* from its item coverage.

Clearly, the authors of NRM2 have thought about the 'props and walings' issue, and it is a shame that an opportunity has been missed.

Cutting off the tops of piles, and preparation of rebar for pile caps and ground beams, has, logically, been moved to Work Section 5: *Excavating and filling* (5.20.1 refers) as this work is not usually carried out by piling contractors.

6.10.11   Underpinning

Work Section 8: *Underpinning* is unrecognisable compared with SMM7.

Gone is the detailed measurement of what might be called 'traditional' underpinning to be replaced with a single item, 8.1: *Underpinning*. At Level 1, descriptors can be chosen for *foundations*, *walls* and *bases*, and, at Level 2, scope for a description of the work is provided.

A possible BQ description for a traditional underpinning item is illustrated in Table 6.26.

Table 6.26   Underpinning – 1.

		Qty	Unit	Rate	£	p
	Underpinning					
	Underpinning to existing basement; Class C30 concrete foundation; One brick thick common brickwork in CM 1:3; Pinned to existing foundation; Backfill excavation with Type 2 Sub-base					
	Underpinning					
A	Foundations; Class C30 concrete	82	m			
B	Walls; One brick thick common bricks in CM 1:3	82	m			

However, the rules of measurement are not exactly crystal clear as to what these items actually mean because, whilst the *deemed included* list is a lengthy 12 items, and comprises temporary support, excavation, earthwork support and backfilling, the Level 1 descriptors (*foundations*, *walls* and *bases*) have no accompanying notes or item coverage guidance. Therefore,

- Is the foundation item meant to include all the *deemed included* items?
- Is the wall item meant to include some of them (e.g. cutting away existing foundations/footings, preparing the underside of existing work)?

An alternative way of describing the same item of work is shown in Table 6.27, which may be considered more satisfactory in terms of the *deemed included* items but may be more difficult to value.

Table 6.27   Underpinning – 2.

					£	p
	8 UNDERPINNING					
	Underpinning					
	Underpinning to existing basement; Class C30 concrete foundation; One brick thick common brickwork in CM 1:3; Pinned to existing foundation; Backfill excavation with Type 2 Sub-base					
A	foundations; 2 deep; 1.5 m maximum width	48	m			

Risk Issue

It may seem trivial, but when it comes to valuing work in progress and, especially, variations, it is important to understand what is included in the contractor's rates.

Of relevance is Note 1 in the 'general rules' which says that extensive underpinning work *may be measured separately in accordance with the rules of the relevant trades or Work Sections*, provided that the work is described as '*in underpinning*'. Underpinning with mini-piling would come under this rule.

Strangely, the Note 1 comment is repeated underneath 8.1.*, whereafter items for concrete, formwork, reinforcement, brickwork or blockwork and tanking are provided, along with appropriate units.

6.10.12   Diaphragm Walls and Embedded Retaining Walls

In Work Section 9, it is no longer necessary to measure *excavation and disposal* items for diaphragm walls (SMM7 D40.1 refers) as NRM2 includes this in the list of *deemed included* items.

Backfilling is also included in the item coverage, but there is no specific mention of backfilling empty trenches.

Risk Issue

Of particular concern is that guide walls are not measured in NRM2, nor are they mentioned in the *deemed included* list.

This is a glaring omission of a significant cost item.

The quantity surveyor/cost manager may wish to consider two options to deal with this issue:

- Ensure that relevant information is shown on the drawings to indicate the *extent of work* as provided in the general rules to the Work Section.
- Provide information using the Level 2 descriptor: *details and method of construction*.

A means of ensuring clarity to the contractor's pricing is essential as it might be difficult to resist a claim should the item coverage remain unclear.

In this regard, guide walls are measured in CESMM4 Class C.5.8. They are not measured in the MMHW, but MMHW Chapter III: *General directions* states that *The rates and prices entered in the Bill of Quantities shall be deemed to be the full inclusive value of the work* and this includes *Temporary Works*. NRM2 clearly needs some clarification.

6.10.13   In Situ Concrete Works

Not being tied anymore to the CAWS has enabled NRM2 to make substantial changes to the measurement of concrete work.

Firstly, in situ concrete and precast concrete have gone their separate ways into Work Sections 11 and 12/13, respectively.

Next, in situ concrete is not now measured in terms of structural components (e.g. floors, beams, columns, etc.) but is described as horizontal, sloping or vertical work. It is also now clear that *mass concrete* is any un-reinforced concrete not measured elsewhere and that concrete is cast into formwork *unless otherwise described* (using the Level 3 descriptor provided in the classification table).

The notes to Work Section 11: *In situ concrete works* are important as they identify what horizontal, sloping and vertical work consist of. This is because the work section classification is very much 'slimmed down' and is not at all clear. Table 6.28 illustrates what may be found under each class of in situ concrete work.

It must be stressed that the 'mass concrete' classification is not exclusive and that mass concrete may be measured under horizontal, sloping or vertical work where appropriate. For instance, an un-reinforced concrete ramp would be measured under 'sloping work' and not 'mass concrete'.

Table 6.28 Classification of in situ concrete work.

In situ concrete			
Mass concrete	**Horizontal work**	**Sloping work**	**Vertical work**
• Any un-reinforced bulk concrete not measured elsewhere	• Blinding	• Blinding	• Columns, attached columns and column casings
	• Beds • Foundations • Pile caps	• Beds • Slabs • Steps and staircases	• Walls • Retaining walls • Filling to hollow walls
	• Column bases	• Kerbs	• Parapets or upstand beams (where height ≤3 × width)
	• Ground beams • Slabs • Landings • Beams, attached beams and beam casings • Shear heads • Upstands (where height ≤3 × width) • Kerbs • Copings	• Copings • Attached beams • Upstands • Shear heads	

The Work Section 11 notes also indicate that each type of concrete work can be aggregated or measured separately. If separate items are to be given (e.g. for columns, walls and retaining walls), an additional item descriptor to those provided in the classification table will be needed as illustrated in Table 6.29.

Table 6.29 In situ concrete – additional description.

	'In-situ' concrete works			
	Reinforced in-situ concrete: Class C30			
A	Vertical work; <= 300 thick; In structures; Columns	9	m3	
B	Vertical work; <= 300 thick; In structures; Attached columns	2	m3	
C	Vertical work; <= 300 thick; In structures; Walls	18	m3	

Formwork items are classified differently to in situ concrete, and NRM2 reintroduces the idea of structural components in this part of Work Section 11. Consequently, formwork is measured variously to sides of **foundations**, to soffits of **horizontal work**, to sides and soffits of attached **beams** and to sides of isolated **columns**.

Whilst more concise than SMM7, with a number of minor items now confined to a *deemed included* list, the formwork section of NRM2 still, surprisingly, requires plain and suspended wall kickers to be measured, whilst other kickers are *deemed included*.

6.10.14   Structural Metalwork

Work Section 15: *Structural metalwork*, which includes structural steelwork and structural aluminium work, has some major changes compared with SMM7 G10–G12.

Taking structural steelwork as an example, the NRM2 classification has rationalised and simplified the provisions of SMM7, but framed members are still subdivided into fabrication and erection. Isolated members are still measured separately but have been assimilated more sensibly in the work section.

The weight classification has changed – there are now four levels instead of three as previously – and the classification of members has been rationalised at Level 3. The identification of castellated, tapered, curved and hollow members has been relegated to the *Notes, comments and glossary* column, and so, where applicable, NRM2 Paragraph 3.3.3.3 applies. There is no place for portal frames, but Note 7 does refer to *compound fabrications*.

Fittings were included in the mass of steel in SMM7, but they are now measured separately, stating whether they are to framed or isolated members. Measurement of fittings is either by calculated weight (15.5.1) or percentage (15.5.2). No distinction is made between fittings as part of fabrication and those fixed on-site.

> **Risk Issue**
>
> If a percentage is used, this will need to be marked PROVISIONAL in accordance with NRM2 Paragraph 3.3.8.1 and remeasured on completion as per Paragraph 3.3.8.2.
>   The 20% rule (Paragraph 3.3.8.2) will thus apply.

Measurement of steelwork is subdivided as before into *fabrication* and *permanent erection on-site*, but this is where the similarity to SMM7 ends, as Table 6.30 demonstrates:

- Trial erection has been afforded a completely separate status from the erection of framed members (15.13.1) but is only measured when NOT at the contractor's discretion.
- Permanent erection is now deemed to include delivery to site.
- Permanent erection is also deemed include all **specified** operations subsequent to fabrication.
- The site drilling of holes and site welding are not deemed included and must be assumed to qualify as specified operations subsequent to fabrication.

A simplified example of the measurement of structural steelwork is provided in Figure 6.9 which also illustrates (a) framed members, (b–d) typical fittings and (e) holding down bolts.

Profiled Metal Decking

This is an interesting topic that merits some detailed comparison with SMM7.

Profiled metal decking is essentially a means of supporting a cast in situ concrete suspended slab during construction. It remains in place once the concrete has set and, to all intents and purposes, acts as 'permanent formwork' in the same way that proprietary open mesh expanded metal sheets do.

Profiled metal decking may be attached to the structural steel frame by shot firing or using self-tapping screws or may be welded through the deck to the steelwork with shear stud connectors.

From a measurement perspective, the method of fixing profiled metal decking is crucial because certain methods do not create a floor that is a composite structural union between the concrete slab and the steel frame and others (e.g. shear stud connectors) do.

Table 6.30 Structural steelwork – framed.

SMM7		NRM2			
	Rules	**Works and materials deemed included**		**Mandatory information to be provided**	
D1	Fabrication includes all operations up to and including delivery to site	3	Permanent erection is deemed to include all **specified** operations subsequent to fabrication including delivery to site	2	Specification describing fabrication, welding, testing, erection and everything else necessary to complete the installation
D5	Erection includes all operations subsequent to fabrication			**NB:** From the above, it would appear that welding and testing are as much <u>not</u> part of fabrication as erection	
M1	The mass of framing includes all components and fittings of the same material		An allowance for fittings is separately measured, either by weight or a percentage		

NB:
It is assumed that 'fabrication' would normally include:
- Cutting steel members to length, including splay cuts
- Drilling holes for shop and site fixing
- Attaching cleats, brackets, etc. by bolting or welding (but see NB above)

Summary			
	Fabrication	**Transport**	**Erection**
SMM7	G10 1.* Fabrication	G10.2.1 Trial erection	G10.2.2 Permanent erection
NRM2	15.1.*.*.* Fabrication	15.13.* Trial erection	15.2.*.*.* Permanent erection

In SMM7, permanent formwork is a measured in two places:

- E20: *Formwork to in situ concrete.*
- G10: *Structural steel framing.*

Therefore,

- When permanent formwork to slabs is *structurally integral with the framing*, it is measured under G10.3.1, with the method of fixing stated.
- But when permanent formwork is simply *designed to remain in position*, it is measured in accordance with E20.8.*.*.4.

This is fairly straightforward, but SMM7 also provides the option of measuring profiled metal decking under G30: *Metal profiled sheet decking* with no test as to whether it is fixed

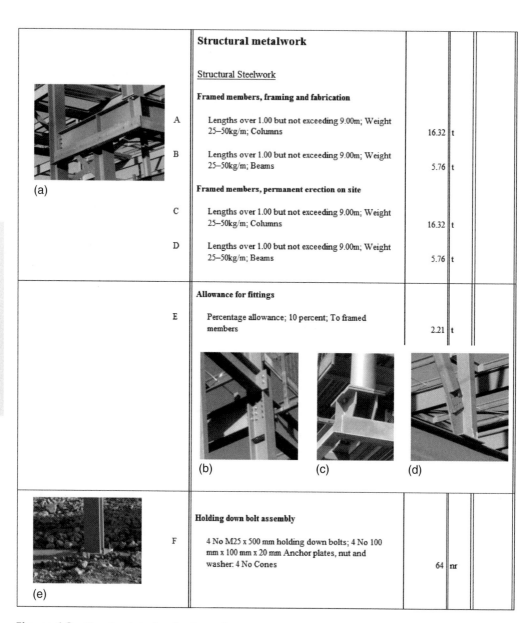

Figure 6.9   Structural steelwork – framed.

in a way that it will be *structurally integral with the framing.* On-site or off-site labours, such as holes and notches, are also measured under G30.3.*.*.

The three situations are illustrated in Figure 6.10.

Thankfully, this confusing issue has been simplified in NRM2, which measures permanent formwork under Work Section 11.15.*.* (General note 4 refers), but permanent formwork that is integral with the framing is described as *profiled metal decking* and measured under Work Section 15.8.*: *Structural metalwork.* The options are illustrated in Table 6.31.

There is an added complication, however, in the form of 15.8.* Note 1 which states that:

- Profiled metal decking *is only measured here* (i.e. Work Section 15.8.*) *when it forms part of the structural steel package otherwise it would be measured in accordance with the rules for permanent formwork* (i.e. 11.15.*.*).

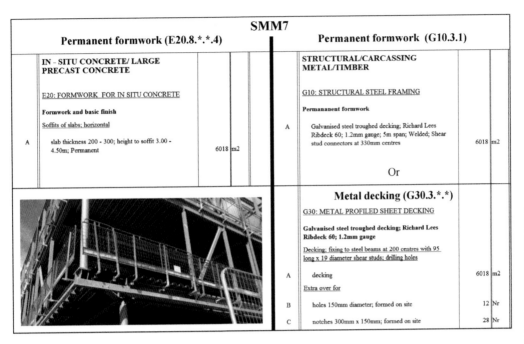

Figure 6.10   Profiled decking/permanent formwork – SMM7.

Table 6.31   Profiled decking/permanent formwork – NRM2.

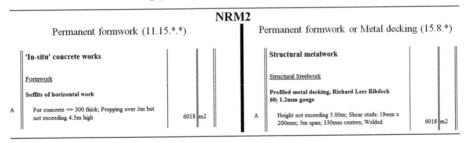

A further issue is that edge trims to profiled metal decking are measured extra over item 15.8.* but are not joined with Note 1.

Risk Issue

In a traditional tendering situation, the bill compiler cannot possibly make the judgement as to how each tenderer will package the work as there are several options:

- Include the profiled metal decking with the structural steelwork work package.
- Separately package the profiled metal decking to a specialist installer.
- Include the profiled metal decking in the concreting work package with/without the through-deck welding work.
- Install the profiled metal decking with own labour and sublet the through-deck welding work.

6.10.15   Precast/Composite Concrete

The use of the forward slash (/) – also called slant, oblique or solidus – in the title of Work Section 12 is a carry-over from SMM7 (E60: Precast/Composite concrete decking).

Without being over-pedantic, the use of this symbol in SMM7 indicates a choice between precast and composite concrete decking and, therefore, may be taken to have the meaning 'or'. In NRM2, however, precast concrete decking is measured under Work Section 13, and precast/composite concrete decking is measured under Work Section 12. Consequently, the (/) symbol is more likely to indicate composite concrete construction that includes an element of precast concrete rather than a choice between the two.

The term 'composite construction' is used to describe any method of construction that combines two or more dissimilar materials to form an element of construction such as a wall, partition, decking or floor.

In Work Section 12, a composite concrete floor comprising precast concrete planks and in situ concrete topping incorporating either fibre or mesh fabric reinforcement would be coded as 12.2.1.1.

Risk Issue

The Level 2 code (2) should be used because there is nowhere else in NRM2 to provide a description of the in situ concrete element:

- The Work Section 11 'deemed included' preamble does not mention in situ concrete.
- Work Section 11: *In situ concrete works* does not cover composite work.
- The notes to Work Section 11.2 – *Horizontal work* refer to coffered and troughed slabs but not to composite slabs.

6.10.16   Suspended Ceilings

Work Section 30: *Suspended ceilings* largely follows the layout of SMM7 K40: *Suspended ceilings* but with some additions and omissions from the list of measured items.

Risk Issue

A glaring omission from Work Section 30 is the lack of a *deemed included* rule to cover the suspension system for suspended ceilings.

This is *deemed included* in K40 Coverage Rule C1, but there is no similar rule in NRM2. *Mandatory information to be provided* requirements in NRM2 include details of the *construction of suspension framing and systems*, but this is not an item coverage rule.

This issue may need to be clarified with a 'notwithstanding' provision in the bill of quantities.

Alternatively, it could be argued that a suspended ceiling cannot, by definition, be 'suspended' without a suspension system, and therefore, this constitutes an *implied included* rather than a *deemed included* item.

6.10.17   Drainage Below Ground

Work Section 34: *Drainage below ground* is significantly different to the equivalent R12: *Drainage below ground* in SMM7.

Drain Runs

The first notable change is that 'drain runs' are now measured as composite items and are no longer measured in detail, that is, with excavation measured separately from beds, pipes, etc.

From the point of view of coding the take-off, this work section is very badly organised as discussed in detail in Section 6.8.6 of this chapter. Notwithstanding this, library-based software houses have taken the pragmatic, and helpful, view of creating a hierarchical structure to the formulation of item descriptions for 'drain runs' as illustrated in Table 6.32.

Table 6.32   Formulating item descriptions for 'drain runs'.

					£	p
34 DRAINAGE BELOW GROUND						
Storm water drain systems						
Drain runs						
A	average trench depth 1.50 m; push fit joints; pipe bedding and or surround; granular; Type 1 sub-base backfill; below ground water level	27	m			

Risk Issue

A fundamental flaw in the classification table is that neither the item to be described (34.1.*: *Drain runs*) nor the list of *work and materials deemed included* mentions excavation. The deemed included list includes earthwork support, trimming of excavations, backfilling and disposal but not excavation. (NB: The same problem applies to chambers under 34.6–10.*)

This is to be contrasted with SMM7 which employs the same item coverage list but also uses the phrase *excavating trenches* in the first column of the classification table and measures *Excavation* to chambers.

Consequently, unless 'excavation' is to be considered an implied term of Work Section 34, this issue needs to be clarified, perhaps in a preamble to the bill of quantities.

Depth classifications for drain runs are now in 500 mm increments with the average depth for each drain run being calculated without reference to the maximum depth. A sensible clarification is that measurements are now taken between the external faces of manholes.

Risk Issue

However, there is no requirement to state a commencing surface for excavation which, under SMM7, had to be stated where more than 250 mm *below existing ground level* (R12.1.*.*.1).

Once again, perhaps a preamble note would clear this up.

A number of extra over items are provided in Work Section 34, including extra over for breaking out hard materials. Note 2 of 34.2 defines 'hard material' as:

> any hard material which is of such a size, position or consistency that it can only be removed by special plant or explosives.

This definition differs from the definition of 'rock' in Work Section 5: *Excavating and filling*, in that it refers to the broader category of 'hard material' and also makes no reference to the use of *wedges* or *rock hammers* as a 'test of hardness'. It should be noted that SMM7 defined *rock* but not hard material.

Risk Issue

The logic for defining 'hard material' in this way, rather than using the 'rock' definition in Work Section 5, is hard to fathom.

The authors of NRM2 chose not to define the likes of concrete and masonry as 'hard material' in Work Section 5, perhaps on the basis that there is usually no argument in determining the presence of such materials in excavations, so why make the distinction in the drainage work section?

A further curiosity is the dependence on *special plant or explosives* as the 'test of hardness'.

Most contractors would use excavator-mounted hydraulic breakers where possible, which would classify as *rock hammers*, but *special plant* is much more difficult to define (and agree with the contractor!), and the use of *explosives* is rare and often specifically prohibited.

At this point, it might be prudent to "*ask the audience or 'phone a friend*" or, alternatively, take care of the issue (once again!) in a preamble.

The width of trenches for the purpose of 'extra over' items is to be taken in the designed width of the trench bed or, where there is no bed, the nominal pipe diameter +300 mm. The minimum width of 500 mm only applies where there is no bed, unlike SMM7, where the minimum applies in both cases.

Chambers

Chambers are now measured as composite items in NRM2, and the arduous chore of measuring all the components thereof is reduced to the less arduous, but still time-consuming, chore of providing a detailed item description.

Risk Issue

The item descriptors, and the 'general' item coverage for chambers, exclude excavation for chambers as discussed earlier, and whilst 'extra over' for *breaking out hard materials* is measured, there is no definition in the side notes to Work Section 34.12.

Rocker pipes are deemed included in the items for chambers (34.6–11 Note 2), but, quite bizarrely, step irons are measured separately under 'Sundries' (34.13.1.1).

Just why the authors of NRM2 chose not to measure manholes and other chambers by reference to standard construction details, thereby avoiding the need for detailed descriptions, is puzzling to say the least, but, then again, this entire work section is a bit of a disaster!

NB: Homework for authors of NRM2:

Compare and contrast the provisions of NRM2 Work Section 34 with Classes I – L of CESMM4.

6.10.18   Builder's Work in Connection with Mechanical, Electrical and Transportation Installations

Apart from being a welcome addition to the rules of measurement, this 'Cinderella' work section, unfortunately, merits attention only for its shortcomings.

Risk Issue

- No *deemed include* list in the 'general' rules.
- No mention of 'excavation' in the measurement of underground service runs.
- No item coverage for chambers.
- No list of 'sundries' to chambers.

On reaching Work Section 41, perhaps the authors of NRM2 had 'run out of steam'!

Part 2

Notes

1. *Pacific Associates v Baxter* [1990] Q.B. 993 (CA).
2. That is, *NRM2: Tabulated rules of measurement for building works.*
3. This is thought to be a mistake and should read *six*.
4. http://www.designingbuildings.co.uk/wiki/Comparison_of_SMM7_with_NRM2 (accessed 26 March 2015).

References

Crotty, R. (2012) *The Impact of Building Information Modelling*, SPON Press, Abingdon.

Lee, S., Trench, T. and Willis, A. (2014) *Willis's Elements of Quantity Surveying*, 12th Edition, Wiley-Blackwell, Chichester.

Ministry of Justice (2012) *Pre-Action Protocol for Construction and Engineering Disputes*. https://www.justice.gov.uk/courts/procedure-rules/civil/protocol/prot_ced (accessed 27 April 2015).

Ostrowski, S.D.C. (2013) *Measurement Using the New Rules of Measurement*, Wiley-Blackwell, Chichester.

Patten, B. (2003) *Professional Negligence in Construction*, Routledge.

Pittman, J. (2003) *Building Information Modelling: Current Challenges and Future Directions*, SPON Press, New York.

The Chartered Institute of Building (2009) *Code of Estimating Practice*, 7th Edition, Wiley-Blackwell, Chichester.

Chapter 7
Civil Engineering Standard Method of Measurement

The Civil Engineering Standard Method of Measurement (CESMM)[1] is now in its fourth edition, published in 2012, and this has retained the benefits of previous editions whilst bringing the document up to date and broadening its appeal both to UK users and to the international construction market.

CESMM4 is approved by its sponsors – the Institution of Civil Engineers and the Civil Engineering Contractors Association – *for use in works of civil engineering construction* and *may be used with any conditions of contract for civil engineering work **that includes measurement*** (CESMM4 – Preface).

This needs to be read in conjunction with General Principles 2.1 and 2.2 which state that CESMM4 is also intended to include *simple building works incidental to civil engineering works* (see also Class Z).

Previous editions of CESMM were closely aligned with the ICE Conditions of Contract (now the Infrastructure Conditions of Contract (ICC) – Measurement Version), but the latest edition represents a significant departure from this arrangement to the effect that CESMM4 is now:

- Contract neutral.
- (Largely) National standard neutral.

7.1 Contract neutral

Unlike CESMM3, CESMM4 may be used in conjunction with any form of contract without amendment or any contractual arrangement that includes quantities or approximate quantities or requires the work to be measured at some point. This could include:

- Lump sum contracts where any increase or decrease in the actual quantities is treated as a variation.
- Measure and value (admeasurement) contracts where the final quantities are determined on completion of the work.
- Target contracts where a priced bill of quantities is used to establish the 'target' price which is then compared to actual cost in order to determine the eventual pain/gain or profit share, if any.

Managing Measurement Risk in Building and Civil Engineering, First Edition. Peter Williams.
© 2016 John Wiley & Sons, Ltd. Published 2016 by John Wiley & Sons, Ltd.

Standard forms of contract that could be used in such circumstances include:

- JCT 2011 SBC/Q and SBC/AQ.
- ICC – Measurement Version.
- NEC3 Options B and D.
- FIDIC Construction Contract or Short Form of Contract.
- Overseas 'local' contracts that include provision for measurement.

In order to achieve contract neutrality, CESMM4, Paragraph 5.6, requires a schedule to be included in the Preamble to the Bill of Quantities which ensures compatibility between the method of measurement and the conditions of contractor used for a particular project. The schedule is intended to provide a definition of terms used in CESMM4 in relation to the conditions of contract employed for the project (see later under Section 7.7: *Preparation of the Bill of Quantities*).

7.2   National standard neutral

In order to widen the appeal of CESMM4, reference to British Standards or other UK specifications in the completed bills of quantities has been kept to a minimum, thereby allowing other countries to use their own standards more easily. In order to do this, it is necessary to establish exactly which standards the contractor is expected to work to on the drawings and/or in the specification for the job.

There are two exceptions to this national standard neutrality principle where the drafting committee found it impossible to avoid using UK standards:

- The types of concrete mixes referred to in CESMM4 are those of the relevant British Standards.
- The specification of work items for road construction is based on the Specification for Highway Works.

Where the bill compiler wishes to use different specifications for these items, CESMM4, Paragraph 5.6, requires this information to be included in a schedule in the Preamble to the Bill of Quantities (see later under Section 7.7).

7.3   Section 1: Definitions

Changes from CESMM3

- References to the ICE Conditions of Contract, numbered clauses and British Standards have been omitted
- Now refers to *contract administrator* and not *the engineer* (Paragraph 1.3)
- *The contract administrator may be the employer, his agent or representative*

Important definitions are contained in CESMM4 Section 1: *Definitions* including:

1.2	Work	This includes *work to be carried out, goods, materials and services to be supplied, and the liabilities, obligations and risks to be undertaken by the contractor under the contract.*

The word 'work' used in CESMM4 is not to be confused with the word 'Works' used in the conditions of contract (e.g. Infrastructure Conditions, NEC3 or JCT). 'Work' relates to the effort and resources expended, and the risks and responsibilities undertaken, by the contractor in order to complete the structure or physical asset described in the contract (the 'Works'). The reason for the distinction is that the method of measurement describes *work commonly encountered in civil engineering* (CESMM4 Paragraph 3.1), and under the contract, the contractor is responsible for delivering the 'Works' on or before the completion date.

1.4	Expressly required	This expression means *shown on the drawings, described in the specification or instructed by the contract administrator pursuant to the contract.*

The importance of this phrase is that the contractor will only be paid for work that falls within the definition. Therefore, if sheet piling is required, but is not indicated on the drawings or otherwise specified, the contractor will not be paid for the work unless the contract administrator specifically instructs its use during the contract.

CESMM4 Paragraph 2.6 states that *all work which is expressly required should be covered in the Bill of Quantities,* and so, if the design requires the use of sheet piling, even if only as temporary works, it must be measured and paid for.

1.5	Bill of quantities	A list of items giving brief identifying descriptions and estimated quantities of the work comprised in a contract.

This definition emphasises that the BQ does not determine the kind or quality of work required in a contract; it is the drawings and specification that do this job. The bill of quantities merely describes the work involved and provides an estimated quantity for the contractor to price having taken into account the detail stated in these other documents. This is to be contrasted with other contracts (e.g. JCT 2011 SBC/Q) where the Contract Bills do describe the nature or extent of the works required.

1.6	Daywork	…*The method of valuing work on the basis of time spent by the operatives, the materials used and the plant employed.*

Daywork is work that may have to be measured and valued at some point in a contract but cannot be determined at the outset. It is not defined on the drawings or in the specification but may be *expressly required* by the contract administrator who will issue specific instructions for this work to be carried out.

1.8	Original surface	…*The surface of the ground before any work is carried out.*

CESMM4 refers to surfaces rather than levels in order to describe where work, such as excavation, commences and finishes (c.f. NRM2). Before any work is started, the 'original surface' is what can be seen in its undisturbed state (see Figure 7.1).

1.9	Final surface	…*The surface indicated on the drawings to which excavation is to be carried out.*

Part 2

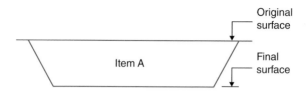

Figure 7.1   CESMM surfaces – 1.

The 'final surface' indicates where excavation work is to finish. Any excavation work below this surface is likely to be for the removal of unsuitable material such as soft spots or contamination (see Figure 7.1).

1.10	Commencing surface
1.11	Excavated surface

The definitions of 1.10, *Commencing surface*, and 1.11, *Excavated surface*, are each in two parts, the first of which is straightforward to follow.

1.10	**Commencing surface**	*...In relation to **an item** in a Bill of Quantities, the surface of the ground before any work covered by the item has been carried out.*

In this case, the definition refers to the surface at which work in relation to the item begins. This could be the original surface or the excavated surface of a preceding item(s) of work.

Figure 7.2 illustrates that the commencing surface may be stated:

a)   In relation to specific items (A, general excavation; B, pile caps; C, ground anchors; and D, piles).

b)   In relation to a change in ground conditions (e.g. top of rock strata).

1.11	**Excavated surface**	*...In relation to **an item** in a Bill of Quantities, the surface to which excavation included in the work covered by the items is to be carried out.*

Here, the definition refers to the surface at which work in relation to the item ends. This could be an intermediate surface if there is further excavation to follow, or it could be the final surface. This is illustrated in Figure 7.2.

The second parts of the definitions of 1.10, *Commencing surface*, and 1.11, *Excavated surface*, are less easy to follow.

1.10	**Commencing surface**	*...In relation to **a group of items** in a Bill of Quantities, for work in different materials in an excavation or a bored, drilled or driven hole, the surface of the ground before any work covered by any item in the group has been carried out.*

1.11	**Excavated surface**	*...In relation to **a group of items** in a Bill of Quantities, for excavation in different materials, the surface to which excavation included in the work covered by any item in the group is to be carried out.*

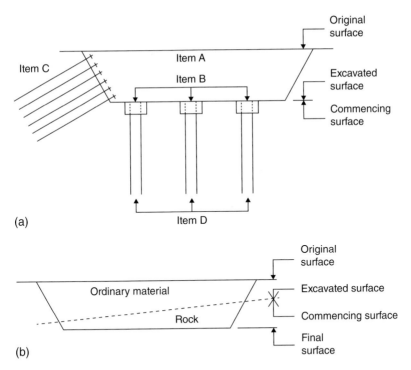

Figure 7.2   CESMM surfaces – 2. (a) Surfaces relating to items of work and (b) surfaces relating to ground conditions.

Evidently, the commencing surface is where the work relating to **any item in the group** starts, and the excavated surface is the point where the work relating to **any item in the group** ends.

The commencing surface for a diaphragm wall, therefore, is the surface where the preliminary trench and guide walls start and not where the diaphragm wall excavation starts which will normally be at a lower level. There is no distinction, therefore, between the first two items which will be carried out by a normal excavator and the excavation carried out by the diaphragm walling rig. Similarly, the excavated surface will be the bottom of the diaphragm wall.

Where the two definitions refer to *a group of items in a Bill of Quantities*, this phrase is not defined in CESMM4 and could therefore mean:

- Any group of items in a BQ defined by the bill compiler (e.g. a group of drainage items under Class I: *Pipework – Pipes* comprising the same pipe specification and diameter in the same location but with different depths to invert).
- A specific group of items in a BQ defined by the CESMM4 classification tables (e.g. Class C: *Diaphragm walls* consisting of guide walls, excavation, concrete, rebar, etc.).

The reference to *work in different materials* is emphasising that different commencing and excavated surfaces do not have to be specified where excavation or drilling is passing through different types of ground. In a diaphragm wall excavation, therefore, it is not necessary to state where rock excavation (if any) commences or finishes albeit that, if there is rock to be excavated, this still has to be measured.

The phrase *in an excavation or a bored, drilled or driven hole* means that the definition of commencing surface refers to:

- Any type of excavation which could be to a diaphragm wall, cutting, foundation, general excavation or a trench, etc.
- Work that is not in an excavation but is a bored hole (e.g. bored pile), a drilled hole (e.g. a ground/rock anchor) or a driven hole (e.g. a driven pile or where pipes are driven into the ground (i.e. not drilled) in order to facilitate grouting).

It should be noted that the definition of 'excavated surface' refers only to excavation and not to boring, drilling or driving. This is common sense because there could be additional excavation beyond an excavated surface but not in the case of a bored, drilled or driven hole which, in each case, will only have a final surface for obvious reasons.

7.4   Section 2: General principles

Changes from CESMM3

None

7.4.1   Title application and extent

Paragraph 2.1 explains that CESMM4 is the 'official' abbreviation to use and that the method of measurement is intended for civil engineering works and simple building works incidental to civil engineering work.

In the CESMM, *simple building works* means just that – simple! Foundations, walls, floors and even roofing could be measured using CESMM4, and there is provision for items such as joinery, doors, windows, plastering, simple building services and the like. Where there is a need for anything more complex, the bill compiler has a number of choices:

Alternative 1
Include a provisional sum in the bill of quantities in respect of the building works required (CESMM4, Paragraph 5.18).

Alternative 2
Measure the building work in a separate bill (e.g. called 'building works') in the bill of quantities using an alternative method of measurement (CESMM4, Paragraph 5.4, refers).

Alternative 3
Consider a procurement method more suited to multifaceted work than a traditional engineer-designed single contract.

When it comes to a £50 million railway station refurbishment in a large city centre, comprising complex roof replacement, giant atrium, refurbished platforms, enlargement of pedestrian concourses handling 150 000 passengers per day, 15 lifts and 30 escalators, CESMM4 would clearly be inadequate to deal with the building work involved with the civil engineering elements of such a project.

With a traditional procurement route, such a project would cause considerable problems for the bill compiler.

Risk issue

Alternative 1

The problem with any provisional work to be undertaken by the contractor is establishing the correct rate for valuing the work when it is carried out.

It is very likely that there will be no analogous rates in the bill of quantities, and so rates will have to be negotiated in order to establish the value of the work in progress. These will not be competitive market rates, and thus, the employer will be paying a premium for this work.

Alternatively, the contractor could be asked to supply a quotation for the work albeit that this would not be in competition either.

Alternative 2

The Preamble to the Bill of Quantities will have to state which method of measurement has been used to prepare the quantities (e.g. SMM7, NRM2, etc.) and any amendments thereto (if any).

In this case, careful attention will have to be paid to the Appendix to the conditions of contract or the Contract Data (NEC3) which normally state that only one method of measurement shall apply to the contract. If this issue is not addressed, the problem will be that the bill of quantities will contain quantities that have not been measured in accordance with the method of measurement stated in the contract.

Alternative 3

Framework contract

The civil engineering and building work content could be separated into individual work packages and tendered separately within the framework. Each work package could utilise the same form of contract if desired, but the stated method of measurement would be CESMM4 and SMM7/NRM2, respectively. For a very large contract, there could be a number of such discrete packages.

For this procurement arrangement, it would be advisable to appoint a construction manager to manage the interface between the packages. This could be either the civils package contractor or the building package contractor depending upon the respective size/importance of the packages. Alternatively, a separate managing contractor could be appointed; this would be a necessary appointment for a very large project. In any event, a principal contractor would have to be appointed to undertake the statutory CDM role which could be one of the contractors or, possibly, the employer.

Management contract or construction management

Separate work packages could be devised for the civil engineering and building work using the most appropriate form of contract and method of measurement for each package. The packages would be coordinated by the management contractor/construction manager who could also be appointed to the CDM role unless undertaken by the employer.

Design and build

The risk would no longer reside with the employer as it would be the contractor's decision as to how the work would be carried out/subcontracted. In this case, each work package could be let using any number of different subcontract forms along with the appropriate method of measurement.

7.4.2   Object of CESMM4

It would be easy to skip over this small paragraph in CESMM4. It looks quite innocuous but is, in fact, there for good and sufficient reasons.

Paragraph 2.3 establishes two facts concerning the object of CESMM4 which *is to set forth the procedure according to which*:

1.   *The Bill of Quantities shall be prepared* and *priced* and
2.   *the quantities of work expressed and measured.*

Despite being poorly phrased, this paragraph emphasises that, where CESMM4 is stated as the method of measurement in the contract:

- The Bill of Quantities shall have been prepared in accordance therewith.
- The contractor's pricing of the Bill of Quantities will be construed as being in accordance with the relevant provisions of the method of measurement as regards:
  - What the contractor has included in a particular rate irrespective of the actual pricing.
  - How general items and method-related charges, etc. have been priced irrespective of how the contractor has actually priced these items.
- The quantities of work in the Bill of Quantities shall be described in accordance with the method of measurement.
- The quantities of work in the Bill of Quantities shall be measured in accordance with the rules in the method of measurement.

Risk issue

Failure to describe or measure items in accordance with CESMM4 is technically a breach of contract but may not be sufficiently important so as to go to the root of the contract. Nevertheless, any deviation from CESMM4 in the bill of quantities must be dealt with in accordance with the provisions of the contract, and where there is no such provision, a claim for damages may be valid.

In the valuation of work in progress and the valuation of variations shall be based on the item descriptions and item coverage rules as expressed in the method of measurement irrespective of how the contractor has actually priced the items. Therefore, an item of excavation shall be deemed to include an appropriate allowance for working space and earthwork support notwithstanding that the contractor may have priced such work elsewhere in the bill of quantities.

7.4.3   Objects of the bill of quantities

Paragraphs 2.4 and 2.5 emphasise the importance of the Bill of Quantities with regard to the financial control of civil engineering projects with the proviso that the BQ should be *as simple and brief as possible.*

All work *expressly required* should be included in the Bill of Quantities (Paragraph 2.6) but only to the extent that it is required to be measured by the provisions of CESMM4.

Risk issue

Where work is included on the drawings or in the specification, then it is *expressly required* and must be measured in the bill of quantities. Where work is not *expressly required* but is nevertheless indispensable for the technical realisation and the safe execution of the design, then the contractor has to allow for such work in his rates and prices.

In this regard, the contractor must be fully aware of the item coverage rules in CESMM4 because this is where work that is indispensable but not expressly required is to be found.

Paragraph 2.7 states that it is the Work Classification system of CESMM4 that determines the successful realisation of the objects of the Bill of Quantities.

7.5   Section 3: Application of the work classification

Changes from CESMM3

- Reference to BS 4449 omitted (Paragraph 3.3)

Section 3 is more than explanatory, and the use of the word *shall* in Paragraphs 3.1–3.6 and 3.9 and 3.10 elevates these paragraphs to the status of rules to be followed. Paragraph 3.3 (final paragraph) is advisory (*should*), and the remaining paragraphs (3.7, 3.8 and 3.11) are explanatory.

7.5.1   Item descriptions

Section 3 of CESMM4 explains how the Work Classification is to be applied. This is the heart of CESMM4 where *work commonly encountered in civil engineering contracts* is arranged into 26 'classes' A–Z.

Class A: *General items* deals with items commonly known as 'preliminaries', and Classes B–Z represent items of work to be measured. Paragraph 3.1 goes on to explain how each class is arranged, and this is illustrated in Table 7.1.

7.5.2   Mode of description

In Paragraphs 3.1 and 3.4, the rule is that work item descriptions shall be made up of one descriptive feature from each of the three divisions. This enables work items to be kept short and succinct, and this is emphasised in Paragraph 3.2 which states that *item descriptions for permanent works shall generally identify the component of the works and not the tasks to be carried out.*

As a consequence, unnecessary embellishment of descriptions is avoided so that an item such as *formwork, fair finish, plane vertical, width exceeding 1.22 m* is kept free of phrases such as 'make, erect and strike' which is a task as opposed to a component of the Works.

Inexplicably, this rule is repeated in Paragraph 3.3 but using the less emphatic word *should* for some unknown reason.

It should be noted that there is no rule in CESMM4 that requires the bill compiler to use the phraseology of the Work Classification so long as work item descriptions comply with Paragraph 3.1 and are kept brief in accordance with Paragraph 3.3.

7.5.3   Separate items

The items included in a bill of quantities should be individual and discrete, and in this regard, CESMM4, Paragraph 3.4, requires that each item *shall* exhibit no more than one descriptive feature from each of the three divisions of the Work Classification.

This requirement ties in with the item coding and numbering protocols in Section 4.

7.5.4   Units of measurement

The Work Classification determines the unit of measurement that shall apply to each item. The units of measurement are to be found variously in divisions one, two and three, but the rule in

Table 7.1 Work classification.

Left-hand page			Right-hand page			
First division	Second division	Third division	Measurement rules	Definition rules	Coverage rules	Additional description rules
• Each division has up to eight descriptive features • Each BQ item is made up of one descriptive feature from each division			Rules dealing with how work items should be measured	Rules that provide an explanation of words and phrases used in the work classification and in the BQ	Rules that define what is deemed to be included in an item of measured work	Rules that determine how an item of work should be described in addition to the descriptive features required in the first, second and third divisions

Example

Class J: Pipework – fittings and valves

First division	Second division	Third division	Measurement rules	Definition rules	Coverage rules	Additional description rules
Type of pipe e.g. Concrete pipe fittings	Type of fitting e.g. Bends	Diameter of pipe e.g. 200–300 mm	**M1** *Pipe fittings* comprising backdrops to manholes shall be included in the items for manholes measured in Class K	**D2** A *straight special* is a length of pipe either cut to length or made to order	**C3** Items for *straight specials* shall be deemed to include cutting	**A4** *Fittings* to pipework not in trenches shall be so described

Paragraph 3.5 nevertheless limits the units used to those stated in the Work Classification irrespective of which division they appear in.

The unit of measurement attached to a descriptive feature *shall* apply to all items wherein that particular descriptive feature appears.

7.5.5   Measurement rules

The right-hand page of the Work Classification is where the various rules needed to measure and describe civil engineering work are to be found. Included therein are 'measurement rules'.

The measurement rules set out:

- *The conditions under which work **shall** be measured.*
- *The method by which the quantities **shall** be computed.*

This is where rules are to be found that govern what is to be measured or not measured in particular circumstances. For instance, Class G: *Concrete ancillaries* Measurement Rule **M1** states that an item of *formwork shall be measured for surfaces of in situ concrete which require temporary support during casting unless otherwise stated in CESMM.*

The measurement rules also state how the quantities are to be calculated. For example, Class G: *Concrete ancillaries* Measurement Rule **M7** says that *the mass of steel reinforcement shall be taken as 0.785 kg/m for each 100 mm² of cross section (7.85 t/m³).*

Therefore, a 4 m length of 20 mm diameter rebar shall be deemed to weigh:

$$0.785 \times \pi R^2 \times \text{length} = 0.785 \times 3.142 \times 10^2 \times 4\,\text{m}$$
$$= 986.59\,\text{kg}$$
$$= \textbf{0.99 tonne}$$

This is irrespective of what the rebar actually weighs.

7.5.6   Definition rules

Some of the words and phrases used in the CESMM4 Work Classification in relation to certain classes of work require clarification to be universally and unequivocally understood.

This is the role of the 'definition rules' which help the bill compiler to understand the extent and limits of such work when using the Work Classification and also help the estimator and other users of the bill of quantities that results from the measurement process.

For example, 'diaphragm walls' is an expression used in the Class C: *Geotechnical and other specialist processes* of the Work Classification wherein Definition Rule D2 explains that:

Diaphragm walls are walls constructed using bentonite slurry or other support fluids.

Consequently, if a wall is to be constructed in any other way it is not a diaphragm wall and so, when the expression 'diaphragm walls' appears in the bill of quantities, everyone has a clear understanding of what is meant thanks to Definition Rule D2.

Another example may be taken from Class G: *Concrete ancillaries* where Definition Rule D2 states that:

Formwork shall be deemed to be for plane areas and to exceed 1.22 m wide, unless otherwise stated.

This means that formwork shall be assumed to be flat (i.e. not curved) and wider than a standard sheet of plywood even though it might be vertical, horizontal, sloping or battered. As a result of the definition rule, the word 'formwork' has a clear meaning unless there is some other statement to the contrary.

Part 2

7.5.7   Coverage rules

The idea behind CESMM4 is to create a bill of quantities that can be useful for the cost management of civil engineering projects whilst at the same time providing a simple and succinct document containing enough detail to enable contractors to price the work correctly, when read in conjunction with the drawings and specification.

Inevitably, the price of succinctness is a reduction in the number of measured items in the BQ which, to comprehensively cover all the work required, need to include work that is necessary but not measured. As a result, the coverage rules play an important role in clarifying what is deemed included in specific items where there might otherwise be doubt.

In this regard, it should be noted from CESMM4, Paragraph 3.8, that *a coverage rule*:

- *Does not state all the work covered by an item.*
- *Does not preclude any of the work stated being covered by a Method-Related Charge.*

Coverage rules do not necessarily define everything that might be required in an item of work because:

- Some coverage rules appear above the double line in the CESMM4 Work Classification and therefore apply to all items in a particular class of work; these general coverage rules have to be read in conjunction with the specific coverage rules that apply to individual classes of work.
- *The exact nature and extent of the work is to be ascertained from the Drawings, Specification and Conditions of Contract, as the case may be, read in conjunction with the Work Classification* (CESMM4, Paragraph 5.12).
- All work that is *expressly required* is shown on the drawings, described in the specification or instructed by the contract administrator, and it should be clear from this as to what work the contractor needs to carry out.
- The conditions of contract may further define the contractor's general obligations with regard to the provision of resources, whether temporary or permanent, *as far as they may be reasonably inferred from the Contract* (e.g. ICC – Measurement Version Clause 8).
- The conditions of contract may state that *the contractor shall be deemed to have satisfied himself as to the correctness and sufficiency of the Accepted Contract Amount* which, unless otherwise stated, is deemed to include *all things necessary for the proper execution and completion of the Works* (FIDIC Red Book Clause 4.11).
- Admittedly, on a narrow construction, English common law provides that the contractor will supply everything that is *indispensably necessary* to complete a project even though it might not be measured or specified.

Even though a coverage rule for a class of work states what is deemed included in that class of work, coverage of the work required for a particular item(s) is not necessarily confined to the measured item as it may be the case that the whole or part of that class of work is included as a method-related item in Class A: *General items*.

The possibility that some or all of the work deemed included in a measured item by virtue of a coverage rule is actually priced elsewhere is anticipated in Paragraph 3.8. This possibility would have to be taken into account when it comes to the re-evaluation of BQ rates or the valuation of variations to the contract.

7.5.8   Additional description rules

The nature of methods of measurement and the process of building item descriptions do not represent a precise science, and it is easy for the bill compiler to unconsciously misrepresent an item of work such that the contractor's estimator is left 'in the dark' as to its meaning.

Even when the method of measurement is followed faithfully, there are occasions when judgement suggests that more information is required to describe an item fully, and this eventuality is anticipated by CESMM4, Paragraph 3.9. This allows for more description to be given than is required in Paragraph 3.1 *where required by any provision of Section 5 of CESMM4 or by any applicable additional description rule in the Work Classification.*

Paragraph 3.9 also says that *where additional description is given, a separate item shall be given for each component of work exhibiting a different additional feature.* Therefore, where the composition and materials of bridge bearings of similar type are different, separate items shall be given in accordance with *Additional Description Rule* **A3** of Class N: *Miscellaneous metalwork.*

Notwithstanding the provisions of Paragraph 3.9, Paragraph 5.14 also says that additional description shall be given where strict application of the Work Classification would be inadequate.

Following CESMM4, Paragraph 3.10, additional description rules are sometimes used to override the requirements of a descriptive feature of the Work Classification.

Therefore, the nominal bores of pipes in **CLASS I: PIPEWORK – PIPES** shall be given in the item description instead of the range given in the **SECOND DIVISION** following the requirement of **ADDITIONAL DESCRIPTION RULE A2**. This is shown in Table 7.2.

Part 2

Table 7.2   Additional description rules.

FIRST DIVISION		SECOND DIVISION		THIRD DIVISION	
1	Clay pipes	1	Nominal bore: not exceeding 200 mm	1	Not in trenches	
2	Concrete pipes	2	200-300 mm	2	In trenches, depth:	Not exceeding 1.5 m
3	Iron pipes	3	300-600 mm	3		1.5–2 m
4	Steel pipes	4	600-900 mm	4		2–2.5 m
5	Polyvinyl chloride pipes	5	900-1200 mm	5		2.5–3 m
6	Glass reinforced plastic pipes	6	1200-1500 mm	6		3–3.5 m
7	High density polyethylene pipes	7	1500-1800 mm	7		3.5–4 m
8	Medium density polyethylene pipes	8	Exceeding 1800 mm	8		exceeding 4 m

MEASUREMENT RULES	DEFINITION RULES	COVERAGE RULES	ADDITIONAL DESCRIPTION RULES
			A2 The materials, joint types, nominal bores and lining requirements of pipes shall be stated in item descriptions.

EXAMPLE
Ref   Item
E224   Concrete pipes; **225 mm diameter**; In trenches depth 2–2.5 m

Similarly, the nominal thickness of walls in **CLASS U: BRICKWORK AND MASONRY** shall be given instead of the range listed in the **SECOND DIVISION**.

7.5.9   Applicability of rules

Albeit not a rule in itself, Paragraph 3.11 is nevertheless important because it states that rules printed above a double line on the right-hand page of the Work Classification apply to all items in the class.

Rules printed below the double line only apply to the groups of items that appear on the left-hand page <u>within the corresponding single horizontal lines</u> when read across right to left.

7.6   Section 4: Coding and numbering of items

Changes from CESMM3

None

Section 4 of CESMM4 provides a simple to follow item numbering system that can be used to:

- Create unique reference codes for all measured items.
- Code measured items where there is no appropriate feature in the Work Classification.
- Code measured items where there is no applicable division in the Work Classification.
- Provide a unique code for measured items with additional descriptive features that would otherwise be indistinguishable from similar items.
- Identify individual items in the bill of quantities, if desired.

7.6.1   Coding

Paragraph 4.1 explains that the CESMM4 coding system is alphanumeric with a reference letter identifying the class of work to which the item belongs followed by up to three numbers denoting the relevant descriptive feature from each of the three divisions of the Work Classification.

Where more than one descriptive feature from any of the three divisions of the Work Classification applies to a measured item, an asterisk (*) is used (Paragraph 4.2 refers).

An example of how the coding system works is given in Figure 7.3 where it can be seen that the item of concrete pipes, nominal bore 300–600 mm, in trenches depth 1.5–2 m is coded I233 as per Class I: *Pipework – Pipes*, that is, the Class (I) followed by a code from each of the three divisions.

The item would not be billed with this description because the Class I, Additional Description Rule A2, requires the nominal bore to be given in the description (see CESMM4, Paragraph 3.10, and also Table 7.3).

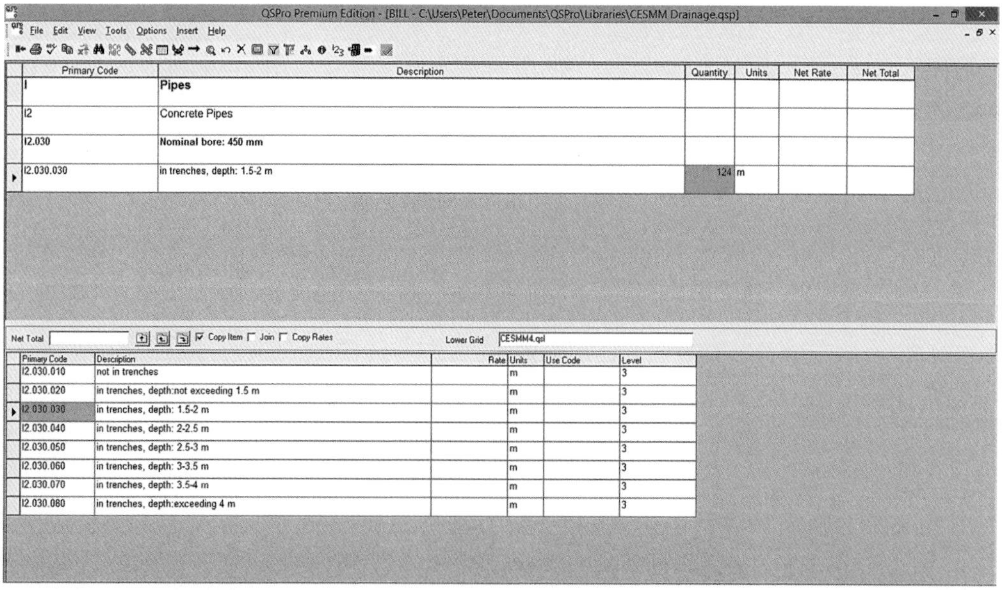

Figure 7.3   Coding.

Table 7.3   Billing with additional description.

Pipes							
Concrete Pipes							
Nominal bore: 450 mm							
A	in trenches, depth: 1.5-2 m		124 m				

CESMM4 shows spaces rather than full stops to separate the characters in an item code except where a prefix and/or a suffix code is required. There is no rule governing this, and the choice is a personal one, so long as unique items are given a unique code when 'taking-off'.

The use of full stops becomes important for contracts where there are several parts to the bill of quantities, or where the BQ items are to be referenced with the CESMM4 codes.

Consequently:

- Items coded I 2 3 3, which appear in more than one part of the BQ, would need to be distinguished, one from the other, by use of a **prefix** denoting the bill number (e.g. item I 2 3 3 in Bill No 3 could be coded 3.I 2 3 3).
- Where additional description is given to an item, the item code would attract a **suffix** (e.g. where an item I 2 3 3 is to be excavated by hand, additional description would be given in accordance with Additional Description Rule A7, and the item could be coded I 2 3 3.1 to distinguish it from other similar items excavated by machine).

Should both a prefix and suffix be used, an item could equally be coded as 3.I 2 2 3.1 as 3.I. 2.2.3.1, so long as both codes are unique.

7.6.2   Item numbers

Once measured items have been coded, Paragraph 4.3 provides that the same code can be used to identify the relevant item in the bill of quantities. This is purely optional and any other referencing may be used instead (e.g. the traditional page number and item letter alphanumeric code). Should the CESMM4 codes be used, the BQ items should be listed in ascending order (i.e. 1–9).

Risk issue

Paragraph 4.4 emphasises that there is no contractual significance attached to the code number used in the bill of quantities and that the code number is not part of the item description, presumably to avoid potential conflict where an item has been described correctly but coded wrongly and vice versa.

7.6.3   Coding of unclassified items

Should a measured item of work possess a feature that is not listed in the Work Classification, the feature receives the code number 9 because the maximum number of features in any division is 8 (CESMM4, Paragraph 4.5, refers).

If it arises that a division of the Work Classification does not apply to a particular item, or where there are less than three divisions that apply to the item, then the code 0 (zero) shall be applied in accordance with Paragraph 4.6.

7.6.4   Numbering of items with additional description

Where a measured item requires additional description, as prescribed by an additional description rule in the Work Classification or by Paragraph 14 of Section 5, the item in question will need to be coded in order to distinguish it from other similar items which do not have the additional descriptive feature.

Accordingly, Paragraph 4.7 requires that a suffix code is given to the item(s) in question.

In Class M: *Structural metalwork*, Additional Description Rule A6 requires that *item descriptions for erection shall separately identify and locate separate bridges and structural frames and, where appropriate, parts of bridges or frames.*

Therefore, the erection item for the steel bridge beams for two grade-separated highway bridges at an interchange would be separately identified in the take-off and in the bill of quantities as illustrated in Figure 7.4.

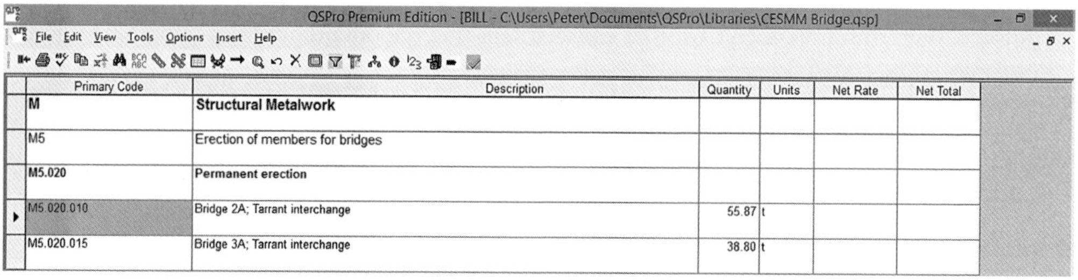

Primary Code	Description	Quantity	Units	Net Rate	Net Total
M	Structural Metalwork				
M5	Erection of members for bridges				
M5.020	Permanent erection				
M5.020.010	Bridge 2A; Tarrant interchange	55.87	t		
M5.020.015	Bridge 3A; Tarrant interchange	38.80	t		

	Structural Metalwork		
	<u>Erection of members for bridges</u>		
	Permanent erection		
A	Bridge 2A; Tarrant interchange	55.87	t
B	Bridge 3A; Tarrant interchange	38.80	t

Figure 7.4   Coding with additional description.

In Figure 7.4, it can be seen that the Class M codes from the first and second divisions have been assigned to each bridge (there is no third division descriptive feature) and that the software has assigned a further discrete code to each item.

It should be noted that CESMM4 only requires the suffix code to be given if the bill of quantities items are to be coded with Work Classification code numbers. Therefore, the BQ items in Figure 7.4 are coded with a simple alpha code, not the CESMM code, but the items would comply with Rule A6 because each bridge is separately identified and located in the items.

The quantity take-off is coded as part of the audit trail back to the original dimensions.

7.7   Section 5: Preparation of the Bill of Quantities

Changes from CESMM3

- Method of measurement now contract and (generally) specification neutral (Paragraph 5.6)
- Schedule (of terms) to be included in the Preamble to the Bill of Quantities (Paragraph 5.6)
- Reference to a *standard schedule* instead of the FCEC Schedule of Dayworks (Paragraph 5.7)
- Provisional sums now to be for *defined work* (Paragraph 5.18)
- CESMM3, Paragraph 5.22: *Form and setting out* of the Bill of Quantities seemingly omitted in error
- There shall be *an item* for a general contingency in the Grand Summary instead of *a Provisional Sum* (Paragraph 5.25)

Section 5 of CESMM4 sets out rules to determine the structure, order and content of the bill of quantities in order that the quantification of civil engineering work based on this method of measurement may assume a uniform, consistent and familiar look.

7.7.1   Measurement of completed work

It is important to note that the measurement of <u>completed</u> work <u>shall also</u> comply with the provisions of Section 5.

Of particular significance in this regard is Paragraph 5.11 which requires that **all work shall be** *itemized and **all** the items **shall be** described in accordance with the Work Classification* but that *further itemization and additional description **may be** provided* should *special methods of construction or considerations of cost* arise.

Risk issue

It is clear, therefore, that the admeasurement of completed work, and the measurement of variations to the contract, shall comply with the provisions of CESMM4 and that other methods of measurement, or *ad hoc* methods, cannot be used instead.

7.7.2   Sections of the bill of quantities

In keeping with the objective to standardise presentation, Paragraph 5.2 prescribes that *the Bill of Quantities shall be divided into the following sections*:

a) List of principal quantities.
b) Preamble.
c) Daywork Schedule.
d) Work items (grouped into parts).
e) Grand Summary.

The Daywork Schedule is optional as Paragraph 5.7 refers to the Daywork Schedule, *if any*.

7.7.3   List of principal quantities

The provision of a list of principal quantities, usually in the specification, was always fairly common prior to CESMM which has simply formalised custom and practice.

Part 2

The presence of principal quantities in the bill of quantities is *solely to assist tenderers* by providing an indication of the *scale and character* of the proposed works *prior to the examination of **the remainder of the Bill of Quantities and the other contractual documents on which their tenders will be based***. The list is provided in order to give the estimator, in particular, a 'feel' for what is involved in the project.

> ### Risk issue
>
> How effective this disclaimer might be in the event that the list of principal quantities creates a false impression of what is involved in the works, or proves to be misleading in terms of the contractor's pricing of the job, might be open to question.
>
> It is clear, nonetheless, that it is the ***remainder*** of the *Bill of Quantities*, along with other documentation, which forms the basis of the contractor's tender offer and not the list of principal quantities.

The contents of the list is purely subjective – there are no rules to guide the bill compiler in determining which items to include or which items might be of most importance or significance to the contractor.

7.7.4   Preamble

The Preamble to the Bill of Quantities has three important functions:

1. To state where CESMM4 has been amended or where another method of measurement has been adopted in preparing the bill of quantities.
2. To provide a definition of rock.
3. To provide a 'schedule' to define terms used in CESMM4 in the context of the conditions of contract used for the project.

1. *Other Methods of Measurement*

Section 5, Paragraph 5.4, requires that the Preamble shall state which other methods of measurement (if any) have been used in preparing the bill of quantities. This includes amendments to CESMM4 as well as other standard methods such as SMM7 or NRM2. Consequently, where work is to be measured, and the standard version of CESMM4 is not used, this shall be stated in the Preamble.

It should be noted that, where bills of quantities are used with NEC3 (Options B and D), amendments to the chosen method of measurement are to be stated in Contract Data Part 1.

CESMM4, Paragraph 5.4, makes reference to two sets of circumstances where CESMM4 might be abbreviated:

- Contractor-designed work where the contractor is asked to design a particular part(s) of the works in accordance with stipulated employer's requirements.
- Other circumstances such as where the contractor is given a choice between alternative construction methods.

In both cases, the bill compiler will provide items in the BQ for the contractor to price but will not use the entire extent of the CESMM4 Work Classification to do so. This will avoid being prescriptive about what the contractor is pricing and enable the contractor to make the choice as to what precisely is covered by the items in question. The BQ items provided will ensure that suitable rates are provided for admeasurement purposes and for the valuation of variations. The extent of the work affected by all amendments shall be stated in the Preamble.

Typical examples of contractor design in civil engineering include:

- Geotechnical work including ground anchors, diaphragm walls and vibroflotation.
- Piling.
- Roads and paving (highways, concrete runways, etc.).
- Rail track work.
- Sewer and water main renovation.

2. *Definition of Rock*
Importantly, Paragraph 5.5 requires a definition of 'rock' to be provided in the Preamble where the work includes excavation (e.g. diaphragm walls, earthworks, drainage), boring (e.g. ground anchors, bored piling) or driving (e.g. driven piles). CESMM4 is silent on the form that the definition shall take which could be:

- A geological definition of rock 'strata' making reference to borehole details supplied with the tender documents.
- A definition of rock 'deposits' where materials excavated from a designated 'zone' or part of the works (shown on the drawings) would be classed as 'rock' irrespective of the material actually found on the site.
- A definition which relates to how the material is to be removed (e.g. by using explosives, wedges, rock hammers or 'special' plant).

3. *Schedule of Terms*
A significant difference between CESMM3 and its successor is that CESMM4 is now contract neutral and can be used with any preferred conditions of contract. Consequently, the references that were made to the ICE Conditions of Contract in CESMM3 do not appear in CESMM4.

This means that terminology that would otherwise have been defined in the ICE Conditions of Contract, the default contract in CESMM3, must now be made explicit in the bill of quantities on the basis that different forms of contract may be used in conjunction with CESMM4 and that words and phrases used in CESMM4 may have different meanings in different forms of contract.

In order to achieve compatibility between the conditions of contract and CESMM4, Paragraph 5.6 requires that a schedule **must be** included in the Preamble listing words and phrases used in CESMM4 along with the clause number and term in the conditions of contract that defines the particular word or phrase used in CESMM4 in the language of the contract. The schedule has no specific title in CESMM4 which, presumably, the bill compiler is at liberty to choose (e.g. schedule of terms).

CESMM4 provides a *pro forma* schedule which lists CESMM4 terms that need to be cross-referenced to an appropriate clause in the conditions of contract. There is a space in the schedule for the bill compiler to insert the equivalent term and clause number from the conditions of contract so that the CESMM4 term can be understood in relation to the conditions of contract. For example, *Daywork Schedule* in CESMM4 Paragraph 5.2 is equivalent to *Shorter Schedule of Cost Components* in Clause 11.2(22) in ECC Option B.

This schedule must be replicated in the Preamble to the Bill of Quantities and may be added to as required, such as when CESMM4 is amended for any reason. Put in simple terms, the schedule provides words or phrases that can be used in place of those used in CESMM4 without creating a conflict between the method of measurement and the conditions of contract.

7.7.5   Daywork schedule

A contractual provision for payment on a Daywork basis is frequently needed on civil engineering projects so that unexpected work that cannot be valued at analogous or fair rates can be paid for on the basis of resources used (see Chapter 4).

Part 2

Such a contractual provision is usually made in two places:

- In the form of contract where the principles that apply to the valuation of variations are established.
- In the bill of quantities where:
  - Provisional items are included as a contingency should payment on a daywork basis be instructed by the contract administrator.
  - Competitive daywork rates are established which are agreed by the parties and reflect the 'going rate' for daywork in the industry at the time of tender.

CESMM4, Paragraph 5.7, provides three methods for including a Daywork Schedule in the Preamble to the Bill of Quantities, none of which are compulsory:

1. A list of the various classes of labour and plant and types of materials is provided which each tendering contractor prices. The list is accompanied by *a statement of the conditions under which the Contractor shall be paid for work executed on a Daywork basis.*
2. A statement that the contractor will be paid Daywork in accordance with the rates and prices stated in *the standard schedule included in the contract* to which the percentages quoted by the contractor at tender stage will be added or deducted.
3. By inference, no provision for a Daywork Schedule at all – *the Daywork Schedule **if any** shall comprise....*

Under Option 1, the *statement of the conditions under which the Contractor shall be paid for work executed on a Daywork basis* included in the bill of quantities must reflect what is said in the conditions of contract which normally contains an express condition as to how claims and variations are to be valued. There should be no conflict between the statement and the conditions.

Under Option 2 in CESMM3, the standard schedule was the *Schedules of Daywork carried out incidental to Contract Work* issued by the Federation of Civil Engineering Contractors.[2] Being contract neutral, CESMM4 does not specify which, if any, standard schedule shall be used as there are several options available.

Paragraph 5.8 suggests that provisional sums *may be given* for work to be executed on a day-work basis listed under the separate headings of labour, materials, plant and other charges. Where Option 2 is used, adjustment items shall be given for each category so that tenderers may price a percentage addition/deduction to the standard schedule. The price inserted in the priced bill of quantities shall be the Provisional Sum for each category ± the quoted percentage.

7.7.6   Work items

CESMM4, Paragraphs 5.9–5.23, concern how the main body of the bill of quantities is to be arranged and provide a small number of general rules that shall apply to work items.

Division of the bill of quantities into parts

Remembering that the bill of quantities must be divided into 'sections' as specified in Paragraph 5.2, and that 'work items' is one of those sections, places Paragraph 5.9 in the correct context. 'Work items' should not be confused with the 'Work Classification' from which the 'work items' may be drawn.

Paragraph 5.2 provides that the bill of quantities **may be** divided into numbered parts in order to distinguish work to be carried out in circumstances that may be *likely to give rise to different methods of construction or considerations of cost*. There is no prescription as to how this might be done, and it is the skill and judgement of the bill compiler that will dictate whether this is done well or badly. The main consideration is to help the contractor to price the work accurately, and to programme activities realistically, without being put to the trouble and expense of isolating the quantities of work requiring special consideration.

It should be noted that Paragraph 5.9 states that items in each separate part of the Bill of Quantities *shall be arranged in the general order of the Work Classification.*

If the nature of the project has no special characteristics requiring subdivision of the Bill of Quantities into 'parts', then it may be subdivided into the applicable classes of the Work Classification. The BQ for an enabling works contract may therefore be subdivided as follows:

- Section A: List of principal quantities.
- Section B: Preamble.
- Section C: Daywork schedule.
- Section D: Work items:
  - Part 1: General items (Class A).
  - Part 2:
    - Geotechnical and other specialist processes (Class C).
    - Demolition and site clearance (Class D).
    - Earthworks (Class E).
    - Piles (Class P).
    - Piling ancillaries (Class Q).
- Section E: Grand summary.

However, a land drainage scheme, where some of the work is to be carried out in tidal conditions and where differences in location, access and construction methods characterise parts of the work, will need to be subdivided into parts because of the nature of the work entailed. In this case, each part of the BQ should be subdivided into the relevant classes of the Work Classification:

- Section A: List of principal quantities.
- Section B: Preamble.
- Section C: Daywork schedule.
- Section D: Work items:
  - Part 1: General items (Class A).
  - Part 2: Sea outfall (tidal):
    - Earthworks (Class E).
    - In situ concrete (Class F).
    - Concrete ancillaries (Class G).
  - Part 3: Drainage and headwalls:
    - Pipework – Pipes (Class I).
    - Pipework – Fittings and valves (Class J).
    - Pipework – Manholes and pipework ancillaries (Class K).
    - Etc.
  - Part 4: Drainage ditches:
    - Earthworks (Class E).
    - Precast concrete (Class H).
    - Etc.
  - Part 5: Pumping station:
    - Excavation (Class E).
    - In situ concrete.
    - Concrete ancillaries (Class G).
    - Precast concrete (Class H).
    - Etc.
- Section E: Grand summary

Other examples where work may need to be subdivided into parts include:

- Work carried out beyond the site boundary.
- Highway work – separation of main carriageway, side roads and structures.
- Airport work – separation of work carried out 'airside'.
- Sewer work – separation of work carried out in 'live' sewers.

Other work will be too large and/or complex for a single bill of quantities and may require non-traditional procurement methods and contract documentation reflecting the size, complexity and contractor design input into such work:

- Rail work:
  May involve station buildings as part of a city centre redevelopment scheme, new and replacement rail track work carried out under measured term contracts, under-/overbridges involving line possessions, multibillion £ electrification framework contracts.
- Power stations:
  A large power station project costing several billion pounds will comprise numerous separate elements, each of which will probably be subdivided into individual contracts. Access roadworks, wharves and jetties, railway works, cooling towers, heavy foundations and piling, structural steel frames, superstructures and ancillary buildings will all be significant contracts in their own right, and all will be procured via separate contracts using different procurement methods.

Should bills of quantities be considered appropriate for any element of such projects, whether supplied by the employer, by contractors or by subcontractors, and if the bills of quantities are prepared using CESMM4 as the basis for measurement, then the rules applying to the subdivision of bills of quantities will need to be followed in order to be compliant.

Headings and subheadings

Where the bill of quantities is divided into parts, Paragraph 5.10 requires that each part *shall be given a heading* and groups of items within parts *shall be given subheadings*. Care needs to be exercised in deciding what the text of these headings and subheadings will be because Paragraph 5.10 stipulates that they *shall be read as part of the item descriptions to which they apply*.

Referring to the example given Section 7.7.6 previously for a land drainage scheme, Part 3: *Drainage and headwalls* could be a heading, and *Pipework – Pipes (Class I)*, *Pipework – Fittings and valves (Class J)* and *Pipework – Manholes and pipework ancillaries (Class K)* could be subheadings.

Paragraph 5.10 also provides rules to ensure that the work included under headings and subheadings shall be clearly defined by underlining the <u>description</u> column and by repeating the headings and subheadings on subsequent pages of the BQ.

Extent of itemisation and description

Barnes (1977) considers that Paragraph 5.11 is one of the most important in CESMM because it allows the bill compiler the freedom to elaborate item descriptions and to separate items beyond the strict limitations of CESMM.

In order to understand this view, the words in Paragraph 5.11 merit careful attention:

- All work **shall be** itemized, and the items **shall be** described in accordance with the Work Classification.
- Further **itemization** and **additional description** *may be* provided under circumstances where *special methods of construction or considerations of cost* may arise.

However, the bill compiler has a choice if it is felt that the Work Classification is inadequate in some way and that *further itemisation **and** additional description* might help the contractor in particular circumstances.

Risk issue

The bill compiler has no choice but to follow the Work Classification. Therefore, provided the bill of quantities is prepared in strict accordance with the Work Classification, there can be no claim by the contractor that the work has been itemised or described incorrectly or misleadingly.

Before exploring the possibility of providing further itemisation *and* additional description, the bill compiler should scrutinise the coverage and additional description rules in the Work Classification to make sure that the issue is not already covered.

Risk issue

Provided that the additional itemisation and description is supplementary to the Work Classification, then again there is no valid claim if the contractor feels that the work has not been correctly billed.

It should be noted, however, that the word 'and' appears between *further itemisation **and** additional description*, not 'and/or'.

In other words, on a strict interpretation, if further itemisation is given, then additional description must also be given.

Descriptions

Paragraph 5.12 acknowledges that BQ descriptions *shall identify the work covered by the respective items* but also places the onus firmly on the contractor to ascertain *the exact nature and extent of the work* from the Drawings, Specification and Conditions of Contract, read in conjunction with the Work Classification. In effect, this means that the bill of quantities does not fully describe and accurately represent the work required but identifies it well enough that further detail may be obtained by scrutinising other documents.

Risk issue

This paragraph may have been overlooked by the review committee because it is identical to that in CESMM3 and resonates strongly with the sufficiency of tender provisions of the ICE Conditions of Contract (now ICC – Measurement Version).

The wording is also 'old fashioned' and sits awkwardly with the phraseology used in NEC3, for instance.

Care would be needed in using CESMM4 with NEC3 to make sure that such words and phrases were clearly defined in terms of the contract. The Paragraph 5.6 schedule (of terms) could be used for this purpose.

Paragraph 5.13 allows that item descriptions do not have to be fully in accordance with the Work Classification provided that any missing descriptive detail is signposted to a Drawing or Specification clause. This permits item descriptions to retain their brevity whilst, at the same time, making sure that the contractor can find the information needed for pricing in some other document.

The converse to this is Paragraph 5.14 which requires that additional description *shall be given* where application of the Work Classification does not identify individual items of work sufficiently clearly. Such additional description shall refer to the location of the work, or to other physical features of it, that may be shown on the Drawings or be described in the Specification.

Ranges of dimensions

In some cases, the Work Classification provides a range of dimensions, one of which may be chosen as a descriptive feature of an item, for example, H 1 2 4: Precast concrete beams, length 5–7 m, mass 1–2 t.

Normally, individual items would be listed under this heading distinguished by a mark or type number, but where the precast beams are all the same size, Paragraph 5.15 permits the actual dimensions to be stated rather than the Work Classification ranges.

Prime cost items

The idea behind P C Items is to provide a sum of money in the contract bills which can be expended during the contract in order to pay for:

- Specialist work to be carried out on the site.
- Specialist work to be carried out off-site.
- Design work in connection therewith.

The sum of money included in the BQ is the estimated cost of the work which is later replaced in the final account by the actual cost.

The usual procedure is that a specialist contractor is subsequently chosen or 'nominated' by the employer/contract administrator and engaged as a subcontractor by the main contractor, subject to the right of reasonable objection. The nominated subcontractor is thus in direct contract with the main contractor (not the employer) who provides on-site facilities to the subcontractor. These attendances (or 'labours') are priced by the main contractor in the BQ against specially provided items along with an additional item for main contractor's profit and overheads ('other charges and profit').

Paragraph 5.16(a)(i) explains that, where the nominated subcontractor is to carry out work on-site, 'labours' only include use of the main contractor's facilities, light and power, disposal of rubbish and space for cabins, etc., whereas Paragraph 5.16(a)(ii) provides that unloading, storing and hoisting materials, etc. are included where the work is carried out off-site (i.e. where the nominated subcontractor is effectively a supplier). It should be noted that the CESMM4 definition of 'labours' prevails only *in the absence of any express provision in the contract to the contrary*.

Where specific rather than general attendances are required for nominated subcontract work on-site, such as material handling, crane hire, etc., it is advisable to insert a BQ item for 'special labours' to be priced by the main contractor.

Allowances for Prime Cost Items are to be included in General Items (Class A) which provides for *labours* and *special labours* in connection with P C Items. Rule M8 applies.

Rule A6 stipulates that the item description for P C Items shall identify the work included, and Rule A7 stipulates that the special labours required shall be stated in the item description. Table 7.4 illustrates these principles.

There is no special wording or way of describing work covered by P C Items in CESMM4.

Risk issue

It is surprising that CESMM4 retains the concept of including P C Items in the Bill of Quantities as its use is somewhat outmoded in modern-day procurement practice.

The preference nowadays appears to be for employers to avoid the risk of nomination, especially where subcontractor design is involved, and to pass this on to the contractor either via a list of 'preferred' subcontractors or by using full or partial contractor design.

Table 7.4   P C items – 1.

	General Items				
	Nominated Sub-Contractors which include work on the Site				
A	Prime Cost Item; Supply and install circular cast iron flap valve; 2m diameter; Fixing to RC headwall; Including fixing bolts; Estuary outfall; Tidal	sum			
B	Labours	sum			
C	Special labours; Offloading, handling and storing; Setting out; Lifting into position; Safe working platform	sum			
D	Other charges and profit	sum			

However, Paragraphs 5.16(a)(i) and (ii) make it clear that such sums may relate to work carried out <u>off-site</u> as well as <u>on the site</u>, and as such, the implication is that separate items would be provided in order to distinguish between the two situations. This is illustrated in Table 7.5.

Table 7.5   P C items – 2.

	General Items				
	Nominated Sub-Contractors which include work on the Site				
A	Prime Cost Item; Electrical installation and pumping equipment	sum			
B	Labours	sum			
C	Special labours	sum			
D	Other charges and profit	sum			
	Nominated Sub-Contractors which do not include work on the Site				
E	Prime Cost Item; Pre-stressed concrete bridge beams	sum			
F	Labours	sum			
G	Special labours	sum			
H	Other charges and profit	sum			

Part 2

In Paragraph 5.17, goods, materials or services supplied to the main contractor by a nominated subcontractor must be identified in the relevant item description or in the heading to the relevant Prime Cost Item. This would be the case, for instance, where the contractor is to fix materials supplied by the nominated subcontractor.

Risk issue

It is important to remember that the facility to nominate subcontractors must be mirrored in the conditions of contract and express provision is made for this in the ICC – Measurement Version (Clauses 58 and 59) and in FIDIC 'Red Book' (1999) (Clause 5).

Nomination is not, however, provided for in JCT contracts or in the NEC3.

Provisional sums

All construction projects need a contingency allowance in order to provide for the unexpected, and it is sensible to include this in the bill of quantities <u>and</u> provide the contract administrator with the express contractual power to spend the money, if need be. This allowance usually takes the form of a general contingency and a list of provisional items.

Paragraph 5.18, which recognises the need for such an allowance, deals specifically with Provisional Sums and distinguishes between *provisional sums for defined work* and the *general contingency* (see also Paragraph 5.25 – not 5.26 as stated in 5.18).

Defined work is where *the scope of the work cannot be completely **designed*** but *the scope can be **defined***. Provisional sums for defined work are included in General Items (Class A.4.2), whereas, presumably, anything else will appear with or be included in the general contingency which *shall be given in the Grand Summary*. Table 7.6 provides an example of work that can be defined but cannot be designed until work commences on-site.

Table 7.6 Provisional sums.

General Items				
	Provisional Sums			
	Provisional Sums - Defined Work			
A	Excavation and disposal to create new wetland habitat for phragmites; Allow for careful removal and deposition of plants; Approximate area 1000 m² in North East corner of site. Shape and depth of individual lagoons to be determined on site by the Engineer	sum		

Phragmites = common reed

Class A coverage rule C2 confirms that the tenderer shall be deemed to have made appropriate allowances in his General Items for the programming, planning and pricing of defined provisional work.

The quantities included in a CESMM bill of quantities are recognised as being *estimated quantities* in Section 1: *Definition 1.5*. Consequently, there is no provision in CESMM for approximate or provisional quantities where the scope of part of the works is uncertain. The only recourse is for the bill compiler to include a provisional sum for defined work or to include an appropriate allowance in the general contingency.

This creates a problem on-site with respect to agreeing rates for provisional work or for contingency spending, and therefore, some bill compilers create bill items that may not be used, or inflate the quantities in 'legitimate' items, in order to establish rates and prices should the need arise and create a 'hidden' contingency at the same time.

Paragraph 5.18 makes the point that quantities *shall not* be increased in order to provide for contingencies. Notwithstanding this, it is common practice to do so and then, hopefully, show a saving when the work is admeasured.

Risk issue

Now that CESMM4 is contract neutral, inflating quantities could be a dangerous strategy because:

- The contract used could easily be a lump sum contract.
- There could be hidden problems with costly variations.
- There could be a clause in the contract entitling the contractor to a re-rate should the actual quantities be less than those estimated beyond a prescribed threshold.

Quantities

Paragraph 5.19 makes the usual stipulation that *quantities shall be computed net using dimensions from the Drawings* unless:

- There is a measurement rule to the contrary.
- Directed otherwise by the contract.

No allowance shall be made for bulking, shrinkage or waste, and the bill complier is at liberty to round quantities up or down and use fractional quantities where necessary, but not to more than one place of decimals.

This paragraph should be read in conjunction with Paragraph 5.1 which states that *appropriate provisions of this section* (i.e. Section 5) *shall also apply to the measurement of completed work.*

Pursuant to Paragraphs 5.1 and 5.19, therefore, the quantities of completed work shall also be measured in the same way as those in the Bill of Quantities:

- Net.
- Using dimensions from the drawings.

Consequently, when it comes to the admeasurement of work, it is a fallacy to imagine that work shall be physically measured on-site unless there are no drawings available, revised and up to date, from which to measure.

Units of measurement

Paragraph 5.20 prosaically lists the units of measurement that shall be used with CESMM4, together with their abbreviations, which do not bear repeating here.

Work affected by water

In civil engineering work, the presence of water is frequently a major risk factor. This might be a river or stream, canal, lake or tidal water such as the sea or an estuary.

Part 2

Paragraph 5.21 requires such bodies of *open water* to be identified in the Preamble whether they are on or bounding the site. Also, they *shall be* referenced to a drawing where the boundaries and surface level of such bodies of open water are indicated. Where the boundaries and surface levels fluctuate, the range of fluctuation *shall* also be indicated which Barnes (1977) considers should be the mean low and high water levels (ordinary spring tides) of the surface of tidal waters.

The presence of groundwater is excluded from the Paragraph 5.21 requirements, and Barnes (1977) considers that there are difficulties in distinguishing everything that will be affected by water. He also considers that the bill complier will have to exercise judgement as to the extent to which the contractor is alerted to the potential effects of the presence of water.

Risk issue

The issue of water is an interesting one as there is no provision in CESMM4 for measuring excavation below groundwater level or in running water or for the disposal of groundwater as there is in NRM2, for instance.

In such circumstances, the contractor must turn to the conditions of contract for relief if it is felt that the physical conditions encountered could not *reasonably have been foreseen by an experienced contractor* (ICE/ICC – Measurement Version, Clause 12) or merit the notification of a compensation event (Clause 61.3) under the NEC3.

Ground and excavation levels

Paragraph 5.22 applies to items of work involving excavation, boring or driving and requires that:

- Each item description shall state the Commencing Surface where this is not the Original Surface.
- Each <u>excavation item</u> shall state the Excavated Surface provided that this is not the Final Surface.
- Depths of excavation stated in accordance with the Work Classification shall be measured from the Commencing Surface to the Excavated Surface.

The object of these rules is to expand on Section 1: *Definitions* and ensure that the contractor is clear where such work starts and finishes. If there are intermediate stages in an excavation, the excavated surface of one stage becomes the commencing surface of anther, but there is no requirement to relate these surfaces to a level. Practical phrases such as '1 m below original level' or 'top of rock' are sufficiently clear for the purposes of this rule.

Paragraph 5.23 is incongruous under this heading and appears to be left over after the deletion of Paragraph 5.22 in CESMM3 (Form and setting of bills of quantities) which does not appear in CESMM4. The paragraph refers to the totalling of bill pages, carrying totals to a Part Summary and carrying the part summaries to the Grand Summary.

7.7.7   Grand Summary

The Grand Summary brings together all the various part summaries of the Bill of Quantities which shall be listed together with their respective monetary totals (Paragraph 5.24).

Underneath the total of the part summaries, three items shall be included:

General contingency allowance

Provided that the employer agrees, a contingency sum shall be included as a reserve fund to cover items of expenditure not provided for in the bill of quantities. There is no rule in Paragraph 5.25

governing the amount to be included, except that it shall be an item, not a percentage, and it may be expended in full or in part or not at all depending on the circumstances that arise on-site.

Adjustment item

The adjustment item is provided so that the contractor may conveniently make last-minute changes to the tender total without altering the rates and/or prices in the Bill of Quantities. Paragraphs 6.4 and 6.5 explain how the adjustment item works with respect to interim payments, retention and contract price fluctuations:

For interim payments, a *pro rata* addition or deduction* is paid/deducted in instalments, but not exceeding the total of the adjustment item, in the proportion:

Adjustment Item	*Less*	Retention (if any)
Total of BQ ± Adjustment Item		

Total of the priced bill of quantities

The monetary total of the part summaries, contingency allowance (if any) and adjustment item represents the contractor's tender total.

If the form of contract used is the ICC – Measurement Version or NEC3 Option B (priced contract with bill of quantities), then the tender total will be subject to admeasurement.

Should JCT 2011 SBC/Q or NEC3 Option A (priced contract with activity schedule) be used, however, the tender total will be a lump sum subject to adjustment only as prescribed in the contract.

7.8   Section 6: Completion, pricing and use of the Bill of Quantities

Changes from CESMM3

- Reference to pound sterling omitted. Now refers to *currency of the contract* (Paragraph 6.1)
- Now refers to *interim payments, interim certificates* and *completion of the works* rather than ICE contract references (Paragraph 6.4)
- Now refers to a *contract price fluctuations* clause rather than the UK-centric 'Baxter formula' references (Paragraph 6.5)

7.8.1   Insertion of rates and prices

Although there is no guidance on how the bill of quantities pages should be set out, as there was in CESMM3 (Paragraph 5.22: *Form and setting*), CESMM4 Paragraph 6.1 nonetheless prescribes that the *rates and prices shall be inserted in the rate column*. This follows CESMM3 albeit the general convention in the industry is to insert rates in the rate column and the prices (i.e. sums of money such as general items and method-related charges) in the extension column (£-p).

Paragraph 5.22 refers specifically to *the currency of the contract* (CESMM3 did not) which hints at the internationalisation of CESMM4. The prevailing currency will be that stated in the conditions of contract:

- NEC3 – Contract Data Part 1.
- FIDIC – Appendix to Tender.

In contracts where there is no provision for stating the currency of the contract (e.g. ICC – Measurement Version), care would be needed to make sure this was included somewhere in the contract documents (e.g. ICC – Measurement Version, Clause 72: *Special conditions*).

Attention may also have to be paid to circumstances where payment for work in progress is to be made in more than one currency.

7.8.2   Parts to be totalled

Paragraph 6.2 states that *each part of the Bill of Quantities shall be totalled and carried to the Grand Summary*. This is a (not quite) verbatim repetition of Paragraph 5.23.

This presupposes that the bill of quantities is subdivided into parts, which is optional pursuant to Paragraph 5.9, albeit Paragraph 5.2(d) suggests that the work items shall be *grouped into parts*.

Some work items could be grouped (e.g. Classes F and G or Classes I–L), but these provisions are less than clear.

7.8.3   Adjustment item

The tendering period is a time of intense activity for the contractor, especially on large projects when a team of estimators, buyers, quantity surveyors, planners, construction managers, senior managers and directors will undoubtedly be involved. Towards the end of the process, many aspects of the tender come together rapidly – alternative methods of construction are considered, the pretender programme is finalised, risk issues are assessed, cheaper quotes for materials and temporary works are received, last-minute subcontract quotations are submitted, etc.

By this time, the bill of quantities will have been priced, rates will have been grossed up (i.e. overheads and profit added to direct costs), preliminaries will have been priced, and the contractor will be close to submitting the finalised bid. Nevertheless, final adjustments will be made to the tender to take account of last-minute economies and changes of mind, and the directors will wish to reflect on the competition and how much they want to win the contract.

With this in mind, adjustments may have to be made to the priced bill of quantities, and this will involve either making changes to the rates and prices or finding a place in the BQ to make an amendment so that the priced BQ totals to the tender figure that the directors decide upon.

Anyone who has been an estimator will know how manic the 'last knockings' of a tender submission are, and it was with some relief that the first edition of CESMM introduced the idea of an adjustment item in the Grand Summary. This has been retained in subsequent editions.

Paragraph 6.3 provides for an Adjustment Item to be given in the Grand Summary and gives the tenderer the option of pricing a lump sum addition or deduction to the total of the priced bill of quantities against it.

This conveniently gives the contractor the chance to quickly adjust the BQ total to the tender total that has been decided on but, at the same time, creates complications with regard to post-contract matters such as payment and the valuation of variations.

If, for example, the contractor has made a last-minute deduction to the BQ total, interim payments must be reduced accordingly in order to avoid overpayment to the contractor, and vice versa. The adjustment made to the interim payment shall be a proportionate adjustment made prior to the deduction of retention. Any positive or negative balance that is left of the adjustment item, once contract completion is reached, shall be added to or deducted from monies due.

Ross and Williams (2013) give a worked example of how an adjustment item works in practice (reproduced in Table 7.7).

Table 7.7   Adjustment item.

Adjustment item at tender stage

Ref	Item	Quant	Unit	Rate	£	p
	GRAND SUMMARY					
A	Class E: Earthworks				146 973	96
B	Class F: In-situ concrete				49 334	72
C	Class G: Concrete ancillaries				9 874	21
D	Class P: Piles				98 652	98
E	Class Q: Piling ancillaries				13 788	67
F	Class U: Brickwork, blockwork and masonry				17 463	22
G	Class W: Waterproofing				6 998	44
	Bills of Quantities Total				343 086	20
	Adjustment item				*12 196*	*20*
	TENDER TOTAL				**330 890**	**00**

Adjustment of interim valuation:

		£	p	£	p
• Paid or deducted in instalments	Gross interim valuation			104 746	96
• Adjusted in proportion to the total of the bills of quantities before the addition/deduction of the Adjustment Item	Adjustment: £12 196.20 x £343 086. 20	104 746	96	(3723	60)
• Adjusted before deducted of retention monies				101 023	36
	Retention 3 %			3 030	70
• Adjustments shall not exceed the total of the Adjustment Item	**Certified value**			**97 992**	**66**

Reproduced by kind permission of the authors.

Paragraph 6.5 further requires that suitable adjustment shall be made in respect of the adjustment item to any payments due to the contractor in the event that a contract price fluctuations clause is applicable to the contract.

7.9   Section 7: Method-related charges

Changes from CESMM3

- Reference to ICE clauses omitted. Method-related charges are deemed to be **prices** (i.e. not rates) for the purposes of *valuing changes to the works and revisions to rates as a consequence of a change in quantity arising from admeasurement* (Paragraph 7.6)
- Reference to ICE clauses omitted. Method-related charges shall now be paid *in accordance with the contract* (Paragraph 7.7)
- Reference to ICE clause omitted. Changes in method shall not result in an increase or decrease in the amount of a method-related charge unless a change in method *has been ordered by the contract administrator*, in which case *the provisions of the contract in valuing changes shall apply* (Paragraph 7.8)

The first edition of CESMM pioneered the concept of separating quantity-related and method-related costs in a bill of quantities based on CIRIA (1977) Research Project 34. The concept has been imitated in other methods of measurement since.

Method-related charges remain an integral feature of CESMM4 which is testament to the vision of the original authors and to the logic and practicality of the idea.

7.9.1   Definitions

Paragraph 7.1 distinguishes between **time-related** and **fixed** method-related charges and defines a method-related charge as:

- *The sum for an item inserted in the Bill of Quantities by a tenderer in accordance with Paragraph 7.2*

A method-related charge can, therefore, be described as:

- A time-related charge *proportional to the length of time taken to execute the work.*
- A fixed charge *which is not a time-related charge.*

Risk issue

A time-related charge is taken to be proportional to the time taken to execute the work to which it relates and not 'the Works' (i.e. the contract).

For interim payment purposes, therefore, the contract administrator must judge the progress the contractor has made compared with the agreed programme when deciding on the proportion of the time-related charge to be paid.

For example:	
Time-related charge	= £10 000
Activity duration	= 10 weeks
Time expired	= 6 weeks
Actual progress	= 5 weeks
Interim payment	$\frac{5}{10}$ × £10 000
	= £5 000

7.9.2   Insertion by a tenderer

Method-related charges are intended to cover <u>items of work</u> that the **contractor** deems necessary in order to construct and complete <u>the Works</u>, which is his contractual obligation. Users of NEC3 will be familiar with the distinction between 'work' and 'Works'.

All that the bill compiler has to do is provide a place in the bill of quantities for the tenderer to insert the item(s) that are desired, but he/she is not required to 'second guess' what the contractor's working methods will be. Some bill compilers, nevertheless, like to structure the Method-Related Charges section using the generic headings, others don't bother. A sample BQ page without the headings is shown in Table 7.8.

It will be noted from Table 7.8 that a preamble has been included that is neither required nor suggested by the method of measurement. A preamble is a personal preference but serves to direct the tenderer's attention to the provisions of Class A.3 and, also, to remind the estimator (at a busy period) not to forget to include the method-related charges in the BQ total.

In order to help the contractor, generic headings are provided in CESMM4 Class A: *General items* under A3: *Method-related charges*, but there is no compulsion for the contractor to use all, or indeed any, of these headings.

Table 7.8   Method-related charges.

General Items			
Method-Related Charges			
Tenderers are invited to enter below such Method-Related Charges, if any, as are deemed appropriate; Method-Related Charges should be generally in accordance with Class A of the CESMM4 Work Classification but may cover items of work other than those set out in Class A			

In this respect, Paragraph 7.2 explains that a tenderer is free to:

- Insert items into the BQ for method-related charges which
- Relate to the way that the contractor intends to carry out the work and
- Are intended to cover items of work that *are not to be considered* as proportional to the quantities of other items
- The cost of which are not included in the rates and prices for other items.

Tenderers are equally free *not* to insert method-related charges into the bill of quantities, and as many or as few of such items as are considered necessary may be inserted.

The words '*not to be considered*' are underlined because this emphasises, along with Paragraph 7.6, that method-related charges are not subject to admeasurement and will not be changed if the quantities of related items increase or reduce. There is a caveat to this, however, as discussed in Section 7.9.6.

In the days that CESMM was directly linked to the ICE conditions, contractors could be assured that:

- Changes in quantities are not variations.
- Method-related charges are not subject to admeasurement unless a variation is instructed.
- Money priced against method-related items is a safe repository, not only for non-quantity-related costs but also for margin moved from other items in order to front-load tenders, unless there is a variation affecting the work concerned.

This would still be the case if the ICC – Measurement Version was to be used, because the ICE and ICC conditions are essentially the same; the FIDIC (Red Book) operates similarly to the ICE conditions as regards variations to the contract.

Risk issue

Now that CESMM4 is contract neutral, the method of measurement could be used with conditions of contract that are not the same as ICE, ICC or FIDIC:

- CESMM4 could be used with JCT 2011 SBC/Q which recognises changes in quantities as a variation.
- Under NEC3 ECC (Black Book) Option A, all prices are lump sums and are not remeasurable. However, if there is a change to the Works Information which has the effect of reducing Defined Cost, the prices (including method-related charges) are reduced.

Part 2

7.9.3   Itemisation

Paragraph 7.3 suggests that method-related charges should follow the Work Classification order in Class A. This provides a structure for the General Items that the bill compiler may choose to provide in the bill of quantities. Alternatively, it may be left for tenderers to 'write in' the detail as necessary. If the Class A.3 structure is not fit for a particular purpose, tenderers can add their own items.

In any event, time-related and fixed charges should be distinguished although there is no obligation to do so. Contractors often like to split up method-related charges in this way as fixed charges such as mobilisation costs, installing cofferdams and providing haul roads represent 'early money' payments to help reduce capital lock-up in projects.

7.9.4   Description

Paragraph 7.4 is a 'rule' in that method-related items *shall be fully described so as to*:

- *define precisely the extent of the work covered.*
- *identify the resources to be used.*
- identify *the particular items of Permanent Works or Temporary Works, if any,* that the method-related item relates to.

These obligations rest firmly at the door of the tenderer (not the bill compiler) because:

- It is the tenderer who 'writes in' the method-related items.
- Unlike other items in the Work Classification, there is nothing on the drawings or in the specification that amplifies the BQ description.
- The clearer the item description, the easier it will be for the contract administrator to value work in progress and variations to the contract which is obviously crucial to how much the contractor will be paid.

7.9.5   Contractor not bound to adopt method

It is frequently the case that contractors change their minds as to how to carry out the work on-site once the contract has been awarded. The involvement of construction management staff will be influential in this, and it is no reflection on the estimating team that a different method is adopted to that which the tender is based on.

In common with all other items in a priced bill of quantities, the moneyed items are purely 'allowances' or budgets, and Paragraph 7.5 acknowledges the contractor's right to spend this money in any way deemed appropriate.

Consequently, a method-related charge for a cofferdam does not preclude the contractor being paid this sum if he decides to 'batter' the excavation nor will the contractor be refused payment if he chooses ready-mixed concrete instead of using the site batching plant priced in the method-related charge in the bill of quantities.

The only thing that will change in each of these instances is the cash flow for each method-related charge. In the first case, payment could be proportional to other excavation and filling works, and in the second case, it could be proportional to the quantity of concrete placed but subject to the provisions of CESMM4, Paragraph 7.8 (see Section 7.9.8).

In both cases, the full amount of the method-related charge will be paid on completion, barring variations.

7.9.6   Charges not to be measured

As discussed in Section 7.9.2, CESMM4 Paragraph 7.6 stipulates that method-related charges *shall not be subject to admeasurement*.

In Paragraph 7.1, a Method-Related Charge is defined as *a sum for an item inserted in the Bill of Quantities* and therefore is not a rate or a price by definition. Consequently, if a contractor prices a method-related charge for a cofferdam, he will be paid the same sum of money irrespective of whether more or less or no sheet piling at all is actually needed.

Risk issue

The point with Method-Related Charges is that the contractor carries the risk attached to his chosen method of working.

If he finds a better, quicker, cheaper method, or if he is lucky with the weather or ground conditions, then he benefits from the 'upside' risk. The 'downside' is that if things go 'pear-shaped', the contractor foots the bill.

Notwithstanding this, Paragraph 7.6 explains that method-related charges *shall be deemed to be prices*, and therefore can be adjusted, in certain circumstances:

- For the purpose of *valuing changes to the works*.
- For making appropriate *revisions to rates* where the process of *admeasurement* results in a change of quantity of items that are not themselves method-related charges.

Therefore:

- If, through the process of admeasurement, rates and prices in the bill of quantities are rendered unreasonable or inapplicable and should be reviewed, this review would include associated method-related charges.
- If the characteristics of measured work items change significantly, or if the conditions under which such work is to be carried out are not as described in the contract, method-related charges could be changed accordingly.
- If the risk associated with any work changes, then the accompanying method-related charges would be reassessed.
- If a variation changes the work envisaged, method-related charges could be reviewed.
- If work is omitted from the contract, any associated method-related charge would not be payable.

Risk issue

It should be noted that the term 'admeasurement' is peculiar to the ICC – Measurement Version but that other standard contracts use words and phrases that essentially mean the same thing.

Therefore, in a measure and value contract, the admeasurement process will take place, if not in name, in order to identify **differences** in the estimated and final quantities of work for the purpose of arriving at a fair valuation of work done.

If this process indicates that the BQ rates should be revisited in the circumstances, then method-related charges will be part of the review in order that a fair valuation is determined.

Part 2

7.9.7 Payment

CESMM4, Paragraph 7.7, requires that method-related charges *shall be certified and paid in accordance with the contract*. This is a change in the wording of CESMM3 which, of course, makes reference to the ICE Conditions of Contract.

The significance of the new wording relates to:

- Who prepares the valuation of work in progress – under some contracts, it is the contractor, and in others, it is the contract administrator or project manager (NEC3 ECC).
- Whether or not the contractor is paid for incomplete items of work (he is not under NEC3 ECC Option A).
- Whether retention is deducted from the valuation or not (this is a secondary option under NEC3 ECC).
- The contractual timing requirements for the issue of certificates and payments.

In order to highlight that the certification and payment régimes vary from contract to contract, a statement to the effect that method-related charges **shall be certified and paid in accordance with the contract** *shall* appear in the Preamble to the Bill of Quantities.

7.9.8 Payment when method not adopted

Under the provisions of Paragraph 7.5, the contractor is not bound to adopt the method of working that he has described in the item that he has priced as a method-related charge in the bill of quantities. The contractor is, nonetheless, entitled to be paid for a method-related charge in such circumstances with the proviso that the work in question has been satisfactorily carried out (Paragraph 7.8).

The issue raised in Paragraph 7.8 is one of payment for method-related charges where the contractor does not adopt the method of working stated in the item description. Two methods of payment are proposed:

1. The contract administrator and the contractor agree the amount(s), timing and trigger points for a number of instalment payments. For example:
    a) Installation/mobilisation (one-off payment), maintenance (monthly) and removal/making good (one-off payment).
    b) In proportion (*pro rata*) to the completed <u>associated</u> permanent works.
2. Failing such agreement, the method-related charge *shall be treated as if it were an addition to the Adjustment Item* and shall be paid to the contractor by way of payments in interim certificates.

Risk issue

The operation of Paragraph 7.8 should be carefully considered in relation to the prevailing conditions of contract in order avoid possible conflict.

For instance, payment under NEC3 ECC Option B (priced contract with bill of quantities), Clause 11.2.28, is assessed by the project manager on the basis of:

- The quantity of work completed.
- A proportion of each lump sum (which could include method-related charges).

However:

- The lump sum is apportioned on the basis of the completed amount of work <u>covered by the lump sum</u> and NOT the completed amount of associated [permanent] works.
- The project manager is obliged to *consider* any payment application by the contractor but is not obliged to take any notice of it or to discuss it with the contractor (Clause 50.4).

If Option B is used, consideration should be given to amending CESMM4, Paragraph 7.8, so as to avoid possible conflict with the conditions of contract.

This could be done by including a statement in Contract Data Part 1, in the space provided, to the effect that the method of measurement has been amended in the manner considered appropriate in the circumstances.

Paragraph 7.8 concludes by emphasising that the contractor is entitled to be paid the full amount of the method-related charge, despite not adopting the stated method of working, unless the change in method has been instructed by the contract administrator. In this event, a variation/change order/compensation event would arise, and this would be valued according to the provisions of the conditions of contract relating thereto.

7.10   Work classification

The Work Classification is the heart of CESMM4 and comprises 26 classes of work, listed alphabetically, including preliminaries (Class A).

Application of the Work Classification has already been discussed in Section 7.5, but there are two additional issues that deserve attention.

7.10.1   'Included' and 'excluded'

Each class is provided with a title, beneath which is a heading in bold text of inclusions and exclusions, followed by the detailed work classification, in three divisions, and the various measurement, definition and other rules.

The heading is important because it states the type of work that is <u>included</u> in each class and also what is <u>not included</u> in each class. Consequently, if a type of work is not included in the heading to the class, then, by definition, it is excluded either:

- Because it is included in another class or
- It is not provided for in the method of measurement.

Take ground freezing – a geotechnical process for altering the properties of soil – as an example.

The technique basically entails the circulation of a cryogenic fluid, such as brine or liquid nitrogen, through a system of pipes drilled into the ground from the surface. Pore water is thereby converted into ice, and pumping continues until the particular design thickness is reached. Once this has been achieved, the freeze plant is operated at a reduced rate in order to maintain the ground in its 'frozen' state.

Geotechnical processes carried out from the ground surface are specifically **excluded** from Class T: *Tunnelling* and specifically **included** in Class C: *Geotechnical and other specialist processes* as stated in the Class T heading. Class C includes *geotechnical processes for altering the properties of soils and rocks* because this is expressly stated in the Class C heading.

Looking at Class C, however, ground freezing is not specifically listed in the Work Classification, and the only drilling items provided are for grouting work (freezing is not grouting).

CESMM4, therefore, does not provide a work classification for ground freezing, and the bill compiler is left with two choices:

1. Adapt the method of measurement in a suitable way, and include a preamble, so that measured items for ground freezing can be included in the bill of quantities.
2. Include a provisional sum for defined work in Class A: *General items*.

7.10.2   Additional description rules

Work is classified in CESMM4 according to the Work Classification, and bill of quantities items are developed from the descriptive features provided in the First, Second and Third Divisions. This is done in conjunction with the measurement rules and definition rules, and contractors are able to understand what is included in the various items by reference to the coverage rules.

The Work Classification is very brief, however, and the resulting item descriptions may not always convey the full extent of the work represented in the measured item in the bill of quantities. This is the role of the Additional Description Rules.

The importance of Additional Description Rules cannot be overemphasised as they 'flesh out' the otherwise limited descriptive information provided in the first three divisions of the method of measurement. Consequently, a great deal of the bill compiler's time is occupied with incorporating additional description into the various BQ items. Additional coding may also be required.

The provision of additional description is, however, not limited to the additional description rules provided in each class. It is also influenced by the bill compiler's judgement in interpreting the requirements of CESMM4, Paragraph 5.11, which states that *further itemisation and description may be provided* should circumstances arise that require *special methods of construction or considerations of cost.*

Risk issue

Failure to observe the Additional Description Rules is equivalent to failing to describe an item correctly and, as such, might constitute a misrepresentation of the work to be priced in the contractor's tender.

7.10.3   Non-standard work

Barnes (1977) raises the issue of *non-standard work* with reference to Class Q: *Piling ancillaries* in the context of pile extraction that is not included in the Work Classification.

By this authority, it seems that the bill compiler has some licence to expand the ambit of CESMM4 not envisaged by Paragraph 5.4.

Whilst Paragraph 5.4 gives tacit approval to amending CESMM4, this would necessitate a statement being given in the Preamble, whilst the discretion implied by Barnes (1977) would appear to suggest that a measured item marked *non-standard work* would suffice.

7.11   Class A: General items

Changes from CESMM3

- Insurance of the Works and third-party insurance no longer listed in the Second Division
- A 1 2 0 *insurances* – no longer any subdivision for different classes of insurance
- C1 revised – there are no longer any *deemed to be included* provisions
- C1 must be defined in the schedule of specific clauses under Section 5: Paragraph 5.6
- A 1 3 0 *parent company guarantee* is new (M2 refers)
- *Specified requirements – contract administrators* replace *engineer's staff;* no longer any reference to *staff*
- A 4 2 0 – Provisional Sums are now for *defined work* (M7, D2, C2 and A5 refer) as well as Daywork

Class A is the 'preliminaries' section of CESMM4.

The unit of measurement in Class A is discretionary, but Measurement Rule M1 suggests that it *may be the sum*. In the case of a *parent company guarantee* (A 1 3 0), Rule M2 states that *an item may be given*.

There are six categories in the First Division of Class A:

1. Contractual requirements		
2. Specified requirements		
3. Method-related charges		See also Section 7.9
4. Provisional sums		
5. Nominated subcontracts which include work on the site	}	See Section 7.7.6
6. Nominated subcontracts which do not include work on the site		

7.11.1   Contractual requirements

CESMM4 has rationalised the Second Division descriptive features of *contractual requirements* and now includes reference to *parent company guarantee* as well as a *performance bond*.

Measurement Rule M2 confirms that a parent company guarantee is an optional provision where a tenderer may be part of a larger company or group.

Risk issue

The choice between a parent company guarantee and a performance bond, provided by a bank or insurance company, requires careful consideration as each provides a different remedy.

A parent company guarantee should ensure that the contract is satisfactorily completed, subject to the continuing solvency of the parent company, whereas a performance bond simply provides a guarantee of a payment, up to a defined limit, in the event of the contractor's default.

The Second Division now only refers to *insurances*, and there is no longer any subdivision into classes of insurance such as works insurance, third-party insurance (see Table 7.9). This is probably sensible because different forms of contract use different phraseology when referring to the types of insurance cover and indemnity limits required.

Coverage Rule C1, which relates to insurances, contains no item coverage information, but the meaning of this rule must be defined in the 'schedule of specific clauses' which is now required under Section 5: Paragraph 5.6.

Table 7.9   Insurances.

	General Items				
	Contractual requirements				
A	Insurances	sum			
B	Parent company guarantee	sum			

Part 2

7.11.2 Specified requirements

Specified requirements are measured in accordance with Rule M1, but where the value of any item is *to be ascertained and determined by admeasurement*, a <u>quantity</u> shall be given for each item in accordance with Rule M3.

Admeasurement is a process for measuring changes during the works, and therefore, bill compilers need to be alert to any specified requirements that might be varied during the contract so that a quantity can be included.

On any contract of substance, facilities are needed for the employer's representatives on-site. Formerly, CESMM provided for facilities for the *engineer's staff*, but the fourth edition, being contract neutral, has different wording. The term *'contract administrators'* is now employed, in the plural, as a collective term for the employer's team. The word 'staff' is no longer used, presumably because not all of the employer's representatives are necessarily directly employed.

Class A 2: *Specified requirements* are defined in Rule D1 as **work** *other than the permanent works* that is *expressly stated in the contract*. As there are no Class A coverage rules, it must be taken that the word *work* implies that *specified requirements* shall be fully functioning, and this is to some extent dealt with in Rule A2 which refers to *their continuing operation or maintenance*:

- Rule A2 requires that item descriptions for *specified requirements* shall *distinguish between the* **establishment** *and* **removal** *of services or facilities and their continuing operation or maintenance*.
- Additional Description Rule A1 states that *specified requirements* needed *after the date for completion* shall be so described in the relevant items.

Risk issue

There is no definition of *completion* in CESMM4, and consequently, any specified requirement that is to be maintained after completion must be correctly referenced to the intended meaning of 'completion'.

'Completion' has several meanings depending upon the conditions of contract used, all of which use different terminologies.

Standard contracts tend to view completion in two stages:

- An 'initial' completion and then, when defects have been corrected:
- A 'final' completion.

The satisfactory conclusion of a defined period after 'initial' completion, usually 6 or 12 months, brings the contract to an end:

Contract	'Initial' completion	'Final' completion
ICE	Substantial completion	Defects correction period
FIDIC	Completion (subject to passing the tests on completion)	Defects notification period
ECC	Completion date	Defects correction period
JCT	Practical completion	Rectification period

Item descriptions should be drafted carefully so that the precise meaning of 'completion' is conveyed and tenderers may price the relevant item accurately.

Specified requirements include accommodation and the provision by the contractor of services and facilities for the contract administrators. In Table 7.10, it can be seen that additional description has been provided for 'accommodation' pursuant to Rules A1 and A2.

Table 7.10   Specified requirements.

	Specified requirements		
	Accommodation for the contract administrators		
A	Offices; Establishment	sum	
B	Offices; Maintenance	sum	
C	Offices; Maintenance during Defects Correction Period	sum	
D	Offices; Removal	sum	
	Services for the contract administrators		
E	Transport vehicles; 2 nr 4 x 4 vehicles as per Specification	sum	

7.11.3   Specified requirements: Temporary works

Specified requirements include testing and temporary works, such as traffic diversions, cofferdams and dewatering.

The temporary works categorised under Class A 2 7 * differ from those listed under Class A 3 5 * and A 3 6 * in that they are specified as being required and are not at the contractor's discretion. As such, these items will be listed in the bill of quantities, along with any required additional description, whereas the contractor's method-related charges are to be inserted by the tenderer.

Additional description for *Specified requirements – Temporary works –* is required under Rule A1 (post-completion requirements) and Rule A2 (establishment, maintenance and removal).

Where *specified requirements* are to be subject to *admeasurement*, a quantity shall be given for each item in accordance with Rule M3. This also provides scope for variations during the works, and any item likely to be subject to change will require a quantity against the item. There might, for instance, be a specified requirement for dewatering to a specific zone of the site, or the specification may contain particular provisions regarding compressed air working pressures, each of which might need to be varied.

This is illustrated in Table 7.11, which also shows how locational information, whilst not specifically required for A 2 7 * items, is provided pursuant to Paragraph 5.14.

7.11.4   Method-related charges

Pursuant to Measurement Rule M4, *Method-related charges (if any) shall be inserted by the tenderer.* This topic has been discussed, at length, in Section 7.9 of this chapter.

Part 2

Table 7.11   Specified requirements – admeasurement.

	Temporary Works		
A	Traffic diversions; Tarrant Road; Establishment		sum
B	Traffic diversions; Maintenance		sum
C	Traffic diversions; Removal		sum
D	Bridges; Temporary; 2-way traffic: River Winchet; Establishment		sum
E	Bridges; Temporary; 2-way traffic: River Winchet; Maintenance		sum
F	Bridges; Temporary; 2-way traffic: River Winchet; Removal		sum
G	De-watering; Wellpoint; Chainage 1+200 to 1+600; Establishment	10400	m2
H	De-watering; Wellpoint; Chainage 1+200 to 1+600; Maintenance	10400	m2
J	De-watering; Wellpoint; Chainage 1+200 to 1+600; Removal	10400	m2

Additional description is required under Rule A4 for *Method-related charges* in order to distinguish between fixed and time-related charges.

There should be no reason why tenderers would not observe this requirement because it is beneficial to contractors for all sorts of post-contract reasons. However, where the rule has not been respected, this could be picked up in the pre-contract checking of tenders and insisted upon by the employer prior to entering into a contract.

There is no guarantee, however, that any method-related charges, or other rates and/or prices for that matter, will not be priced with 'commercial opportunity' in mind. Contractors are alert to opportunities to enhance the value of variations/compensation events and ever mindful of the need to reduce negative cash flow on contracts!

7.12   Class B: Ground investigation

Changes from CESMM3

None

Ground investigation work is usually carried out in advance of a contract so that borehole and trial pit logs may be supplied to tenderers.

In this respect, Class B: *Ground investigation* may be used as part of the procurement process for a 'stand-alone' ground investigation contract, perhaps in conjunction with a suitable form of contract such as the ICC Ground Investigation Version.

Ground investigation items are deemed to include the preparation and submission of records and results in accordance with Coverage Rule C1, but the preparation of analysis reports is regarded as a matter for a separate item under Class B8: *Professional services*. Alternatively, analysis of ground investigation records and results may be undertaken under a separate contract for professional services such as the NEC3 Professional Services Contract.

Where ground investigation is required during a construction contract, it is normal to ask the contractor to excavate trial pits and trenches in order to verify ground conditions or to locate underground services.

Trial pits and trenches warrant two measured items:

1. An enumerated (nr) item stating the maximum depth in depth bands (B 1 * *) with additional description giving the minimum plan area at the bottom of the pit or trench and, for locating services, the maximum length of the trench (Rule A1).
2. An item, in linear metres, measuring the depth of pits and trenches stating the minimum plan area at the bottom of the pit or trench or, in the case of work to locate services, the maximum length of the trench.

The BQ items for 10 nr trenches to locate existing services are shown in Table 7.12.

Table 7.12   Ground investigation.

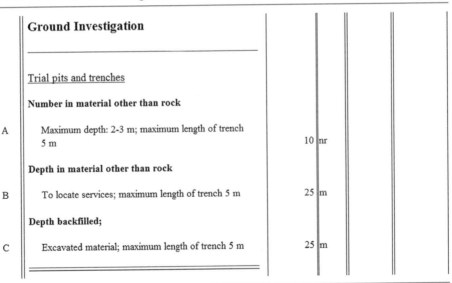

	Ground Investigation		
	Trial pits and trenches		
	Number in material other than rock		
A	Maximum depth: 2-3 m; maximum length of trench 5 m	10	nr
	Depth in material other than rock		
B	To locate services; maximum length of trench 5 m	25	m
	Depth backfilled;		
C	Excavated material; maximum length of trench 5 m	25	m

The alternative to items measured in the contract bills is for the contract administrator to instruct the work to be carried out on a Daywork basis (ICC/ICE contracts) or to value it as a compensation event using the Shorter Schedule of Cost Components (ECC contracts).

7.13   Class C: Geotechnical and other specialist processes

Changes from CESMM3

- *Grout holes* and *grout materials and injection* combined (First Division)
- *Materials* now C 4 5 *
- **NB**:
  There is a mistake in the Second Division of C 4: *grout holes, materials and injections.* C 4 5 * should be separated from C 4 1–4 with a solid line. The Third Division list of grouting materials (1–6) belongs to C 4 5 which requires descriptive features in order to complete the itemisation of C 4 5 * *materials* (CESMM3 refers)
- BS 4449 deleted from C 5 5 * and C 5 6 * (diaphragm walls)
- D3 – reference to BS 4449 deleted
- *Ground anchorages* is now *ground reinforcement* (D4 and A8 refer)
- C 8 * * *Vibroflotation* added

Items for geotechnical work now include ground reinforcement and vibroflotation as well as ground anchors and diaphragm walls, etc.

For all work in this Class, the Commencing Surface shall be stated where this is not the Original Surface (Section 5, Paragraph 5.22), and this shall be used for the admeasurement of completed work (Rule M1). All items in Class C are deemed to include disposal of excavated material and the removal of dead services (Rule C1).

7.13.1   Ground anchors

The measurement of ground/rock anchors is a little involved as there is a multiplicity of items to be measured for this work in Class C:

1. Drilling for grout holes and injection pipes (C 1–3 * *).
2. Items that define the extent of the work involved:
   a)   Number of holes (C 4 1 0)*.
   b)   Number of stages (C4 2 0).
   c)   Water pressure tests (C 4 3–4 0).
3. The supply of grout materials (C 4 5 *).
4. Injection of the grout holes with a choice of grout types (C 4 6 *).
5. Supply, installation and stressing of tendons, testing and constructing anchor heads, etc. (C 6 * *).

1. Drilling for grouting is described as either through *rock or artificial hard material,* or through material other than rock or artificial hard material (i.e. normal ground), and measured in linear metres. Where both ground conditions are present, a quantity is presumably required for each. The drilling inclination is given in the Second Division and the depth of drilling in the Third Division within bands (not exceeding 5 m, 5–10 m, etc.). The diameter of holes is to be given in item descriptions (Rule A1).
2. Additional enumerated items are given for the number of holes, for the number of drilling stages and for carrying out single or double water pressure tests. To some extent, these items reflect the costs of re-mobilising the drilling rig from position to position and moving it around the site.
3. The supply of grouting materials is measured by mass (t), but there is no rule that determines how the quantities shall be calculated for the bill of quantities (e.g. a nominal

mass per m³) or how they shall be verified for admeasurement (e.g. supply of delivery tickets, goods received sheet).

4. Injection is measured by the number required (C 4 6 *), with a separate item for the grout material. The dry materials are measured under C 4 5 in conjunction with a list of cement, fillers and chemicals in the Third Division (features 2–5). The enumerated items represent the cost of preparation, mixing and delivery of grout to the injection point.

5. The number of ground anchorages (nr), and the total length installed (m), either in material other than rock or artificial hard material or in material which includes rock or artificial hard material, are measured in C 6 * *. The measured items shall include details of the location and composition of the anchorages as well as working load and other testing details.

A practical example of the measurement of ground anchors is provided in Chapter 15.

7.13.2 Diaphragm walls

Diaphragm walls are measured in Class C 5 * * and are defined as *walls constructed using bentonite slurry or other support fluids* (Rule D2). They are categorised in the Second Division according to:

- Excavation.
- Concrete.
- Reinforcement.
- Joints and guide walls.

http://www.youtube.com/watch?v=aMyLEpEM9Hg

Excavation is measured in *rock, artificial hard material* or *material other than rock or artificial hard material*, in m³. Depth measurement is taken from the adopted Commencing Surface (Rule M1), and excavation items are *deemed to include preparation and upholding the sides of excavation* (Rule C2). The nature of the material to be excavated (e.g. mass concrete) shall be stated for excavation items described as artificial hard material (Rule A5 refers).

There is no item for preparing the surfaces of diaphragm walls to receive other works (e.g. brickwork, copings, capping beams) as this is deemed included in concrete items (Rule C3 applies).

Concrete is measured in m³ from the required cut-off levels with volumes calculated in accordance with Measurement Rules M1 and M2 of Class F and, of course, with CESMM4, Paragraph 5.19 (i.e. net dimensions taken from the drawings). Item descriptions for excavation <u>and</u> concrete shall state the wall thickness (Rule A4).

Diaphragm walls are constructed by excavating a series of alternate rectangular panels, filled with bentonite and then rebar and concrete, with stop ends forming the key for the subsequent 'infill' panels. This forms a continuous wall which may include water bars cast into place using the stop end formwork.

The on-plan configuration of diaphragm walls can be complex, which increases the complexity of work and adds cost. However, there is no requirement to identify this in item descriptions.

Reinforced concrete guide walls, typically 1–1.5 m deep, are constructed either side of diaphragm walls in order to provide a template for wall excavation and panel layout, to provide earthwork support to the top of the trench and to perform a number of ancillary functions, such as restraint for end-stops, hanging rebar cages, supporting 'tremie' pipes and so on. In practice, pre-trenching is often carried out to remove shallow obstructions and facilitate construction of the guide walls. This is not recognised in the method of measurement and would come under the heading of 'contractor method'.

Guide walls are measured in linear metres to each side of the diaphragm wall (Rule M12). There is no coverage rule for the measurement of guide walls.

Part 2

Figure 7.5 and Table 7.13 show sample take-off and bill items for a diaphragm wall to a 30 m × 20 m basement 21 m deep with rock head at 3 m above the Final Surface.

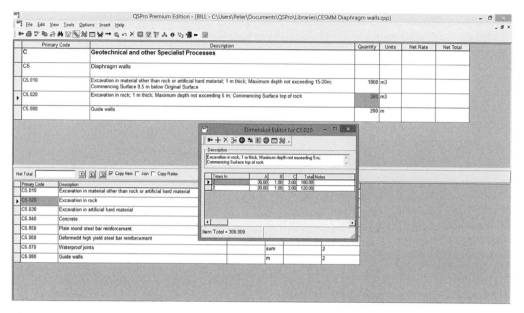

Figure 7.5   Diaphragm wall.

Table 7.13   Diaphragm wall.

Geotechnical and other Specialist Processes		
Diaphragm walls		
A	Excavation in material other than rock or artificial hard material; 1 m thick; Maximum depth not exceeding 15-20m; Commencing Surface 0.5 m below Original Surface	1800 m3
B	Excavation in rock; 1 m thick; Maximum depth not exceeding 5 m; Commencing Surface top of rock	300 m3
C	Guide walls	200 m

7.13.3   Vibroflotation

Vibroflotation is a new First Division descriptive feature in CESMM4 with, disappointingly, little to distinguish it. This geotechnical process is defined as follows:

Compaction of granular soils by depth vibrators is known as Vibro Compaction. The method is also known as Vibroflotation.[3]

The method involves the introduction of a 'vibroprobe' into the ground on a predetermined grid, and a combination of vibration, water and air compacts the subsoil. A certain amount of regrading of the commencing surface is required as part of the process.

The number of probes is enumerated (nr) in the Second Division, and the depths are measured according to wide bandings in the Third Division (e.g. not exceeding 5 m, 5–10 m, 10–20 m, etc.). There is no requirement to state a particular depth within the bands, and there are no rules of any description, other than the general rules that apply to the whole of Class C.

Presumably, any performance criteria or testing requirements will be provided in the specification or on the drawings although nothing is required to be stated in item descriptions, an example of which can be seen in Table 7.14.

Table 7.14   Vibroflotation.

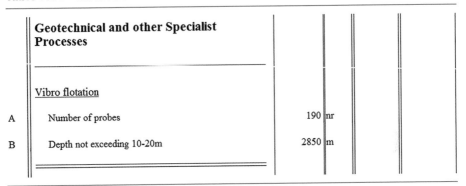

	Geotechnical and other Specialist Processes		
	Vibro flotation		
A	Number of probes	190	nr
B	Depth not exceeding 10-20m	2850	m

7.14   Class D: Demolition and site clearance

Changes from CESMM3

- *General site clearance* is now *site clearance*
- *General clearance* and *invasive plant species* added to the Second Division (D2, C3 and A3 refer)
- Trees – girths revised (D 2 * 0)
- Stumps – diameters revised (D 3 * 0)

7.14.1   Site clearance

For site clearance, there are two potential measured items in Class D:

- D 1 1 0 – General clearance.
- D 1 2 0 – Invasive plant species.

The first is measured in hectares (ha) and the second in m². In Additional Description Rule A2, the area for general clearance must be identified if this is not the total area of the site. If nothing is stated in the item description, then the area of removal of invasive plant species will effectively be 'extra over' general clearance.

General site clearance includes everything *expressly required to be cleared* except for those objects *for which separate items are given* in the bill of quantities as required by Class D (Definition Rule D1 refers).

This would appear to indicate, therefore, that somewhere in the tender documents there is a drawing or schedule that particularises everything that is not included in the bill of quantities.

Thus, trees of less than 500 mm girth, which are not measured individually, must be *expressly* detailed in order for removal to be deemed included in the general site clearance item.

CESMM4 includes a new category of site clearance – *invasive plant species*. Definition Rule D2 states that such species are those *whose control is governed by legislation* and requires treatment *by herbicidal or chemical process*.[4] Japanese knotweed, giant hogweed and Himalayan balsam are well-known examples in the United Kingdom.

There are several ways to deal with such plants, and the bill compiler should take advice as to the most appropriate control measure in each case as the method of treatment must be stated in the item description as well as the type of vegetation (Rule A3). The removal of invasive plant species by digging can require extensive excavation – 500 mm or more for some species and up to 4 m for others – and disposal to a licensed tip by a licensed carrier.

Disposal of material arising from site clearance is deemed included in measured items by virtue of Coverage Rule C1, but Rule C3 also emphasises that items *shall include for the disposal of any vegetable matter remaining after treatment* where herbicidal or other treatment has been applied prior to removal. Rule C3 is a coverage rule, and thus, no additional description is required.

> **Risk issue**
>
> An item for the site clearance of invasive plant species should ring alarm bells for contractors because the additional description and item coverage rules could infer extensive excavation and costly disposal to a seemingly innocuous item.

7.14.2   Trees and stumps

Trees less than 500 mm girth are not measured, but those that are larger than this are enumerated stating the girth as 500 mm to 2 m and exceeding 2 m measured 1 m above ground level (Rule D3). Items shall include the removal of stumps if required (Rule C4). Tree stumps are measured by diameter (not girth) of less than or exceeding 1 m.

In both cases, holes remaining after removal that require backfilling warrant a description of the nature of the backfill in the measured item (Rule A4).

7.14.3   Buildings and other structures

The demolition and removal of buildings and other structures are measured by the *sum*.

The identity of buildings or other structures shall be given in item descriptions (Rule A5), and the Second Division list of descriptive features provided implies that a predominant material shall be stated or, alternatively, that a suitable statement will clarify that there is no predominant material present. The volume shall be stated, according to the Third Division categories, which Rule D4 deems shall be *the approximate volume occupied*.

The phrase *volume occupied* is not defined but, presumably, means the volume occupied by the building or structure which, again presumably, is the volume given by the product of its external dimensions.

Definition Rule D4 specifies that the clearance of buildings and other structures are classified in such a way as to exclude *any volume* below the Original Surface, that is, *the surface of the ground before any work has been carried out* (CESMM4, Paragraph 1.8). This, presumably, includes foundations, basements and the like.

Therefore, the removal of a ground floor slab above existing ground level would be included in the demolition item, and a similar object at or below ground level would be measured as removal of *artificial hard material* in Class E. However, a ground floor slab that is partially above and partially below the existing ground would not fall clearly into either category and would require a decision by the bill compiler as to where in the bill of quantities to include this work.

Underground tanks are not categorised but, presumably, come under the heading of *other structures* of stated material (e.g. metal) and volume range as per D 5 4 *.

7.14.4   Pipelines

The Class D heading specifically excludes *articles, objects, obstructions and materials at or below the Original Surface*, and therefore, items for the removal of pipelines must only include those above the Original Surface.

This is to some extent confirmed in Coverage Rule C5, which deems that the demolition and removal of supports are included in the measured items. Pipelines within buildings are only measured where the nominal bore exceeds 300 mm (Rule M1 refers).

7.15   Class E: Earthworks

Changes from CESMM3

- Introduction of *controlled and hazardous material* in the Second Division of excavation items at E * 6 * and E * 7 * (D4 and A7 refer)
- Also in *Excavation ancillaries* at E 5 * 5 (A11 refers)
- *High energy impact compaction of general fill* has been introduced under E6; the Second Division numbering appears wrong and should be 5 not 1 in order to create a discrete item code (M22 and A17 refer)
- In Measurement Rule M7 and Additional Description Rule A2, reference to Paragraph 5.20 should read 5.21 (Work affected by water)

7.15.1   Excavation

Barnes (1977) explains that the principle behind Class E is that excavation items should cover:

- One type of excavation (dredging, cuttings, foundations, general).
- In one type of material (topsoil, rock, material other than topsoil, rock or artificial hard material, rock and so on).

NB: The work classification, and the rules of Class E, should be read in conjunction with Section 1: *Definitions* and with Paragraph 5.22.

Earthwork depths are measured from the Commencing Surface, which shall be stated in item descriptions where this is not the Original Surface (Rule A4), but Rule A4 also provides that the Excavated Surface shall be identified in item descriptions where this is not the Final Surface.

The bill compiler needs to be aware of any circumstances in the contract where excavation is *expressly required* to be carried out in stages, as an item for each stage shall be given according to Measurement Rule M5. The phrase *expressly required* could refer to a clause in the specification, to a requirement on a drawing or to a specific preamble where a particular sequence of work is required.

Clarity is of crucial importance in the item descriptions for earthworks, and the contractor needs to fully understand how the work has been measured. Paragraph 5.22 is crystal clear that:

- Where excavation begins below the Original Surface, the Commencing Surface shall be stated:
  - This could be an Excavated Surface or
  - The Final Surface.

Part 2

Therefore:

- Earthworks that follow topsoil removal commence from the underside of the topsoil which is the Excavated Surface of the topsoil item.
- Excavation below the Final Surface is, *ipso facto*, not an excavation item under E 1–4 and would be measured as an earthworks ancillary (E 5 6 0).

This logic is illustrated in Figure 7.6 in relation to the relevant bill of quantities items.

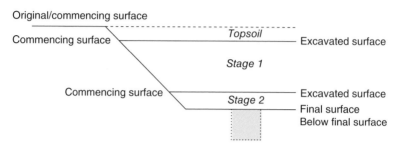

Figure 7.6   Excavation in stages.

Table 7.15 (which should be read in conjunction with Figure 7.6) shows the relevant bill of quantities items for earthworks carried out in stages compared to the items that would appear if work were not required to be done in stages.

Table 7.15   Excavation in stages/not in stages.

	Excavation in stages				Excavation not in stages	
	Earthworks				**Earthworks**	
	Excavation for cuttings				Excavation for cuttings	
	Topsoil				Topsoil	
A	generally	m3		A	generally	m3
	Material other than topsoil, rock or artificial hard material				Material other than topsoil, rock or artificial hard material; Commencing Surface underside of topsoil	
B	Commencing surface underside of topsoil; Excavated Surface 300 mm above formation	m3		B	generally	m3
C	Commencing Surface 300 mm above formation	m3			Excavation ancillaries	
	Excavation ancillaries				Excavation of material below the Final Surface and replacement with acceptable material	
	Excavation of material below the Final Surface and replacement acceptable material			C	generally	m3
D	generally	m3				

In common with other classes of CESMM4, the Commencing Surface adopted for preparation of the bill of quantities shall also be used for admeasurement of the completed work.

Depth is classified in bands according to the Third Division descriptive features, but this classification <u>only applies</u> to foundations and general excavation and <u>not to</u> dredging and cuttings.

For foundations and general excavation, the depth bandings refer to the total depth of the excavation and not to the different materials within the excavation (e.g. topsoil, material other than topsoil, rock or artificial hard material, etc.). Common good practice, however, is to alert tenderers to the presence of rock within a specific excavation, despite there being a measured item for rock, and to describe the excavations accordingly.

Table 7.16 illustrates that a general excavation, maximum depth 2–5 m, requires the removal of topsoil (normally at the surface) and that rock is measured to a maximum depth of 2 m from the point where the maximum depth of the 'normal' dig finishes.

Table 7.16   Excavation of rock.

	Earthworks				
	General excavation				
	Topsoil				
A	maximum depth: 2-5m		m3		
	Material other than topsoil, rock or artificial hard material				
B	maximum depth: 2-5m; Commencing Surface underside of topsoil; Excavated surface top of rock		m3		
	Rock				
C	maximum depth: 1-2m; Commencing surface top of rock		m3		

Quite frequently, in practice, there will be an interface between different classifications of excavation. There may, for instance, be a general excavation, or a cutting for a road, where there is also a structure, such as a bridge, gantry foundation or retaining wall which has a foundation to be excavated. The bill compiler needs to make a decision on how to measure the work so that tenderers may be clearly informed as to what to price.

This situation is illustrated in Figure 7.7, where it can be seen that the distinction is made between bulk excavation and excavation for foundations by using:

a)   A 'payment line' on the relevant drawing, which could then be referenced in the item description.

b)   The top of the second-stage excavation (i.e. the layer of earth specified to be left in place as protection for the formation) which could be used as a Commencing Surface for the foundation excavation item.

In Table 7.17, suggested method of billing is shown for excavation items relating to the crib retaining wall shown in Figure 7.7. This is to be constructed within a general excavation, and so a demarcation is needed between the bulk excavation and that for the foundation to the wall. In passing, it will be noted that Item A contains locational information in accordance with Additional Description Rule A3.

The excavation of rock and artificial hard material is provided for in the Second Division of Class E but there are no particular rules dealing with rock and artificial hard material in Class E, except for Measurement Rule M8; this limits the measurement of isolated volumes of each material which must be at least $1 \, m^3$ or a minimum of $0.25 \, m^3$, where the width of the excavation is less than 2 m.

Part 2

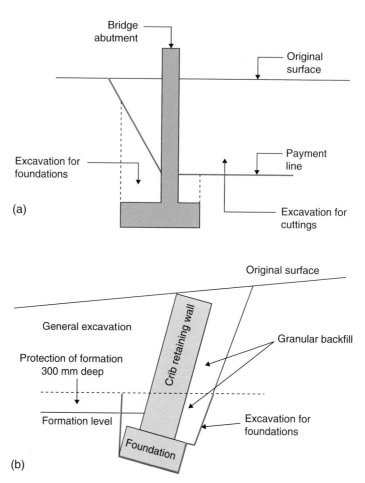

Figure 7.7   Demarcation between types of excavation. (a) Demarcation – payment line. (b) Demarcation – earthwork stages.

Artificial hard material must be described (*stated*) in item descriptions and separately billed as *exposed at the Commencing Surface* or *not exposed at the Commencing Surface*. Rock is not defined in CESMM4. In both cases, the Third Division depth categories are applicable only to general excavation and excavation for foundations.

Risk issue

The *Commencing Surface* is not necessarily the *Original Surface* and could equally be an *Excavated Surface*. Consequently, artificial hard material *exposed at the Commencing Surface* could be below ground.

The removal of such material could be a completely different proposition to, say, removing concrete slabs at ground level, and the cost of removal will, in all probability, be equivalent to that of *artificial hard material* **not** *exposed at the Commencing Surface*.

Bill compilers should pay careful attention, therefore, to ensuring that item descriptions are clear, and contractors, no doubt, will be alert to the possibility of a misrepresentation of the item description.

Table 7.17   Billing of excavation items for crib retaining wall.

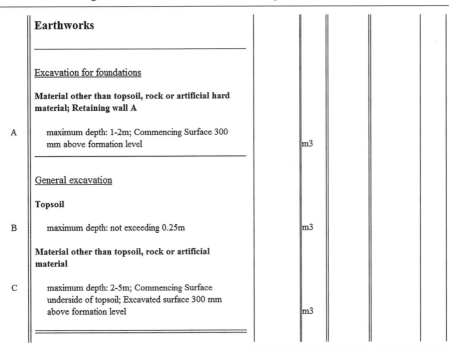

	Earthworks				
	Excavation for foundations				
	Material other than topsoil, rock or artificial hard material; Retaining wall A				
A	maximum depth: 1-2m; Commencing Surface 300 mm above formation level	m3			
	General excavation				
	Topsoil				
B	maximum depth: not exceeding 0.25m	m3			
	Material other than topsoil, rock or artificial material				
C	maximum depth: 2-5m; Commencing Surface underside of topsoil; Excavated surface 300 mm above formation level	m3			

Table 7.18 illustrates an item for the removal of reinforced concrete exposed at the Commencing Surface with respect to the foundation excavation for the crib retaining wall shown in Figure 7.7. The Commencing Surface is the Excavated Surface of the general excavation which is considerably below original ground level.

Table 7.18   Artificial hard material.

	Earthworks				
	Excavation for foundations				
	Reinforced concrete exposed at the Commencing Surface				
A	maximum depth: 1-2m; Commencing Surface 300 mm above formation level	m3			

Two further descriptive features have been added to the Second Division in order to provide for the excavation of hazardous material. This is defined in Definition Rule D4 as *material whose excavation and disposal is governed by legislation* which the Work classification describes as *Controlled and hazardous material*.

It should be noted that Rule D4 includes *invasive plant species* in the definition of controlled and hazardous material. This provision should, therefore, be read in conjunction with Class D: *Demolition and site clearance* which also provides a Second Division descriptive feature for *invasive plant species*.

Risk issue

In Class D, Rule A3 requires item descriptions to include the *type of vegetation* and the *method of treatment*, and Rule C3 deems the *disposal of any vegetable matter remaining after treatment* to be included.

As one method of treatment of invasive plant species is removal by excavation, consideration needs to be given as to whether this should be dealt with in Class D or Class E.

In Class E, only Additional Description Rule A7 refers to *controlled and hazardous material*, with the requirement that *the nature of the material* shall be stated in item descriptions.

In view of the fact that Class D excludes the removal of materials below the Original Surface, it would seem to make sense to include surface treatment and removal of vegetation in Class D and any excavation requirement in Class E. The method of measurement is less than clear though.

Excavation items are deemed to include earthwork support, working space requirements and the removal of dead services (Coverage Rule C1 applies).

Risk issue

Coverage Rule C2 provides that items for excavation shall also *include the removal of existing pipes of **any** material or diameter.* There is no further detail given, and no mention is made of the removal of manholes and other chambers.

Presumably, the intention of Rule C2 is that the removal of drain pipes, ducts, culverts and the like are deemed included to the extent that they are present in an excavation, whether a general excavation, a foundation excavation or a cutting, etc.

Where existing drainage installations, statutory services and the like are known and require cutting off and removal or diversion, it must be assumed that a provisional sum for defined work would be included in the bill of quantities for such work. Alternatively, the work would be instructed as a variation (ICC/ICE conditions) or a compensation event (ECC).

Not included in excavation items is disposal which is classed in E 5 3 * *Excavation ancillaries*.

Disposal is deemed to be disposal off-site *unless otherwise stated* (Rule D5), and any double handling of excavated material must be *expressly required* to warrant measurement (Rule M14). Material to be disposed of on the site shall be stated in the item descriptions for *disposal of excavated material* (i.e. not in the excavation item), in accordance with Rule A10.

The volume of disposal is determined by the difference between *the total net volume of the excavation* and *the net volume of excavated material used for filling* (Rule M12).

Risk issue

Double handling of excavated material is determined by the *void formed in the temporary stockpile from which the material is removed* (Rule M13).

This is a site measurement as opposed to a quantity *computed net using dimensions from the Drawings*, and as such, Rule M13 must be considered a derogation of the rule stated in CESMM4, Paragraph 5.19.

7.15.2   Dredging

Dredging normally involves the removal and disposal of unwanted material submerged below a body of water. Whilst recognised as a special subdivision of earthworks (E 1 * *), CESMM4 offers no definition of what is meant by the word 'dredging', except:

- Measurement Rule M3 – where excavation is classed as *by dredging* in the bill of quantities, it *shall be* [ad]*measured as by dredging irrespective of the method of excavation* adopted by the contractor on-site.

Additional Description Rules A1 and A2 are clearly of crucial importance in describing dredging work, and not only should the body of open water be identified in the BQ Preamble, and similarly identified in item descriptions, but also the location and limits of excavation by dredging should be stated, unless this would be obvious.

There are many instances where *excavation by dredging* may be clearly classified as such – land reclamation, removal of silt from rivers and estuaries, cleaning of canals and ditches, creation of trenches in the seabed to accommodate pipelines, etc.

In other cases, the distinction is less clear, as excavation below bodies of water may be required without meriting the sobriquet of 'dredging'. The removal of unacceptable material from within cofferdams in rivers, estuaries and harbours, for instance, may equally be classed as *excavation by dredging*.

Dredging to remove silt is measured as an excavation ancillary but only to the extent that (a) it is *expressly required* and (b) that the silt to be removed is that which has accumulated *after the Final Surface has been reached*. In all other cases, E 1 * * would apply.

The Work Classification requires that the quantity of *excavation by dredging* shall be measured in m³, and Measurement Rule M4 states that this shall be determined *from soundings* unless otherwise stated. Soundings are normally carried out using ultrasonic echo sounders in order to determine (a) the depth of a given point beneath the surface of a body of water and (b) the resulting depth after dredging work has been carried out.

Disposal is measured in accordance with E 5 3 *, but Rule A8 requires that disposal of excavation by dredging shall be distinguished in item descriptions as is the case for other excavation ancillaries.

Where the scope of dredging work is extensive, or not undertaken as work ancillary to civil engineering work, consideration might be given to formulating a 'dredging-only' contract. In this event, the usual conditions of contract might not be considered suitable, and the FIDIC standard Form of Contract for Dredging and Reclamation Works (2006) – the 'Dredgers Contract' or Blue Book – could be a more appropriate procurement choice.

The topic of dredging is discussed further in Chapter 9.

7.15.3   Filling

Filling is measured to structures, embankments, in general areas and to stated depth or thickness, in a range of stated materials, according to E 6 * *. It is deemed to include compaction (Rule C4).

The presumption is that filling material is to be *non-selected excavated material*, excluding topsoil or rock, unless otherwise stated (Rule D7). Filling with *excavated rock* is not measured unless expressly required, in which case the location of the work must be stated in item descriptions (Rule D8).

Bulk filling is not classed as *to stated depth or thickness* notwithstanding any specification requirements for the compaction of fill in particular layers.

In practice, there will be instances where there is an interface between filling to structures and filling to embankments. In the case of a bridge, for instance, the embankment is usually constructed in two stages. Stage 1 is brought up whilst the bridge abutment is being constructed, and when this is finished, the embankment is completed behind the structure.

Consequently, Stage 2 of the embankment could be viewed as part of the embankment, but equally, it could be argued that there should be an item for filling to structures as the two operations are quite different and could be carried out using different plant. This situation is illustrated in Figure 7.8.

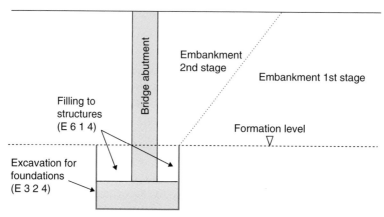

Figure 7.8   Filling to structures.

The resolution of this dilemma lies in Measurement Rule M16. This states that an item of filling around completed structures shall only be measured when the volume filled has also been measured as excavation in accordance with Measurement Rule M6. Rule M6 determines that the volume measured for the excavation of a structure shall be that which is *occupied by or vertically above* the structure.

Consequently, an item of filling to structures can only be measured as indicated in Figure 7.8 because, in all probability, the *excavation for foundations* item will have been measured from formation level (or from the top of the protection layer specified).

Imported fill must be identified in item descriptions, stating the material, in accordance with Rule A12. There is no rule for determining the volume of imported fill, but referring to Rule M12, it must be given as the difference between *the total net volume of the excavation* and *the net volume of* [acceptable/suitable] *excavated material used for filling*.

Risk issue

Where additional fill is required because of the settlement of underlying material or where fill has penetrated underlying material, the contractor is responsible for the first 75 mm of the 'ground loss' (Rule M18).

This is not an easy item to measure, even on-site, as the total 'ground loss' can only be derived from the total volume of fill less the volume of fill shown on the drawings. Even if this could be determined, a 'rough' calculation would have to be made to separate out the contractor's risk contribution (i.e. the first 75 mm of penetration).

If the fill is imported, then delivery records would be helpful, but determining the actual volume of excavated material would require very careful records of which excavations the fill came from, especially if some excavated material is to be disposed of off-site.

For rock filling in soft areas, the **volume** shall be measured *in the transport vehicles at the place of deposition*. Similarly, where fill deposited below water cannot be quantified, it shall be measured in the same way.

Risk issue

Measuring the **volume** of fill in a waggon is unrealistic because the material is in an uncompacted state. Even if a volumetric measurement were possible, it would be necessary to check every vehicle because the loads carried vary.

Where such circumstances may arise, Rules M20 and M21 should perhaps be revisited and arrangements made for quantities to be determined by weight.

A new item has been added to *Filling* in the form of *High energy impact compaction. The surface area to be treated* is measured in m², where the treatment is to existing ground (Rule M22), but where selected excavated material or imported material is to be compacted using this method of compaction, it is measured in m³. Item descriptions *shall state the type of compaction* (Rule A17) such as dynamic compaction using Cam (3-sided) or Pentagonal (5-sided) shaped impact drums.

A typical item is shown in Table 7.19.

Table 7.19   High energy impact compaction.

				£	p
CLASS E: EARTHWORKS					
Filling					
High energy impact compaction general fill; Dynamic compaction					
A existing ground	10032	m2			

Risk issue

There is no rule as to how selected excavated material, or imported material, which is to be subjected to *high energy impact compaction*, is to be measured.

Imported material is delivered by the tonne, and selected excavated material, unless designated as double handling (Rule M13), is a bulked material of indeterminate volume.

7.15.4   Water

Where water affects the works, this is at the contractor's risk, and there is no entitlement to a measured item for dealing with groundwater, or that from bodies of open water such as rivers and canals.

Paragraph 5.21 requires that bodies of open water shall be identified in the Preamble to the Bill of Quantities. A reference to a drawing shall also be given (presumably in the Preamble) showing the levels and boundaries of the water and the anticipated range of any fluctuations in levels.

The issue of groundwater raises the 'thorny' problem of borehole data and its reliability and also the spectre of contractual claims and compensation events should the data prove unreliable. This does not alter, however, the contractor's obligation to keep excavations free from groundwater which will normally be specified in the contract (e.g. Specification for Highway Works, Clause 602).

Having said all that, there <u>is</u> an entitlement in Class E to a quantity for excavation *below a body of open water* pursuant to Measurement Rule M7. This is not a requirement of the Work Classification but emanates from Additional Description Rule A2.

The volume measured is specified in Rule M7 as:

- *The volume below water when the water surface is at the level (or the higher level of fluctuation if applicable) shown on the drawing...*

and Rule A2 requires that:

- The item description *shall identify the body of water.*

Figure 7.9 illustrates a typical situation where a bridge pier is to be constructed either side of a body of open water.

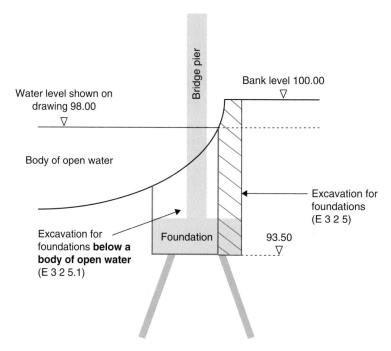

Figure 7.9   Excavation below a body of open water.

It can be seen that part of the excavation, whilst affected by the body of open water, is not below the water level shown on the drawing. Consequently, the shaded portion of the excavation cannot be described as *below a body of open water*. This means that two items are required for the same excavation, one with additional description to alert tenderers to the fact that some of the excavation work will take place below the indicated water level of a named body of open water.

The resulting bill of quantities items are shown in Table 7.20.

With reference to Table 7.20, it should be noted that there is no similar rule in Class P: *Piles* to those of Rules M7 and A2 in Class E. However, the requirements of Paragraph 5.21 must be respected, and it might be considered prudent to include the information referred to in Paragraph 5.21 in any relevant item for piling.

Table 7.20   Billing of excavation below a body of open water.

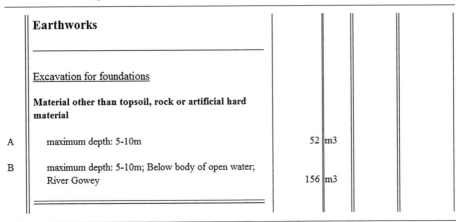

	Earthworks				
	Excavation for foundations				
	Material other than topsoil, rock or artificial hard material				
A	maximum depth: 5-10m	52	m3		
B	maximum depth: 5-10m; Below body of open water; River Gowey	156	m3		

7.16   Class F: In situ concrete

Changes from CESMM3

- Provision of concrete:
  - *Standard mix* omitted
  - The word *mix* now largely replaced by *concrete*
  - Concrete classified to BS 8500 and BS EN 206-1 (D1, A1 and A2 refer)
  - New Additional Description Rules A3–A7
  - F 1 * * is *designed concrete* where the designer is responsible for ensuring compliance with standards (D2 refers)
  - F 2 * * is designated concrete where the concrete producer is responsible for the design (D3 refers)
  - F 3 * * is *standardised prescribed concrete* for simple structural non-reinforced applications (D4 refers); replaces old *standard mix* but mix designations remain the same in the Third Division (e.g. ST1, ST2, etc.)
  - F 4 * * is *prescribed mix* (although the definition rule refers to *concrete*) where the designer is responsible for the mix design (D5 refers)
  - F 5 * * is *proprietary concrete* for concrete that is *outside normal performance criteria* (D6 refers)
  - Now only three maximum aggregate sizes in the Third Division
- Placing of concrete:
  - New Definition Rules (D12 and D13) and Additional Description Rule (A11) relating to *sprayed concrete*
  - Sprayed *concrete* to be described as *other concrete forms* (D12 refers)
  - The horizontal line across the top of *other concrete forms* should have been extended into the Third Division. It makes no sense to describe *sprayed concrete* by cross-sectional area (also, Rule A11 refers to thickness not cross-sectional area)
  - The NOTE on page 40, which refers to Paragraph 5.10, should read 5.11

Notwithstanding substantial changes to the specification of concrete materials, Class F retains the distinction between the *provision of concrete* and the *placing of concrete* in the Work Classification.

Part 2

The objective of this approach is to comply with CESMM4, Paragraph 2.5, which emphasises the need to distinguish between items of *work of the same nature*, carried out in different circumstances or locations, which *may give rise to different considerations of cost*.

All concrete, both supply and placing, is measured in m^3, but there is a degree of 'approximation' to the calculation of volumes. Measurement Rule M4 includes a list of features, such as cast-in components, rebates, holes, etc., that are <u>not excluded</u> from the measured volume, and Rule M2 <u>excludes</u> nibs and external splays from the volume where <0.1 m^2 in cross section (i.e. 100 mm × 100 mm).

There are no item coverage rules in Class F, and great reliance is placed on Class A: *Method-related charges* to cater for major cost items involved in the provision and placing of concrete.

The intention is to derive a set of rates that are more transparent so that changes in concrete specification, or the conditions in which work is carried out, are more easily accommodated in the valuation of variations or compensation events. This is not to say that method-related charges are 'ring-fenced' and immune from admeasurement – far from it.

Class F relies heavily on Paragraphs 5.11: *Extent of itemisation and description*, and the NOTE on page 40 of CESMM4 emphasises that, where the location of concrete members may impact the method and rate of placing, this *may be stated* in the relevant item descriptions. Height above or below ground, size and rate of pour, mass per m^3 of rebar, access limitations and so on can have considerable impact on concrete placing costs, and CESMM4 encourages bill compilers to think about such issues whilst, at the same time, not trying to 'second guess' the contractor's working method.

Additional Description Rule A8 is a useful one for estimators because it is made clear that concrete is presumed to be filled into formwork unless *expressly required to be placed against an excavated surface*. The only exception to this rule is in the case of blinding.

Blinding is measured in m^3 but Definition Rule D8 states that the thickness used for classification *shall be the minimum thickness*.

Risk issue

The minimum thickness of concrete in the Third Division of the Work classification is <150 mm. This is not especially helpful in relation to blinding where the waste factor due to 'ground loss' can be considerable.

A ground loss allowance of 12.5 mm in blinding 75 mm thick is approximately 17%, but this can vary between 8% and 25% for thicknesses of 150 mm and 50 mm, respectively.

A rule requiring the precise thickness of blinding would have been sensible.

Concrete features are described in the Third Division by thickness, except that columns, beams and steel casings are described by cross-sectional area. In each case, there is no requirement to state the precise thickness or cross-sectional area.

Figure 7.10 illustrates three situations covered by the rules in Class F:

a) Rule M3 requires that attached columns shall be measured as walls, and Rule D9 states that the thickness shall exclude the additional thickness of the column.
b) In the case of attached beams, these shall be measured as slabs with the thickness stated as the slab thickness (Rules M4 and D9 apply).
c) Sprayed concrete is described as *other concrete forms* (Rule D12), and Rule A11 requires that item descriptions shall state:

(i)   The specification of the concrete.
(ii) Whether it is reinforced.
(iii) The minimum thickness.

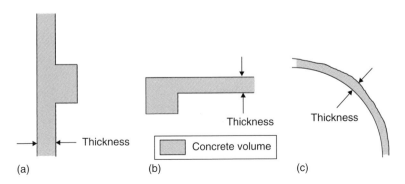

Figure 7.10   Concrete thicknesses. (a) Attached column, (b) attached beam and (c) sprayed concrete.

Reinforcement added to sprayed concrete, such as synthetic or steel fibres, is not classed as reinforcement (Class G 5).

CESMM4 distinguishes between 'sprayed concrete' in Class F: *In situ concrete* and in Class T: *Tunnels*, which measures sprayed concrete in connection with *in situ linings* to tunnels, shafts and other cavities. Another differentiating factor is that sprayed concrete in tunnels is measured in m² and that in Class F is to be given in m³.

Whilst the method of measurement is silent on the distinction, engineers will understand that 'sprayed concrete', variously known as *gunite*, *shot concrete* and *shotcrete*, is used for many purposes. Canal linings, reservoirs and dams, tunnel linings, diaphragm walls, piled wall facings and sea and river walls are amongst its many applications in addition to tunnelling using the New Austrian Tunnelling Method (NATM) (Sprayed Concrete Association).

Presumably *sprayed concrete* in Class F is intended to mean sprayed concrete other than that used in the NATM tunnelling process. Confusingly, Class F, Additional Description Rule A11, makes reference to *sprayed concrete* **support**, which is the very function of sprayed concrete in tunnel linings.

In any event, the method of measurement treats sprayed concrete differently in Classes F and T.

In Class T: *Tunnels*, Measurement Rules M2 and M5 state that in situ linings are measured either to *payment lines* or to the net dimensions shown on the drawings. There are no such rules in Class F, albeit Rule A11 requires the minimum thickness of *sprayed concrete* **support** to be stated in the item descriptions. In this case, it can only be assumed that the requirement does not apply to sprayed concrete in other applications.

Class F and Class T item descriptions are compared in Table 7.21.

Thicknesses of concrete in Class F are given in ranges and are not precise thicknesses. Therefore, similar work in the same location would be aggregated into one item provided the thickness of the various items is in the chosen range of thicknesses. The same principle applies to concrete components whose thickness varies but nonetheless remain within the dimension range (e.g. a tapering wall).

Where all work in an item has the same thickness, then the precise thickness, and not the range, would be given following the rule in Paragraph 5.15.

Table 7.21   Sprayed concrete in Class F and Class T.

	In Situ Concrete		
	Placing of concrete - Reinforced		
	Other concrete forms		
A	Sprayed concrete support; Grade 40; Reinforced; Minimum thickness 50 mm	m3	
	Tunnels		
	In situ lining to tunnels		
	Sprayed concrete primary; Grade 40; Reinforced		
B	Diameter: 6 - 7 m; 50 mm thick	m²	
	Sprayed concrete secondary; Grade 40; Reinforced		
C	Diameter: 6 - 7 m; 200 mm thick	m²	

7.17   Class G: Concrete ancillaries

Changes from CESMM3

- G 6 * – reference to BS numbers omitted
- Old D6 omitted
- A9 – BS references omitted
- Footnote to page 43 omitted (i.e. permitting formwork to separate surfaces of concrete components of constant cross section to be measured by length as one item)
- However, G * 8 * and A5 relating to the measurement of formwork *for concrete components of constant cross section* remain
- Footnote to page 45, allowing *similar inserts which vary in size* to be combined within size ranges, omitted

The suffix 'ancillaries' in no way demeans the importance of Class G, which deals with the measurement of formwork, rebar, joints, post-tensioning and concrete accessories such as surface treatments, inserts and grouting base plates and the like. In keeping with their importance, the rules of measurement in Class G are well developed and understood, but certain items are nonetheless worthy of discussion in the context of this book.

7.17.1   Formwork

Barnes (1977) regards 'formwork' as temporary works treated as if it were permanent works. This has long been the tradition in UK methods of measurement, and almost paradoxically, contractors are not only entitled to measured items in the bill of quantities for formwork, but they are also entitled to be paid for it even if it is not used or needed.

This is not to say that formwork is always measured, even if needed, notwithstanding Measurement Rule M1 which clarifies that, *where surfaces of in situ concrete...require temporary support*, formwork *shall be measured*. Consequently, Measurement Rule M1 must be read in conjunction with Rule M2 which provides the list of exceptions. These include:

- Edges of blinding <0.2 m wide.
- Temporary surfaces formed at the contractor's discretion.
- Surfaces of concrete *expressly required* to be cast against an excavated surface.
- Surfaces of concrete cast against an excavated surfaces <45° to the horizontal.

Formwork is measured according to four categories of finish, five categories of orientation and five width ranges. It is surprising, in a 'neutral' method of measurement, that two of the width categories, G * 4 and 5, remain geared to UK plywood sizes (1.22 m width). Many other countries, which are truly 'metric', have standard widths of 1.2 m or 1.25 m, and thus, the Third Division classification makes little sense. In any event, formwork is rarely the precise dimension of the finished concrete, and contractors are more than capable of working out formwork uses without the help of a method of measurement.

More to the point, the limitations of Class G: *Concrete ancillaries* relate to the fact that measurements are aggregated irrespective of the concrete element in question. This means that, on many occasions, tenderers are forced to remeasure the billed quantities in order to disaggregate formwork to the various structural elements. With a bridge, for instance, Class G does not separate G 1 4 * (formwork, rough finish, vertical) to abutment foundations, walls, wing walls, etc., and separate items are only measured should there be a requirement for a distinguishing 'finish' to the concrete.

Additional Description Rule A1 requires that *formwork left in* shall be so described. However, no distinction is made between formwork that <u>has to be</u> left in, because of the design or form of construction, and <u>permanent</u> formwork, that is, formwork that is designed to be left in (e.g. soffit formwork between bridge I-beams). In either case, reliance must be placed on CESMM4, Paragraph 5.11, in order to clarify such instances.

For concrete components of constant cross section, G * 8 *, there is a more useful Third Division classification relating to beams, columns, walls, etc.

In this case, the formwork required to create a concrete component may be measured and billed in linear metres, identifying the type of component in accordance with the Third Division classification, rather than having to measure the individual surfaces in linear metres or m².

The provisions of the Work Classification are to be read in conjunction with Additional Description Rule A5, which states that *the principal cross-sectional dimensions* are to be stated in the item descriptions, along with any other identifying feature.

Figure 7.11 illustrates (a) an in situ concrete column and (b) a beam, with the dim sheet attached, which shows:

- The width classification for the formwork relevant to each component.
- The dimensions for each (i.e. the column + the <u>vertical</u> sides of the beam).
- The total area of formwork measured.

Table 7.22 shows the completed BQ item (G 2 4 4).

A better way of measuring these components is by describing them as *components of constant cross section* (G 2 8 1 and G 2 8 2), as illustrated in Table 7.23, which compares the two methods of billing the same items. The snag here is that the item for the beam includes formwork to both the <u>vertical sides</u> and the <u>soffit</u> of the beam, and the estimator would have to perform a separate calculation in order to segregate the two types of formwork which would each be priced differently.

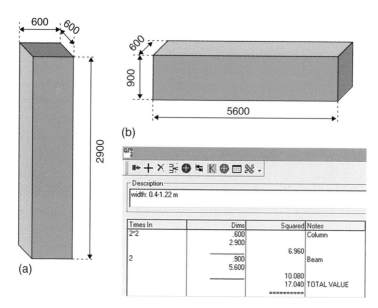

Figure 7.11   Width classification – 1. (a) Column and (b) beam.

Table 7.22   Width classification – 2.

	Concrete Ancillaries		
	Formwork: fair finish		
	Plane vertical		
A	width: 0.4-1.22 m	17	m²

Risk issue

The billed item in Table 7.22 is not particularly helpful for pricing as not only will the formwork for each component be constructed differently, but the two components are in different planes with different support/falsework requirements.

If there were to be a number of these components in a structure, the tenderer would have to either (a) remeasure the formwork to derive quantities for both components or (b) split the billed quantities by proportion.

In the 'pressure pot' of tendering, method (a) could lead to mistakes/loss of valuable time and method (b) would be inaccurate.

Either way, this adds risk to the tender.

Of much more practical use in CESMM3 was the footnote to page 43, which permitted the separate surfaces of formwork, for components of constant cross section, to be measured by length as one item instead of by area. This meant that the sides and soffits of beams could each be

Table 7.23   Components of constant cross section.

Concrete Ancillaries					
	Formwork: fair finish				
	Plane vertical				
A	width: 0.4-1.22 m	17 m²			

NB:
This item is for formwork to the column + the <u>vertical</u> sides of the beam.

Alternative description

For concrete components of constant cross-section					
B	beams; 600 mm x 900 mm	6 m			
C	columns; 600 mm x 600 mm	3 m			

NB:
Formwork to the <u>vertical</u> sides of the beam is measured separately from the formwork to the column.

given as separate items. This was particularly useful for attached beams and the like and permitted the various widths of formwork to be measured separately as illustrated in Figure 7.12.

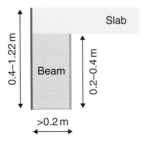

Figure 7.12   Separate surfaces of formwork.

Unfortunately, the page 43 footnote is no longer included in the fourth edition of CESMM. Whether this is by accident or design is known only to the authors of the method of measurement!

7.17.2   Reinforcement

Class G provides for three issues of importance in an otherwise featureless work classification for rebar:

- For bar reinforcement exceeding 12 m, before bending, lengths shall be stated in item descriptions to the next higher multiple of 3 m – Rule A7.
- Items for rebar are deemed to include supporting reinforcement (chairs) – Rule C1.
- Supports to top reinforcement are to be included in the quantity of rebar measured – Rule M8.
- The areas of fabric reinforcement measured are exclusive of laps – Rule M9.

7.17.3   Joints

Measurement Rule M10 stipulates that joints shall be measured only where expressly required. Joints that require support during concrete casting are classed as *formed joints*, whereas other joints are *open surface joints* (D7 refers). Consequently, a horizontal joint in a concrete wall is an *open surface joint*, unless it is a construction joint, when it shall not be measured. A vertical joint in a floor slab is a *formed joint*.

Formwork to joints is deemed included in the item (Rule C3).

7.17.4   Post-tensioned prestressing

Concrete members that require post-tensioning on-site may be cast in situ or precast, and this is recognised in the Second Division of G 7 * *. The casting of in situ concrete components is measured in Class F: *In situ concrete*, and the supply of precast concrete components is measured in Class H: *Precast concrete*.

Coverage Rule C5 states that items for prestressing shall include ducts and grouting, and Rule A12 requires details of the component (to be stressed), tendons and anchorages to be stated in item descriptions. The jacking process is measured separately under G 7 5 0.

7.17.5   Concrete accessories

A variety of in situ concrete finishes is measured in G 8 including trowelling and applied finishes. Surface treatments of already formed surfaces, such as bush hammering, are also included.

Granolithic finishes are deemed to include materials, joints and formwork and surface treatment (Rule C6).

Linear and other inserts in in situ concrete seemingly cannot be combined within a size range owing to the omission of the footnote on page 45 of CESMM3.

7.18   Class H: Precast concrete

Changes from CESMM3

None

Class H: *Precast concrete* is chiefly concerned with the supply and fixing of large concrete units manufactured off-site, such as bridge beams, bridge deck slabs, subways, culverts, etc.

Precast concrete units are defined as *cast other than in their final position* (Rule D2), but where concrete items are precast <u>on-site</u>, Definition Rule D3 determines that they shall be measured as *in situ concrete* where:

- The reason for precasting is <u>not</u> to obtain multiple uses of formwork.
- The nature of the work is characteristic of in situ concrete albeit the cast unit has to be moved into its final position.

Figure 7.13 illustrates the point, where a bridge is to be constructed to one side of a railway line and then slid into its final position during a line possession.

The bridge would be measured as in situ concrete, with items given in Class A for the temporary works associated with the slide shown in Figure 7.13 (Rule D3 refers). Rule D3 implies that the

Figure 7.13   Bridge slide.

bill compiler is responsible for inserting the Class A Method-related charge, whereas Class A Rule M5 states that it is the tenderer's responsibility in accordance with CESMM4 Section 7: *Method-related charges*.

   Class H 6 0 * and H 7 0 0 respectively distinguish between *segmental units* and *units for subways, culverts and ducts*. There are no Second and Third Division features with either, but whilst H 7 0 0 is fairly clear, there is no direction as to what is meant by H 6 0 *: *Segmental units*.

Segmental units are clearly not *units for subways, culverts and ducts*, despite being 'segmental', and thus, it must be assumed that H 6 0 * refers to sectional retaining walls, crib-type retaining walls, precast concrete overfilled arch structures, etc., but not concrete, masonry or artificial stone blocks which are measured under Class U.

Equally, however, H 6 0 * could refer to segmental precast concrete bridge units. These items are very large, normally of box girder construction, with a widened top flange, which are stressed together with wire tendons and/or Macalloy bars to form the completed structure. The top flange forms the full width of the carriageway, and they are usually cast on, or close to, the construction site.

The reasoning to class segmental bridge units as 'precast concrete' is that they are cast on or near to the site because of their size and weight rather than as a function of formwork cost efficiency.

Segmental units are classified in the Third Division according to which range their individual mass falls in (D1 also refers) and are enumerated (nr). However, Rule A4 overrides this by requiring the cross-section type and principal dimensions to be stated. The Third Division classification does not apply to *units for subways, culverts and ducts*, however, but the provisions of Rule A4 apply equally to both.

Units for subways, culverts and ducts, on the other hand, are measured in linear metres with the length being calculated as *the total length of identical units* (Rule M1 refers).

Part 2

Risk issue

Precast concrete structures, such as box culverts, subways, overfilled arch bridges and short-span underbridges are of 'proprietary' manufacture, and thus, for public sector projects, this raises the issue of EU-inspired legislation that prohibits the creation of a 'barrier to trade' (see Chapter 8).

The MMHW deals with this issue in a somewhat convoluted way by using 'designated outlines' to define the location, but not the specification, of proprietary structures so that they may be designed by the contractor.

There is no similar provision in CESMM4, and bill compilers should be aware of the need to comply with legislation by not allowing one manufacturer an unfair advantage over others. Consideration might be given, therefore, to avoiding cross-section types and principal dimensions of proprietary structures where this information might identify, or be deemed to favour, a particular manufacturer.

7.19   Class I: Pipework – pipes

Changes from CESMM3

- *Addition to Measurement Rule M5 making reference to pipes in dual trenches*
- **NB**:

  Coverage Rules C1 and C2 in CESMM3 **have been omitted** in CESMM4
  These are key coverage rules that determine what is deemed included in item descriptions in Class I
  This must be a mistake[1] as neither Class I nor Class L 2 * * make sense without C1 and C2
  The heading beneath the class title says what is included and excluded, but this does not have the status of a coverage rule

[1] This book is written on the basis that Coverage Rules C1 and C2 apply:

C1	C2
Items for pipes shall be deemed to include the supply of all materials by the contractor unless otherwise stated. Items shall be deemed to include pipe cutting	Items for pipes *in trenches* shall be deemed to include excavation, preparation of surfaces, disposal of excavated material, upholding sides of excavation, backfilling and removal of dead services except to the extent that such work is included in Classes J, K and L

Source: CESMM3

The three divisions of Class I provide for the provision, laying and jointing of pipes, including those described as *not in trenches*, and for excavation and backfilling of pipe trenches.

Coverage Rule C1 confirms that the contractor's rates shall be deemed to include for the supply of all materials, but this is tempered with the exclusion of work included in Classes J, K, L and Y in the Class I heading. Consequently, Class I items do not include for pipe bedding, for special backfills or for the supply of pipe fittings or for the reinstatement of trenches, etc. This is emphasised in Coverage Rule C2.

In the Second Division, ranges of nominal bore are provided, but this is overridden by Additional Description Rule A2 which states that the nominal bores of pipes shall be stated in item descriptions.

The Commencing Surface for excavation shall be identified in items, except where this is the same as the Original Surface, but in both cases, the commencing surface adopted for preparation of the bill of quantities shall be adopted for admeasurement (Rule M1).

There are various approaches to measuring the depth of pipe trenches:

- NRM2 – the average depth between manholes irrespective of maximum depth
- MMHW – the arithmetic mean of depths measured every 10 m and terminal (i.e. short) lengths

The CESMM approach is quite different to either of these.

Depths of pipe trenches are categorised in the Third Division in depth bands – not exceeding 1.5 m, 1.5–2 m, 2–2.5 m, etc. – and are measured from the Commencing Surface to pipe invert (Rule D1). Average depths are not used in CESMM.

Consequently, for any given length of pipe between two manholes, the length(s) of pipe occupying each depth category or zone shall be measured and identified in the bill items. This is illustrated in Figure 7.14.

From Figure 7.14, it can be seen that the first 8 m of pipe, starting from the outfall end, lies within the 2–2.5 m depth zone and the next 27 m falls within the 2.5–3 m band and so on. The depths of pipe in the run from manholes 4–5 are all determined in this way, and it is by this process that the pipe runs are categorised. This is shown in Figure 7.15, which displays the dim sheet for the 2.5–3 m depth category, that is, a total of 57 m. Another 28 m lies in the 2–2.5 m depth band as shown.

The item descriptions and quantities are billed as shown in Table 7.24, where it can be seen that:

- The location of the group of items is indicated (Class I, Rule A1, refers).
- The nominal bore, the materials and pipe joint type is stated.

When measuring and describing pipework, it is vital to appreciate the role of Additional Description Rule A1 which is the means by which important information is conveyed to tenderers:

a)   The location of pipework.
b)   The type of pipework, in each item or group of items.

The whole point of the CESMM approach is to alert tenderers to risk, so that appropriate allowances can be made in the tender.

Part 2

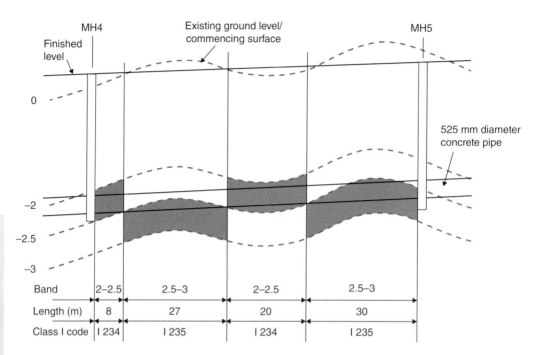

Figure 7.14   Pipework depth categories – 1.

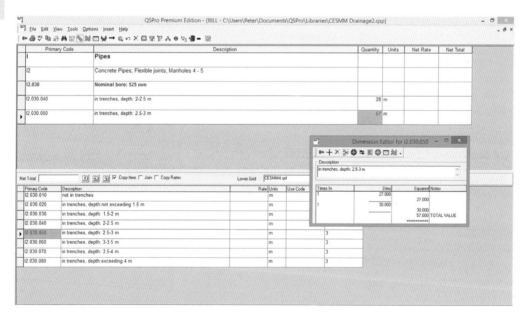

Figure 7.15   Pipework depth categories – 2.

Different types of pipework, in different locations, will affect access requirements, working methods, choice of plant, temporary works and output of the drainage gang. The bill complier must exercise considerable judgement in deciding how to comply with Rule A1, as the method of measurement provides no guidance.

Table 7.24   Pipework depth categories – 3.

Pipes		
Concrete Pipes; Flexible joints; Manholes 4 - 5		
Nominal bore: 525 mm		
A in trenches, depth: 2-2.5 m	28	m
B in trenches, depth: 2.5-3 m	57	m

Grouping work in live sewers, drain runs subject to tidal influence, work in live carriageways, gully connections and branches for future development are just some of the many ways to individualise item descriptions and help the estimator to price the work accurately.

Class I items, whilst composite in nature, are not complete items. Pipework supports and protection, for example, are measured elsewhere (L * * *), for instance, as illustrated in Table 7.25.

Table 7.25   Pipework supports and protection – 1.

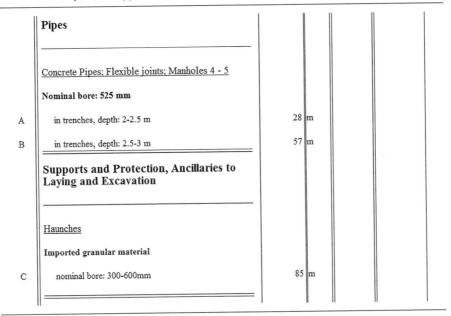

Pipes		
Concrete Pipes; Flexible joints; Manholes 4 - 5		
Nominal bore: 525 mm		
A in trenches, depth: 2-2.5 m	28	m
B in trenches, depth: 2.5-3 m	57	m
Supports and Protection, Ancillaries to Laying and Excavation		
Haunches		
Imported granular material		
C nominal bore: 300-600mm	85	m

Items for pipe bedding, haunches and surrounds are required to state:

- The material used (Rule A3).
- Depth of beds (Rule A3).
- The nominal internal diameter of the pipe (Rule D3), notwithstanding the Third Division classification.

The rule requiring the depth of beds to be stated is not clear and not very helpful either. It could be taken to mean the depth of beds <u>only</u> (L 3 * *) or the depth of the beds beneath haunches (L 4 * *) and surrounds (L 5 * *) <u>as well as</u> the depth of beds (L 3 * *). This part of Rule A3 is somewhat academic, however, as estimators will invariably search for standard details on the drawings, or make reference to other standard details, for the bedding requirements for different types and diameters of pipes.

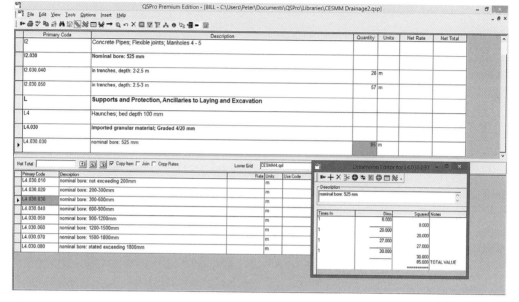

Figure 7.16   Pipework supports and protection – 2.

It can be seen from Figure 7.16 that the pipe support and protection item is 'anded-on' from the foregoing pipework items and that the billed item includes the type of material, the bed thickness and the nominal pipe diameter as illustrated in Table 7.26.

Table 7.26   Pipework supports and protection – 3.

	Pipes		
	Concrete Pipes; Flexible joints; Manholes 4 - 5		
	Nominal bore: 525 mm		
A	in trenches, depth: 2-2.5 m	28	m
B	in trenches, depth: 2.5-3 m	57	m
	Supports and Protection, Ancillaries to Laying and Excavation		
	Haunches; bed depth 100 mm		
	Imported granular material; Graded 4/20 mm		
C	nominal bore: 525 mm	85	m

Backfilling of pipe trenches with material other than excavated material is measurable but not in Class I. Pursuant to Class I, Rule M2:

- Filling of French and rubble drains shall be measured in Class K.
- Backfilling with material other than that excavated from trenches shall be measured according to Class L.

However, backfilling of pipe trenches <u>above or below</u> the Final Surface shall only be measured where it is *expressly required* (Class L, Rule M7). The words *expressly required* can have two meanings:

a)   Shown on the drawings as being required.
b)   Instructed by the contract administrator during construction.

Risk issue

The difference between (a) and (b) is that admeasurement of items shown in the contract documents would similarly *be computed net using dimensions from the Drawings* (CESMM4, Paragraph 5.19), whilst items instructed by the contract administrator would be variations or compensation events and would, therefore, be subject to the contract conditions as regards the valuation of these items.

Items for backfilling above the Final Surface with concrete, or other stated material, and for excavation and filling below the Final Surface, are measured in m³ and according to Class L, Rules M4–7 and D1, the latter determining the nominal trench width used to calculate the volumes.

The volume of backfilling above the Final Surface, excavation being included in Class I items, is calculated by multiplying together *the average depth and length of the material … backfilled and the nominal trench width*. Note reference to the *average* depth and not the depth categories in Class I.

Whilst there is no rule, it would seem logical to measure the average depth from the Commencing Surface to the top of pipe bedding/haunch/surround. Strictly speaking, for beds and haunches, this would necessitate a deduction for the space occupied by the pipe, but there is no measurement rule requiring this, save for CESMM4, Paragraph 5.19, which states that *the quantities shall be computed net … from the Drawings*.

Sample BQ items for Class L backfilling to pipe trenches are provided in Table 7.27.

Table 7.27   Backfilling trenches – 1.

	Supports and Protection, Ancillaries to Laying and Excavation		
	Extras to excavation and backfilling		
	In pipe trenches		
C	backfilling above the Final Surface with Clause 803 Type 1 Unbound Mixture	43	m3
D	excavation of natural material below the Final Surface and backfilling with concrete ST1	5	m3

Part 2

The dim sheet for the Class L 1 1 6 item (backfilling with sub-base) is shown in Figure 7.17 where it can be seen that an estimated length of 18 linear metres of backfill (from a total of 28 m + 57 m = 85 m) is *expressly required*. This has been multiplied by the nominal trench width (525 mm + 500 mm (see Rule D1) = 1.025 m) and the average depth to the top of the pipe haunching.

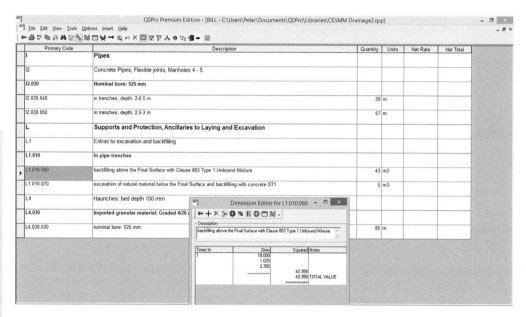

Figure 7.17   Backfilling trenches – 2.

The measurement of excavation and filling below the Final Surface is a straightforward calculation of length × nominal trench width × <u>average</u> depth (Class L, Rule M4). This is basically a provision for soft spots in trenches and, as such, must be considered 'provisional'.

Risk issue

There is no concept of 'provisional' quantities in CESMM4 but, clearly, a measured item for soft spots cannot be viewed as being 'estimated' in the same sense as other measured items in the Bill of Quantities (CESMM4, Section 1, Definition 1.5, refers).

In circumstances where (a) there are no soft spots found or (b) the quantity allowed is proved to be significantly inadequate, a contractual issue may arise as to whether the method of measurement has been respected and/or whether the quantities have been misrepresented in the bill of quantities.

Contractors will, no doubt, view Class L items L 1 1 7–8 with suspicion and price them accordingly!

7.20   Class J: Pipework – fittings and valves

Changes from CESMM3

- Additional Description Rule A1 – reference to British Standard specifications omitted

Class J provides for the measurement of fittings and valves.

Pipe fittings are accorded seven First Division features and eight in the Second Division. *Valves and penstocks* are measured under J 8 * 0.

These items are 'supply and fix' as determined by Coverage Rule C1. However, the length of pipes (measured in Class I) are measured over fittings and valves (Class I, Rule M3) with the consequence that fittings and valves are effectively 'extra over' the pipes to which they relate.

Straight specials (J * 8 *) or 'rocker pipes' are defined in Class J, Rule D2, as being a length of pipe *cut to length or made to order*. They are not measured unless *expressly required* in the contract. Straight specials do not occur where the contractor is obliged to cut pipes to accommodate pipe runs that are not multiples of standard pipe lengths (Rule M2), but where they are specified, items are *deemed to include cutting* (Rule C3).

In common with Class I, the nominal bore of pipe fittings shall be stated in item descriptions in place of the nominal bore ranges (e.g. 200–300 mm) given in the Third Division of the classification.

For manholes and other chambers with a backdrop, Rule M1 states that the associated pipe fittings are included with the Class K item (K 1 * * and K 2 * *).

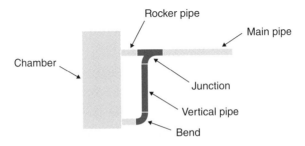

Figure 7.18   Backdrop.

However, Figure 7.18 indicates that the junction on the main pipeline at the top of the backdrop is part of both the backdrop and the pipeline. In this case, Class I, Rule M3, states that the *lengths of pipes in trenches shall … exclude lengths occupied by pipes and fittings comprising backdrops to manholes*. Thus, the length of the junction would be excluded from the length of pipe between manholes.

7.21   Class K: Pipework – Manholes and Pipework Ancillaries

Changes from CESMM3

None

7.21.1   Manholes, chambers and gullies

Manholes, chambers and gullies are provided for in Class K, along with French drains and the like, ducts, metal culverts, crossings and reinstatements.

Manholes and other chambers are enumerated and described according to their type (with or without backdrop) and depth within the Second and Third Divisions. In practice, it is likely that standard details of different chambers will be provided, and where this is the case, reference numbers must be included in item descriptions according to Additional Description Rule A1.

Depths are measured from the tops of covers to invert, or tops of base slabs in catchpits and the like. There is no reference to a Commencing Surface for these items, but as excavation is likely to represent a relatively small proportion of the cost, normal practice would be to assume the full depth for the volume calculation when pricing such work.

7.21.2   Piped french and rubble drains

Piped French and rubble drains are measured in two Classes:

a) **Class I** provides for excavation and pipe laying (M1) (Table 7.28 refers).
    The location or type of pipework shall be stated according to Class I, Additional Description Rule A1, and this is stated in Table 7.28 as 'French drains'.

Table 7.28   Class I – Piped French and rubble drains.

	Pipes					
	Clay Pipes					
	Nominal bore 150 mm; French drains					
A	in trenches, depth: not exceeding 1.5 m	227	m			

b) Filling is measured in m³ in **Class K4**, with details of the filling material given in the description (A4) (see Table 7.29).

Table 7.29   Class K – Piped French and rubble drains.

	Pipework - Manholes and pipework ancillaries					
	French drains, rubble drains, ditches and trenches					
	Filling French and rubble drains with graded material					
C	Type B 20/40 mm graded filter media	176	m3			

There is no measurement rule in Class K for calculating the volume of filling to French drains and the like.

However, it would be difficult to argue against the application of rules provided in Class L which, *inter alia*, define the nominal trench width. For a 150 mm diameter French drain, the nominal trench width would be 150 mm + 500 mm = 650 mm, as illustrated in Figure 7.19.

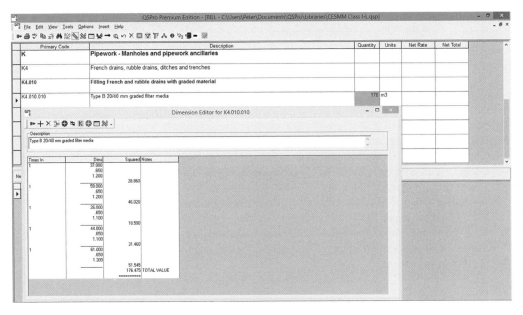

Figure 7.19   Filling to piped French and rubble drains.

Risk issue

The separation of the excavation and filling of French/rubble drains creates a problem for the estimator when there is more than one diameter of piped trench.

This is because it is very difficult to relate the volume of filling to the length of piped trench in order to calculate the additional excavation and waste factor (i.e. the additional excavation, disposal and filling for a trench of a given depth and nominal width) due to the width of the excavator bucket.

The waste factor for filling a trench 1 m deep (150 mm diameter pipe) is given below and also compared to that for a 225 mm diameter pipe.

Pipe diameter mm	Bucket width	Nominal width	Waste calculation
150	750 mm	150 mm + 500 mm = 650 mm	$\dfrac{750 - 650}{650} \times 100 = \mathbf{15.38\%}$
225	900 mm	225 mm + 500 mm = 725 mm	$\dfrac{900 - 725}{725} \times 100 = \mathbf{24.14\%}$

In Class K, piped French drains and rubble drains are distinguished from un-piped rubble drains (K43*). Such items are measured in linear metres, with a stated cross-sectional area, and are deemed to include excavation, earthwork support, disposal, etc. in accordance with Coverage Rule C1 (Table 7.30 refers).

The filling of un-piped rubble drains, however, is measured under K410 or K420 in m³.

Table 7.30 Un-piped rubble drains.

	Trenches for unpiped rubble drains					
B	cross-sectional area: 0.75 - 1 m²		174	m		

7.21.3   Vee ditches

Vee ditches are classed as *lined* (K 4 7 *) or *unlined* (K 4 6 *) and are measured in linear metres within a stated cross-sectional area.

Whether or not vee ditches are lined, *the cross-sectional area shall be measured to the Excavated Surface* (Rule M2), that is, the underside of any lining. Details of any linings are to be stated in the item description (Rule A5) which shall include the materials and dimensions of the linings. As illustrated in Figure 7.20, this can involve several different linings, sub-bases and membranes in the one item.

Figure 7.20   Lined vee ditches.

Risk issue

In the absence of any coverage rule in Class K, it can only be assumed that the items for lined ditches shall include the lining itself and any membranes.

The excavation, disposal and preparation of surfaces are deemed included by Coverage Rule C1.

7.21.4   Ducts and metal culverts

Ducts and metal culverts are measured in linear metres in accordance with the rules for pipes in Class I (Class K, Rule M3, refers). Such items are measured as *in trenches* or *not in trenches* with those in trenches classified according to the same depth banding as Class I: *Pipework – Pipes*.

Additional Description Rule A6 stipulates that the rules of Class I (presumably the additional description rules thereof) shall apply to ducts and metal culverts, and therefore, pipe diameters and other details shall be given in item descriptions.

According to Definition Rule D1, the lengths of multi-pipe ducts are calculated along a centre line which is taken as being a line equidistant from *the inside faces of the outer pipe walls*. This may not represent the true centre line of the trench.

7.21.5   Crossings and reinstatement

Where pipelines cross specified physical features, a measurable item is created for *crossings* (K 6 * *).

Road and rail crossings are not listed in the Second Division features because:

a)   Road crossings and the like are measured under K 7 1–4 *.
b)   Rail crossings would normally be subject to *special pipe laying methods* and would thus be measured under L 2 * *.

Items for crossings of the physical features listed in the Second Division include reinstatements unless otherwise stated (Coverage Rule C6) but:

▪   Crossings of streams <1 m wide shall not be measured (M5).
▪   Where linings are broken through any reinstatement shall be stated in the item description (A7).

Crossings of rivers, streams and canals are classified according to their widths measured at the water level shown on the drawings (D6).

Reinstatements are measured for pipes, ducts and metal culverts (M6) in linear metres (K 7 * *).

The length of reinstatement is measured along the centre line (Class K, Rule M7), and lengths include manholes and chambers (M7). The centre line for multiple pipes, ducts and metal culverts is defined in Class K, Rule D1, as *the line equidistant between the inside faces of the outer pipe walls*, as illustrated for a three-way duct in Figure 7.21.

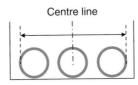

Figure 7.21   Multiple pipes, ducts and metal culverts.

Logic would seem to imply, therefore, that reinstatements are measured along the centre lines of pipes, ducts and metal culverts irrespective of the number of pipes in the trench. This makes practical sense but regard must also be paid to Class K, Rules M3 and D5, which state that Class I 'rules' (NB: measurement, definition, coverage and additional description rules, <u>not</u> the Work Classification) shall apply to ducts and metal culverts. This would seem to imply that the centre line measurement is solely a mean length but that each pipe in the trench is nevertheless measured.

Therefore, for three 100 mm pipes in a trench, the quantity of pipework is 3 × the length along their centre line (the outer pipes may be longer or shorter than each other due to the angle or radius of the trench).

Consequently, the measurement of reinstatement is less than clear. One the one hand, Class K, Definition Rule D1, provides a rule regarding the centre line for multiple pipes, ducts or culverts, but on the other hand, Class I 'rules' also apply to pipes, ducts and metal culverts which state that *the length of the pipe*, not the length of the <u>trench</u>, shall be measured where multiple pipes are *expressly required*.

Therefore, where there is more than one pipe in a trench, care needs to be taken with measuring associated reinstatements, because Rule D7 states that the classification of bore in the Third Division is to be taken as the *maximum nominal distance between the **inside** faces of the **outer** walls of the pipe, duct or culvert to be installed*.

This definition rule is clearly written in the singular (e.g. *the pipe*) with the implication that a reinstatement item is measurable for <u>each pipe</u> in a dual trench or <u>each pipe</u> in a multiway duct. This is emphasised in Class I, Measurement Rule M3, which states that *where more than one pipe is expressly required to be laid in one trench*, the length measured *shall be the length of **the** pipe* (i.e. <u>each</u> pipe).

Risk issue

As there is no provision for stating the width of reinstatements, save to include the bore of the pipe as per Definition Rule D7, or any clear rule as to how reinstatements are to be measured for trenches with multiple pipes, contractors would be wise to verify the quantities stated in the BQ with the drawings.

The danger is that they may 'overprice' these items when there is more than one pipe in a trench, and in a city centre sewer replacement scheme, the extent of reinstatement could be considerable.

7.22   Class L: Pipework – supports and protection, ancillaries to laying and excavation

Changes from CESMM3

None

Class L provides a work classification for special pipe laying methods, such as thrust boring and pipe jacking, and for the provision of supports and protection to pipes, ducts and metal culverts, as well as for cost significant items incidental to such work, including the excavation of rock and the excavation and filling of soft spots.

Coverage Rule C1 determines that items of work in Class L shall include excavation, disposal, earthwork support, backfilling, etc., except where such work is included in:

- Classes I, J and K.
- Extras to excavation measured in Class L.

Consider the item for pipe jacking in Table 7.31.

This item is measured in Class I (I 2 3 1), and thus, Coverage Rules C1 and C2 must be consulted to see what is deemed included as regards materials and excavation, supports to excavation, disposal, etc.

As I 2 3 1 is described as *not in trenches*, Coverage Rule C2 does not apply (C2 only applies to pipes <u>in trenches</u>), and thus, the item coverage in Class L must now be referred to.

C3 states that items for *special pipe laying methods* shall be deemed to include, *inter alia*, *work associated with special pipe laying methods* not included with the Class I item. As this

Table 7.31   Pipe jacking – 1.

	Pipes				
	Concrete Jacking Pipes; Steel collar and elastomeric sealing gasket				
	Nominal bore: 600 mm				
A	Pipe jacking beneath A492; MH3 - MH4; not in trenches		24	m	

has been described as *not in trenches*, it is regarded in C3 as an item for *provision, laying and jointing of pipes* and thus does not include excavation work.

The item must, therefore, be completed with a Class L item for *special pipe laying methods* as illustrated in Table 7.32.

Table 7.32   Pipe jacking – 2.

	Pipes				
	Concrete Jacking Pipes; Steel collar and elastomeric sealing gasket				
	Nominal bore: 600 mm				
A	Pipe jacking beneath A492; MH3 - MH4; not in trenches		24	m	
	Supports and Protection, Ancillaries to Laying and Excavation				
	Special pipe laying methods				
	Pipe jacking; MH3 - MH4				
B	nominal bore: 600mm		24	m	

With regard to *special pipe laying methods*, Class L, Measurement Rule M9, requires that such items shall only be measured where *expressly required*. This is echoed in Class I, Definition Rule D2.

It is likely that, in most cases, the decision as to whether to install pipes in heading, by pipe jacking or by thrust boring will be taken by the contractor – perhaps with some design responsibility – and not the bill compiler. In this event, the work would not be measured as *not in trenches* (I * * 1), but additional description (Class I, Rule A1) would be given to alert the tenderer to the possibility that special methods could/should be considered.

The items of *extras to excavation and backfilling* (L 1 * *) are for the measurement of items of work which are additional to the items measured in Class I and in Class L 2 * *: *Special pipe laying methods*. They fall into two groups:

1. Items L 1 * 1–6.
2. Items L 1 * 7–8.

Items in the first group – excavation of rock, backfilling above the Final Surface, etc. – may be considered as 'extra over' the descriptive features in the Second Division (e.g. pipe trenches, manholes, pipe jacking) because they occur <u>within the confines</u> of the already measured items in Class I and Class L 2 * *.

Items in the second group – excavation and filling below the Final Surface – are 'full value' items because they are measured <u>outside the confines</u> of the items already measured in Class I and Class L 2 * *.

There is no definition of 'rock' in CESMM4, but when defined in the Preamble in accordance with Paragraph 5.5, such material shall be measured when expected or encountered in pipe trenches, manholes, headings, thrust boring and pipe jacking.

The volume measured is determined by Measurement Rules M4–M6 and Definition Rule D1:

Volume		Nominal trench width
M4	**In pipe trenches:** The *average depth × average length ×* **nominal** *trench width of the material removed or backfilled*	D1 ▪ *The maximum* **nominal** *distance between the inside faces of the outer pipe walls*[1] *(d) + 500 mm where d = <1 m*
		▪ *The maximum* **nominal** *distance between the inside faces of the outer pipe walls*[1] *(d) + 750 mm where d < 1 m*
M5	**In manholes or other chambers:** The maximum plan area of the manhole or other chamber	[1]This convoluted definition is needed in case there is more than one pipe in the trench.
M6	**In headings, thrust boring and pipe jacking:** Length × internal cross-sectional area of the pipe	

In pipe trenches, the use of the word 'nominal' is important because it means 'notional' and not 'actual'. Consequently, the quantities that might appear in the bill of quantities are based on a drawing or standard detail, and the quantities measured during admeasurement are not actual save that the average length and average depth are determined on-site.

In manholes and other chambers, the maximum plan area is not defined but is likely to be taken as including any concrete surround but not the space occupied by formwork.

In headings, thrust boring and pipe jacking, the wall thickness of the pipe is not taken into account when determining the volume.

Risk issue

For the contractor, the volumes given in accordance with M4–6 and D1 represent an under-measure, and due allowance needs to be made for additional excavation, disposal and filling:

- In pipe trenches:
  - When the bucket width of the excavator is greater than the nominal width of the trench
  - When the excavation process results in a wider trench excavation as it invariably does
  - When earthwork support creates an excavation wider than the nominal trench width
- In manholes and other chambers:
  - For the space occupied by formwork and earthwork support and for working space
- In headings, thrust boring and pipe jacking:
  - For the volume of the annulus of the pipe
  - For 'overbreak'
  - For the volume of specification tolerances

7.23   Class M: Structural metalwork

Changes from CESMM3

- Measurement Rule M6 amended to omit reference to BS 4360

Class M is the repository of fabricated structural steelwork where fabrication is measured separately from erection. Both are measured in tonnes, and both are categorised as members for bridges, members for frames and other members.

There is no specific mention of structural steelwork for buildings in Class M, and neither is Class M referred to in Class Z: *Simple building works incidental to civil engineering works.* There are some descriptive features listed in the Second Division that would suffice for a portal frame or simple trussed roof steel structure, but the list is not as comprehensive as that found in NRM2. In this regard, it may be necessary to adopt an alternative method of measurement for such works pursuant to CESMM4, Paragraph 5.4.

With the exception of plates and flats, fittings are not measured, but weld fillets, bolts, nuts, washers, rivets and protective coatings shall be included in the mass of members measured (Measurement Rule M4). Exceptionally, site bolts are measured under the erection items M 5–7, and items are deemed to include for supply and delivery to site as per Coverage Rule C3.

Anchorages and holding down bolts are measured items (M 3–4 8 0), but only for fabrication, with anchorages and holding down bolts being measured by the number of complete assemblies and enumerated (nr).

Additional Description Rule A6 provides that items for erection shall identify separate bridges and structural frames in descriptions and, *where appropriate*, parts of bridges or frames.

Surface treatments such as metal spraying, galvanising and painting carried out <u>off-site</u> are measured under M 8 * 0, with the materials and number of applications stated in the item description. Surface treatments carried out <u>on-site</u> are measured under Class V: *Painting.*

Part 2

7.24   Class N: Miscellaneous metalwork

Changes from CESMM3

- C5 – items for *bridge bearings* to include for grouting beneath plates
- A3 – item descriptions for *bridge bearings* to state composition and materials of the bearing
- A4 – item descriptions for *tanks* to state the principal dimensions

Class N is for 'odds and sods' of metalwork not measured in other classes, and this is recognised in a long list of exclusions in the class heading. This is not to minimise the importance of items measured in this Class as the likes of bridge bearings, bridge parapets, stairways and handrails do not come cheaply.

The 'odd' nature of items in this Class is recognised in the First Division, where there are no groupings for the descriptive features offered in the Second Division.

For items of miscellaneous metalwork, great reliance is placed on Additional Description Rule A1 as well as upon drawings or standard details, to portray items that would otherwise require lengthy bill descriptions. This emphasises that, perhaps more than anywhere else in the method of measurement, civil engineering bills of quantities are not intended to fully describe the work in a project but that the BQ must be read in conjunction with the other documents that make up the contract to appreciate the full extent of the work involved.

Of particular note in Class N is the measurement of walings (N 2 5 0) and tie rods (N 2 4 0) in connection with piling work; the former is measured in linear metres and tie rods by number (nr) with items for tie rods deemed to include concrete, reinforcement and joints (Coverage Rule C4 refers).

Tanks – a new group of descriptive features in the Second Division – are to be enumerated within eight volume bands, with Additional Description Rule A4 requiring that the principal dimensions are to be stated. This is in addition to Rule A1 which requires full specification and other information to be provided as well.

New Coverage Rule C5 stipulates that grouting beneath bridge bearing plates is deemed included in item descriptions.

7.25   Class O: Timber

Changes from CESMM3

- None

Class O: *Timber* specifically excludes timber used for formwork, piling, rail work, tunnels and simple building works, etc. and thus is restricted to timber required for constructing jetties, marina decking, gangways, footbridges and the like. In fact, timber components (O 1–2 * *) are measured separately from timber decking (O 3–4 * 0), and each has a different unit of measurement.

Timber components are measured in linear metres with decking measured in m².

It would seem prudent to identify the structure to which the work measured in Class O relates albeit this is not required in the Work Classification. For instance, it would be helpful to tenderers to know that a measured quantity of timber components and decking relates to a particular jetty in a specific location, and thus, the provisions of CESMM4, Paragraph 5.11, might be considered appropriate.

Notwithstanding the Second Division categories, the nominal cross-sectional dimensions of timber components shall be given in item descriptions (Additional Description Rule A1), along with type and grade of timber, treatment details and surface finish (sawn, wrot).

Openings and holes are not deducted from the area measured for decking unless <u>each</u> opening/hole exceeds $0.5\,m^2$ (Measurement Rule M2). This includes gaps between decking boards, if specified.

Therefore, in a given area of decking 3 m wide with 12 mm gaps, the net effect of Rule M2 is that no deduction would be made as <u>each gap</u> is less than the threshold $(3\,m \times 0.012\,mm = 0.036\,m^2)$. The deduction is not cumulative but is for <u>each</u> opening or hole.

No labours (fixing, cutting, boring, etc.) are measured to timber work in accordance with Coverage Rule C1, but fixings, such as straps, bolts, etc., are enumerated under O 5 * 0.

7.26   Class P: Piles

Changes from CESMM3

- *Preformed prestressed concrete piles* omitted from the First Division
- *A7* – item descriptions for *preformed concrete piles* to state if they are *prestressed*
- New First Division item *P8*** for *stone columns* (C2 and A15 refer)

The description of work items for cast-in-place and preformed piling relies on the classification provided in the Third Division of Class P: *Piles*:

Bored/driven* cast-in-place piles		Preformed concrete piles	
■ Number of piles	nr	■ Number of piles of stated length	nr
■ Concreted length	m	■ Driven depth	m
■ Depth bored or driven	m		

*Using the FRANKI® pile system.

Bored or driven piles are reinforced and filled with in situ concrete, and preformed piles are precast off-site.

Groups of bored/driven piles (i.e. the same type/materials, diameter/cross section and location) are required by Measurement Rule M2 to be billed as follows:

- An item for the number of piles.
- An item for the concreted length.
- An item for the total depth bored or driven.

This is illustrated in Table 7.33.

In Table 7.33, the concreted length is measured according to Rule M3, this being the measurement from the *cut-off levels* to the *toe levels* expressly required. It will also be noted that:

- The maximum bored/driven depth is stated according to Definition Rule D1; this is the maximum depth of any pile in a group of piles.
- The structure to be supported is identified (Additional Description Rule A3).
- The *Commencing surface* is stated (also A3).

Part 2

Table 7.33   Piles – 1.

	Bored cast in place concrete piles; Retaining wall B; Commencing surface 1m above Final surface			
	Diameter: 450 mm			
A	number of piles	18	nr	
B	concreted length	162	m	
C	depth bored or driven; maximum depth 12m	180	m	

For preformed piles, Rule M4 requires:

- An item for the number of piles.
- An item for the total depth driven.

This is illustrated in Table 7.34.

Table 7.34   Piles – 2.

	Preformed concrete piles; prestressed: Bridge 2; Commencing surface 300 mm above Final surface			
	Cross-sectional area: 0.05 - 0.1 m²			
A	number of piles; length 15m	14	nr	
B	depth driven	196	m	

The depth of pile driven is not necessarily given by the number of piles × length but is a function of the head and toe levels stipulated in the contract. It will also be noted in Table 7.34 that the preformed concrete piles are described as *prestressed* as required by Additional Description Rule A7, and the details required by Additional Description Rule A3 are also given in the item.

The Commencing Surface for the purpose of measurement is defined in Measurement Rule M1 as *the surface at which boring or driving is expected to begin.*

This requires the bill compiler to make a judgement as to where piling work is to start, which might not be the actual commencing surface on-site. Measurement Rule M1 further provides, however, that the measurement of completed work shall adopt the same commencing surface so that the contractor is not disadvantaged in terms of the final measured quantities, except to the extent that the choice of commencing level is conditioned by the working method adopted.

There is no 'deemed included' provision, or item coverage rule, for the supply of piles or the supply of concrete for cast-in-place piles. There is only the implication, in Additional Description Rule A1, that item descriptions shall include details of the *materials of which piles are composed.* Reinforcement to piling, on the other hand, is measured in Class Q: *Piling ancillaries.*

The new Class P category of *stone columns*, for foundations constructed using the vibro-displacement/vibro-replacement method, merits two measured items – one for the number of columns (P 8 * 1) and another for the total length (P 8 * 2). Additional Description Rule A15 requires that the diameter of stone columns shall be stated in item descriptions, notwithstanding the eight Second Division diameter ranges provided.

The same rules apply to stone columns as to piles:

- Commencing surface to be stated A3.
- Depth measurement – commencing surface to toe (M1).
- Identify structure to be supported (A3).
- Disposal of excavated material deemed included (C1).

Additionally, Coverage Rule C2 states that items for stone columns shall include for boring. Table 7.35 illustrates a typical bill item.

Table 7.35   Stone columns.

	Stone Columns; Marina gangway raft foundation; Commencing surface 2m below Original surface		
	Diameter 600mm		
C	Number of columns	80	Nr
D	Length of columns	184	m

The provision of an item for the number of piles and stone columns recognises the fixed cost of establishing and re-establishing the piling rig at each pile location. However, the mobilisation of the piling work, bringing plant to site and later removal, transporting testing equipment and constructing pile mats, etc. would be best priced as method-related charges in Class A.

Carrying out pile tests is measured by number (nr) according to the type of test and the test load under Q 8 * *.

It should be noted that there is no similar rule in Class P: *Piles* to that of Rules M7 and A2 in Class E regarding additional description for piling carried out *below a body of open water*. However, the requirements of Paragraph 5.21 must, nevertheless, be respected, and good practice suggests that including the information referred to in Paragraph 5.21 in any relevant item for piling would help tenderers with the pricing of such items.

This view is confirmed in by the Federation of Piling Specialists (2007), who recommend that the item descriptions for piling and embedded retaining walls should identify work affected by bodies of water and also include relevant information on fluctuations in tide levels.

7.27   Class Q: Piling ancillaries

Changes from CESMM3

- Definition Rule D2 – reference to BS 4449 omitted

Class Q: *Piling ancillaries* accompanies Class P: *Piles* and provides for all the supplementary work associated with piling such as cutting off surplus lengths of piles, supply and fixing of rebar to cast-in-place piles, provision for dealing with obstructions and pile testing, etc.

Risk issue

The work in this Class shall be measured only *where it is expressly required*, and therefore, unless shown on the drawings or expressly or impliedly included in the specification or elsewhere in the contract, the contractor will not be paid for any additional work carried out (Measurement Rule M1 refers).
There is an exception to this in the case of *backfilling empty bore for cast-in-place concrete piles.*

The pile types in this Class mirror those of Class P except that pile diameters are not required in lieu of the Third Division descriptive features, except those for *enlarged bases for bored piles* whose diameter shall be given (Rule A1).

Items for piling ancillaries are *deemed to include disposal of surplus excavated materials* unless otherwise stated (Rule C1 refers).

Where driven piles are to be extended, two items are provided:

- An enumerated item (nr) for preparing the pile to be extended (Q 3 4 *, concrete, or Q 4 4 *, timber).
- A linear metre item for supplying the pile extension (e.g. concrete piles Q 3 5 * <3 m or Q 3 6 * >3 m).

The enumerated item is ideal for pricing the mobilisation costs associated with extending piles already driven to a provisional set, but driving extended piles is added to the length measurement in Class P by virtue of Measurement Rule M5.

The completion of piles to receive ground beams, pile caps, etc. is measured in two items. Cutting off surplus lengths of piles is measured linearly at Q 1 7 * and Q 3 7 *, and preparing the heads of piles is itemised under Q 1 8 * and Q 3 8 * for concrete piles. Disposal is deemed included in both items as per Coverage Rule C1.

7.28   Class R: Roads and pavings

Changes from CESMM3

- Reference to *DTp specified* replaced with *Manual of Contract Documents for Highway Works,* Volume 1 Specification for Highway Works in Definition Rule D1
- Classification of road building materials changed substantially:
  - R1 *unbound sub-base* items 1–5 in the Second Division;
    - **NB**: *Geotextiles* (R 1 6 *) and *additional depth of stated material* (R 1 7 *) remain from CESMM3
  - R2 *Cement and other hydraulically bound pavements* items 1–8 in the Second Division (A4 also refers)
  - R3 *Bituminous bound pavements* items 1–8 in the Second Division (A5 also refers)
  - R4 *Bituminous bound pavements* items 1–6 in the Second Division including *EME2 base and binder course* high performance asphalt (R 4 2 0), *porous asphalt* (R 4 4 0) and *cold recycled bound material* (R 4 6 0 – A6 also refers)
    - **NB[1]**: *Regulating course of stated material* (R 4 8 0) remains from CESMM3
    - **NB[2]**: solid line across the Third Division excludes depth categories above; this must be an unintended error
  - R 4 8 0 – new item for *cold milling/planing* (misspelt in CESMM4) in the Second Division (C2 also refers)

- R5 *Concrete pavements* substantially changed:
  - Pavement classification changed (Second Division items 1–5)
  - New Rule A7 relating to *wet lean concrete*
  - R 5 6 * reference to BS 4483 deleted (A8 also refers)
  - Revised additional description rule (now numbered A8) is incomplete
  - R 5 6 * *Other fabric reinforcement* deleted from the Second Division
  - R 5 7 * reference to BS 4449 deleted
  - R 5 8 0 wording changed to *separation and waterproof membranes*
- *Joints in concrete pavements*:
  - Two new categories in (1) *transverse joints* and (5) *longitudinal joints*
  - *Butt joints* omitted
- *Kerbs, channels and edgings* now include *footways and paved areas* as well:
  - Classification of kerbs channels and edgings rationalised and reference to BS 7263 omitted (new C5 also refers)
  - *In situ concrete kerbs* and *asphalt channels* omitted
  - First Division *light duty pavements* omitted
  - New Second Division list of pavings and surfacing created (R 7 items 5–8), including *grass concrete paving* and *flexible surfacing* (R 7 6 0)
  - New C5 stating that *kerbs, channels and edgings* are now deemed to include earthworks and concrete ancillaries as well as beds and backings, reinforcement and so on
  - New C6 relating to *precast concrete natural stone block and clay slabs and pavers* and new A13 requires *material, size and thickness* to be stated
  - New C7 relating to *flexible surfacing and in situ concrete surfacing* along with new A14 requiring *materials and thicknesses* to be stated
  - New C8 relating to *grass concrete pavings*
- *Ancillaries* now includes:
  - An item for *raised rib* markings (R 8 2 6)
  - A new A15 relating to *traffic signs and surface markings* requiring *material, size and type to be stated*; type is used instead of reference to *DoT traffic signs, regulations and general directions*
  - C9 amended to refer to M10 not M9

Class R includes roads, runways and other paved areas as well as kerbs, light duty pavements, traffic signs and road markings.

Light duty pavements are no longer accorded their own First Division 'status', and such work is now measured under R 7 6 0: *Flexible surfacing*.

Risk issue

Flexible surfacing items are *deemed to include for beds* by virtue of Coverage Rule C7.

Depth categories in the Third Division of Class R are subject to Additional Description Rule A1 which requires that item descriptions shall *state the depth of each course or slab*.

A new Second Division category of *cold milling and planing* has been added at R 4 7 0 which is deemed to include for disposal of arisings (C2) with, presumably, the depth stated in the Third Division (see NB[2] previously).

Risk issue

There is no requirement to state the location of cold milling/planing despite this being of crucial importance in the contractor's choice of plant and, therefore, his pricing.

Despite the inclusion of cold planing, it remains a mystery why CESMM4 provides no other road maintenance and repair items, such as the removal and replacement of kerbs and edgings and adjustment of precast concrete slabs.

Risk issue

Kerbs, channels and edgings are deemed to include beds and backings, reinforcement and so on (C4) but are now also deemed to include earthworks and concrete ancillaries by virtue of the new C5.

Under *ancillaries*, both illuminated and non-illuminated *traffic signs* are enumerated and hence require a comprehensive coverage rule (C9) that includes a long list of everything required for such items other than ducts and electrical work. Ducts are measured under Class K: *Ducts*, and electrical work may be given as a Prime Cost Item in Class A.

7.29   Class S: Rail track

Changes from CESMM3

- *Track foundations*:
  - Items for *ballast cleaning, tamping* and *pneumatic ballast injection* added to the Second Division along with accompanying Measurement Rules M3, M4 and M5
  - The *provision of ballast to fill voids* is *deemed included* with *ballast cleaning* (new Coverage Rule C1 refers)
- *Taking up*:
  - *Eutectic strip* now included in *sundries* – Third Division (S 2 6 7)
- *Lifting, packing and slewing* now includes *works to existing track*:
  - Second Division items now exclude *turnouts* which have been moved to the Third Division
  - *Buffer stops* moved to the Third Division S 3 * 5
  - *Spot re-sleepering* has been moved from S 6 * 8 to S 3 3 * and renamed *spot replacement of sleepers*
  - New items for *rail turning* (S 3 4 *), *re-railing* (S 3 5 *) and *stressing rail* (S 3 6 *) added
  - New list of Third Division descriptive features and units of measurement
  - First Division unit of measurement (nr) retained in S 3 * *, but this seems to be an oversight
  - New Measurement Rules M8 (old D3), M9 and M10
  - New D3 relating to *spot replacement of sleepers*
  - New C3 is old C2
  - C4 and C6 concern *spot replacement of sleepers*; they are slightly different but it looks like the intention to combine them was overlooked in the editing
  - New C5 for *re-railing*
- *Supplying* (S 4 * *):
  - *Steel* sleepers added at S 4 7 3
- *Supplying* (S 5 * *):
  - New Second Division heading *turnouts and crossings* and new Third Division list of descriptive features (including *diamond crossings*) provided along with associated measurement units
  - Second Division measurement unit (nr) retained but looks like an oversight
  - New Second Division item for *prefabricated track panels* and Third Division list of descriptive features (S 5 3 *)
  - New Second Division item for *prefabricated turnouts and crossings* and Third Division list of descriptive features (S 5 4 *)

- ○ New Additional Description Rules A15, A16 (old A10 with reference to *diamond crossings* removed); A18 (old A12) and A19 are virtually the same except for the addition of the word *type* in A19 (looks like an editing issue)
- There are now two First Division items for laying S 6 * * and S 7 * *
- S 6 * * – *Laying with bearers on ballast*:
  - ○ This is the old S 6 * *
  - ○ *Welded joints* is now S 6 * 6 (not S 6 * 7)
  - ○ *Spot re-sleepering* moved to S 3 3 *
  - ○ New M17, D4 and C14 relating to laying rails direct to *concrete slab track*
  - ○ New A20 and A21 (old A13 and A14 slightly amended)
  - ○ New A23 and A25
  - ○ A24 is old A16, but instead of *type and length*, reference is made to *type and weight*
- (S 7 * *) – *Laying*:
  - ○ Sundries moved from CESMM3 S 6 8 * with *static sander* (S 7 5 8) and *eutectic strip* (S 7 5 9) added
  - ○ New Second Division items for *slab track, prefabricated track panels, prefabricated turnouts,* etc.
  - ○ New accompanying Third Division descriptive features and units of measurement
  - ○ New A24 (old A16 with weight added instead of length)
  - ○ New A25
  - ○ New A26 (old A18)

Class S in CESMM4 introduces significant changes compared to its predecessor CESMM3 reflecting the 'high-tech' and highly mechanised rail infrastructure engineering techniques employed worldwide.

The need to measure railway work may arise in a variety of circumstances:

- Laying of new permanent way.
- Replacement of railway lines.
- Quarries and mines.
- Power stations and steelworks.

Rail track work is highly industrialised and mechanised, and there is a variety of track laying methods available, depending upon the length of track required, its access, location and complexity in terms of diamond crossings and turnouts, etc.

Rail track work involves both new track laying and work to existing track.

For new rail track work, materials are often provided by the employer, delivered to site, hence the separation of *supplying* from *laying* for new track work in Class S.

However, the measurement of track foundations is measured on a 'supply and fix' basis (S 1 * 0) with ballast measured in m³ but separated into bottom ballast (beneath the sleepers) and top ballast (between the sleepers, but with no deduction of the volume occupied by the sleepers).

In Class S, work to existing track is classified as (a) *taking up* and (b) *lifting, packing and slewing*, the former relating to the removal of track and the latter to the adjustment or realignment of track to be retained. Treatment of existing ballast, such as cleaning and tamping, is also measured, and cleaning is *deemed to include* the provision of ballast to fill voids (Coverage Rule C1 refers).

In some instances, track to be slewed or realigned will require some sleepers to be replaced, and this is provided for in item S 3 3 *, *Spot replacement of sleepers*. The item coverage includes opening out the old sleepers and packing and boxing in the new sleepers with ballast (Rule C3 refers).

The measurement of rail track, both new and existing, is taken along the centre line of the track and includes two rails, but the laying of check, guard and conductor rails is along one rail only.

Part 2

However, care needs to be taken with turnouts and diamond crossings because:

- The measurement of work to existing track under S 2 * * and S 3 * * <u>excludes</u> turnouts and diamond crossings (Rules M6 and M9 refer).
- The lengths of check, guard and conductor rails under S 2 4 * and S 2 5 * <u>excludes</u> turnouts and diamond crossings (Rule M7 refers).
- The lengths of laying new track is <u>inclusive</u> of the lengths occupied by turnouts and diamond crossings (Rule M14 refers).

The measurement of slab track ('ballastless track') and prefabricated track panels, with timber or concrete sleepers, are measured along the centre line of the track (two rails) including the lengths occupied by turnouts and diamond crossings.

7.30   Class T: Tunnels

Changes from CESMM3

None

Class T: *Tunnels* excludes pipe laying by pipe jacking, thrust boring and the like which is measured under Classes I–L. 'Cut and cover' tunnels are measured under Class E, Class F and other appropriate classes of the method of measurement.

The level of risk in tunnelling operations is recognised in CESMM4, and a considerable burden of financial risk is retained by the employer:

- Both temporary and permanent support shall be measured (Rule M1); this is quite different to other classes where the contractor is deemed to include earthwork support (*supports and stabilisation* is measured under T 8 * *).
- *Supports and stabilisation* also includes a measured item for *forward probing*, whereas the risk of determining prevailing ground conditions lies with the contractor in other classes.
- Work in compressed air shall be so described as an additional description to relevant items (Rule A1 refers).
- Ground freezing, jet grouting, chemical grouting, cement grouting and other soil stabilisation methods are not measured in Class T, or elsewhere, but may be provided for via a provisional sum for defined work in Class A.

Both in situ and preformed tunnel linings are measured under the First Division descriptive features T 2–7, but where complex concrete shapes are required, these *may be* measured under Classes F and G (footnote to CESMM4, page 88, refers).

In situ linings are described in the Second Division as 'sprayed concrete'. This is a generic term for a number of different applications of the process of spraying a mixture of cement and fine aggregate, under pressure, onto a variety of natural and other surfaces in order to carry out structural repairs, fire protection to structural steelwork, tunnel and refractory linings and for lining swimming pools, reservoirs, and a variety of complex-shaped structures such as domes and shells.

The term 'sprayed concrete' is used specifically in Class T which provides two descriptive features in the Second Division – primary and secondary sprayed concrete. This is an implied reference to NATM which relies on an initial thin application of sprayed concrete to the excavated surface followed by a second, thicker, application.

Table 7.36 illustrates a theoretical application of the Work Classification.

Table 7.36   Class T sprayed concrete.

	Tunnels				
	In situ lining to tunnels				
	Sprayed concrete primary; Grade 40; Reinforced				
A	Diameter: 6 - 7 m; 50 mm thick	m²			
	Sprayed concrete secondary; Grade 40; Reinforced				
B	Diameter: 6 - 7 m; 200 mm thick	m²			

It will be noted that the item descriptions in Table 7.36 include the words 'primary' and 'secondary', and therefore, it must be assumed that where there is a 'primary' lining a 'secondary' lining follows by implication.

However, NATM is not the only application of sprayed concrete in tunnel linings, and where a single application is required, or where complex shapes are involved, Classes F and G should be used (the NOTE on page 88 of CESMM4 refers).

Risk issue

The measurement of sprayed concrete in Class T: *Tunnels* is subject to different rules and units of measurement compared to Class F: *In situ concrete* and Class G: *Concrete ancillaries*.

In Class T, the volume for excavation of tunnels is calculated according to a *payment line* shown on the drawings; otherwise, the net dimensions of the excavation are to be used (Rule M2). Payment lines are normally taken to the outside of the tunnel lining, perhaps with a tolerance allowance.

A contractor risk item for overbreak is accommodated by an item of *excavated surfaces* (T 1 7 0 and T 1 8 0) measured in m². This is the area of the payment surface defined by the payment lines on the drawings but, otherwise, is the net area excavated (Rule M4). CESMM4 does not define the word 'overbreak'.

However, Attewell (1995) makes the point that the term *overbreak* should be restricted to rock as, whilst soils can deform, they cannot fracture or 'break'. In this sense, the term *over-excavation* would be more appropriate when referring to soil or degraded rock. Attewell (1995) also suggests that any additional excavation beyond the payment line is down to contractor method or preference for an oversized tunnel, perhaps for technical reasons.

Attewell (1995) further explains that the excavation of soil pockets or zones of rock degraded to soil consistency in rock tunnels, beyond the payment line, must be carefully considered in terms of who carries the risk. However, Class T, Measurement Rule M2, stipulates that this would classify as *excavation of other cavities* and, as such, would be measurable under T 1 5 * (for rock) and T 1 6 * (for other stated material) in m³, with the diameter stated in accordance with the Third Division.

Filling within tunnels, including *other cavities*, is included in Class E: *Excavation* (see Class T heading of exclusions). However, there are no specific items in Class E relating to tunnelling or in Class F: *In situ concrete* for filling.

Coverage Rule C1 establishes that *excavated surfaces* items, along with other excavation items, are deemed to include off-site disposal and removal of dead services.

7.31 Class U: Brickwork, blockwork and masonry

Changes from CESMM3

- Unit of measurement for walls exceeding 1 m thick changed to m² (was m³)
- This may not be intentional as Measurement Rule M2 still refers to *volumes*

Measurement Rule M1 requires that *each skin* of cavity construction *shall be measured*, and hence, cavity walls are not regarded as 'composite'. Additional Description Rule A4 requires *cavity or composite construction* to be stated in the item description for walls, facing to concrete, casings and so on, and Rule A5 provides that wall thickness shall be given.

Rule M3 requires that *mean dimensions* shall be used to calculate areas and volumes in this class of work, including the heights of *columns* and *piers*, and surface features, such as band courses and corbels, shall be measured by their mean length (Rule M4 refers). *Inter alia*, this implies that centre lines of brickwork, blockwork and masonry shall be used for quantity take-off.

The area for *fixings and ties* to cavity work and the like shall be given under U * 8 6, but the area measured shall be *the smaller of the two areas measured* for cavity brickwork, masonry, etc., that is, not the centre line of the wall.

7.32 Class V: Painting

Changes from CESMM3

- Reference to lead in V 1 * * omitted

Painting is straightforwardly measured as the area covered, with no deductions for holes or openings <5 m². Narrow widths are measured in linear metres.

Less easy to interpret is the phrase *isolated groups of surfaces*, which is used in several places in Class V.

Measurement Rule M3 states that items for *isolated groups of surfaces* arise when surfaces of different shapes or dimensions are measured separately if they satisfy the Definition Rule D1 test that each group does not exceed 6 m² in total. Additional Description Rule A3 requires the work and its location to be stated in the item description.

The point of this provision is to provide the tenderer with a count (nr) of small areas of work whose location falls outside the main body of work and therefore has a different cost implication. *Isolated groups of surfaces* of different materials are separately measured.

Notwithstanding the foregoing, the footnote to CESMM4, page 94, states that painting to areas having **the same** shape and dimensions may be measured by number as *isolated groups of surfaces* instead of by area or length (narrow widths).

Risk issue

As there is no definition rule to this footnote, nor any reference to other definition rules, such isolated areas could well be substantial.

It would, therefore, be incumbent on the tenderer to establish such areas from the drawings in order to determine their quantities for pricing purposes.

7.33   Class W: Waterproofing

Changes from CESMM3

- W 6 * * – *Sheet linings membrane* added

Waterproofing, protective layers and sheet membranes, etc. are straightforwardly measured as the area covered, with no deductions for holes or openings <5 m². Narrow widths are measured in linear metres.

Less easy to interpret is the phrase *isolated groups of surfaces*, which is used in several places in Class W.

Measurement Rule M4 states that items for *isolated groups of surfaces* arise when surfaces of **different** shapes or dimensions are measured separately if they satisfy the Definition Rule D2 test that each group does not exceed 6 m² in total. Additional Description Rule A2 requires the work and its location to be stated in the item description.

The point of this provision is to provide the tenderer with a count (nr) of small areas of work whose location falls outside the main body of work and therefore has a different cost implication. *Isolated groups of surfaces* of different materials are separately measured.

Notwithstanding the foregoing, the footnote to CESMM4, page 96, states that waterproofing to areas having **the same** shape and dimensions may be measured by number as *isolated groups of surfaces* instead of by area or length (narrow widths).

Risk issue

As there is no definition rule to this footnote, nor any reference to other definition rules, such isolated areas could well be substantial.

It would, therefore, be incumbent on the tenderer to establish such areas from the drawings in order to determine their quantities for pricing purposes.

7.34   Class X: Miscellaneous work

Changes from CESMM3

- X 1 8 * *Fences – Road restraint system* replaces *metal crash barriers*
- *Gates and stiles* – Second Division rationalised
- *Drainage to structures above ground* – *Asbestos cement* deleted from the Second Division
- A 5 0 0 – *Open cell block systems* added; measured by (nr) with details to be given as per new A7 (*open cell block systems* are not listed in the Class X heading)

Part 2

Items for *fences*, which includes *road restraint systems*, shall be deemed to include excavation, disposal, backfilling, concrete, formwork and reinforcement, etc. (Rule C1) as well as posts (Rule C2). Details of the type of fence and its foundations shall be given as additional description (Rule A2).

The distinction between *box* and *mattress* gabions is made in Rule D4 according to whether they exceed 300 mm or not. Rule D4 also deems that filling shall be imported unless otherwise stated.

7.34.1   Open cell block systems

It is surprising that Class X 5 0 0 – *Open cell block systems* – does not provide a definition of what is meant by the term. Open cell blocks are commonly used in a wide variety of applications, such as grassed parking areas, revetments, retaining walls and so on, and all sorts of different shapes, sizes and configurations are available.

There is no clear guidance, however, and a most unsatisfactory unit of measurement, for the billing of such items. It would have been most helpful had there been a list of Second and Third Division features and units of measurement.

Therefore, the bill compiler is left with the dilemma of choosing whether to measure open cell blocks as:

a) Precast concrete under H 6 0 0: *Segmental units* or
b) Under X 5 0 0: *Open cell block systems*.

Both are enumerated (nr) but:

a) There are no measurement rules for segmental units in Class H although the bill compiler is helped by Additional Description Rules A1 and A2.
b) Additional Description Rule A7 of Class X requires the descriptive features of such work to be stated.
c) In neither case is there provision to describe open cell block systems adequately.

Risk issue

There is no suitable method of measurement for a substantial gravity retaining wall within CESMM4.

7.35   Class Y: Sewer and water main renovation and ancillary works

Changes from CESMM3

■ None

Thanks to 'no-dig' or trenchless technology, a great deal of sewer renovation can be carried out without the disruptive effect of open excavation.

Where excavation is *expressly required*, Coverage Rule C1 deems that relevant Class Y items are *deemed to include* earthwork support, backfilling, removal of dead services, etc., and item descriptions are to state the depth of excavation in 1 m stages pursuant to Additional Description

Rule A4. Crossings, reinstatements and extras to excavation and backfilling, etc. are to be measured according to Classes K and L (Rule M2 refers).

Where manual or remotely controlled methods are *expressly required*, this shall be stated in item descriptions (Rule A3 refers).

Additional description given for Class Y work shall include provision of sewer dimensions and profiles (Rule A2).

7.36   Class Z: Simple building works incidental to civil engineering works

Changes from CESMM3

- D8 – *taps* are classed as *pipe fittings*
- A23 and A27 – reference to British Standards omitted
- A30 and A31 – reference to British Standards omitted

Class Z is more comprehensive than it would appear, because building work such as in situ concrete, drainage, ducts, brickwork, etc. would be covered by the relevant Class in other parts of the method of measurement, as noted in the heading on page 105 of CESMM4.

Some types of building work may be 'hidden away' in certain cases, such as *drainage to structures above ground* in Class X: *Miscellaneous work*, which provides itemisation for guttering, downpipes, fittings, etc. (X 2 * *).

Due to the detailed nature of building work, it is likely that there will be a considerable degree of additional description required in Class Z and items should be coded with an appropriate suffix where required.

When measuring suspended ceilings, bill compilers familiar with SMM7 should note that both primary support systems and edge trims are deemed included in Z 4 5 * (Coverage Rule C6 refers).

Notes

1. *The headings in this Chapter generally follow those of CESMM4.*
2. Now published by the Civil Engineering Contractors Association (CECA).
3. http://www.vibroflotation.com/Vibro/vibroflotation_fr.nsf/site/Vibro-Compaction.Effects-and-Test (accessed 7 April 2015) and https://www.youtube.com/watch?v=_v10uOev78U (accessed 7 April 2015).
4. https://www.gov.uk/japanese-knotweed-giant-hogweed-and-other-invasive-plants (accessed April 7, 2015).5.

References

Attewell, P.B. (1995) *Tunnelling Contracts and Site Investigation*, E&FN Spon, London.

Barnes, M. (1977) *Measurement in Contract Control*, Thomas Telford, London.

CIRIA (1977) CIRIA Research Project 34.

Federation of Piling Specialists (2007) *Measurement of Piling and Embedded Retaining Wall Work*, Federation of Piling Specialists, Beckenham.

FIDIC Conditions of Contract for Construction (First Ed. 1999).

Ross, A. and Williams, P. (2013) *Financial Management in Construction Contracting*, Wiley-Blackwell, Chichester.

Sprayed Concrete Association. http://www.sca.org.uk/ (accessed April 7, 2015).

Chapter 8
Method of Measurement for Highway Works

In the United Kingdom, responsibility for looking after the construction and maintenance of motorways and major trunk roads devolves to a number of 'overseeing departments'. In England, the Department for Transport (DfT) is the ministerial department responsible which operates via a government-owned company, namely, Highways England (formerly the Highways Agency). Wales, Scotland and Northern Ireland have their own authorities.

The 4300 mile arterial road network in England[1] is supplemented by other roads which are managed by local authorities which become the 'overseeing organisation' in the event that they use the Manual of Contract Documents for Highway Works (MCHW), and its accompanying Method of Measurement, for their highway projects.

8.1 Manual of Contract Documents for Highway Works

The MCHW[2] consists of seven substantial 'loose-leaf' ring binders:

- Volume 0: Model Contract Document for Major Works and Implementation Requirements.
- Volume 1: Specification for Highway Works (SHW).
- Volume 2: Notes for Guidance on the Specification for Highway Works.
- Volume 3: Highway Construction Details (HCD).
- Volume 4: Bills of Quantities for Highway Works.
- Volume 5: Contract Documents for Specialist Activities.
- Volume 6: Departmental Standards and Advice Notes on Contract Documentation and Site Supervision.

Volumes 1–4 were originally published by HMSO and Volumes 0 and 5 by the Department of Transport (DTp) on behalf of the then overseeing departments (now Highways England, Transport Scotland, the Welsh Assembly Government or the Department for Regional Development according to who is responsible for the looking after the specific project).

Each volume has its own schedule of Registration of Amendments which allows changes to be made to the documents as needed and thereby avoids the need to publish new editions of the documents from time to time as was hitherto the case.

Managing Measurement Risk in Building and Civil Engineering, First Edition. Peter Williams.
© 2016 John Wiley & Sons, Ltd. Published 2016 by John Wiley & Sons, Ltd.

As the MCHW is used for public sector projects that have to be advertised in the European Union (EU) Journal, amendments are subject to Directive 98/34/EC whereby member states are required to notify the European Commission of proposed changes to their technical regulations and standards at draft stage. This process places the proposed amendments in an EU Notification standstill period during which other member states may examine the drafts and react to them. Failing objections, the draft changes may be published after a 3-month wait.

The main driving force behind publication of the MCHW in 1991, and the Seventh Edition of the Specification in particular, was the need to remove 'barriers to trade' in accordance with Notification Directive 83/89 of the European Commission. In practical terms, this means that particular products or suppliers cannot be prescribed in specifications or on drawings as this might preclude the procurement of equivalent products or suppliers from EU member states. Additional reasoning was the need to make the highway documentation more user-friendly within a controlled system that recognised the need to move towards greater quality assurance both in design offices and on-site (Money and Hodgson, 1992).

8.1.1   Volume 0: Model contract document for major works and implementation requirements

Volume 0 contains typical contract documents and instructions for tendering for new trunk road contracts and for improvement and maintenance schemes in the United Kingdom and Northern Ireland. Originally based on the ICE Conditions of Contracts, the preferred contractual basis for such projects is now the New Engineering Contract (NEC) (both the Engineering and Construction Contract (ECC) and the Short Form).

A variety of procurement routes are available including full and partial contractor design, target contracts with pain/gain provisions and the use of the ECC Secondary Option Partnering Clause (X12). The preferred pricing document is the activity schedule, but bills of quantities (BQ) are envisaged in certain circumstances.

In this context, an unusual feature of the model contract is the option of **risk transfer in BQ** for projects that can be reasonably accurately quantified in advance and where the likelihood of remeasurement is low.

This feature is facilitated by use of a 'Z clause' (Z36), but subject to the proviso that *the project sponsor needs to be satisfied that the bill fully represents a true reflection of the project, even to the extent of erring on the generous side when fixing quantities. There is no intention of catching out a contractor by deflating quantities. The motivation behind the clause is to stop claims and to achieve certainty of price outcome.*

There are some 38 'Z clauses' in Volume 0, some of which are marked (M), meaning that they are mandatory. Other non-mandatory Z clauses include provisions for:

- Named subcontractors.
- Retention bonds.
- Bonus for early completion.
- Lane rental charges for late completion.
- Incentivised railway possessions.
- Value engineering.

The model contract document contains extensive project administration arrangements and pro forma documentation.

Whilst Highways England prefers to use the NEC (main and short forms) for its projects, Mitchell (2014) suggests that this is not necessarily the case with local authority clients who may still prefer to use the ICE Conditions (now the Infrastructure Conditions of Contract).

8.1.2   Volume 1: Specification for Highway Works

Along with the Highway Construction Details, the Specification is the heart of the Manual of Contract Documents.

The Seventh Edition of the Specification for Highway Works (SHW) is designed to be 'contract neutral' for use with a variety of procurement methods and forms of contract. The usual practice of including contract-specific terms, such as engineer's directions and/or approvals, has been avoided in line with the 'neutrality' policy of the Specification.

Despite the 'contract neutrality' of the SHW, British Standards (BS) and British Standard Codes of Practice may be incorporated in contracts unless they have been superseded by a European Standard (BS EN) or a harmonised European Standard (hEN).

hENs allow products to be given a CE marking, but, from 1 July 2013 under the Construction Products Regulation (CPR) 2011, it is mandatory for manufacturers to apply a CE marking to any products covered by a hEN or European Technical Assessment (ETA).[3] This UK legislation has been enacted as a result of EU Construction Products Regulation (EU) No 305/2011, a follow-on from the **Construction Products Directive (CPD) (Council Directive 89/106/EEC)**, which is now repealed.

8.1.3   Volume 2: Notes for guidance on the specification for highway works

The Notes for Guidance are purely intended to assist users in the interpretation and use of the SHW, but, unlike the Specification, are not intended to be incorporated into the contract. The Introduction to the Notes for Guidance emphasises the 'contract neutrality' of the SHW and that, other than when the ICE/ICC Conditions of Contract are used, the engineer's approval of workmanship and materials is not required.

Mitchell (2014) questions the extent to which the Notes for Guidance are used in practice and observes that they are often misapplied or ignored altogether.

8.1.4   Volume 3: Highway Construction Details

The drawings are of vital importance to the proper working of the MCHW because this is where the contractor has to look to see exactly what is expected in the contract. It is not good enough to simply look at the Specification and MMHW item coverages because these are generic and are designed to cover all of the work that might be required on any highway project. Clearly, not all projects will require every possible specification requirement.

The drawings in question consist of:

1. Project-specific drawings, including general layouts, plans and long sections, for the particular project in question.
2. Drawings/details contained in Volume 3: *Highway Construction Details* (HCD). This shows standard details for fencing, pipe beds and surrounds, manholes, gullies, carriageway pavement details and so on, but it must be emphasised that these standard details are also generic and each and every standard detail does not necessarily apply to each and every project.

Consequently, a specific contract may not require all the work covered by the Specification or the HCD or the items listed in the MMHW item coverage, and the precise requirements of an individual project can only be discovered by looking at the project drawings.

Taking a drainage item, for example, the item coverage for headwalls under Series 500: *Headwalls and Outfall Works* includes brickwork, copings, string courses and the like, but if the drawings only show headwalls as being concrete, then there is no requirement for the contractor to price for any brickwork. Similarly, specification clauses under Series 1800: *Structural Steelwork* will not be relevant if there is no requirement for structural steelwork shown on the drawings.

Part 2

8.1.5   Volume 4: Bills of quantities for highway works

Volume 4 is slightly different to other volumes in the MCHW in that it is not a homogeneous document but rather is made up of three interrelated sections that come together to enable BQ to be prepared:

- Section 1 – Method of Measurement for Highway Works (MMHW).
- Section 2 – Notes for Guidance on the Method of Measurement for Highway Works.
- Section 3 – Library of Standard Item Descriptions for Highway Works (LSID).

Section 1 provides the rules of measurement and Section 3 the means of converting measurements into meaningful descriptions of the work required. Section 2 assists the bill compiler in interpreting the requirements of the method of measurement. However, the MMHW is somewhat idiosyncratic compared with other standard methods such as SMM7, NRM2 and the Civil Engineering Standard Method of Measurement (CESMM4) as there is no requirement to:

- Include specification information in the item descriptions.
- Make reference to drawn information.
- Make reference to other data such as subsoil reports or groundwater levels.

The ensuing BQ is purely a 'neutral' list of quantified work items that must be interpreted by the contractor in conjunction with Volume 1: SHW, Volume 3: HCD and the project-specific drawings. All of these documents are intended to be used in conjunction with the Model Contract Document for Highway Works in order to arrive at a complete set of procurement documents.

Figure 8.1 illustrates how particular components of the MCHW, along with the scheme drawings, come together to form a BQ.

Each component part is integral to the process, and both the bill compiler and the contractor's estimator need to cross-reference the BQ to each of them in order to acquire the complete picture of what the contract requires.

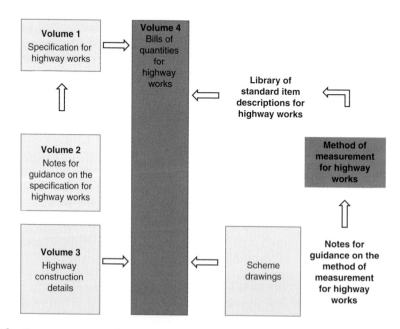

Figure 8.1   Documents required to compile a bill of quantities for highway works.

The Library of Standard Item Descriptions (LSID) is also an idiosyncratic feature of Volume 4 as the library consists of a number of 'root narratives' with numbered inserts which enable standard item descriptions to be compiled by selecting appropriate descriptions from numbered lists of variables.

In ordinary language, this means that the library has lots of standard item descriptions with numbered gaps and the gaps are filled in by selecting the appropriate description from a list according to the number of the insert. The consequence of this is that a number of similar (but nonetheless unique) item descriptions can be compiled from the one 'root narrative' which, in turn, means that the item descriptions so derived will be the same whoever compiles the description.

The 'root narrative' approach to item description ensures consistency in BQ production as, apart from 'rogue items', there is no room for individuality in describing the items. BQ may be produced manually or by using a computer package such as CATO or QSPro. The workings of 'root narratives' is explained in detail later in this chapter.

According to Money and Hodgson (1992), the MMHW was *poorly understood and often mis-used by those measuring and supervising contracts*, and this is the reason why the Notes for Guidance were reintroduced in the 1991 edition. Mitchell (2014) is also critical of the lack of understanding of users of the MMHW, and having had some experience of running practitioner training courses on the subject, this author has some sympathy with his views.

The Notes for Guidance are not particularly extensive but are sufficient to assist the bill compiler (or contractor) with some important matters of interpretation and correct practice.

8.1.6   Volume 5: Contract documents for specialist activities

Volume 5 comprises model contract documentation, specification requirements (including performance specifications), notes for guidance and methods of measurement for specialist activities in connection with highway works. This includes:

- Geodetic surveys (aerial surveys and digital mapping).
- Ground investigation.
- Mechanical and electrical (M&E) installations in road tunnels, moveable bridges and bridge access gantries.
- Trenchless installation of highway drainage and service ducts.
- Closed-circuit TV (CCTV) surveys of highway drainage systems.

The methods of measurement for ground investigation work and trenchless drainage and service ducts follow the principles of the main MMHW (Volume 4 of the MCHW), whereas other specialist activities are measured more simply. There is a model BQ for geodetic surveys, and a sample BQ to help bill compilers with the measurement of M&E work, but no complex measurement rules for either.

The nature of the work dealt with in Volume 5 is beyond the scope of this book.

8.1.7   Volume 6: Departmental standards and advice notes on contract documentation and site supervision

This volume contains standards and advice notes concerning such matters as the implementation of the Construction (Design and Management) Regulations 2015, the Use of Substances Hazardous to Health in Highway Construction, the New Roads and Street Works Act 1991, etc. These topics are beyond the scope of this book.

Part 2

8.2 Design manual for roads and bridges

As stated elsewhere in this book, the fundamental starting point for calculating quantities is something from which to measure. At BQ production stage, this usually means a set of drawings – either hard copy or digital – or a model.

The drawings for highway projects are derived from two basic sources:

1. Project-specific drawings.
2. Highway Construction Details (HCD).

The HCD are contained in Volume 3 of the MCHW, whereas the project-specific drawings come from either:

1. Engineer design.
2. Contractor design.
3. Both.

Although not part of the MCHW, the Design Manual for Roads and Bridges (DMRB) is a *companion manual* intended to guide designers of highway projects, whether they be client-engaged engineers or contractors.

Originally published as a series of separate documents by the various UK overseeing departments (e.g. the then Highways Agency in England), the DMRB was first introduced in 1992 as a comprehensive manual of loose-leaf volumes containing current standards, advice notes, references and other documents relevant to the design of trunk roads. The manual comprises 16 volumes (Volumes 0–15) dealing with design issues such as highway structures, geotechnics and drainage, road geometry and environmental assessment, etc. Some of its features include the following:

- The DMRB represents good practice that is intended for use on trunk road schemes.
- The manual may be used or adapted by highway authorities for local roads and bridges projects.
- Some documents may be restricted where not applicable to Scotland, Wales or Northern Ireland.
- Some documents may be restricted to a particular overseeing organisation.
- Designers may relax the specified design standards where thought necessary, but such relaxations must be justified to the overseeing organisation.
- Referencing to the MCHW and, in particular, to the SHW and HCD.

As well as setting standards, the DMRB provides guidance in the form of advice notes, and there is also an extensive system of administrative and technical procedures dealing with document control and the issue of new and amended documents in order that a rigorous system of quality control is maintained.

8.3 Highways England procurement

Highways England is in the business of procuring a wide variety of highway projects from relatively small highway maintenance schemes to very large and often complex improvement schemes involving technical and logistical problems not associated with 'run-of-the-mill' construction work. River and rail crossings are often involved as well as complex earthworks, ground stabilisation and geotechnical work, large bridges and other structures.

Due to the size and intrinsic value of many highway schemes, procurement of such projects warrants careful consideration and an appropriate balance of risk between the parties befitting the constraints and degree of difficulty involved.

Within the confines of the MCHW, overseeing organisations, such as local highway authorities, are at liberty to choose procurement methodologies appropriate to their own projects. For major projects costing more than £20 million, Highways England uses a variety of NEC3 contracts for project delivery within a framework arrangement. A procurement strategy is developed for each project on a case-by-case basis, and the most appropriate contract model is chosen on that basis.

In this context, the procurement strategies employed by the Highways England include:

- Frameworks.
- Early contractor involvement (ECI).
- Design and build (D&B).
- Individual (discrete) contracts.
- Managing Agent Contractor (MAC and TechMAC).
- Private finance.

8.3.1   Frameworks

Frameworks are used to procure works, goods and services over a prescribed period of time. They are generally used where there are either known repeat requirements in developed, competitive markets or where there are variable demands and requirements that are best suited to a flexible procurement arrangement.

As far as Highways England is concerned, Major Projects from 2014 are procured on the basis of a *Collaborative Delivery Framework* of pre-qualified 'suppliers' of professional design, engineering and construction services. This replaces the previous *Major Projects Framework*.

The framework comprises four separate lots, one for engineering design services and three for small-, medium- and large-value construction projects, with a total value of £4–5 billion over 4 years and a possible 2-year extension[4]:

Collaborative Delivery Framework		
Lot 1	Professional design and engineering services	8–12 consultants
Lot 2	Medium-value construction works (<£25 million)	3–5 contractors
Lot 3A	High-value construction works (£25–100 million)	4–6 contractors
Lot 3B	High-value construction works (£100–450 million)	4–5 contractors

8.3.2   Early Contractor Involvement (ECI)

The Highways England procurement strategy accommodates both ECI and late contractor involvement where it is felt that input by a contractor prior to submission to the planning process would be beneficial. Late contractor involvement permits a contractor to be hired closer to construction commencement but nevertheless before submission of the scheme to the planning process.

When a contractor is hired after planning consents have been obtained or when planning consents are not required, an NEC3-based D&B contract is used.[5]

8.3.3   Design and Build (D&B)

D&B is used for substantial projects, beyond the threshold of Agency Frameworks (see Section 8.3.1), but where the ECI approach is deemed inappropriate. Larger highway renewal schemes may warrant this approach and also those where most design decisions are fixed in advance.

Part 2

Design innovation is encouraged in order to generate time and cost savings and improved quality and value based on prior experiential learning.

8.3.4   Individual (discrete) contracts

Projects beyond the scope of frameworks, MACs and TechMACs, or where there is a specific need or requirement, are procured on the basis of an individual or discrete contract.

Such projects may involve a degree of contractor design, and the emphasis will be on delivering the appropriate balance of risk and value.

8.3.5   Managing Agent Contractor (MAC)

The preferred procurement route for maintenance works over £0.5 million is through 4-year works asset-management frameworks. The main contractual vehicle is the MAC (and TechMAC) contract on the basis of either:

- Managed works – where a number of main civil engineering contractors are given a task order for each scheme or
- Construction management – where several specialist contractors are appointed to complete a single scheme on the basis of task orders with the MAC acting as supervisor and principal contractor[6].

8.3.6   Private finance

Private finance contracts are the preserve of high-value strategic projects according to government policy using design, build, finance and operate (DBFO) contracts. Such projects are beyond the scope of the MCHW.

8.4   Measurement implications of procurement choices

The preference of Highways England is to use the NEC ECC, with its various options, and the NEC Short Form of Contract, as provided for in Volume 0: *Model Contract Document for Major Works and Implementation Requirements*.

As a consequence, the pricing documents available for use on highway schemes are as follows:

Contract	Type of contract	Pricing document
• ECC Option A	Lump sum	Activity schedule
• ECC Option B	Measure and value	Bill of quantities
• ECC Option C	Target	Activity schedule
• ECC Option D	Target	Bill of quantities
• Engineering and Construction Short Contract	Lump sum	Price list – unquantified
• Engineering and Construction Short Contract	Measure and value	Price list – quantified

Under NEC3, the use of an activity schedule/unquantified price list implies a lump sum contract, and a BQ/quantified price list implies a measure and value contract.

8.4.1   Activity schedules

Appendix H of Part 7 of MCHW Volume 0: *Model Contract Document for Major Works and Implementation Requirements* provides an illustrative activity schedule for highway works. This is shown in Table 8.1, but *Tenderers are responsible for inserting activity descriptions and stage payment requirements.*

The use of the term 'stage payment requirements' is interesting as it resonates with the ECC *Guidance notes* which state that:

- Individual activities on the activity schedule are paid for when completed.
- Grouped activities are paid for when the whole group of activities is completed which *is how the ECC provides for milestone or stage payments.*

It could be interpreted, therefore, that the invitation to tenderers to state their *stage payment requirements* in Volume 0 of the MCHW is a tacit relaxation of the provisions of the ECC. This would make a great deal of sense because the sort of activities listed in the illustrative activity schedule, including site establishment and compilation of the health and safety file, are activities that would normally take several months, if not years, to complete and could otherwise pose a considerable cash flow problem for the contractor.

Table 8.1   Illustrative activity schedule.

No.	Activity	Price
1	Establish site	
2	Topsoil strip area A	
3	Topsoil strip area B	
4	Culverts	
5	Headwalls for culverts	
6	Earthworks for slip roads	
7	Earthworks for main line	
8	Drainage for slip roads	
9	Drainage for main line	
10	Environmental bund	
11	Ducts for slip roads	
12	Ducts for main line	
13	Milling	
14	Capping	
15	Sub-base	
16	Road base	
17	Basecourse	
18	Wearing course	
19	White lines	
20	Traffic lights	
21	As-constructed information	
22	Health and safety file	
	Lump sum fixed price (£)	

Source: Manual of Contract Documents for Highway Works – Volume 0.

When used in conjunction with the ECC, the illustrative activity schedule shown in Table 8.1 would be problematic. As discussed in detail in Chapter 4, the activity schedule should reflect

Part 2

the accepted programme, but the detail provided in the illustrative activity schedule falls a long way short of the level of detail expected in the ECC.

For highway works, the contractor's programme will usually be presented in the form of a time–chainage diagram, which is more illustrative of the 'linear' nature of roadworks projects than the more ubiquitous linked bar chart. An example of a time–chainage diagram is given in Figure 8.2.

8.4.2   Bills of quantities (BQ)

BQ are traditionally used for the procurement of highway works, and the resulting contract is usually on a measure and value (or admeasurement) basis. Suitable forms of contract include:

- ICC – Measurement Version.
- ECC Option B.
- Engineering and Construction Short Contract.
- FIDIC 1999.
- ECC Option D (for a target contract based on a BQ).

The basis of the quantities for admeasurement contracts is the MMHW which places the measurement risk with the employing authority and the constructional risk with the contractor.

Consequently, if the quantities of work measured in the BQ are less than the final quantities, the overseeing organisation will 'foot the bill'. If, on the other hand, the work proves more difficult than expected, this risk lies with the contractor, but only up to the point where an experienced contractor could not have anticipated the circumstances prevailing.

Where there is some certainty about the quantity of work required, the overseeing organisation might choose to employ the Z36: *Risk transfer in bill of quantities* secondary option clause, where the quantity risk passes from the employer to the contractor.

8.5   Contractual arrangements

In keeping with other construction projects, highway schemes require an integrated and coherent infrastructure of documents in order to bring about an effective contract. The contract documents will be slightly different depending upon whether a traditional engineer-designed project is employed or whether the ECC is preferred. Table 8.2 illustrates typical contract documents for the ICE/ICC Conditions of Contract compared with their NEC equivalents.

Despite the different vocabulary between ICE/ICC and NEC contracts, it can be seen from Table 8.2 that, in each case, the specification and drawings, together with a pricing document of some sort, are employed. These documents are central to any contract, but the difference with the MCHW is the reliance on **standard documents** and especially:

- Volume 0 (Model Contract Document (MCD)).
- Volume 1 (Specification).
- Volume 3 (Highway Construction Details).
- Volume 4 (Method of Measurement).

Risk Issue

The documents contained within the MCHW are not contract documents. The *ensemble* is a means of creating a contract, but the various MCHW documents must be included in the contract by specific reference, that is, in preambles, and by ensuring that the method of measurement (i.e. the MMHW) is stated in the contract.

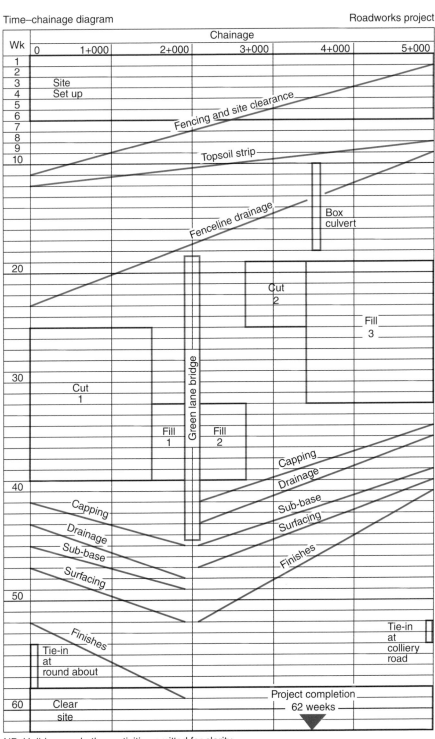

NB: Holidays and other activities omitted for clarity

Figure 8.2   Time–chainage diagram.
Figure 21.4 on page 479 of *Construction Planning, Programming and Control* by Cooke and Williams.

Part 2

Table 8.2 Comparative contract documents.

ICE/ICC	NEC equivalent
The contract conditions	• Core clauses (ECC or Short Form) • Appropriate main and secondary option clauses
The Contract Appendices Parts 1 and 2	• Contract Data Part 1 (completed by the employer) • Contract Data Part 2 (completed by the contractor)
Form of agreement (optional)	Form of agreement (optional)
The form of tender	Form of tender (not an NEC document)
Formal acceptance of the contractor's tender offer	Formal acceptance of the contractor's tender offer
The priced bill of quantities	Priced activity schedule or bill of quantities
The specification	Works Information
The drawings	Site Information

An important point to remember is that the documents contained within the MCHW have the status of 'Standards', and therefore, *any variation or waiving of a requirement contained within a MCHW document shall be regarded as a Departure from Standard.*[7]

Consequently, for contracts based on the MCHW, there must be no divergence from the standard documents unless the reason for any change is approved by the overseeing organisation. In this context, Highways England must also be consulted for information regarding suitable contracts where D&B is contemplated, and other UK overseeing organisations, including local authorities, must endorse forms of contract to be used on any highway projects under their control.

As might be imagined, there is a system within the MCHW framework for incorporating changes to the standard documents. The reasoning for a strict and hierarchical system is twofold:

- Highway projects are usually public sector projects, and as such, the overseeing organisations need to exercise a suitable degree of control.
- Clarity and consistency in the contract documents produced are necessary especially considering the need to avoid barriers to trade for contractors and suppliers from the EU.

The Introduction to Volume 0 Section 1 Part 2 DG 01/08 of the DMRB also contains formal protocols for dealing with departures from design standards.

Amendments to the MCHW are issued from time to time, and subscribers receive loose-leaf replacement pages for the various volumes as appropriate. Amendments to Volume 2 (NG SHW) will include a *Schedule of Pages and Relevant Publication Dates* for Volume 1 (SHW), and amendments to Volume 4 will include a *Schedule of Pages and Relevant Publication Dates* relating to the MMHW. Each of these schedules is to be incorporated into the contract where the overseeing organisation deems that the contract should include such amendments. Amendments to Volume 3 will similarly include a *Schedule of Pages and Relevant Publication Dates* relating to the HCD.

The MCD for the ECC and the Short Form is to be found in MCHW Volume 0 Section 1 Part 7. This provides extensive procurement documentation, including Z clauses and guidance notes, as well as various pro formas:

- Form of tender.
- Instructions for tendering.
- Illustrative activity schedule.
- Contract Acceptance Letter and Contract Agreement.
- Contract Data Parts One and Two.
- Dual-envelope tendering arrangements.
- Quality/price tender assessment and scoring provisions.

Risk Issue

The illustrative activity schedule, which is intended to form part of the Contract Data Part 2 (which is completed by the contractor) is populated with a list of 22 very broad activities such as establish site, culverts, earthworks for main line, drainage for main line, capping, sub-base, etc.

In keeping with the philosophy of the ECC, the lump sum items priced in an activity schedule become due to the contractor only when the activity is completed. If the activity is part of a group of activities, then the entire group has to be completed before payment becomes due.[8]

8.6   Specification for Highway Works

Most standard methods of measurement not only quantify the work to be done by the contractor but also describe the nature and quality of work required. Where this is the case, BQ item descriptions may be supplemented by either a bespoke or standard specification (e.g. the Specification for the Water Industry). The MMHW does not work like this as BQ based on this method of measurement contain no specification information.

Unlike other standard methods of measurement, the MMHW operates in a highly structured and controlled infrastructure along with other interlinking documents. Central to these documents is the Specification for Highway Works (SHW).

8.6.1   Introduction

Along with the drawings, the SHW is of vital importance in a highway project as it is central to a number of important stages in the administration of a project:

- Preparation of BQ in accordance with the MMHW.
- Preparation of activity schedules by tendering contractors.
- Pricing the BQ/activity schedule at tender stage.
- Interpreting the requirements of the contract as work on-site progresses.
- The administration of variations to the contract.
- The compilation and administration of contractual claims.

Before continuing, it is essential to understand that:

- The SHW is a 'Standard' and as such should not be altered without authority from the relevant overseeing organisation.
- Unlike projects in the private sector, the SHW cannot be 'bespoked' in any way that the compiler wishes.
- Where it is desired to make the Specification 'contract specific', this must be done by using the system of appendices prescribed (this is explained in detail in MCHW Volume 2: *Notes for Guidance on the Specification for Highway Works*).
- The SHW is incorporated into a contract by reference, and thus, any alterations to the 'standard' must be made clear in the contract documents.

Part 2

8.6.2   Structure of SHW

In its 'standard' form, the SHW consists of:

- 29 numbered series

  The numbered series consists of an Introduction (Series 000) and Preliminaries (Series 100) together with 27 other series such as Fencing (Series 300), Drainage and Service Ducts (Series 500), Earthworks (Series 600), Special Structures (Series 2500) and Maintenance Painting of Steelwork (Series 5000). Each series describes the quality of materials and workmanship required by the contract. Series 2200 is not used.

- 8 lettered appendices (A–H)

  The lettered appendices (A–G) to the SHW contain standard information and requirements relating to quality management, product certification and other approvals, whilst Appendix H specifies four categories of quality records (Categories A–D) that must be submitted to the overseeing organisation before, during and after completion of construction activities. All the lettered appendices are national requirements and, with the exception of Appendix F (publications referred to in the Specification), are intended to be used unaltered.

Additional and substitute clauses that appear in the numbered series or lettered appendices only apply to the overseeing organisations in Scotland, Wales and Northern Ireland, and these clauses are suffixed TS, NAW and NI for each country, respectively. These clauses replace each 'standard' Specification clause with the equivalent non-suffixed number. For instance, 601TS: *Classification, Definitions and Uses of Earthworks Materials* applies to projects in Scotland and replaces the 'standard' Series 601 specification. Additional and substitute clauses are deemed to apply unless otherwise stated in Appendix 0/5 (see later reference to this appendix).

8.6.3   Numbered appendices

Additional to the 'standard' Specification structure are the contract-specific numbered appendices which are employed to tailor the Specification to the particular contract. There are two types of contract-specific numbered appendices:

- The so-called 'zero-series' numbered appendices (0/1–0/5) which contain contract-specific information and requirements:
  - Appendix 0/1 is for incorporation of additional, substitute or cancelled clauses.
  - Appendix 0/2 is for incorporation of minor alterations to existing clauses, tables and figures and any alterations to Appendix F.
  - Appendix 0/3 contains a complete list of the contract-specific numbered appendices included in the contract:
    - List A contains those referred to in the national Specification.
    - List B contains any additional numbered appendices devised by the compiler relating to Appendices 0/1 and 0/2.
  - Appendix 0/4 contains a list of the drawings in the contract including those from the HCD.
  - Appendix 0/5 relates to special national alterations of the overseeing organisation of Scotland, Wales or Northern Ireland.
- Numbered appendices such as Appendix 1/10 and 1/11 which refer to contractor design of permanent structures and to contractor design of temporary works and temporary structures, respectively.
- Other numbered appendices beginning with 1/1 that are referenced to the appropriate numbered series. Therefore, numbered appendix 16/5 relates to Series 1600: *Driven Cast-in-Place Piles*.

The SHW Notes for Guidance suggest that the information contained in the contract-specific numbered appendices 0/1 to 0/5 and 1/1 onwards should be bound in one or more volumes in the tender/contract documents with a clear title indicating what they relate to.

8.6.4   Contractual issues

The SHW has been drafted to be 'contract neutral' in order that a wide variety of procurement methods and forms of contract may be used. Consequently, terms specific to particular forms of contract have been avoided, and no references are made to approvals and/or directions by parties (such as the engineer) that might appear in particular forms of contract.

For a typical highway project, the Specification is incorporated into the contract by specific reference, but it is not reproduced as a contract document either in whole or in part. Consequently, tendering contractors must have access to a copy of the SHW in order to be able to price the job – available online at http://www.dft.gov.uk/ha/standards/mchw/ free of charge. However, even if a contractor has the latest published version of the SHW, this is not necessarily up to date or concurrent with the specific requirements of a particular contract. This is because:

- Amendments to the SHW may be issued between the last publication date and the date of any specific contract.
- There may be contract-specific changes to the SHW for a particular project.

In order to ensure that contractors are pricing a tender based on the latest relevant Specification clauses, the *Schedule of Pages and Relevant Publication Dates* issued with the most recent update must be included in the contract. This will usually be done as a preamble, and the Notes for Guidance (NG Series 000) provide a *Standard Preamble to the Specification* in order to assist compilers to fulfil this requirement.

The *Schedule of Pages and Relevant Publication Dates* lists the publication date of each page of the SHW applicable to the contract, and the current schedule is included in Series 000 of the SHW. The Schedule of Pages should be bound into the contract documents, with the contract-specific numbered appendices, following the Preamble to the Specification.

The contract-specific numbered appendices to the Specification are drawn up for the contract by the compiler and are also included in the documents. Cross reference should be made in the contract-specific numbered appendices to any relevant drawing number(s) that might be appropriate.

It is important to note that the contract-specific Numbered Appendices 1/1 onwards should only be used to extend the information in Specification clauses, tables or figures and *not* to change them and they must not be used to alter the 'standard' SHW clauses. Some numbered appendices are partially drawn up by the complier and completed by the contractor, whilst others must be completed by the contractor and returned to the overseeing organisation.

Appendix 0/3 contains a complete list of numbered appendices to be included in the contract. This list is in two parts and comprises national specifications (List A), and numbered appendices (List B) put together by the bill compiler.

Part 2

Risk Issue

Series 005/1 is important in terms of measurement and pricing in that it states that the thickness of items specified are intended to be the finished or compacted thickness.

8.7   Method of Measurement for Highway Works

As can be seen in Part 8.6 of this chapter, the SHW is much more than a 'static' standard specification, the like of which are common in the construction industry. The SHW is not only regularly amended via the *Schedule of Pages and Relevant Publication Dates* system, but it is also capable of being adapted to be 'contract specific', albeit within controlled limits, using the system of numbered appendices.

The same system of publishing a *Schedule of Pages and Relevant Publication Dates* applies to the MMHW. There is no system of numbered appendices, but great reliance is placed on the Preambles to Bill of Quantities wherein may be found any amendments to the Method of Measurement.

Risk Issue

Where applicable, the Method of Measurement for Highway Works will be the stated method of measurement in the contract. This information will be found in the Contract Appendix (ICE/ICC conditions) or in the Contract Data (NEC contracts).

However, the contract will not state the relevant edition of the MMHW (e.g. unlike CESMM, Fourth Edition) as this information is found the *Schedule of Pages and Relevant Publication Dates*.

There are downsides to this MCHW system because:

- Tendering contractors have to be alert to published changes to the 'standard' documents in their possession.
- It takes time for tenderers to scrutinise the Schedule of Pages to ensure that they are using the correct specification information.
- The tender period is a high-pressure time for contractors, and it is easy to overlook important changes to the specification.
- Accurate pricing of the BQ is heavily dependent upon having the correct specification requirements to hand.

Risk Issue

Checking published changes to the SHW is particularly important as the Method of Measurement for Highway Works does not provide for the inclusion of specification information in the item descriptions.

The importance of MCHW Volume 1: *SHW* and Volume 3: *HCD* to Volume 4 MMHW cannot be overemphasised.

8.7.1   MMHW structure and contents

The MMHW is to be found in Section 1 of Volume 4 of the MCHW – *Bills of Quantities for Highway Works*. It comprises four chapters:

(i)   Definitions.
(ii) General Principles.
(iii) Preparation of Bill of Quantities.
      Preambles to Bill of Quantities.
      Schedules of Pages and Relevant Publication Dates.
(iv) Units and Method of Measurement.

BQ prepared under the MMHW are organised along traditional lines:

- Preambles.
- Main measured Work Sections.
- PC and provisional sums (where used).
- Final summary.

However, unlike other methods of measurement that tend to subdivided into 'Work Sections', the various measured sections of the MMHW are categorised into 'series' such as Series 500: *Drainage and Service Ducts* and Series 600: *Earthworks*. The various 'series' are to be found in Chapter IV of the MMHW.

The importance of the Preambles to the Bill of Quantities cannot be overstated. This is where some very important directions, rules and information are to be found that are of central importance to the contractor's pricing of the tender and to the interpretation of issues that might arise during construction of the works.

8.7.2   Preparation of BQ

Format and contents

It should not come as a complete surprise that the format and contents of BQ prepared using the MMHW is strictly controlled. This even extends to the format of BQ pages which must follow a standard layout as determined by Implementation Standard SD 3/92 (Annex A). If a particular word-processing package is unable to achieve the correct layout, approval to use a different layout must be sought from the relevant overseeing organisation.

Without going into unnecessary detail, BQ are structured using three levels as illustrated in Table 8.3. This structure is delineated in Table 1 of Chapter III of the MMHW wherein the main divisions (Level 1) of the BQ are listed as:

  (i)   Preliminaries.
 (ii)   Roadworks.
(iii)   Structures.
(iv)   Structures Where a Choice of Designs is Offered.
 (v)   Structures Designed by the Contractor.
(vi)   Service Areas.
(vii)   Maintenance Compounds.
(viii)   Accommodation Works.
 (ix)   Works for Statutory or Other Bodies.
 (x)   Daywork.
(xi)   PC and Provisional Sums.

Each level is further subdivided as necessary, for example:

(ii)   Roadworks:
   ◦   Roadworks General.
   ◦   Main Carriageway.
   ◦   Interchanges.
   ◦   Side Roads.
   ◦   Signs, Motorway Communications and Lighting.
   ◦   Landscape and Ecology.

It can be seen that this layout is completely different to that used in CESMM4, for example, which employs a classification system based on classes of *work commonly encountered in civil engineering* which are then subdivided into three divisions. The classification system in the

MMHW is almost 'elemental' in nature albeit that the itemisation system used in the various series is perhaps a combination of the two approaches.

Table 8.3   Structure of bills of quantities.

Level 1 Division		Level 2 Construction heading	Level 3 MMHW series heading	
Division	**Subdivision**			
(iii) Structures	Structure in form of bridge or viaduct; name or reference	Special preliminaries		
		Piling	1600	Piling and embedded retaining walls
		Substructure – end supports	500	Drainage and service ducts
			600	Earthworks
			1100	Kerbs, footways and paved areas
			1700	Structural concrete
			1800	Structural steelwork
			1900	Protection of steelwork against corrosion
			2300	Bridge expansion joints and sealing of gaps
			2400	Brickwork, blockwork and stonework
		Substructure – intermediate supports Substructure – main spans Substructure – approach spans		As for end supports

Quantities

Quantities are presented in the eventual BQ in whole units except for those items measured in tonnes or hectares which are given to three places of decimals.

Special preliminary items

A peculiar feature of the MMHW is the use of the term 'Special Preliminary Items'. These items are included in the BQ purely at the discretion of the overseeing organisation, but, unusually, there is no provision for them in the method of measurement. Additionally, there is no obligation to include such items in the BQ, and, in any event, their inclusion or exclusion does not relieve the contractor from his obligations under the contract.

The point of Special Preliminary Items is that, for a particular contract, the cost of preliminaries associated with certain work may be considered as disproportionate to the value of the work in question. Consequently, even though such work is not required to be measured separately, a case may be made that the contractor's preliminaries associated with that work merit special provision in the BQ. Should they be included, Special Preliminary Items must be *adequately covered in the documentation*, and the compiler must also ensure that a suitable item coverage is included.

Alternative types of pavement

It has long been the case with the MMHW (and its predecessor the MMRB) that contractors are offered a choice as to the type of pavement that they would prefer to employ. This was always a question of the relative economics of different pavement solutions, and this led to the existence of both macadam and concrete road surfaces on Britain's motorway network. In the 1970s especially, some contractors preferred concrete carriageways as they owned the necessary 'concrete trains' used for their construction and others preferred to sublet the surfacing work to 'tarmac' subcontractors. The eventual decision as to the pavement to be used was the overseeing organisation, but tenderers would invariably price the contract so that significant savings would emanate from their preferred pavement type.

Structures where a choice of design is offered

On occasion, the contractor is given the choice between pricing his tender on the basis of a structure designed by the overseeing organisation or one of his own designs. This may be for a small bridge, underpass or culvert, for example.

In such cases, the BQ must contain a bill for each alternative both prepared in accordance with the method of measurement. The difference between the bills is that one is measured and itemised in accordance with the appropriate series of the MMHW with the other, containing a single item only, being prepared in accordance with Series 2500: *Special Structures*. If the contractor decides to design the structure himself, then he prices the bill with the single item. Irrespective of the contractor's choice, only one of these bills may be used to compile the total of all the bills that make up the tender figure.

Structures designed by the contractor

In circumstances where the overseeing organisation wishes the contractor to design a structure(s), the drawings will show a designated outline of where the work is located. The contractor will price all the work included within the designated outline with the exception of any work scheduled by the overseeing organisation as *not to be included*.

For such work, the BQ will contain a bill with a single item measured in accordance with Series 2500: *Special Structures*. Work not included in the single item is to be included in other bills prepared by the overseeing organisation.

Landscape and ecology

This type of work, covered by Series 3000, includes all sorts of seeding, planting, pruning and maintenance of hedgerows and trees along with the management of wildlife species, aquatic plants and wildlife and the construction of wildlife tunnels and underpasses and so on.

Where required by the contract, this work is to be measured in a separate bill within the Roadworks bill. Payment for some of this work will be made in stages, and this is to be made clear in the contract documents.

Part 2

8.7.3   Definitions

MMHW Chapter I is a 'must-read' chapter because it contains definitions of terms used in the MMHW that may either be unusual or not the normally accepted meaning. For instance:

- **Definition 1(c)**
  'Bill of Quantities' *means a list of items giving **brief identifying descriptions** and **estimated quantities** of work comprised in the execution of the works to be performed.*
  The BQ does not, therefore, describe the nature and extent of the work required which is the function of the drawings and the specification.
- **Definition 1(d)**
  Items designated 'Provisional' *are items for which the quantities of work to be executed **cannot be determined with the same degree of accuracy as other items** but for which it is deemed necessary to make provision.*
  Therefore, despite the fact that the BQ items are ***estimated quantities***, and thus subject to admeasurement, provisional quantities are deemed to be less accurate than 'estimated' quantities.
- **Definition 1(h)**
  'Hard Material' means the following:

  (i)   *material so designated in the Preambles to Bill of Quantities; and/or*
  (ii)   *material which requires the use of blasting, breakers or splitters for its removal but excluding individual masses less than $0.20\,m^3$;*

  This definition has changed over the years but is of profound significance as it determines whether or not the contractor will be paid additional monies for excavating materials that might be called 'more difficult than the norm'. The definition is further developed in MMHW Chapter III – *Preambles to Bill of Quantities*.
- **Definition 1(k)**
  'Designated Outline' means the designated outline shown on the Drawings.
  This unusual term is used in the MCHW to delineate parts of the works, such as structures or culverts, which are to be designed by the contractor or where a choice of designs is offered for contractors to choose from. Designated Outlines have implications as regards both measurement and pricing. This term should be distinguished from the term 'Earthworks Outline' which is defined in Series 600: *Earthworks*.

8.7.4   General principles

In order that tenderers are able to prepare a reliable and accurate tender and that the employer may depend upon the completeness of the price without fear of claims and cost overrun on the contract, certain basic principles are set out in Chapter II of the MMHW.

The first of these principles is that the MMHW is used for highway contracts using any form of contract. Secondly, unlike other standard methods of measurement, the nature and extent of the work to be performed is not contained in the BQ but is to be ascertained by referring to the drawings, specification and conditions of contract. Thirdly, the work covered by the respective items in the BQ is identified both by the various BQ subheadings and item descriptions **and** by the list of matters contained within the relevant 'Item coverage' marginal heading in the appropriate series of the method of measurement.

A further important principle is that tenderers are provided with full details of the construction requirements of the contract, which, in practice, means that:

- The tender documents are based upon a completed design.
- The design work is based on a complete site investigation.
- The full extent of work required by the contract is shown on the tender drawings.

The various parties to highway contracts will view these ideals with a degree of cynicism as they are rarely achieved in practice. Mitchell (2014) is especially critical of the standard of tender documentation and of the *cavalier approach* of compilers to their work.

The MMHW is based on the SHW and on the HCD published as Volumes 1 and 3, respectively, of the MCHW. However, amendments to the MMHW are sometimes required where additions or amendments to the SHW or the HCD are not adequately covered by the prevailing provisions of the method of measurement. Such amendments are accommodated in the Preambles to the Bill of Quantities as provided in MMHW Chapter III.

As far as BQ item descriptions are concerned, the guiding principle is that they are to be compounded from features contained in one or more of the groups listed in each series of Chapter IV of the Method of Measurement. The groups and features are to be found under the relevant marginal heading 'Itemisation'. *An item description may contain Features from as many Groups as necessary to identify the work required, but may include only one Feature from any one Group.*

8.7.5   Preambles to bill of quantities

Whilst the MMHW does not contain any specific 'rules of measurement', the 'standard' preambles that appear under the heading *Preambles to Bill of Quantities* in Chapter III come very close to being 'rules' without using that particular word.

> **Risk Issue**
>
> Being incorporated into the bill of quantities, the preambles assume contractual importance because the bill of quantities is usually a contract document.

Chapter III: *Preparation of Bill of Quantities* of the MMHW makes particular reference to the heading *Preambles to Bill of Quantities* which contains a number of matters that must be specifically included in the BQ as preambles. This includes several directions and statements that are important with regard to the contractor's pricing of the tender. In addition to the 'standard' preambles, any additional matters must be included as 'numbered' preambles, and any amendments to the method of measurement must also be listed.

There are 19 'standard' preambles with number 20 reserved for listing any amendments to the method of measurement. Some of the more important preambles are considered below:

Preamble 1

This states that the method of measurement shall be the MMHW, but, as previously mentioned, reference has to be made to the *Schedules of Pages and Relevant Publication Dates* to discover the method of measurement page number and date of publication that shall apply to the contract.

Preamble 2

Preamble 2 states that *the nature and extent of the work is to be ascertained by reference to the Drawings, Specification and Conditions of Contract*. This emphasises that the measured items in the BQ is purely a quantified list of items of work to be done and is not a complete description of the contractor's obligations under the contract.

Part 2

Also in Preamble 2 is the first of what might be called 'measurement and/or coverage rules'. This informs the contractor as to what is deemed to be included in the rates and prices. This includes those that might usually be expected but also others that merit careful attention:

- Usual items
  - Labour, materials and plant including associated oncosts and the cost of fixing, erection and placing in position, waste, testing and checking, etc.
  - Establishment charges, overheads and profit although the distinction between establishment charges and overheads is not made.
- Unusual items
  - Significantly, temporary works are deemed included in the rates and prices because there is no separate provision in the MMHW (unlike CESMM).
  - The impact of anything *set forth or reasonably implied* in the tender documents on the phasing of the works or any element thereof.

Risk Issue

The MMHW makes no provision for method-related charges, either in the preliminaries or elsewhere. Such work is deemed included in the rates and prices.

Whilst Series 100: *Preliminaries* contains relatively few items, most contractors price temporary works and method-related charges here as there is both more opportunity to influence cash flow by doing so and such tactics reduce the risk of underpayment should the admeasurement process not work advantageously.

It is clear that Preamble 2 has implications for the contractor's programme, but specific constraints or requirements of the contract must be stated in Appendix 1/13 of the Specification.

Preamble 3

In this preamble under the heading of 'Measurement', the stipulation is made that *work shall be computed net from the dimensions stated in the Contract* unless otherwise indicated in the method of measurement. It is further stated that any work affected by 'pavements' shall be measured on the basis of the thinnest pavement construction. Other 'rules' are stipulated that relate to alternative specified designs, materials and options within types of pavement where the contractor is given a choice.

Preamble 4

This requires a rate or price to be stated against each item in the BQ. The use of 'nil' or 'included' is thus not acceptable, but a rate of £0.00 would be.

Preamble 6

This preamble raises the 'thorny' issue of public utilities such as gas, electricity and telephones and their infrastructure. The perennial question is, 'who is responsible for finding them and dealing with their support and protection during the works?'

In this preamble, the contractor is clearly reminded that it is his obligation under the contract despite any information that might be furnished in the tender documents. Therefore, whilst this information *is believed to be correct*, it is nonetheless down to the contractor to locate and deal with any existing services and to arrange with the appropriate utility company where existing services need to be interrupted. The preamble is silent on the issue of diverting any existing services that the contractor might have located during the works.

Preamble 9

Preamble 9 is important because it deals with the issue of water – both tidal and non-tidal open water. Dealing with existing flows of water and sewage is covered in Preamble 10.

Unlike CESMM, where excavation below a body of open water is measured, under the MMHW it is not. Additionally, no provision is made in the MMHW for the separate pricing of sheet piling, cofferdams and the like, whereas under CESMM the contractor has the opportunity to price a separate method-related charge.

Risk Issue

Contractors should be aware that their rates and prices are deemed to include *taking measures required* to work within and below tidal water and non-tidal open water as well as undertaking any investigations into water boundaries, levels or tidal and non-tidal ranges.

Preamble 10

Here, the MMHW states that the contractor must include for *taking measures to deal* with existing flows of water and sewage and the like in the rates and prices. This would include diverting flows, sandbagging and pumping as necessary. Again, this preamble is to be contrasted with CESMM where method-related charges would be provided in the BQ.

Preamble 13

Of prime significance in the *Preambles to Bill of Quantities* is Preamble 13 as this concerns the often contentious issue of Hard Material – formerly 'rock'.

Hard material is measured 'extra over' (EO) excavation because of the relative cost of removing such material, if encountered. Hard material is only measured in Series 300: *Fencing*, 500: *Drainage and Service Ducts* and 600: *Earthworks* although it is included in the item coverage in a number of other series.

Risk Issue

In series where the excavation of hard material is included in item coverages, but is not measured in that series in the bill of quantities, the contractor's rates and prices must include for the excavation of hard material where its presence can be reasonably inferred from the contract documents.

This is not to say that the contractor will not be paid for excavation in hard material if unexpectedly encountered but that there will be no measured item and no admeasurement.

Where the contractor believes that entitlement is due the matter will be subject to a 'claim' or 'compensation event'. The terms of the contract, and whether or not the contract documents were clear, will be key, and a great deal will also depend upon the ground investigation reports and the contractor's own site investigations.

Albeit hard material is subject to admeasurement in Series 300, 500 and 600, it is incumbent upon the overseeing organisation to ensure that the quantities of hard material included in the BQ represent, as accurately as possible, the circumstances that the contractor is likely to encounter as this has a direct impact upon his choice of construction method and pricing.

Part 2

Hard Material is defined in MMHW Chapter I *Definitions* (Definition 1(h)) which states that it is:

(i)   *material so designated in the Preambles to Bill of Quantities; and/or*

(ii)   *material which requires the use of blasting, breakers or splitters for its removal but excluding individual masses less than 0.20 m^3;*

The definition of Hard Material in Definition 1(h)(i) is expanded in Preamble 13 to include:

a)   *....... strata[9];*
b)   *those deposits designated by limits shown on the Drawings;*
c)   *existing pavements, footways, paved areas (but excluding unbound materials) and foundations in masses in excess of 0.20 m^3.*

a)   Strata will be identified as granite, chalk, limestone and so on according to perusal of the ground investigation data.
b)   Deposits designated by limits shown on the drawings will be interpolated from borehole records which will enable an 'assumed line' of hard material to be drawn.
c)   Existing pavements, footways, paved areas, etc., whether buried or at surface level, should be shown on the drawings or, alternatively, will be obvious once encountered on-site.

This topic is further developed later in this chapter.

Preamble 15

Where parts of the permanent works are to be designed by the contractor, the rates and prices shall include for the costs of incorporating the design into the works. This includes the cost of design preparation, drawings and other data, certification, testing and sampling, etc. The contractor also has to allow for the time and cost implications of submissions of designs for approval and for any modifications and resubmissions necessary.

Preamble 16

For each priced BQ containing a single item with respect to structures to be designed by the contractor, the contractor must subsequently prepare a *priced schedule of quantities* for the structure in question showing quantities, rates and prices totalling to the single figure stated in the priced BQ submitted at tender stage. The contractor's priced schedule must be prepared in accordance with the MMHW.

The reason for this requirement is so that the post-contract issues of payment and valuation of variations may be correctly administered and for no other reason.

The same rules apply should the contractor decide to design a structure himself as opposed to one designed by the overseeing organisation.

The contractor shall include in the priced schedule of quantities all the parts of the works within the designated outline with the exception of those designed and scheduled by the overseeing organisation as not to be included.

Preamble 18

Highway works are subject to a great deal of testing both to verify workmanship, goods and materials incorporated into the permanent works and to test the permanent works in order to prove the overseeing organisation's design. The contractor's rates and prices are to include such tests as determined by Preamble 2(x).

Testing of existing structures and other investigative works shall be individually measured within the relevant series.

Should procedural trials, trial panels and trial areas be required in advance of the permanent works, they shall be included in the item coverage for the relevant series. The trial erection of structural steelwork is measured separately in accordance with Series 1800.

Preamble 19

Where works of Landscape and Ecology are required, the rates and prices in the BQ for new planting, seeding and turfing under Series 3000 shall include for all post-planting maintenance work required by the contract documents.

A *Staged Payments Schedule* is to be inserted in the Bill of Quantities immediately preceding the collection page for Landscape and Ecology that shall be used for assessing payments due. This provision is intended to reflect the scope and duration of the planting and post-planting requirements of the contract which differ from the normal run of highway works.

Preamble 20

Pursuant to Paragraph 1(b) of Chapter II *General Principles* of the MMHW, additions or amendments to the SHW or the HCD which are not adequately covered by the Method of Measurement require suitable amendments to be made. Preamble 20 is reserved for this purpose in order to bring such amendments to the Method of Measurement into a particular contract. The MMHW is thus deemed to be amended in accordance with the pages that appear in this preamble.

Part 2

8.7.6   Chapter IV: Units and method of measurement

Volume 4 of the Manual of Contract Documents – *Bills of Quantities for Highway Works* – consists of three sections. The first two – *MMHW* and *Notes for Guidance on the Method of Measurement for Highway Works* – must be read in conjunction with each other in order to understand how the method of measurement is intended to work, and the third, *LSID*, is indispensable when forming descriptions of the work to be measured.

This is an unusual arrangement as most standard methods of measurement, to a large extent, combine all three in the one document. The CESMM, for instance, combines Volume 4 Sections 1 and 3 in the three divisions of each class of measured work and in the additional description rules, whilst Section 2 is to be found in the accompanying measurement and definition rules.

To a certain extent, this structure creates a problem with regard to defining the prevailing Method of Measurement in the Appendix to the Form of Tender (ICC – Measurement Version) or Contract Data (ECC with Options B and D) as the Method of Measurement is separate from the other two sections whilst, at the same time, being entirely reliant upon each of them. This distinction is made clear in Volume 4 Section 1 Chapter III: *Preambles to Bill of Quantities* which states that:

> The Bill of Quantities has been prepared in accordance with the Method of Measurement for Highway Works published by The Stationery Office as **Section 1** of Volume 4 of the Manual of Contract Documents for Highway Works.

From this, it is evident that Sections 2 and 3 are not intended to be contractual, and Mitchell (2014) confirms this view.

Risk Issue

Disputes concerning interpretations of the MMHW or the correct formulation of item descriptions could well arise between overseeing organisations and contractors who may find themselves disagreeing over whether or not non-contractual documents should be relied upon to clarify a 'grey area' of measurement detail.

8.7.7   Series

Rather than the usual 'work sections', the various measured sections in Chapter IV of the MMHW are categorised into 'series'. These series are numbered to coincide with the numbering system employed in Volume 1: *SHW*.

Series 100: *Preliminaries* is followed by Series 200: *Site Clearance* and culminates with Series 5000: *Maintenance Painting of Steelwork*. Other series include Series 500: *Drainage and Service Ducts*, Series 600: *Earthworks* and Series 1600: *Piling and Embedded Retaining Walls*.

Some series, such as 800, 900 and 1000, are 'not taken up' which means that the equivalent series in the SHW are not measured. This is because these series are purely 'specification' series and do not contain work items as such. Other series 'not taken up' are 2200, which is 'not used' in the SHW and 2600: *Miscellaneous* which contains specifications for bedding mortar, types of concrete and fencing which are either measured in other series or are included within 'item coverages' in other series.

Each series has a title (e.g. Series 1200: *Traffic Signs and Road Markings*) followed by a number of main headings which represent work to be measured (e.g. Traffic Signs, Remove from Store and Re-erect Traffic Signs, Road Markings and Road Studs, etc.). Each main heading has three subheadings (marginal headings) – that is, Units, Measurement and Itemisation. Under the marginal heading 'Itemisation', there is a list of separate items that shall be provided in the BQ. These items are categorised into discrete 'groups', and each group has at least one feature.

8.7.8   Groups

Each series in the MMHW is hierarchical which means that, for instance, *Traffic Signs* is a subset of Series 1200: *Traffic Signs and Road Markings*. *Traffic Signs* is then divided into seven groups (I–VII) which enable item descriptions to be built up for different types of traffic signs of different sizes and having different types of support structure. Each group has one or more features, but an item description can only contain one feature from each group.

8.7.9   Features

Group I of *Traffic Signs* has two features:

1. *Permanent traffic signs.*
2. *Prescribed temporary traffic signs.*

These two features could clearly not be combined in the same item description as a traffic sign could not be both permanent and temporary!

Within the marginal heading *Permanent traffic signs* is the 'item coverage' for that feature which lists what is deemed to be included in a BQ item for *Permanent traffic signs*.

8.7.10   Item coverage

Item coverage is a list of items that are deemed to be included in any item description to which they relate. This is illustrated in Table 8.4 which shows the item coverage for a Series 600: *Earthworks* extra over item of excavation in hard material.

Table 8.4   Item coverage.

Extra over excavation for excavation in hard material	**23** The items for extra over excavation for excavation in Hard Material shall, in accordance with the Preambles to Bill of Quantities General Directions, include for:
Item coverage	a. Preliminary site trials of blasting b. Blasting, splitting, breaking and the like c. Cutting through reinforcement d. Saw cutting and trimming e. Removal of existing paved areas by course or layer, cleaning surfaces, milling or planing, stepping out and treatment to bottoms of foundations

Part 2

Risk Issue

It is important to remember that all the items in the item coverage 'list' are not necessarily required for a specific item on a specific contract. This is because it is the specification and drawings for that contract which determine the work required for that contract. This, in turn, determines the items that are deemed to be drawn from the list for that specific item of work.

Consequently, for the example shown in Table 8.4, the contractor must allow in his rates for cutting through reinforcement but only when the presence of reinforced concrete is indicated in the borehole logs or on the drawings.

It will be noted that the item coverages of the various series of the MMHW do not include the provision of labour, materials, plant, overheads and profit, etc.

Consequently, item coverages must be read in conjunction with MMHW Chapter III: *Preambles to Bill of Quantities*, General Directions, Paragraph 2, which provides a generic item coverage that applies to all items. This paragraph, coupled with a specific item coverage, makes the contractor's rates 'fully inclusive' of labour, material, plant, overheads and profit, etc.

Risk Issue

General Directions Paragraph 2 also includes Temporary Works and the effect on the phasing of the works or any element of the works *to the extent set forth or reasonably implied in the documents on which the tender is based*.

'Temporary Works' is not defined in paragraph 1 of MMHW Chapter I: *Definitions*, and care needs to be exercised when interpreting the specification, drawings and conditions of contract for a specific project as to what exactly the contractor is to deemed to provide.

The provision and removal of haul roads and the order and sequencing of operations, for instance, are not included in the item coverage for earthworks neither are there any method-related items measured in the Series 100: *Preliminaries* in the MMHW.

8.7.11 Written short item coverage

In a number of series of the MMHW, the item coverage is 'written short' either in whole or in part to varying extents. This means that the item coverage for a particular item of work refers to the item coverage of another item elsewhere for the detail as to what is deemed to be included in the item in question.

An instance of this appears in Series 1600: *Piling and Embedded Retaining Walls*, Paragraph 28, with respect to the item coverage for reinforcement for cast-in-place piles which shall include for:

a) Reinforcement (as Series 1700 Paragraph 26).
b) Bending projecting reinforcement.

The written short item (a) is expanded in Series 1700: *Structural Concrete* Paragraph 26 which states that items for reinforcement shall include for:

a) Cleaning, cutting and bending.
b) Binding with wire or other material.
c) Supports, cover blocks and spacers (except for steel bar supports to reinforcement where shown on the drawings).
d) Extra fabric reinforcement at laps.
e) Welding.
f) Mechanical connections.

Risk Issue

The principle that an inclusion in the item coverage only applies where stated in the specification or shown on the drawings applies to written short item coverages as well.

8.7.12 Units of measurement

The unit of measurement to be used for specific work items is given under the marginal heading 'Units' for each subheading of each series. Therefore, *Traffic Signs*, being a subheading of Series 1200: *Traffic Signs and Road Markings*, has a given unit of measurement which is stated as 'number'.

Approved units of measurement are given in Chapter III: *Preparation of Bill of Quantities* of Volume 4 of the MCHW which also lists the abbreviations to be used. Most units and abbreviations are those commonly used in other methods of measurement, but some are more unusual because they relate specifically to roadworks (e.g. vehicle week/v. week and vehicle day/v. day).

Chapter III also requires that *Quantities shall be expressed in whole numbers except for units of measurement of tonnes and hectares in which case the quantities shall be to three decimal places.*

8.7.13 Measurement rules

Unlike other standard methods of measurement, the MMHW does not have extensive 'measurement rules' in the sense that a great deal of prescription is given to what shall and shall not be measured and how. Such measurement rules as there are in the MMHW are to be found under the marginal heading 'Measurement' in each series, and they tend to be simple in nature. For example:

▪ *The measurement of precast copings, capping units, plinths and the like shall be the measurement along the centre line.*

- *The measurement shall be the area of formwork which is in contact with the finished concrete but measured over the face of openings of 1 m² or less and features described in (c) below.*

Whilst some series have fairly extensive 'measurement rules' (e.g. Series 600: *Earthworks*), others are minimalistic as the MMHW tends to place greater reliance on 'item coverage' and the drawings in order to convey what is 'ruled in' and 'ruled out' of an item of work. This is backed up by the specification together with any project-specific numbered appendices that might be included in the documentation.

There may well be dangers in this approach to measurement from a claims point of view and Mitchell (2014) is critical that a fundamental principle of measurement – clarity – is lost by virtue of the lack of measurement rules. Others might argue that less adversarial approaches to contracting mitigate against the need for extensive measurement rules and that a more 'crude' approach to measurement is the future for the 'BIM world' of a modern construction industry.

8.7.14   Relationship with contract

In common with other methods of measurement, the MMHW is not a contract document. Standard practice is to incorporate a method of measurement in the contract by specific reference, and this is how it works with the MMHW.

For instance, Clause 57 of the ICC – Measurement Version states that the method of measurement shall be the CESMM Third Edition 1991 *or such later or amended edition thereof as may be stated in the Appendix to the Form of Tender*. In the Appendix – Part 1, footnote 'e' requires *any amendment or modification adopted if different to that stated in clause 57* to be inserted in the space provided.

In the ECC (Options B and D), the method of measurement is that stated in the Contract Data Part 1 along with any amendments thereto.

8.8   Item descriptions

The modern trend in methods of measurement such as NRM2 and CESMM is towards brief item descriptions supplemented by drawings and other documents. This is the approach taken by the MMHW, but brevity is taken to a new level in this method of measurement. The absence of specification information in item descriptions contributes significantly to this because MMHW billed items do not attempt to fully describe the nature and extent of the contract works contrary to the approach taken by other methods of measurement.

MMHW item descriptions are derived in a different way to other approaches to measurement and instead rely upon:

- Groups and features listed in each series of Chapter IV of the MMHW.
- 'Root narratives' taken from the LSID.
- 'Item coverages' listed under each subdivision of each series of the method of measurement.

8.8.1   Groups and features

The first port of call for compiling item descriptions is the marginal heading 'Itemisation' under each subheading of each series of the method of measurement. This is where the 'groups' and 'features' reside which make up the hierarchy or 'family tree' of MMHW item descriptions.

A specific BQ item may comprise any number of groups required, but not more than one feature from each Group may be used in any one item description. In Table 8.5, it can be seen

Table 8.5 Itemisation of headwalls and outfall works.

Group	Feature	
I	1	Headwalls
	2	Revetments
II	1	Different types
III	1	Different materials
IV	1	Pipe not exceeding 100 mm internal diameter
	2	Pipe exceeding 100 mm but not exceeding 300 mm internal diameter
	3	Pipe exceeding 300 mm but not exceeding 600 mm internal diameter
	4	Pipe exceeding 600 mm but not exceeding 900 mm internal diameter

that the itemisation of **Headwalls and Outfall Works** in Series 500: *Drainage and Service Ducts* is categorised in four Groups.

Group I of the itemisation comprises two features, **Headwalls** and **Revetments**. Headwalls are structures used to terminate land drainage pipes and culverts at their inlet and outfall in order to minimise erosion of the banks of watercourses. Revetments are similarly used to line watercourses in order to prevent erosion especially at times of flooding or heavy seasonal rains.

Headwalls come in different shapes and sizes and are constructed from different materials with different diameters of inlet and outlet pipes. Consequently, further itemisation is needed to create a clear and comprehensive item description. This is provided in Groups II–IV.

Whilst the itemisation provided in the method of measurement could be used to construct an item description, it can be seen that the eventual description would lack consistency as each bill compiler would undoubtedly use different phraseology. Standardisation is provided by the LSID which is Section 3 of Volume 4 of the MCHW. The basis of this standardisation is the 'root narrative'.

8.8.2   Root narratives

The LSID is organised in 'series' that correspond to those contained within the MMHW itself. This is a master library which creates directly comparable BQ whether produced by manual billing or computer software packages.

Item descriptions based on the LSID are drawn from standard item descriptions that contain numbered inserts. These standard item descriptions are called 'root narratives'. By using a numbered variable from the appropriate numbered group, unique item descriptions can be produced for all standard work on a highway project.

Within each series of the LSID, the applicable root narratives are firstly listed, and these are then followed by a list of available 'variables'.

The root narrative for a headwall, for instance, is provided by Item 27 of the LSID. Headwalls are a subset of the **Headwalls and Outfall Works** subsection of Series 500: *Drainage and Service Ducts*. An extract from the relevant section of the library is illustrated in Table 8.6. For a revetment, Item 28 would be used.

Table 8.6   Root narrative.

Headwalls and outfall works		
Item	*Root narrative*	*Unit*
27	Headwall 22*23* to pipe 24*	No
28	Revetment 22*23* to pipe 24*	No

Part 2

It can be seen in both cases that the root narratives contain three variables, 22*, 23* and 24*. These are listed in the LSID as Group 'variables' as shown in Table 8.7. Each variable allows the bill compiler to create a standard item description whilst retaining consistency of phraseology. The type (i.e. design configuration) of headwall or revetment will be shown on a drawing or standard detail, for example, Headwall Type A, B, C, etc.

Table 8.7   Group variables.

Group	Variable	
22*	(i) etc.	=[stated type]
23*	(0)	=no entry
	(i)	=in brickwork
	(ii)	=in mass concrete
	(iii)	=in reinforced concrete
	(ii) etc.	=[stated material]
24*	(i)	=not exceeding 100 mm internal diameter
	(ii)	=exceeding 100 mm but not exceeding 300 mm internal diameter
	(iii)	=exceeding 300 mm but not exceeding 600 mm internal
	(iv)	=exceeding 600 mm but not exceeding 900 mm internal diameter

In order to construct a BQ item for a headwall, the root narrative would be **Headwall 22* 23* to pipe 24*** where:

- Variable 22* is the stated type of headwall.
- Variable 23* is the material used to construct the headwall.
- The pipe diameter around which the headwall is to be constructed is given by variable 24*.

Consequently, Headwall Type A, constructed of reinforced concrete with an inlet/outlet pipe diameter of 750 mm, would be described using the standard phraseology of variables 22(i), 23(iii) and 24(iv). The eventual item description would be:

- **Headwall** Type A (**22***).
- In reinforced concrete (**23***).
- **To pipe** exceeding 600 mm but not exceeding 900 mm internal diameter (**24***).

NB: Where the bold type represents the root narrative.

This process is illustrated in Figure 8.3 using the MMHW Library of the QSPro software package.

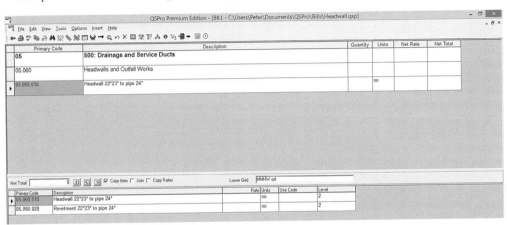

Figure 8.3   Root narratives.

Part 2

The root narrative variables (e.g. 22*) would not appear in the eventual item description which would read:

> Headwall Type A in reinforced concrete to pipe exceeding 600 mm but not exceeding 900 mm internal diameter.

Where a variable is not listed in a group, but would naturally belong there, it may be added to the group and numbered sequentially.

Rogue items – that is, those which cannot be compiled from the standard root narratives in the library – may be drafted using the same principles as those upon which the library is based. Rogue items not included in the library but which consistently recur may be forwarded to the HA for evaluation and possible inclusion in standard amendments that may be issued from time to time.

8.8.3   Item coverage

As previously discussed, item coverage is a list of items that is deemed to be included in any item description to which it relates.

'Item coverage' appears as a marginal heading and normally relates to a specific Group I feature of each subsection of each series under the Itemisation marginal heading. Sometimes, the item coverage applies to all the Group I features of the subsection as is the case with Drainage and Service Ducts in Structures. This is illustrated in Table 8.8 and Table 8.9 which shows the item coverage for **Drainage and Service Ducts in Structures.** Table 8.9 demonstrates that each of the Group I features of this subsection of Series 500: *Drainage and Service Ducts* has the same item coverage.

Table 8.8   Drainage and service ducts in structures.

Group	Feature	
I	1	Drainage
	2	Service ducts
II	1	Substructure – end supports
	2	Substructure – intermediate supports
	3	Superstructure
	4	Reinforced earth structure
	5	Anchored earth structure

Table 8.9   Item coverage for drainage and service ducts in structures.

Drainage and service ducts in structures	**55** The items for drainage and service ducts in structures shall in accordance with the Preambles to Bill of Quantities General Directions include for:
Item coverage	a. Drains, service ducts, filter drains, fin drains and narrow filter drains and connections (as this Series Paragraphs 16, 22, 28 and 32) b. Chambers (as this Series Paragraph 37) c. Gullies (as this Series Paragraph 38) d. Pipework, gullies, downpipes, fittings and the like including brackets, hangers and straps, fixing to or building into the structure e. Making good protective system and waterproofing f. Permeable backing including compaction and supports g. Channels

The item coverages that appear in the MMHW emanate from the SHW and the HCD, but, contrary to common perception, the item coverage does not represent an estimator's 'shopping list'. The item coverage must be read in conjunction with the drawings and specification for each project in order to understand whether all, or just some, of the items listed are to be included in the contractor's rates and prices for that project.

Failure to understand this important principle can lead to considerable inaccuracies in pricing and potential claims that can foster ill feeling and disputes. However, misunderstandings can also arise because item coverages are not fully inclusive as they do not include:

- Items that are contingently and indispensably necessary.
- General obligations set out in the conditions of contract.
- Specific obligations set out in the *Preambles to Bill of Quantities*.

The first two bullet points raise the issue of completion and, in particular, the contractor's general obligation to construct and complete the works. The usual obligation is that the contractor will provide everything necessary for completing the works whether specified in the contract or to be reasonably inferred from it – such as cement in mortar and concrete and fixings for fencing, safety barriers and the like.

Therefore, provided that the rules of the method of measurement have been observed by the bill compiler, the contractor must carry out any work, whether described or not, which is obviously (or **indispensably**) necessary, on a true construction of the contract, to complete the works. Additionally, the contractor will be obliged to carry out anything **contingently** necessary, short of the impossible, in dealing with problems arising from the carrying out of the contract barring, of course, any relief granted by specific contract terms (Loots and Charrett, 2009).

The third bullet point signals the importance of referring to the *Preambles to Bill of Quantities* which is to be found in Chapter III of the Method of Measurement for Highway Works and, in particular, to Paragraph 2 thereof. This is where the contractor is alerted to the items that his rates and prices shall be deemed to include such as labour, materials, plant and equipment, overheads and profit and to his *general obligations, liabilities and risks involved in the execution of the Works*.

Risk Issue

It is vitally important to include the *Preambles to Bill of Quantities* in every contract, irrespective of whether the MMHW is amended or not, because it is via the preambles that 'item coverage' is incorporated into the bill of quantities as demonstrated by the wording of Paragraph 2 referred to above:

In the Bill of Quantities the sub-headings and item descriptions identify the work covered by the respective items, read in conjunction with the matters listed against the relevant marginal headings 'Item coverage' in Chapter IV of the Method of Measurement for Highway Works, these Preambles and the amendments to the Method of Measurement immediately following these Preambles.

When reading item coverages, attention must be paid to any references made to item coverages in other series of the method of measurement. This type of abbreviation, known as 'written short', tempts the reader to think that the 'written short' item coverage applies in its entirety to the original item coverage.

For instance, in Series 500: *Drainage and Service Ducts*, the item coverage for **drainage and service ducts** is provided in Paragraph 16. Within Paragraph 16, the following entry is found:

(j) Formwork (as Series 1700 Paragraph 15).

Reference to Series 1700, Paragraph 15, reveals an item coverage that includes the erection of trial panels, falsework and all sorts of labours and intricate details and finishings. Such requirements have no place in formwork for drainage work, and these particular elements of the item coverage would clearly not apply in practice. As is generally the case with item coverage, reference must always be made to contract-specific drawings and specification amendments to reveal exactly what is required under the contract. A dash of 'common sense' also helps!

8.9   Contractor design

On the face of it, a section on contractor design in a chapter about the Highways Method of Measurement may seem strange as, surely, any requirement for a contractor-designed element in a highway project would imply that the contractor will assume responsibility for preparing the quantities for the contractor-designed element. True, but, in the MCHW, contractor-designed elements have a significant influence on how the tender documents are prepared and, in particular, the tender BQ.

Procurement strategies for the design, construction and maintenance of UK highways have changed significantly since the early 1990s. Roadworks entirely designed by the employer's 'in-house' or consultant engineer have become a thing of the past, and DBFO roads are common. The United Kingdom has a number of privately financed toll bridges and tunnels, and the first toll motorway, the Birmingham Northern Relief Road, was opened in late 2003. Even highway maintenance is largely carried out on the basis of contractor design and delivery using a type of term contract.

There is a wide variety of choice available to overseeing organisations for contractor involvement in the design of highway schemes:

- Full contractor design
  - Contractor design of a highway scheme (e.g. a new bypass).
  - Contractor design within a DBFO scheme (e.g. toll bridge).
  - Contractor design and delivery of maintenance works.
- Partial contractor design
  - Design of proprietary structures within an engineer design (e.g. culvert).
  - Design of proprietary elements within an engineer design (e.g. bridge bearings).
  - Design of 'special structures' as an alternative to a non-proprietary engineer design (e.g. a small-span footbridge).
  - Design of temporary works (e.g. special temporary works such as falsework for major and complex structures, temporary works that are alongside or temporarily support or span live carriageways or railway lines, etc.).
  - Design of temporary structures (e.g. bridge lifting systems, demolition of existing structures, support systems and platforms over or adjacent to highways, railways, watercourses, etc.).

This chapter is concerned with the measurement and billing of partial contractor design because the MCHW, and the Highways Method of Measurement in particular, contain no provisions for full contractor design.

8.9.1   Contractor design involvement

Contractor design in highway construction was slow to 'catch on', largely due to the haphazard introduction of the implementation standards and advice notes needed by participants in order to understand the procedures (Mitchell, 2014), but the 'powers that be' were no doubt prompted by the scathing criticism levelled at the DTp by the National Audit Office in its 1992 report *Contracting for Roads* (National Audit Office, 1992). Amongst a plethora of reproaches were observations concerning high project outturn costs, late completion and the proliferation of contractual claims on highway projects.

In the early days of large highway contracts, there was little or no design involvement on the part of the contractor and no limitation, apart from cost efficiency, on the engineer's design choices. A bridge design, for example, could include any combination of materials and methods for the supports and deck and any choice of structural elements, such as piles, bridge bearings and movement joints. Common practice was for the engineer to specify *Messrs ABC Precast Concrete Driven Piles* or *Messrs Acme Bridge Bearings* 'or other equal and approved', and this gave tenderers the chance to offer an alternative supplier, perhaps at a more competitive price.

This arrangement changed a little over the years, but it was not until 1992 that such practices were deemed to constitute a 'barrier to trade' and had to be stopped pursuant to the Single European Act of 1986. It was at this time that the engineer's hands became tied and the freedom to specify any preferred products or materials was curtailed.

8.9.2   The Single European Act

Highway schemes are normally large and complex projects comprising a wide variety of elements such as carriageways, bridges, culverts, underpasses, safety barriers, street lighting and communications installations, etc. These elements are specified in the various series of the SHW which assists engineers to design using their preferred combination of materials, products and methods.

The aim of the Single European Act was to prevent the specific naming of any 'proprietary' materials and structures in the tender documents in order that specified manufacturers and installers did not have an unfair advantage over their competitors. This meant that discrimination against Member States, whose materials and products could otherwise be eligible for use on highway projects within the EU, was outlawed in favour of a 'neutral' means of specifying and contractor choice. The impact of this was significant on the design, specification and measurement of highway works.

8.9.3   Barriers to trade

A barrier to trade is one where an unfair advantage is given to one product over its competitors or where EU member states are put at a disadvantage, due to the way that a design is presented and specified. It is also a barrier to trade where a design discourages innovation in product development.

Prior to 1992, contractors were able to tender for highway contracts on the basis of 'alternative specified materials' that were both approved and equal in every respect to those specified in the tender documents. This was, and still is, common practice in the construction industry as it provides a competitive element to tenders.

As far as Highways England is concerned, however, the concept of 'alternative specified materials' is thought to constitute a 'barrier to trade' on the basis that it does not comply with the Single European Act. Other public authorities, bound by the same legislation, take a different view of the law, and Mitchell (2014) believes that the Highways England approach amounts to *overkill*.

Nonetheless, the 'bottom line' is that, since 1992, the MCHW has contained complex arrangements to ensure that barriers to trade are avoided and this has made the measurement and billing of certain elements of highway projects quite complex and, in some instances, unworkable according to Money and Hodgson (1992) and Mitchell (2014).

Part 2

Despite such eminent opinion, 'we are where we are', and so, in order to overcome the 'barrier to trade' problem, specifiers (e.g. the employer's engineer) must not only avoid stipulating specific 'proprietary' materials in their design, and leave the choice and design of such materials to the contractor, but also wrestle with complex measurement and billing issues.

8.9.4   Proprietary products

Proprietary products are those manufactured and supplied under a brand name (e.g. Acme Box Culverts) and are defined in the DMRB as *a structure with* CE *marking **or product** with* CE *marking manufactured to a system covered by a patent and/or a registered design.*

The issue with the MCHW is the avoidance of any implication in the engineer's design or specification that a particular proprietary material is preferred. There is no limitation on engineer-designed structures using non-proprietary materials (e.g. reinforced concrete), and the engineer is free to design a structure using non-proprietary materials, even where a proprietary structure could be used, provided that the design choice is made for sound engineering reasons.

8.9.5   SHW Clause 106

The need to avoid 'barriers to trade' prompted the introduction of Clause 106: *Design of the Works by the Contractor* in the 1992 SHW. This signalled an important change as it affected both the specification of materials for highway works and also introduced the idea of contractor design of structures, structural elements and other features where:

- Proprietary materials are habitually used.
- Proprietary materials could be proposed as a suitable alternative to a conventional design.

Consequently, because of the need to avoid barriers to trade (restriction of competition and discouragement of innovation), the choice and design of such products are not specified in the tender documents anymore but are left to the contractor. Just how this may be achieved in practice is not easy as not only are the resulting measurement and billing issues complex but also Highways England does not consider the following methods to be acceptable either:

1. A general arrangement drawing based on a proprietary product with the caveat 'or equivalent' (although other public bodies bound by the same legislation accept this method as legitimate).
2. A detailed design based on non-proprietary materials with the opportunity for the contractor to propose an alternative design based on a proprietary product.

Both methods are considered to be a barrier to trade because:

- Method 1 could imply preference for a particular manufacturer or product which is a 'barrier to trade'.
- Method 2 constitutes a disincentive to the use of proprietary products that could otherwise be employed and therefore also constitutes a 'barrier to trade'. The 'disincentive' is that tenderers may consider the *technical approval procedures* too onerous (Money and Hodgson, 1992).

There is an important caveat to Method 2 in that *where the Engineer considers that a non-proprietary design and a proprietary system are more or less equal in terms of cost and performance, then both possibilities are run in parallel but the tenderer only prices his choice* (Money and Hodgson, 1992).

8.9.6   Avoiding barriers to trade

The need to avoid 'barriers to trade' as and from 1992 created the atmosphere for partial contractor design of highway projects, and the MCHW was drafted in such a fashion that this may be achieved in three distinct ways:

Method 1
    Contractor design of proprietary manufactured structures (e.g. box culverts, retaining walls).
Method 2
    Contractor design of proprietary manufactured structural elements as part of a non-proprietary engineer design (e.g. the bearings and movement joints of a large-span bridge).
Method 3
    Contractor design of structures where there is a choice of designs (e.g. engineer design of a non-proprietary small-span reinforced concrete underbridge and a contractor-designed proprietary alternative).

The measurement and incorporation of partial contractor design in the BQ for each of these methods must be considered separately because, although Methods 1 and 3 are similar, there are differences and Method 2 is quite different and poses considerable problems for the measurement and billing process.

8.9.7   Contractor design elements

From a risk perspective, partial contractor design as part of an engineer design is a completely different proposition compared with a full contractor design. In the latter case, competition will be limited, and risk will be reduced, by using a multistage procurement process. In a competitive tendering environment, on the other hand, it is obvious that the cost of preparing full designs for large elements of an engineer-designed highway project, weighed against the chances of winning the contract, is a risk too far for even the biggest contractors.

Given this situation, therefore, design and construct solutions for highway projects require careful procurement choices by the overseeing organisation, and, consequently, traditionally procured projects will, at most, employ the partial contractor design of relatively small elements of the works. The more usual elements that may be realistically designed by contractors as part of a competitive tender include:

- Structures, such as small bridges and earth retaining structures.
- Buried structures, such as culverts and underpasses.
- Environmental barriers.
- Structural elements such as ground anchors, piles, combined drainage and kerb systems, drainage channels, bridge bearings and bridge expansion joints.

These are referred to in the MCHW as 'Special Structures', and provision is made for them in the Specification and Method of Measurement in Series 2500.

8.9.8   Forms of construction

The type of structure most appropriate for a specific situation on a highway project will be one of the following:

- A uniquely designed structure, based substantially on non-proprietary materials such as reinforced concrete or structural steel, perhaps using some components that may be proprietary products; this would be largely designed by the engineer with some contractor design involvement.

Part 2

- A proprietary manufactured structure, such as a precast concrete culvert selected from a manufacturer's catalogue of products, designed entirely by the contractor.
- A structure whose form of construction could equally be uniquely designed or of proprietary origin; the non-proprietary design would be prepared by the engineer and the proprietary alternative by the contractor.

In order to avoid the risk of discrimination, the designer must be able to demonstrate to the Technical Approval Authority (TAA) that all three types of structure have been considered prior to making a decision. This is done at the outline approval in principle (O/AIP) stage. However, the designer is not obliged to adopt any particular design if, after considering engineering and/ or aesthetic issues, any of the options is considered inappropriate or if one option has clear advantages over the others.

Complications can arise when planning authorities do not approve of a particular design and written justification for their decision needs to be sought.

8.9.9   Technical approval of the design

Technical approval (TA) for a highway structure is a certification process whereby contractor-designed elements are categorised according to the extent of design checking needed.

The process is rigorous and time consuming and involves submitting an O/AIP application for design proposals. Applications to the TAA have to be made before designs can be implemented. The TAA is the overseeing organisation and might be a local county council where the authority is, or is to become, the highway authority, usually assisted by a firm of consulting engineers.

The process follows several distinct stages that the contractor must allow for both in the planning and pricing of the project:

1. Acceptance of the approval in principle (AIP) document which agrees the form of the proposed structure and its principal details together with traffic loadings and the technical standards to be employed. The category of the design check is also established at this stage.
2. Acceptance of a Design Certificate.
3. Acceptance of a Check Certificate.
4. Acceptance of a Construction Compliance Certificate including submission of as constructed drawings and a maintenance manual incorporating the health and safety file.

The 'Standard' for the Technical Approval of Highway Structures, BD2, is contained in Volume 1 Section 1 Part 1 of the DMRB. This standard outlines the procedures necessary for the successful adoption of proposed designs. The procedures, format and terms used in BD2 are intended to be contract neutral and may thus be used with any method of procurement, with the exception of DBFO contracts.

8.9.10   Contractual implications of contractor design

Under the ICC – Measurement Version form of contract, once the contractor's design has been approved, the overseeing organisation becomes responsible for *the integration and co-ordination of the Contractor's design with the rest of the Works* (Clause 7(7)).

This *does not relieve the Contractor of any of his responsibilities under the Contract* (Clause 7(7)) which includes taking *all reasonable skill care and diligence in designing any part of the Permanent Works for which he is responsible* (Clause 8(2)). This is the same standard of design liability as that expected of a professional designer as opposed to the higher, fit for purpose, standard of a design-construct contractor.

Under the ECC form of contract, Clause 21.1 states that the *Contractor* is to design those parts of the works stipulated in the Works Information and this design is to be submitted to the *Project Manager* for acceptance under Clause 21.2. The *Contractor's* design responsibility is normally limited to the professional standard of *reasonable skill and care* by using Option Clause X15, and the *Project Manager's* acceptance of the design does not change the *Contractor's* liability for his design (Clause 14.1).

The contractual requirement for proprietary manufactured structures is included in the contract by means of the O/AIP form which stipulates the criteria that must be applied to the detailed design of a highway structure by the contractor.

8.10   Measurement and billing of contractor-designed elements

The incorporation of contractor-designed elements into the tender documents, and into the BQ in particular, poses certain difficulties that have not entirely been overcome in the MMHW. There are three procedures according to the three ways of incorporating partial contractor design into the contract, that is,

1.   For the design of proprietary manufactured structures.
2.   For the design of proprietary manufactured structural elements.
3.   For structures where there is a choice of designs.

Clearly, in a BQ prepared by the overseeing organisation, a contractor-designed element cannot be measured and billed at tender stage because each tenderer will probably approach the design differently and, in any case, there will likely be a multiplicity of design choices available. Additionally, care must be taken to ensure that the BQ is not drafted in such a way as to create a barrier to trade.

Consequently, where tenderers are to be asked to design and price a proprietary structure or an alternative to an engineer-designed structure, Series 2500: *Special Structures* of the Method of Measurement for Highway Works, the LSID and the Notes for Guidance must all be referred to. The measurement and billing of proprietary structural elements, however, follow the measurement rules, standard item descriptions and guidance in the series relevant to that item. In this case, Series 2500 does not apply.

In order to appreciate just where 'Special Structures' sit in the context of the overall BQ for a project, it must be remembered that a highway scheme is made up of a number of different 'bills'. This means that a separate bill is needed for proprietary structures to be designed by the contractor and another for structures where there is a choice of designs even though both are categorised as Series 2500: *Special Structures*. Each of these separate bills will be further subdivided as necessary (see Chapter 16 for examples).

The measurement and billing of proprietary structures, or alternatives to engineer-designed structures, depends on the use of a **'designated outline'** which delineates the boundaries of the structure in question. There is no designated outline for proprietary structural elements, however, and this raises complexities in the measurement and billing of such items.

8.10.1   Designated outline

The idea of a designated outline is to effectively separate proprietary manufactured structures to be designed by the contractor, and structures where a choice of designs is offered, from other works that must be measured and billed by the overseeing organisation.

A simple example is shown in Figure 8.4 which illustrates a designated outline for a small-span underbridge. Items not to be included in the contractor's design (or price) are listed in a schedule of excluded items.

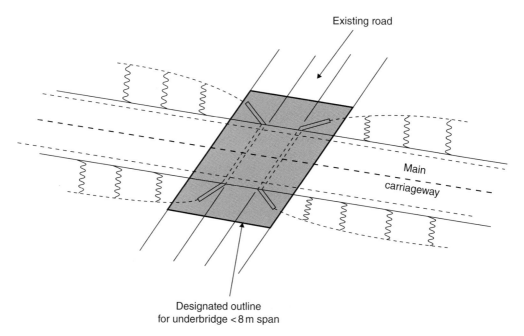

Schedule of excluded items
• Main carriageway drainage
• Service ducts
• Cables
• Pavements
• Safety barriers
• Road markings

Existing road

Main carriageway

Designated outline
for underbridge < 8 m span

Figure 8.4 Designated outline.

The term 'designated outline' is defined, somewhat unhelpfully, in MMHW Chapter I: *Definitions* as:

- *the designated outline shown on the Drawings.*

However, this is followed by a note which, more helpfully, explains that:
A Designated Outline is shown as enclosing each structure to be designed by the Contractor and each structure for which a choice of designs is offered. The Designated Outline delineates the limits of measurement of work to be included for each structure (with the exception of those works scheduled as not to be included).

The *Notes for Guidance on the MMHW* is more expansive and lists several criteria that should be considered when selecting a designated outline:

- It should be given careful consideration.
- It should be clearly defined and fully enclosing.

- It should be sufficiently large to include any non-proprietary design of the overseeing organisation as well as all the possible options that the contractor might submit.

A designated outline is three-dimensional (3D) and, therefore, should define and fully enclose the foundations to structures and any special backfill requirements proposed by the contractor in response to the design requirements stipulated by the overseeing organisation. This may raise measurement and admeasurement issues as discussed later in the chapter.

Although beyond the scope of this book, requirements for the detailing of designated outlines may be found in Departmental Standard SD 4/92 and Appendices 1/10(A) and (B) of the Specification should contain details of the various options for special structures to be designed by the contractor.

Ideally, the designated outline should exclude common items such as pavements, kerbing, safety barriers, service ducts, cables, headwalls and the like. This is not always feasible though, and, if not, a schedule should be provided which lists the features that encroach into the designated outline but are to be excluded from the lump sum item. Such items shall be included elsewhere (e.g. in the Roadworks or other Bill) and priced as such.

Risk Issue

It should be carefully noted that all earthworks within the designated outline should be included in the contractor's price. This is because Paragraph 11 of Series 600: Earthworks specifically states that *Earthworks within Designated Outlines shall not be measured in this Series.*

Where common items such as those referred to above are integral to the structure to be designed, they should not be excluded from the lump sum item, and this should be made clear in the tender documentation.

Where a footbridge or small-span underbridge is to be designed by the contractor, the designated outline should be configured so as to ensure that the foundations are enclosed within it.

Risk Issue

Where the contractor's design requires alterations or variations, either to works outside the designated outline or to existing works, such alterations or variations shall not be included in any admeasurement as prescribed in Paragraph 16 of the Preambles to Bill of Quantities:
Unless specifically stated to the contrary in the contract the measurement of the works affected by the incorporation of the contractor's design shall be based on the tender documents and not on the works as amended and completed to incorporate the contractor's design.

8.10.2   Proprietary manufactured structures

Proprietary manufactured **structures** are essentially relatively simple structures such as box culverts, precast concrete small-span underbridges, crib and gabion walls and the like. Where such structures are required by the contract, they must not be specified in the tender documents but are left for the contractor to make the choice and carry out the design. This avoids the possibility of specifying or implying the adoption of a particular manufacturer.

Part 2

Structures to be designed by the contractor are listed in the tender documentation in Appendix 1/10(A) of the Specification pursuant to Clause 106 of the SHW. Design criteria are stipulated on an O/AIP form *which records the agreed basis and criteria for the detailed design or assessment of a highway structure.* The contractor may then select a product that meets the O/AIP criteria which would then be subject only to checking and technical approval, should the contractor's tender be successful.

This arrangement is detailed in Annex D of BD2 *Design Manual for Roads and Bridges* which provides specific procedures for the approval of such structures. Proprietary manufactured structures may include:

- Various types of culvert.
- Small-span underbridges (up to 8 m span) in precast concrete.
- Various systems of earth retaining structures such as reinforced or anchored earth systems and crib and gabion walls.
- Lighting columns.
- Large sign supports (greater than 7 m high).
- Environmental barriers.

Some of these proprietary products will require Highways England-'type approval' pre-contract where this is not already the case.[10]

Because proprietary manufactured structures to be designed by the contractor cannot be measured and billed in the normal way like drainage, in situ concrete and surfacing, they must be included in the BQ in a special way:

- They must be listed in Appendix 1/10(A).
- They must also be identified and delineated in the contract by means of a **designated outline** shown on the drawings.
- They must be billed separately to other measured work.

Where proprietary structures are to be included in a contract, the following steps must be followed:

- The contractor must be free to choose and design proprietary structures in order to avoid a 'barrier to trade'.
- The location and extent of the work involved in each contractor-designed structure must be delineated on the drawings with a 'designated outline'.
- The designated outline must be extensive enough to include the extremities of the work envisaged including any foundations and special backfill.
- The work involved in proprietary structures must be billed separately from other non-proprietary work.
- Each structure to be designed by the contractor shall have its own BQ comprising a single item.

A worked example of how this might be done may be found in Chapter 16.

8.10.3   Structures where there is a choice of designs

In some situations, contractors will be invited to tender for a highway contract on the basis of providing a contractor design for a structure as an alternative to a design by the overseeing organisation. Consequently, where the overseeing organisation believes that there is little to choose between a proprietary and a non-proprietary design, the MMHW provides for the measurement and billing of such structures in Series 2500: *Special Structures.* Where this is the case, there will be no barrier to trade.

Series 2500 is the same one that is used for the measurement and billing of proprietary structures designed by the contractor, but the arrangements for measurement and billing are significantly different.

Structures that might be designed by the contractor as an alternative to one designed by the engineer are the same as those for proprietary structures to be designed by the contractor, with the difference that the engineer believes there to be little to choose between a proprietary and non-proprietary design. The structures in Series 2500 are:

- Buried structures.
- Earth retaining structures.
- Environmental barriers.
- Underbridges up to 8 m span.
- Footbridges.
- Piped culverts.
- Box culverts.
- Drains exceeding 900 mm internal diameter.
- Other structures.

The basis of this option for contractor design is fundamentally different to that for proprietary structures in that the contractor has choices to make and must decide:

- Whether to offer a contractor design or not (for proprietary structures, there is no choice other than to price the bill items or not submit a tender).
- Whether the time, effort and cost of preparing a design are commercially attractive.
- Whether the cost of preparing an outline design at tender stage is worthwhile considering the attendant risk of not winning the contract.
- Whether a design can be developed that is more cost efficient than that prepared by the overseeing organisation.

Risk Issue

It will be for tendering contractors to weigh these issues against the chances of winning the contract and to decide whether or not it is worth the additional cost of preparing a design in sufficient detail in order to be able to price the work involved into the tender bid.

Overseeing organisations have some choice as regards such designs, but, in the main, choices of materials are constrained by the need to avoid specifying proprietary products which would create a barrier to trade.

In common with proprietary structures designed by the contractor, structure(s) where a choice of designs is offered must be identified on the drawings by means of a **designated outline** that must fully enclose the design and delineate the limits of measurement of work to be included for each structure; the designated outline should define and fully enclose the foundations to structures and any special backfill requirements as before.

The major difference between designs for proprietary structures and designs for non-proprietary engineer designs is in the measurement and billing of the alternatives; this is determined by MMHW Chapter III: *Preambles to Bill of Quantities*, Paragraph 6:

- *where the Contract provides for a structure designed by the Contractor to be constructed as an alternative to the structure which has been designed by the Overseeing Organisation, a separate Bill of Quantities is to be provided for each of the two construction procedures permitted by the Contract.*

Part 2

This means that <u>two</u> BQ are needed for <u>each structure</u> in addition to the main BQ for the remainder of the works, but only one of these additional BQ will be priced and included in the tender total pursuant also to Paragraph 6 which states that:

- *Provision is to be made for only the one Bill of Quantities which relates to the form of construction elected to be constructed by the Contractor to be priced and included in the Tender Total.*

A worked example of how this might operate is provided in Chapter 16.

Which additional BQ are included in the eventual contract depends upon whether:

- The contractor wishes to submit an alternative design for all or any of the structures listed in the appendix.
- An alternative design is more competitive than the engineer-based design.

Should the tender be successful and the contractor design adopted, Paragraph 16, *Structures Designed by the Contractor*, of the MMHW Chapter III: *Preambles to Bill of Quantities*, requires that:

- The contractor shall prepare **a priced schedule of quantities** in respect of each priced BQ comprising a single item for a structure designed by the contractor.
- This priced schedule shall be prepared in accordance with the relevant chapters and series of the MMHW.
- The priced schedule shall be submitted to the overseeing organisation.

Consequently, the contractor must be prepared for the time and cost of preparing a detailed BQ for each 'special structure' to be designed by him and must, furthermore, ensure that:

- The schedule of quantities is prepared according to the Method of Measurement for Highway Works.
- The extended total of the quantities, rates and prices in the priced schedule shall equate to the lump sum submitted at tender stage.

Paragraph 16 further states that the priced schedule of quantities shall only be used for:

- Payment applications.
- Valuation of variations ordered under the contract in connection with structures designed by the contractor.

8.10.4   Proprietary manufactured structural elements

Proprietary manufactured structural **elements** differ from 'Special Structures', whether designed by the contractor or the engineer, because they represent parts of a design and not a complete design in their own right.

Such elements include:

- Ground anchors.
- Piles.
- Combined drainage and kerb systems.
- Drainage channels.
- Bridge bearings.
- Bridge expansion joints.
- Etc.

Once again, the issue with proprietary manufactured structural elements is the avoidance of any implication in the contract that a particular product has preference over others that could equally well be selected. Therefore, structural elements, or other features of a design, cannot be specified where they are based on proprietary systems but must be designed by the contractor or a manufacturer and listed in Appendix 1/10(A) to the specification. This avoids the possibility of creating a 'barrier to trade'.

The procedures associated with proprietary manufactured structural elements are less complicated than for structures as there is no O/AIP requirement and no need to identify a 'designated outline', but the measurement issues created are more difficult to resolve. In fact, Money and Hodgson (1992) raised concerns over 20 years ago and suggested that the (then) DTp *must urgently address this issue and prepare a measurement mechanism for structural elements*. Mitchell (2014) observes that *still nothing has been done*.

The issue of proprietary manufactured structural elements arises in connection with designs prepared by the overseeing organisation. These designs are measured and billed according to the various series of the MMHW, and examples might include:

- The design of highway carriageways where combined drainage and kerb systems are needed.
- The design of a large-span bridge requiring proprietary bridge bearings and expansion joints.
- The design of a non-proprietary retaining wall needing proprietary piled foundations or ground anchors.

In each of these cases (and others), the engineer would need to ensure that the design did not include reference, either express or implied, to any proprietary product which, of course, would create a barrier to trade. Consequently, any requirement for a proprietary product in the design would have to be the contractor's choice, and the contractor would assume responsibility for the design of that part of the structure at tender stage and price accordingly.

The rules for the measurement and billing of such items are not clear in the MMHW, however. Proprietary products are not covered by Series 2500: *Special Structures* nor is there any other specific series dealing with these elements. For some clarification, it is necessary to resort to the *Notes for Guidance on the Specification for Highway Works* which states that:

- *The compiler should ensure that work items and elements based on proprietary products have not been specified in the contract. Such items and elements should be designed by the Contractor, or where appropriate, by the manufacturer and proposed by the Contractor.*[11]
- *Each work item or element for which a design is to be submitted by the Contractor should be listed in contract specific Appendix 1/10(A).*[12]

These requirements are also emphasised in the item coverages of the appropriate series of the MMHW. In Series 600: *Earthworks*, for example, the item coverage for ground anchors is given in Paragraph 132, and this includes:

a) *design;*
b) *provision of data and drawings;*
c) *certificates;*
d) *resubmissions and modifications;*
e) *amendments to the Works;*

Similar item coverages are given for piling (Series 1600), bridge bearings (Series 2100) and expansion joints (Series 2300).

The problem for the bill compiler is that the MMHW requires proprietary manufactured structural elements to be billed in specific units of measurement and not lump sums as with proprietary structures and structures where there is a choice of designs. Consequently, ground anchors and piles are to be measured in linear metres, bridge bearings by number and expansion joints by number with a stated length. It is obviously impossible to 'second-guess' what the contractor's design will be, and thus, the number and length of piles, the number of bridge bearings and the nature and quantity of expansion joints are all unknowns at the billing stage.

The *Notes for Guidance on the Method of Measurement for Highway Works* provide no help to the bill compiler who is basically left to his/her own devices to provide a solution. In the absence of any guidance on the matter, either from Highways England or any other source, suggestions can only be made as to a way forward.

Mitchell (2014) offers the idea that a designated outline could be superimposed on the engineer's design where appropriate with details of the relevant proprietary structural elements left blank. This is illustrated in Figure 8.5.

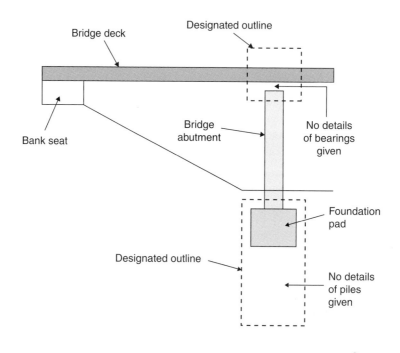

Figure 8.5   Proprietary structural elements. Adapted from Mitchell.

In this illustration, the unit of measurement in the BQ would be a single lump sum with the proviso that:

- The contractor would not be responsible for the design of the bridge deck, pier and pad foundation.
- These items would be listed in a schedule as not included.
- Changes to the engineer's design necessitated by the contractor's design (e.g. rebar) would be accommodated by the relevant series' item coverage which includes *amendments to the Works*.

However, Mitchell (2014) also suggests that there may be problems with this approach that the bill compiler would be unable to surmount, especially with regard to piling, where the depth and rake of the piles would be indeterminable.

A personal view is that measured items could be included in the BQ with the quantities left blank. The quantities derived from the contractor's design could then be completed by the contractor at tender stage in order to arrive at a tender price for the work concerned. Interface problems between the engineer's design and the contractor's design could be dealt with in the relevant item coverages as Mitchell (2014) suggests or as a variation or compensation event.

A worked example may be found in Chapter 16, but it must be emphasised that this is in no way meant to represent 'official' guidance or recommended good practice.

8.11   Measurement of highway works

The measurement of highway works using the Highways Method of Measurement is, *prima facie*, straightforward because fewer items have to be measured compared with other methods of measurement and the items that are measured are 'composite' in nature due to the principle of 'item coverage' that the MMHW employs.

It would be a mistake to think this way because the MMHW is, in fact, complex, both in the context of its relationship with the other documents that make up the MCHW and, as Mitchell (2014) observes, in the context of the *many of the problems associated with the MMHW* [that] *stem from* [the] *lack of proper rules or explanations, resulting in improbable interpretations* on behalf of the compiler.

Unlike other methods of measurement, the various series of the MMHW are not 'stand alone', and some series have to be read in conjunction with others. The influence of Series 600: *Earthworks* is particularly important because not only does the Earthworks Outline define the starting point for the excavation of drainage and other work items but also because it contains definitions that are cross-referenced in other series and item coverages that are 'written short' in other series.

Consideration of all the series in the MMHW in a book of this nature is impossible – it would require a book in its own right. Consequently, some key series are discussed which, hopefully, will assist the reader in interpreting the requirements of other series that have not been included. In this context, it may be helpful to think of a highway in terms of earthworks, roadworks and structures.

Part 2

8.12   Series 100: Preliminaries

Anyone used to SMM7, NRM2 or CESMM4 will find the Series 100: *Preliminaries* section of the MMHW quite alien. There is a fairly short list of preliminaries items for tenderers to price:

- Temporary Accommodation.
- Vehicles for the Overseeing Organisation.
- Communication System for the Overseeing Organisation.
- Operatives for the Overseeing Organisation.
- Information Board.
- Traffic Safety and Management.
- Temporary Diversion for Traffic.
- Recovery Vehicles.
- Progress Photographs.
- Temporary CCTV System for the Monitoring of Traffic.
- Temporary Automatic Speed Camera System for the Enforcement of Mandatory Speed Limits at Roadworks.

There are no familiar items here and no contractual requirements or method-related charges as might be found in other methods of measurement. Mitchell (2014) points an accusing finger at the then Department of the Environment for *refusal to adopt* method-related charges way back in 1974 and for ignoring CIRIA Report 34 which largely underpinned the adoption of method-related charges in the CESMM.

8.12.1   Itemisation of preliminaries

In common with other series of the MMHW, each item in Series 100: *Preliminaries* has paragraphs that specify the units of measurement to be used, how itemisation is to be structured and an item coverage that specifies what each item is deemed to include.

Temporary Accommodation, for example, is measured by the 'item', and the itemisation structure is shown in Table 8.10.

Typical BQ items are illustrated in Table 8.11.

Table 8.10   Temporary accommodation.

Group	Feature	
I	1	Erection
	2	Servicing
	3	Dismantling
II	1	Principal offices for the overseeing organisation
	2	Principal laboratories for the overseeing organisation
	3	Portable offices for the overseeing organisation
	4	Portable laboratories for the overseeing organisation
	5	Offices and messes for the contractor
	6	Stores and workshops for the contractor
III	1	Provided by the overseeing organisation
IV	1	At the place of fabrication or manufacture
V	1	Until completion of the works
	2	After completion of the works

Note: Group IV and V features shall be applied only to items of temporary accommodation for the overseeing organisation.

Table 8.11   Typical BQ items for temporary accommodation.

					£	p
	SERIES 100 - PRELIMINARIES					
	Temporary Accommodation		(NA)			
	Erection of principal offices					
A	for the Engineer	ITEM				
	Erection of offices					
B	for the Contractor	ITEM				
	Servicing of principal offices					
C	for the Engineer	ITEM				
D	for the Contractor	ITEM				
	Dismantling of principal offices					
E	for the Engineer	ITEM				
F	for the Contractor	ITEM				

8.12.2   Pricing and interim payment

Taking the example of servicing the engineer's principal offices, the itemisation of what are effectively time-related charges can cause problems when it comes to interim payment. For instance, should payment be linked to the contract period or to the contractor's agreed programme, which might be quite different?

Another issue with Series 100 is the pricing of temporary works. In other methods of measurement, items are provided where the contractor can include monies that are effectively 'ring-fenced' and largely protected from admeasurement, but in the MMHW no such facility exists. Instead, the contractor must refer to MMHW Chapter III: *Preambles to Bill of Quantities – General Directions* Paragraph 2 which states that *the rates and prices entered in the Bill of Quantities shall be deemed to be the full inclusive value of the work covered by the several items*, and this includes *Temporary Works* (Paragraph 2(v) refers).

Risk Issue

In view of the high fixed cost of installing and removing temporary works, contractors should not risk pricing such items in rates that will be subject to admeasurement.

The pricing of temporary works, at least in the author's experience, is best included in the erection and dismantling items of temporary accommodation for the contractor (see Table 8.11) or such other 'ring-fenced' preliminaries items as may suit the purpose.

The itemisation structure adopted by the MMHW effectively encourages the 'front-loading' of preliminaries items and discourages a balanced approach to the valuation of variations.

8.12.3   Special preliminaries

An exception to the Series 100: *Preliminaries* itemisation rules is where there is a need to include provision for 'Special Preliminary' items in the BQ. Such items are 'contract specific' and therefore do not appear in Chapter IV: *Units and Method of Measurement*. They are included at the discretion of the overseeing organisation.

'Special preliminaries' are intended to be used for temporary works, cofferdams, accesses, advance operations and the like where the work involved is unusual in relation to the measured works and where:

a)   *the magnitude of such work, not separately measured, is such as to be disproportionately high in cost in relation to the measured work with which it is associated; or*
b)   *an operation, not separately measured, is required to be executed far in advance or after the main measured operation to which it relates.*

An example might be the construction of a launching gantry for the erection of the incrementally launched precast concrete deck units for a large bridge or the construction of a casting basin with manufacturing facilities for constructing immersed tube tunnel sections to be later floated into position (https://www.youtube.com/watch?v=GqpZamvvJnU).

The *Notes for Guidance on the Method of Measurement for Highway Works* also suggest that *consideration should be given to include a Special Preliminaries item* where temporary king post walls are to be constructed based on a design by the overseeing organisation.

In such instances, special Series 100: *Preliminaries* items will have to be drafted, complete with their own units of measurement, itemisation structure and item coverage, and a preamble is also required to follow Paragraph 20 of Chapter III: *Preambles to Bill of Quantities* of the MMHW.

Part 2

8.13   Series 600: Earthworks

The importance of earthworks in a highway project, and its impact on other work activities, leads this to be the first of the 'measured work' series to be considered.

The quantities of earthworks for a highway project can run into the millions of cubic metres. For instance, the £485.5 million 44 km long M6 Toll (formerly Birmingham Northern Relief Road) in the United Kingdom required the removal of 1.3 million m^3 of topsoil, 9.2 million m^3 of excavation and 7.5 million m^3 of fill[13] – some 400 000 m^3 of 'muck shifting' per kilometre.

Earthworks operations influence much of what happens on a highway project both in terms of the way that construction work is planned and carried out and in terms of the measurement and billing process. Excavated material has to be moved from one part of the site to another, some has to be imported, some may have to be 'won' from borrow pits near the site, and some may have to be treated in various ways to render it acceptable to use for construction purposes. Haulage distances can be considerable, and there is the added complication of existing roads, rivers, canals and railways to contend with.

The 'earthworks balance' calculation and 'mass haul diagram' are critical in all this, but the presence of structures, such as bridges, retaining walls, culverts and underpasses, breaks up the bulk earthworks and influences its measurement. Contractor-designed structures also have to be considered because the earthworks within the designated outline of such structures are excluded from the bulk earthworks.

8.13.1   Classification of materials

In determining the disposition and quantities of earthworks, highway designers and bill compilers are greatly assisted by modern 3D ground modelling software and will be able to work from geological plans and long sections of the route showing the locations of borehole and trial pit data. Notwithstanding, expert help may be needed to interpret this data as great care is needed to avoid misrepresenting the quantities of acceptable and unacceptable materials in the design and the BQ.

Earthworks materials must be classified in accordance with Clause 601.1 and Table 6/1 of the Specification and any modification thereof determined by contract-specific Appendix 6/1. Despite not being a contract document, bill compilers are also required to pay attention to the classification flow chart at Table NG 6/1 in the *Notes for Guidance on the Specification for Highway Works*, that is, Volume 2 of the MCHW. By a process of elimination, this determines whether excavated material is available for use in the works or is for disposal off-site.

Risk Issue

The classification and measurement of earthworks is not a precise science. Judgements may well be based on reliable data, but borehole information is only representative of the immediate location of the borehole itself, and interpolation of the data is fraught with uncertainty.

Claims for additional payment abound in this area and changes in quantities can at least precipitate requests for tender prices to be 're-rated'.

Typical layouts for earthworks excavation and fill schedules are suggested in Series 600 of the *Notes for Guidance* as (partly) illustrated in Table 8.12. This shows the excavation side of a roadworks schedule which would be repeated for fill of different types. Another schedule would be developed for structures.

Table 8.12 Earthworks schedule (part).

ROADWORKS EARTHWORKS SCHEDULE																
EXCAVATION																
ACCEPTABLE						UNACCEPTABLE										
5A	3		Class 5A Other than Class 3 and		Total Acceptable other than Class 5A (to include processed U1A & U1B material)	U1A		U1B		U2		Total Excavation other than Class 5A	EO Hard Material	Processing of Class U1A	Processing of Class U1A	LOCATION
	Above Earthworks Outline	Below Earthworks Outline	Above Earthworks Outline	Below Earthworks Outline		Above Earthworks Outline	Below Earthworks Outline	Above Earthworks Outline	Below Earthworks Outline	Above Earthworks Outline	Below Earthworks Outline					
1	2	3	4	5	6	7	8	9	10	11	12	13	14	15	16	
																ROADWORKS Main Carriageway 0–500
																Main Carriageway 500–1000
																Interchange
																Side Roads
																Sub total
																ROADWORKS TOTAL

As well as obligations under the Specification, contractors have important contractual responsibilities for ensuring the acceptability of the finished earthworks. Under the ICC – Measurement Version, Clause 36(1) states that *All materials and workmanship shall be of the respective kinds described in the Contract* and ECC Clause 20 requires that the contractor *Provides the works in accordance with the Works Information.*

8.13.2 Principles of measurement

Contractors and earthworks subcontractors should be aware that the way in which earthworks are measured and billed *will often not correspond with the actual quantities on-site* (Money and Hodgson, 1992). Therefore, whilst the MMHW sets out precisely how earthworks shall be measured and what shall be allowed for in the contractor's prices, the complexities of earthworks operations on-site are left for the contractor to sort out. Money and Hodgson (1992) suggest that this is a deliberate policy to simplify and rationalise measurement of the work. Consequently, actual quantities of excavated materials on-site may vary due to:

- Overbreak:
  - From earthworks in Series 600.
  - From excavations in other series.
- Additional excavation for working space.
- Excavation arisings from other series (e.g. fencing, drainage, piling and embedded retaining walls, etc.).
- Acceptable material that is allowed to become unacceptable.
- Settlement beneath embankments.
- Excavations within designated outlines.
- Etc.

Part 2

Compilers are not meant to 'second-guess' how the contractor will go about earth-moving operations, however. For instance, the contractor may opt to render unacceptable material acceptable for use in the works, rather than importing acceptable material. In such cases, the work should be measured as though the unacceptable material had been disposed of and acceptable material, of the class rendered acceptable, imported. In any event, the Preambles to the Bill of Quantities should draw the contractor's attention to the fact that the work will be admeasured as billed and not in accordance with how the contractor chooses to carry out the work.

The contractor has choices as to how to carry out earthworks operations but also has contractual obligations regarding the excavation, selection, handling, weathering, testing and filling of earthworks materials as determined by the SHW. The Specification is also influential in the contractor's choice of plant and whether finished earthworks operations are to be judged on the basis of prescriptive or end-product standards, both of which are used in the SHW.

8.13.3   Earthworks outline

The measurement of earthworks is contingent upon the Earthworks Outline which is defined in Paragraph 1 of Series 600, unless expressly stated otherwise, as:

- *the finished earthworks levels and dimensions (prior to topsoiling) required by the Contract for the construction, where specified, of:*
  a) *carriageway, hard shoulder, hard strip, footway, paved area, central reserve, verge, side slope;*
  b) *sub-base;*
  c) *fill on sub-base material, base and capping;*
  d) *contiguous filter material, lightweight aggregate infill;*
  e) *surface water channels;*
  f) *landscape areas, environmental bunds.*

In all cases of filter drains, except narrow filter drains, the Earthworks Outline shall be the top of the filter material.

Figure 8.6 illustrates the Earthworks Outline in three situations:

a) For a Type 1A flexible carriageway.
b) For a structural foundation.
c) For a retaining wall.

Figure 8.6a shows an edge of pavement detail taken from MCHW Volume 3: *HCD* which indicates the Earthworks Outline as top of capping layer. Where the quality of subsoil is such that a capping layer is not needed, road formation, and therefore the Earthworks Outline, would be the underside of sub-base.

Some situations are not covered in Volume 3, but, in any event, bill compilers need rules as to how the Earthworks Outline is to be used for measurement purposes. Such rules are provided in Series 600 Paragraphs 1–11 of the MMHW which define Earthworks Outline, Existing Ground Level and Subsoil Level as they apply in different situations. Measurement of earthworks is determined by these definitions.

The rule for measuring excavation for structural foundations is derived from MMHW Series 600 Paragraph 15(d), and this is illustrated in Figure 8.6b which shows that the measurement of excavation commences at the underside of blinding and extends up to the Earthworks Outline.

Where there is a structure such as an earth retaining wall, a rule determining where the Earthworks Outline is drawn is provided in MMHW Series 600 Paragraph 5 as illustrated in Figure 8.6c.

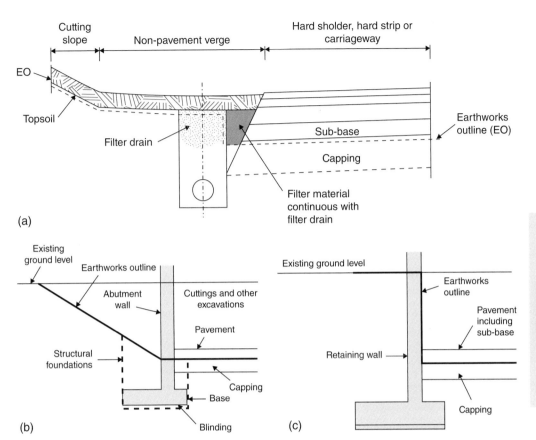

Figure 8.6 Earthworks Outline. (a) Pavement Type 1A (Flexible carriageway), (b) structural foundation and (c) retaining wall.

8.13.4   Earthworks boundaries

Bill compilers need to be aware that the MMHW imposes boundaries on earthworks measurement in addition to the Earthworks Outline and that excavation is categorised in three ways in the MMHW:

1. According to whether it is 'acceptable' or 'unacceptable' as defined by the Specification.
2. According to what type of excavation it is, for example:
    a) Cutting.
    b) Structural foundation.
    c) Foundations for corrugated steel buried structures and the like.
    d) Etc.
3. According to location in the works, for example:
    e) Roadworks.
    f) Structures.

The first two classifications are to be found in the itemisation table in Series 600 Paragraph 15 as illustrated in Table 8.13.

The third categorisation is given in Paragraph 1, Table 1 of Chapter III: *Preparation of Bill of Quantities* of the Method of Measurement which prescribes how the BQ is to be subdivided. This is illustrated in Table 8.14.

Table 8.13 Itemisation of earthworks.

Group	Feature	
I	1	Excavation
II	1	Acceptable material Class 5A
	2	Acceptable material excluding Class 5A
	3	Unacceptable material Class U1A
	4	Unacceptable material Class U1B
	5	Unacceptable material Class U2
III	1	Cutting and other excavation
	2	Structural foundations
	3	Foundations for corrugated steel buried structures and the like
	4	New watercourses
	5	Enlarged watercourses
	6	Intercepting ditches
	7	Clearing abandoned watercourses
	8	Removal of surcharge
	9	Gabion walling and mattresses
	10	Crib walling
	11	Caps to mine working, well, swallow hole and the like
IV	1	0–3 m in depth
	2	0–6 m in depth and so on in steps of 3 m

Table 8.14 Structure of bill of quantities.

BQ division		Construction heading	Series heading No.	Series heading Title	Notes
Roadworks		Roadworks General	600	Earthworks	
Structures	Structure in form of bridge or viaduct; name or reference	Substructure • End supports	600	Earthworks	To include wing walls and paved areas beneath structures
		Substructure • Intermediate supports • Main span • Approach spans	600	Earthworks	To include piers and columns
		Superstructure • Main span • Approach spans • Arch ribs	600	Earthworks	
		Finishings	600	Earthworks	
	Retaining wall, culvert, subway, gantry, large headwall, gabion wall, diaphragm wall, pocket-type reinforced brickwork retaining wall and the like; name or reference	Main construction	600	Earthworks	
		Finishings	600	Earthworks	

In Table 8.14, it can be seen that the billing of earthworks must be subdivided into Roadworks and Structures and that 'structures' are to be billed as either 'bridge or viaduct' or 'retaining wall, culvert, etc.' with an associated name or reference.

The billing of earthworks to bridges or viaducts is further subdivided into substructure, superstructure and finishings, whilst bills for retaining walls and so on are to be billed as 'main construction' and 'finishings'.

In the case of bridges or viaducts, it will be noted that various supports and spans are to be identified. This is illustrated in Figure 8.7a.

8.13.5   Structures

In Table 8.14, it is clear that Substructure End Supports are to be billed separately, but it is not clear, either in the *Method of Measurement* or in the *Notes for Guidance*, whether substructures for other supports should be measured separately or collectively. Common sense would indicate that a judgement would be made according to the extent of such supports.

Exactly what 'Superstructure earthworks' are is not clear either, but it would make sense if this referred to bulk earthworks to structures. This would tie in with the requirement for bulk earthworks to be those above the Earthworks Outline and for Substructure earthworks to be those below the Earthworks Outline.

Figure 8.7b illustrates this interpretation as it applies to an overbridge. Here, it can be seen that the volume of earthworks to be removed is partly in cutting and partly in structural foundations. These items are measured separately as required by Series 600 itemisation structure (MMHW Series 600 Paragraph 16) and are indicated as being, respectively, above and below the Earthworks Outline.

However, where a structure interrupts the earthworks, bill compilers need to make it clear how they have measured the excavation items with regard to the bulk excavation in the Roadworks bill. Mitchell (2014) suggests that this is best done by indicating the limits of bulk earthworks and earthworks to structures on the drawings. This is illustrated in Figure 8.7c.

The rules for distinguishing the bulk earthworks (in cuttings and so on) from excavation for structural foundations are provided in Series 600 Paragraphs 15b(i) and 15d(i) and (ii), respectively. Both rules refer to the Earthworks Outline as the reference point for measurement.

8.13.6   Earthworks within designated outlines

In the case of Series 2500: *Special Structures*, the measurement of earthworks quantities depends upon whether:

1. The contractor is responsible for the design of a proprietary structure.
2. There is a choice of designs – engineer's design.
3. There is a choice of designs – contractor's design.

In all cases, the 'special structure' is distinguished from the remainder of the works by the designated outline for the structure. Within the designated outline for a contractor-designed structure, the contractor is responsible for quantifying the earthworks quantities. For a special structure to be designed by the engineer, the overseeing organisation is responsible for measuring and billing the earthworks quantities.

Risk Issue

Bulk earthworks within the Designated Outline for a Special Structure must be included in the quantities for that structure pursuant to MMHW Chapter III Paragraph 7. This says that *Earthworks within the Designated Outlines shall not be included in the Earthworks Schedules.*

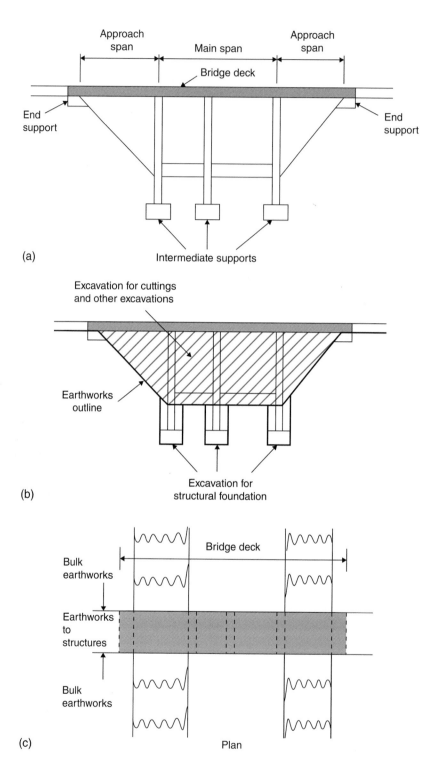

Figure 8.7   Billing of structures. (a) Billing headings, (b) billing of earthworks and (c) earthworks boundaries.

8.13.7 Dealing with water

With regard to the excavation of both topsoil and acceptable and unacceptable material, the item coverages in Series 600 Paragraphs 17 and 18 require the contractor to include in his rates for keeping earthworks free of water.

The measures that the contractor might be expected to take for dealing with water are not specified, or measured, in the method of measurement, but MMHW Chapter III Paragraphs 9 and 10, respectively, spell out that the contractor shall allow for:

- Taking measures required to execute work within and below non-tidal open water or tidal water and any investigations to ascertain actual boundaries, surface levels and ranges affected by non-tidal open water or tidal water (Paragraph 9).
- Taking measures to deal with the existing flow of water, sewage and the like (Paragraph 10).

Risk Issue

Such requirements place onerous obligations on the contractor to which there is no relief in the MMHW.

In circumstances where an experienced contractor could not have anticipated the prevailing circumstances, relief may be sought in the conditions of contract under a Clause 12 Claim (ICC – Measurement Version) or a Compensation Event (ECC Clause 60).

8.13.8 Capping layer

A 'capping' layer is required where the CBR (or other) test reveals that the ground bearing capacity beneath a road is inadequate for the designed traffic loading. If a capping layer is included in the contract, the surface level of the capping becomes the formation level of the road – that is, underside of sub-base – and also the Earthworks Outline.

Capping is not a measured item, and thus, the excavation, deposition and compaction of material to be used as capping are billed under the appropriate feature classification for the various specified classes of acceptable material available in the contract.

Where capping is specified in the contract:

- The void formed to accommodate the capping layer beneath the Earthworks Outline is added to the volume of excavation in *Cutting and other excavation* pursuant to Paragraph 15(b)(i) of Series 600.
- Excavation of structural foundations is measured to the Earthworks Outline according to Series 600 Paragraph 15(d)(i), that is, top of capping layer.

Where capping is to be stabilised with lime or cement, this is to be measured as soil stabilisation (Series 600 Paragraphs 53–56 refer).

8.13.9 Surcharge

The question of 'surcharge' arises where embankments are to be constructed in areas of soft ground. The idea is to put the embankment under load for a period (say, 6–12 months) so that it penetrates the soft ground causing settlement to occur prior to construction of the road pavement, thus avoiding later settlement of the road surface when in use.

Surcharging may be carried out with acceptable material arising on-site or with imported fill, but, being 'temporary', a quantity of material – not necessarily the original quantity of surcharge – is subsequently removed. Where there is a 'loss' of surcharge, this may be due to settlement of the embankment or the contractor's method of working.

The volumes of surcharge material to be placed and removed are given in the earthworks schedules, but only the removal of surcharge is measured as an item in the BQ. However, the tender documents must be suitably explicit such that surcharge requirements, and the likely loss of surcharge material, can be established both in the earthworks balance and to enable tenderers to price the relevant items. The contractor is responsible for the first 75 mm of ground loss under embankments pursuant to Series 600 Paragraphs 13, 33, 45 and 52.

The disposal of any surplus surcharge is not measured separately but is included in the volume of disposal of acceptable material off-site. However, the earthworks balance is initially calculated on the basis that surplus surcharge will be reused on-site whether or not the contractor decides to do so.

The item coverages for filling and compaction items include for dealing with surcharge, but, again, this is not measured in the BQ.

Paragraph 17 of Series 600: *Earthworks* of the *Notes for Guidance on the Method of Measurement for Highway Works* explains in detail how to deal with surcharge and provides worked examples of how to calculate the earthworks balance in three different situations.

8.13.10   Extra over

In the context of a 'base item' measured according to the MMHW, extra over (EO) is a means of measuring any *significant **additional** burden* of work *placed upon the Contractor to undertake **extra work** of much the same nature as the work covered by the base item* (the *Notes for Guidance on the Method of Measurement for Highway Works* Paragraph 2 of Chapters I, II and III refers).

On the face of it, therefore, if *Excavation of acceptable material excluding Class 5A in cutting and other excavation* is the 'base item', an item of *Extra over excavation for excavation in Hard Material in cutting and other excavation* reflects the additional work, over and above normal excavation, for the removal of hard material, such as rock.

The item of *Extra over excavation for excavation in Hard Material* is measured in m³ and, pursuant to Series 600 Paragraph 21, *shall be the volume of Hard Material within the void measured under paragraph 15 of this Series* with a separate item provided for each excavation feature (e.g. cuttings, structural foundations, etc.). This is emphasised in the *Notes for Guidance* which state that *quantities to be billed for the EO items must be in respect of work included with the quantities for the base item.*

All of the foregoing resonates precisely with the commonly understood industry meaning of 'EO' which is that EO items *are not to be priced at the full value of all their labour and materials, as these have to a certain extent already been measured* (Lee et al., 2014) in the base item. Therefore, quite rightly, the item coverage of Series 600 Paragraph 23, *Extra over excavation for excavation in Hard Material*, is limited to the additional costs of excavating hard material, that is:

a)   Preliminary site trials of blasting.
b)   Blasting, splitting, breaking and the like.
c)   Cutting through reinforcement.
d)   Saw cutting and trimming.
e)   Removal of existing paved areas by course or layer, cleaning surfaces, milling or planing, stepping out and treatment to bottoms of foundations.

However, the *Notes for Guidance* state that the item coverage in respect of the quantities for the EO item comprises a summation of item coverages for the base item and the EO item.

Risk Issue

If tenderers follow this guidance, they will be pricing the extra over (EO) item at 'full value' which conflicts with principle of 'EO'.

The reason for this guidance is not clear because the EO item coverage is not 'written short' (i.e. it does not refer to a further detailed item coverage elsewhere) and is a 'stand-alone' measurable item.

However, remembering the *Notes for Guidance on the Method of Measurement for Highway Works* definition of 'EO', this should signal to contractors that the EO rate is to be derived from the 'full value' rate for excavation of hard material *less* the 'full value' rate for normal excavation.

8.13.11   Hard material

The presence of hard material in earthworks can have a significant direct cost implication for contractors, but there may also be a considerable impact on working methods, choice of plant and equipment and upon the planning and sequencing of excavation operations. This is particularly the case when hard material arises unexpectedly or contrary to indications in geological reports.

The measurement of hard material is dealt with in all standard methods of measurement, and each one has its own way of defining exactly what constitutes 'hard material'. At the simplest level, hard material may be categorised as either:

- Naturally occurring hard material, that is, rock

  or
- Artificial hard material such as concrete, brickwork or bound material such as tarmacadam.

The *Notes for Guidance on the Method of Measurement for Highway Works* confirm that the MMHW recognises the cost significance of removing hard material and that such work should be measured *extra over normal excavation*.[14] They also recognise that a consistent approach is needed when dealing with the measurement issues arising from the presence of hard material in order to be fair to both overseeing organisations and to contractors. Such sentiment no doubt arises from the long history of disputes concerning payment for the removal of hard materials, and this partially explains why the definition of hard material has changed over the years.

The *Notes for Guidance* suggest that the current definition should be accepted by all parties and that its inclusion in the contract documents *effectively excludes all other forms of definition*. It is also suggested that the tender documents should clarify what material the contractor is expected to encounter and that the contractor should supplement this with his own inspections where required by the conditions of contract (e.g. ICC Conditions Clause 11(2)).

The *Notes for Guidance* furthermore conclude that the overseeing organisation *should designate which strata or deposits are to be measured as Hard Material* and that bound materials in existing pavements and the like will always be measured as hard material. This is made clear in the MMHW itself where the first part of Definition 1(h) of Chapter I: *Definitions* states that 'Hard Material' is defined as:

(i) *material so designated in the Preambles to Bill of Quantities*

The second part of the definition is that 'Hard Material' is

(ii) *material which requires the use of blasting, breakers or splitters for its removal but excluding individual masses less than 0.20 m³*

In this respect, the *Notes for Guidance* make an important point of clarification that:

- Sub-paragraph (ii) of the definition *outlines the means of determining the volume of Hard Material when circumstances preclude the use of sub-paragraph (i)* albeit that *these circumstances should be rare.*

Consequently, it may be concluded that sub-paragraph (i) is the preferred means of determining the volume of hard material and sub-paragraph (ii) is the 'fallback' definition.

Significantly, however, sub-paragraphs (i) and (ii) are separated by the expression '*and/or*', and this would seem to indicate that 'hard material' is defined as either:

- That which is so designated in the Preambles to Bill of Quantities AND which requires the use of blasting, breakers or splitters for its removal OR
- That which requires the use of blasting, breakers or splitters for its removal

Therefore, this means that:

- Where material is designated as Hard Material in the Bill of Quantities, the material so designated must also require the use of blasting, breakers or splitters for its removal to qualify as 'hard material'.
- Where the designation of hard material is precluded for some reason, 'Hard Material' is defined as that which requires the use of blasting, breakers or splitters for its removal.

This seems fair enough as it would be unreasonable for a contractor to expect payment for the removal of hard material if it were capable of being excavated along with other 'normal' material. The *Notes for Guidance* emphasise this point by stating that where, in the judgement of the overseeing organisation, material is likely to be encountered in bulk excavation that is capable of being removed by *conventional rippers*, then such material *should not be classed as 'hard material'*. An example of this would be using ripper tines attached to the rear of dozers to loosen hard materials for excavation by other plant such as back acters. In coming to this decision, *factors such as the location and extent of the excavation, the size of the project and other limitations* would have to be taken into account.

It must be said that the *Notes for Guidance* are not crystal clear on the subject of hard material, but on a reasonable construction, there is a considerable distinction to be made between the **measurement** of hard material for incorporation in the BQ at tender stage and the **admeasurement** of hard material during construction of the works. This conclusion may be drawn from three specific statements in the *Notes for Guidance*:

1. *Once a strata or deposit has been designated as Hard Material it is not subject to reclassification.*
2. *Where Hard Material is designated by reference to named strata alone the total quantity excavated from within those strata is subject to admeasurement.*
3. *Where deposits are designated by limits shown on the Drawings that volume is measured and paid for as Hard Material.*

From this, it may be concluded that the decision as to what is and what is not hard material at tender stage has a huge impact on the admeasurement of hard material during construction and that this places a considerable burden of risk on the overseeing organisation and a great deal of pressure upon the professional judgement of the bill compiler.

A further conclusion is that once the designation of hard material has been made, that is it – irrespective of the nature of the material encountered. This is confirmed by the *Notes for Guidance* which state that *if the material found during the course of construction is that which was shown at the time of tender, or could be ascertained by the Contractor's pre-tender inspection, then admeasurement should follow the same designations irrespective of the actual hardness of the material.*

8.13.11.1   The measurement of hard material

The whole point of a BQ is to provide the contractor with the most accurate possible calculation of the quantity of work to be carried out on-site. With a completed design, this can be done fairly precisely; otherwise, an approximation has to be made. Hard material, however, is difficult to measure at tender stage because its location and extent is so unpredictable, despite the existence of borehole information.

Consequently, the BQ compiler needs to exercise considerable judgement to determine the quantities of such material for insertion in the BQ. The starting point is usually the borehole logs, but the problem here is that:

- The number of borehole logs available is limited by cost.
- Boreholes or trial pits are only representative of the ground conditions at their precise location.
- Interpolation of borehole and trial pit data is fraught with uncertainty.
- The interpretive data that accompanies ground investigations is an opinion and not a precise statement of fact.
- Geological reports relating to the presence of groundwater can be misleading due to subsequent climatic changes and the time of year.
- The contractor's ability to rip or otherwise dig hard material is dependent on many factors including the location and depth of the excavation and the size, type and condition of earthmoving plant.

The *Notes for Guidance* suggest that the two parts of the definition of hard material should, in general, be compatible. Remembering that the definition states '*and/or*', this means that material designated as 'hard material' in the Preambles to the Bill of Quantities (i.e. the first part of the definition) should be material that is interpreted as requiring the use of blasting, breakers or splitters for its removal (this is the second part of the definition).

Alternatively, if no designation of strata or deposits is made, the '*or*' part of the definition comes into play, that is, the hard material measured in the BQ shall be that which requires blasting, breakers or splitters to remove it. Consequently, the bill compiler should be looking at the borehole logs with a view to finding materials that correspond to the second part of the definition so that they can either be:

- Designated as hard material and measured as such in the BQ or
- Can be measured according to the second part of the definition.

The means of designating hard material for inclusion in the BQ is found in Paragraph 13(c) of the Preambles to the Bill of Quantities which sets out three methods of doing so:

a) Designated strata.
b) Designated deposits with limits shown on the drawings.
c) Existing pavements, footways, paved areas and foundations.

This designation of hard material is purely for measurement purposes and not for the technical classification of such material for incorporation in the works (e.g. as acceptable material for fill).

Risk Issue

The *Notes for Guidance* make it clear that *material similar to that designated as Hard Material in a deposit within defined limits shown on the Drawings*, found elsewhere on-site, will not necessarily be measured as Hard Material.

Naturally occurring hard material is rock of which there are three main types – igneous, sedimentary and metamorphic. Within these classifications is a bewildering array of different materials with different strengths and other characteristics which:

- Can be excavated with normal plant.
- Can be excavated with normal plant but with difficulty or with special teeth or ripping tines (i.e. hard dig).
- Require explosives, pneumatic breakers or other means of stressing to break the rock.

At tender stage, the bill compiler has to decide how to measure such materials in the BQ. This is usually done by a desk study of borehole logs and other information, perhaps supplemented by expert geotechnical opinion, and then a decision has to be made as to how this is to be represented in the contract by identifying:

a) **Designated strata,** which are natural layers of specific materials that are classified as 'hard material' such as granite, limestone, chalk, sandstone, gritstone, etc. Such materials will be identified in the borehole logs, and the relative disposition of such strata can be interpolated from the closest available log(s) as illustrated in Figure 8.8a.

   Consequently, when such strata are found on-site, these will be admeasured as 'hard material'.

b) **Designated deposits,** which are identified on the drawings as locations on-site where the material to be removed is thought to be 'hard', and these locations are indicated by means of identifying lines or 'assumed limits' as shown in Figure 8.9a.

c) **Existing pavements, footways, paved areas and foundations** are, by definition, 'hard material' and do not have to pass the test of needing blasting, breakers or splitters for its removal. Unbound materials, such as sub-base, within existing pavements are not regarded as 'hard material'.

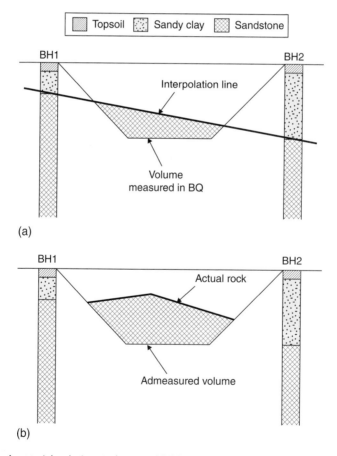

(a)

(b)

Figure 8.8   Hard material – designated strata. (a) Measurement situation and (b) admeasurement situation.

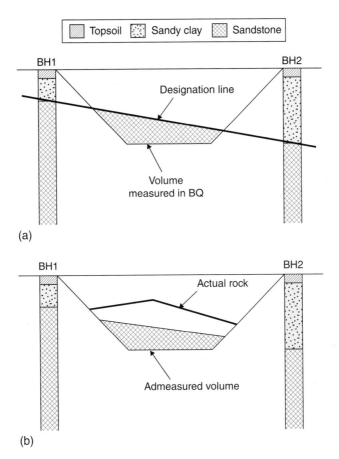

Figure 8.9   Hard material – designated deposits. (a) Measurement situation and (b) admeasurement situation.

At the time of tender, there is no practical difference between the designation of 'strata' and 'deposits' from a measurement perspective as can be seen from Figures 8.8a and 8.9a. In both cases, the disposition of hard material (i.e. rock) is the same as this is interpolated from the same borehole logs.

However, the decision of the overseeing organisation as to how the designation should be made will have a considerable impact when it comes to the admeasurement of the hard materials on-site, as illustrated in Figures 8.8b and 8.9b.

Bound materials in existing pavements and the like will always be measured and admeasured as Hard Material.

8.13.11.2   The admeasurement of hard material

Once construction work begins, the admeasurement of BQ items for *extra over excavation for excavation in Hard Material* becomes as necessity because:

- Traditional highway contracts are measure and value (admeasurement) contracts.
- BQ items are, therefore, approximate.
- BQ quantities of hard material are even more approximate due to the uncertain nature of geological data.
- The 'reality' of site conditions is rarely the same as the 'virtual reality' of drawings and borehole logs.

Accordingly, the means by which hard material is designated in the BQ conditions the approach to the admeasurement of such material because:

- The contractor's tender rates must be established on the basis of a consistent and fair definition of 'hard material' in order to be realistic.
- The admeasurement must be approached in such a way that the tender rates are not rendered inappropriate.
- There must be a means of distinguishing between the ground conditions likely to be encountered on-site and those which are unexpected and likely to give rise to a claim.

Remembering that the MMHW regards existing pavements, footways, paved areas and foundations as, by definition, 'hard material', other hard material (i.e. rock) must be designated with respect to either:

- Specific strata or
- Specific deposits.

This is illustrated by the measurement situations in Figures 8.8a and 8.9a. The distinction between the two designations is subtle but important in terms of admeasurement because the *Notes for Guidance* state that:

1. *Where Hard Material is designated by reference to named **strata** alone the total quantity excavated from within those strata is subject to admeasurement.*
2. *Where **deposits** are designated by limits shown on the Drawings that volume is measured and paid for as Hard Material.*

Before continuing, it must be stressed that material designated as 'hard material' in the Preambles to the Bill of Quantities, whether designated by strata or by deposits, retains its designation for admeasurement purposes. Once designated, hard material is not further subject to the test of whether or not its removal requires the use of blasting, breakers or splitters in the admeasurement situation.

The admeasurement of material otherwise measured in the BQ as 'hard material' (i.e. in series other than Series 600), because its removal is thought to require the use of blasting, breakers or splitters, will always be subject to this test because that is how the material is defined in Definition 1(h)(ii) of the Preambles to Bill of Quantities.

Risk Issue

The contract should contain information known about the existence and extent of hard material, and this includes the whereabouts of any existing buried roads and the like. However, whilst it would appear that the contractor is on 'easy street' with regard to the measurement and admeasurement of hard material, this is not the case.

The problem for contractors is that whilst the material encountered on-site may not be the same as that described in the contract, this may not give rise to a re-rate or a claim. This is because the conditions of contract will usually place a duty on the contractor to make his own interpretation of the likely soil conditions both from geological data supplied by the overseeing organisation and from his own investigations (e.g. arranging permission to dig trial pits at tender stage).

Consequently, any claims would be tempered by the need for the contractor to establish that the prevailing conditions could not have been reasonably anticipated by an experienced contractor.

Assuming that the hard material encountered on-site is the same material as that anticipated at tender stage, but its actual disposition is not as envisaged, then the admeasurement situation is different for hard material designated by strata compared to that designated by assumed limits. This is illustrated in Figures 8.8b and 8.9b where it can be seen that:

- The designated strata is now disposed differently to that envisaged at tender stage and thus will only be measured as hard material according to its actual disposition on-site.
- The designated deposits of hard material will still be admeasured as being bound by the identifying lines on the drawings irrespective of what material is encountered.

This is confirmed by the *Notes for Guidance* which state that:

1. *If the material found during the course of construction is that which was shown at the time of tender, or could be ascertained by the Contractor's pre-tender inspection, then admeasurement should follow the same designations irrespective of the actual hardness of the material.*

Conversely, should the material encountered on-site be different to that envisaged in the tender documents, the contractor would have to argue that such materials could not reasonably have been envisaged and that additional cost had been incurred as a consequence.

This would then give rise to a claim under the conditions of contract, if permitted, as confirmed by the *Notes for Guidance*:

2. *If the material found in the course of construction is not as described in the tender documents or apparent by inspection, the Contractor may raise a claim if permitted under the Conditions of Contract. It will then be for the Contractor to demonstrate that the material could not reasonably have been foreseen and that extra costs had arisen, according to the terms of the Contract.*

Risk Issue

In practice, the disposition of soils can vary widely, and different soils can combine with others, thereby creating an indistinct 'grey area' between strata. Also, hard materials may fragment and merge with other soils.

In such cases, the *Notes for Guidance* suggest that admeasurement of hard material strata is *ascertained by the application of sub-paragraph (ii)* of MMHW Chapter I: *Definitions*.

8.13.12   Soft spots

Having nearly lost a Land Rover, whilst working as a site engineer on the M53 motorway in Cheshire, this author is fully aware of the presence of soft spots in highway projects!

Soft spots and other voids are measured in the MMHW in m^3 under the provisions of Series 600, Paragraphs 61–63, and they are measured separately from the main excavation or filling where the volume:

a) Below structural foundations, foundations for corrugated steel buried structures or in side slopes of cuttings if less than $1\,m^3$.

b) Elsewhere if less than $25\,m^3$.

Separate items are required for the excavation and filling of soft spots, and each is categorised according to whether they are (i) below cuttings or under embankments, (ii) in side slopes or (iii) below structural foundations and foundations for corrugated steel buried structures.

Soft ground above the Earthworks Outline (i.e. formation) is measured as unacceptable material, but, below this level, considerable volumes of such materials can be discovered during earthworks operations – volumes much greater than the thresholds envisaged in the MMHW.

> ## Risk Issue
>
> For contractors, the way that soft spots are dealt with under the method of measurement is less than satisfactory because they can cause disruption to second-stage earthworks operations and may require the contractor to reschedule operations, change working methods or use different plant and equipment to that being used for the bulk muck shifting.
>
> Nonetheless, Series 600 Paragraph 15(b)(i) states that the volume of bulk earthworks includes *the volume of the void formed by the excavation of material below* [the Earthworks] *Outline* and the only recourse is to argue for a 're-rate' should the quantities be sufficiently significant or a misrepresentation should the ground investigation prove erroneous.

8.13.13   *Deposition and compaction of fill*

The deposition and compaction of fill are measured separately in m³ according to Paragraphs 29–33 and 46–52, respectively, of the MMHW.

Excavated material arising from the site, which is to be used for the construction of embankments, reinforced earth structures, anchored earth structures, landscape areas, environmental bunds and fill to structures, is measured as 'deposition of fill' provided that it is:

- Acceptable material.
- Unacceptable material processed to become acceptable.

The quantity of deposition of fill is given by the volume of compacted fill required, calculated in accordance with Series 600, Paragraphs 47–49, *less* the volume of imported fill. Compaction of fill includes the volume of imported fill and deposition of fill.

A distinction is made between the compaction of acceptable material, acceptable material Class 1C and acceptable material Class 6B **according to** Table 6/1: *Acceptable Earthworks Materials* of the SHW. A further distinction is made, pursuant to the itemisation table in Paragraph 51 of Series 600, as to the final destination of compacted fill such as embankments and other areas of fill, strengthened embankments and reinforced earth structures.

The volume of imported acceptable fill is given by the volume of compacted fill, calculated in accordance with Paragraphs 47–49 of Series 600, *less* the volumes of certain acceptable materials and other stated classes of imported acceptable fill.

The quantities and locations of fill are derived from the Roadworks and Structures Earthworks Schedules suggested in the *Notes for Guidance*.

8.13.14   *Imported fill*

Different types of imported fills are categorised in the MMHW according to Paragraph 44 of Series 600 which requires that separate items be given according to fill type and final destination (e.g. embankments, landscape areas, structures, etc.).

Quantities are given in m³ and the item coverage in Paragraph 45(m) stipulates that imported fill resulting from settlement and penetration of landscape areas, environmental bunds and other areas of fill, and from the first 75 mm of settlement and penetration of embankments, shall be included in the contractor's rates.

8.13.15   Disposal

The measured item of 'disposal' is purely for unwanted excavated materials arising from the site categorised according to Paragraph 38 of Series 600:

- Acceptable material excluding Class 5A.
- Acceptable material Class 5A.
- Unacceptable material Class U1A.
- Unacceptable material Class U1B.
- Unacceptable material Class U2.

The volume of disposal is exclusively surplus material arising from the excavation of cuttings, structural foundations, foundations for corrugated steel buried structures and the like, new watercourses, enlarged watercourses, intercepting ditches and so on as itemised in Paragraph 16 of Series 600. This does not include arisings from fencing, drainage or cast-in-place piling, the disposal of which is included in the relevant item coverages of the respective series.

Unlike other methods of measurement, there is no item for the disposal of excavated material on-site. This is classed as deposition of fill.

8.14   Series 500: Drainage and service ducts

The drainage work for a highway project will normally comprise:

- Filter drains (often referred to as 'fenceline' drainage) for cutting off surface and groundwater from land adjacent to the highway, usually consisting of porous/perforated pipework in trenches filled with free draining filter media, carried out 'pre-earthworks'.
- Carriageway 'carrier' drainage for the removal of surface water from the road collected from gullies and drainage channels.
- Piped filter drains, fin drains and narrow filter drains alongside carriageways and, sometimes, in central reservations, filled with suitable filter media.
- Culverts, used to divert existing watercourses under the highway, which may also be considered as 'pre-earthworks' drainage.
- Associated chambers, including catchpits, manholes, gullies and headwalls.
- Lined and unlined ditches.

It should be noted that all drains exceeding 900 mm internal diameter, box culverts, piped culverts and all associated chambers, headwalls, outfall works and concrete bagwork are considered as 'structures' and should be measured in accordance with Series 2500: *Special Structures*.

A further exception is trenches and ducts for street lighting and communications cabling which are measured according to Series 1400: *Electrical Work for Road Lighting and Traffic Signs* and Series 1500: *Motorway Communications*, respectively.

Clearing existing ditches and constructing new and enlarged watercourses and unlined ditches are measured in Series 600: *Earthworks*.

Pre-earthworks drainage follows erection of the permanent fenceline and the topsoil strip over the site. This work, along with culverts, usually commences early in the project, and thus, there may be associated access problems and difficulties delivering materials to specific locations. This is not recognised in the method of measurement as pre-earthworks drainage is not separately itemised in the MMHW.

Drainage arisings – the spoil from drainage excavations – not required for filling trenches may possibly be acceptable as fill elsewhere on the site, or it may be unacceptable and require disposal off-site. This is not a measured item, either in Series 500 or 600, and it is the contractor's responsibility to deal with such arisings as part of the item coverages in Series 500.

Part 2

Decision making as to the suitability of fills is the main contractor's/earthworks subcontractor's responsibility, and the drainage subcontractor will have problems at tender stage deciding what to allow for the disposal of arisings. In view of the expense of haulage, and especially tipping costs, it may be wise to qualify the tender accordingly.

When measuring and billing work in Series 500, it is important to remember that the definitions of Earthworks Outline and other surfaces in Series 600 apply equally well to Series 500.

8.14.1   Principles of measurement

The itemisation of drains and service ducts is determined by Paragraph 15 of Series 500.

Drains are measured separately from service ducts as are pipes of different diameters and design groups. Design groups determine the type of bed/surround to the pipe and the type of backfill, and this information may be found in the MMCHW Volume 3: *HCD*.

Drainage and service ducts are further categorised according to whether they are to be constructed in trench or heading or by pipe jacking or thrust boring and whether they are in side slopes of cuttings or embankments.

Depths to invert are given in 2 m stages with the average depth to invert stated to the nearest 25 mm.

Categorisation is shown in Table 8.15.

Drains and service ducts are measured along their centre lines and between any of the following:

a)   The internal faces of chambers.
b)   The external faces of headwalls.
c)   The intersections of the centre lines at pipe junctions.
d)   The centre of gully gratings (or where no grating is provided, the centre of the gully).
e)   The position of terminations shown in the contract.
f)   The point of change of stage depth.

Depths are measured between invert and existing ground level, but, where the Earthworks Outline is below existing ground level, the depth measurement is taken to the Earthworks Outline.

Table 8.15   Itemisation of drains and service ducts.

Group	Feature	
I	1	Drains
	2	Service ducts
II	1	Different internal diameters
III	1	Depths to invert not exceeding 2 m. The average depth to invert to be stated to the nearest 25 mm
	2	Depths to invert exceeding 2 m but not exceeding 4 m and so on in steps of 2 m. The average depth to invert to be stated to the nearest 25 mm
IV	1	Specified design groups
	2	Particular designs stated in the contract
V	1	Construction in trench
	2	Construction in heading
	3	Construction by jacking or thrust boring
	4	Suspended on discrete supports
VI	1	In side slopes of cuttings or side slopes of embankments

8.14.2   Earthworks outline

The starting point for the measurement of drainage and service ducts in carriageways is the Earthworks Outline. This is defined in Series 600: *Earthworks* Paragraphs 1 to 6 inclusive but applies equally to Series 500.

In practice, the measurement of drainage and the reality of how the work is to be carried out have consequences that need consideration at pricing and pre-contract negotiation stage.

Risk Issue

For drainage subcontractors especially, care needs to be taken with pricing drainage work because the starting level for excavation will invariably NOT be the Earthworks Outline.

In cuttings, the main contractor will leave a protective layer over the formation or may install the capping layer, or part thereof, in order to protect the formation. Consequently, trench depths will be deeper than those measured in the bill of quantities.

The extra cost would be marginal, but, where the drainage subcontractor is responsible for removing arisings, this would be more significant. Common practice, however, is for the earthworks subcontractor to pick up the arisings at a small marginal cost.

In embankments, the converse may be the case, and drainage excavation may commence at levels lower than the Earthworks Outline. This means that trench depths would be shallower than those measured in the bill of quantities.

Whilst that this might be seen as a 'plus', the downside is that the subcontractor would have to return to re-excavate filter drains in order to top up filter media prior to the laying of sub-base.

Drainage arisings that would normally have been picked up by the earthworks subcontractor would now have to be removed by the drainage subcontractor because the bulk earthworks would have finished.

It should be noted that the measurement of narrow filter drains and fin drains differs slightly from that of drains and service ducts. The vertical measurement is from invert to Earthworks Outline depth, but there is only one depth category, that is, not exceeding 1.5 m (Series 500, Paragraphs 25–28 refer).

8.14.3   Types of pavement

The measurement of drainage and service ducts in carriageways is influenced not only by the Earthworks Outline but also by the carriageway construction thickness. In the case of pre-earthworks drainage, existing ground level is used to determine trench depths.

Drainage and service ducts are measured according to the thinnest permissible construction thickness for any of the alternative types of pavement provided for in the contract according to Chapter III 3 (ii) of the MMHW. This is emphasised in the *Notes for Guidance on the Method of Measurement for Highway Works*, Series 500: *Drainage and Service Ducts*, Paragraph 3, which also confirms that *there is no requirement to provide separate drainage Bills of Quantities corresponding with each alternative Type of Pavement*. These rules also apply to the admeasurement of such work, but this is based on the thinnest permissible construction thickness for the alternative chosen by the contractor.

8.14.4   Tabulated billing

An unusual feature of the MMHW is the option to bill drainage work traditionally or using a tabulated layout. However, this is not an option provided by the MMHW itself but rather the *Notes for Guidance on the Method of Measurement for Highway Works* which suggest that the billing of pipe runs of varying diameter and specification may create a lengthy Series 500 bill unless there are non-standard or small quantities when traditional billing *would be best*.

A suggested method of tabulation is provided in the *Notes for Guidance* which helps reduce repetition of item descriptions for pipe runs, manholes and chambers. Tables 8.16 and 8.17 illustrate the principles of tabulation suggested although there is nothing to prevent the bill compiler following another form of tabulation.

Table 8.16   Tabulated billing – drains.

Item	Description				Unit	Quantity	Rate	£	p
	'A' mm internal diameter drain specified design group 'B' in trench depth to invert exceeding 2 m, but not exceeding 4 m, average depth to invert 'C' metres Adjustment on this item for variation greater than 150 mm above or below the average depth of 'C' metres per 25 mm of variation in excess of 150 mm. Rate per metre 'D' (not to be extended)								
	'A' Diametre	'B' Design group	'C' Ave. depth	'D' Adjust. rate					
21	150	6	2.625		m	54			
22	225	7	2.950		m	18			
23	300	7	2.875		m	78			
24	450	8	3.275		m	157			

Note: Adjustment rate 'D' shall apply to both increases and decreases of average depth in excess of 150 mm and will result in either a positive or negative adjustment of the rate.

Table 8.17   Tabulated billing – chambers.

Item	Description						Unit	Quantity	Rate	£	p
	(05/01) Chamber-specified design group 'A' subtype 'B' with 'C' and 'D' and frame depth to invert exceeding 'E' metres but not exceeding 'F' metres										
	'A' Design group	'B' Sub-type	'C' Cover grade	'D' Type	'E' Depth min.	'F' Range max.					
76	2	—	Grade A	Cover	1	2	No	10			
77	3	a	Grade A	Cover	1	2	No	60			
78	3	b	Grade A	Cover	1	2	No	70			
79	3	c	Grade A	Cover	2	3	No	55			

Risk Issue

The *Notes for Guidance* is not a contract document, but there should be no conflict with the MMHW, provided that items are measured and billed strictly in accordance with the method of measurement.

8.14.5   Drainage and service ducts

Items for drains and service ducts, including sewers and piped culverts not exceeding 900 mm diameter, are measured according to Paragraph 15, Groups I–VI of Series 500, and in accordance with the features listed. This means that drains and service ducts, pipes of different diameters and those belonging to different design groups are required to be measured separately as well as pipes constructed by different methods and in different locations.

The item coverage for such items is to be found in Paragraph 16 where it will be noted that:

- Items for excavation, disposal, formwork and protective systems are 'written short'.
- There is a long list of 'deemed to be included items' such as:
  - Fixing draw ropes, removable stoppers, marker blocks and posts in service ducts.
  - Building ends of pipes into headwalls and outfall works.
  - Access shafts to headings and their subsequent reinstatement.
  - Thrust pits and thrust blocks for pipe jacking and their removal on completion.

Risk Issue

Hidden in the depths of Paragraph 16 is item coverage (e) articulated pipes and fittings. This is easily missed and/or misinterpreted because it refers to both (i) articulated pipes and (ii) to pipe fittings generally.

Consequently, pipe bends, junctions/branches and the like along the length of a pipe run are deemed to be included in item coverage 16(e).

The written short items for excavation of both 'acceptable' and 'unacceptable' materials (Series 600 Paragraphs 17–19 refer) include disposal whether or not such material is to be re-used on-site or removed to tips off-site. Not only should the contractor take this into account when calculating the earthworks balance, there is also an interface to manage at pre-subcontract stage between the earthworks and drainage subcontractors as to who carries the responsibility for removal of drainage arisings.

Risk Issue

Drainage subcontractors need to be aware that the removal of drainage arisings will be their responsibility under the subcontract unless undertaken by the earthworks subcontractor.

This is a pre-contract issue, with significant financial implications, that should be resolved before entering into a subcontract.

8.14.6   Drainage and service ducts in structures

Drainage and service ducts in structures are measured separately from 'normal' drainage items according to Series 500, Paragraphs 52–55. Such items should be clearly delineated on the drawings, and the relevant quantities should be scheduled and included either on the drawings or in an appendix.

In the case of structures to be designed by the contractor, and their associated drains and service ducts, they shall be measured in accordance with Series 2500 (Series 500, Paragraph 14 refers).

Part 2

8.14.7   Filter drains

Narrow filter drains and fin drains are categorised separately from filter drains in the MMHW.

Depths of filter drains are calculated as for drains and service ducts with the proviso in Paragraph 19 that the depths of filter drains which have no pipe are measured to the bottom of the trench.

The item coverage for narrow filter drains and fin drains is 'written short' and refers also to the item coverage for filter drains (Series 500, Paragraph 22). This is where the excavation and disposal item coverage is to be found.

Within the itemisation features for filter drains is an item for *Filter material contiguous with filter drains*. This is a separately measured item for additional filter media – detailed in edge of pavement details in MCHW Volume 3: *HCD* – that occupies the void between the filter drain and the carriageway, hard shoulder or hard strip (see Figure 8.6a).

For narrow filter drains (filter media wrapped with a geotextile) and fin drains (a proprietary vertical cored filter membrane wrapped in geotextile), a similar item is measured for *Fill on sub-base material, road base and capping*.

In both cases, the measured item is additional to any filter media deemed included in the drain itself, but the item coverage does not include for any excavation or disposal.

Part 2

Risk Issue

When filter drains and the like are 'topped up' to their final required level, this is usually done sometime after the installation of the drain itself. As such, subsequent earthworks and roadworks operations may have deposited unsuitable materials over the filter drain.

Any subsequent excavation and removal of such materials is not a measured item but must nevertheless be allowed for in the relevant drainage item and in the item for contiguous filter media and fill on sub-base, road base and capping.

8.14.8   Measurement of drains, sewers, piped culverts, etc.

Drains and the like are measured along their centre lines commencing at the outfall or lowest end. Pursuant to Series 500, Paragraph 12, depths are calculated every 10 m except for terminal lengths and lengths shorter than 10 m whose depths shall be calculated at their ends.

In Figure 8.10, a sewer comprising three drain runs is illustrated commencing at the outfall manhole F1 and finishing at manhole F4. Drain run F2–F3 has a terminal length of 7 m and drain run F3–F4 is only 9 m long.

The average depths of the three drain runs are calculated as shown in Table 8.18.

Risk Issue

It should be noted that the arithmetic mean of trench depths is not a weighted mean.

This means that it is possible for a drain run to have wildly fluctuating depths to invert along its length but a relatively low average depth.

This may prove to be significant in terms of the contractor's pricing, including the choice of excavation plant, the rate of forward travel (i.e. output/progress of the drainage gang) and the earthworks support requirements such as the type of temporary works needed, excavation overbreak and additional backfill and/or disposal.

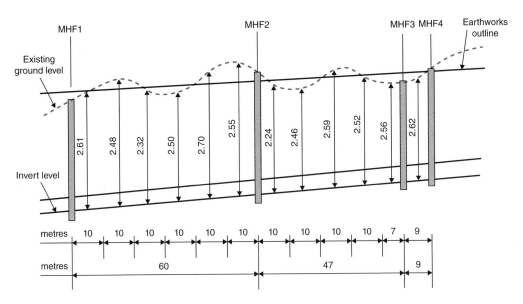

Figure 8.10   Average depths.

Table 8.18   Average depths.

Drain run	Length (m)	Depths (m)	No. of readings	Arithmetic mean (m)	Remarks
F1–F2	60	2.61+2.48+2.32+2.50+ 2.70+2.55=<u>15.16</u>	6	$\dfrac{15.16}{6}$ **= 2.527**	Depth of final 10 m length taken at MH F2
F2–F3	47	2.24+2.46+2.59+2.52+ 2.56=<u>12.37</u>	5	$\dfrac{12.37}{5}$ **= 2.474**	Depth of terminal length taken at MH F3
F3–F4	9	2.62	1	$\dfrac{2.62}{1}$ **= 2.620**	Depth of short length taken at MH F4

8.14.9   Adjustment items

An unusual feature of Series 500 of the MMHW is the provision of an adjustment item to accompany billed items for drains and service ducts and filter drains, but not narrow filter drains and fin drains, as shown in Table 8.19.

The adjustment item recognises that drains and the like will, in all likelihood, be admeasured on completion and that a new rate will have to be established in order to value items where there has been a change in average depth.

Therefore, for each item which includes Group III Feature 1 or 2 (i.e. a drain run where the average depth is stated), *an associated item shall be provided for adjustment of the rate for each 25 mm of difference in excess of 150 mm where the average depth to invert calculated from site measurement varies from that stated in the Bill of Quantities.*

This is illustrated in Figure 8.11.

The adjustment applies to both increases and decreases in the average depth in excess of 150 mm and may result in either a positive or negative adjustment of the BQ rate. A worked example is shown in Table 8.20.

Table 8.19 Adjustment item.

		Qty	Unit	Rate	£	p
	500: Drainage and Service Ducts					
	<u>Drains and Service Ducts (excluding Filter Drains, Narrow Filter Drains and Fin Drains)</u>					
A	300mm diameter drain; specified design group 7; in trench; Depth to invert not exceeding 2 metres average depth to invert 1.875 metres	1076	m			
B	Adjustment on last item for variation greater than 150 mm above or below the average depth of 1.875 metres per 25 mm of variation in excess of 150 mm		m			

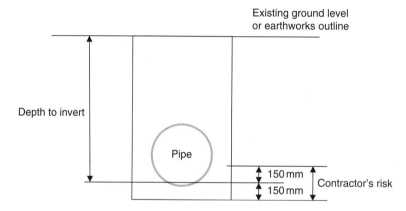

Figure 8.11 Adjustment item.

Table 8.20 Pricing of Adjustment item.

Actual average depth		3.290	
BQ average depth		<u>2.780</u>	
Difference ±	+	0.510	
Less			
Non-adjustable element		<u>0.150</u>	
No. of depth increments	0.025	0.360	14.4 = <u>14</u>
<u>New rate</u>			
Original rate		£37.76	
Adjustment rate	£0.19		
Adjustment factor	14	×	<u>2.66</u>
New rate		**£42.42**	

It will be noted that the adjustment will be based on-**site measurement** and not on revised drawings or drainage schedules.

Risk Issue

The site measurement of drainage can be problematic for a number of reasons:

- The average depth of drains is the arithmetic mean of depths taken every 10 m (or part thereof) measured form the outfall.
- This can only be done when the drainage work is being carried out and before surrounding or backfilling pipes.
- Drainage work is done before, during and after earthworks, and there is no guarantee that there will be an Earthworks Outline to measure from (capping may be incomplete, for instance).
- Drainage is not an exact science and may not necessarily be installed to the precise levels shown on the drawings.
- Site measurements are notoriously inaccurate.
- Remeasurement of drainage, even on a relatively small highway project, is a significant and time-consuming task.

In all practicality, it is unlikely that the adjustment will be based on **site measurement** notwithstanding the provisions of the method of measurement.

8.14.10   Chambers and gullies

Chambers and gullies are itemised in the MMHW (Series 500, Paragraphs 33–38), and the measurement is for the complete chamber or gully. Channels, fittings, benching, building in pipes and fin drain connections are included in the item coverage.

The relevant specified design group must be referred to in the item description which relates to the standard details provided in MCHW Volume 3: *HCD*.

Apart from the design group, such items are distinguished by their depth, stated in 1 m stages, depths being taken from cover level to channel invert or to base slab in the case of catchpits, and by the specification of the cover or grating.

Risk Issue

Standard highway construction details usually require a short length of pipe to be built into the inlet and outlet sides of chambers. These are called 'articulated pipes' in the SHW/MMHW but are commonly referred to as 'rocker pipes'.

Rocker pipes are not measured separately, but, being considerably more expensive than standard length pipes, they demand an 'extra over' allowance to be built into the contractor's tender.

This might be conveniently included in the pricing for each chamber but must be remembered in the 'take-off list' when preparing 'builders' quantities' for such items at tender stage.

8.14.11   Headwalls and outfall works

Headwalls and outfall works to pipes less than 900 mm diameter are measured in accordance with Paragraphs 40–42 of Series 500. For pipes above this diameter, Series 2500: *Special Structures* applies.

Part 2

Headwalls, revetments, etc. are enumerated, and, therefore, tenderers need to carry out additional quantification at tender stage to determine the quantities of excavation, concrete, formwork and brickwork.

Risk Issue

The majority of the item coverage for headwalls and outfall works is 'written short', and special care is needed when pricing such items, particularly if they are in proximity to rivers and tidal waters.

8.14.12   Soft spots

Separate items are required for the excavation of soft spots and other voids, and the filling of soft spots and other voids are classified by the different types of fill specified.

Risk Issue

Measurement of soft spots is the volume of the void *directed to be excavated or filled*, and thus, the contractor would be unwise to undertake such work without an instruction or without confirming a verbal instruction in writing.

There are no limitations as to the volume of soft spots measured under Series 500, as there are under Series 600: *Earthworks*, but there are limitations as to the width of the excavation to be taken:

- For drains, service ducts and filter drains, it is the internal diameter of the pipe plus 600 mm.
- Where there is no pipe, it shall be taken as 600 mm.
- For chambers, gullies and the like, it is the horizontal area of the base slab or, where no base slab is required, the bottom of the excavation.

The item coverage for both excavation and filling of soft spots is 'written short'. For excavation, the relevant paragraphs are 17–19 and 39 of Series 600: *Earthworks* and, for filling, Paragraphs 33 and 52 of Series 600 and Paragraphs 5 and 15 of Series 1700: *Structural Concrete*.

Risk Issue

Excavation of soft spots includes upholding the sides of excavations, dealing with water, disposal and, importantly, overbreak and making good.

8.14.13   Extra over

Notwithstanding the shortcomings of the MMHW approach to EO items, as previously discussed in 8.13.10, items measured as 'EO' in Series 500: Drainage follow the same principles as Series 600: *Earthworks*.

However, whilst the principles are the same, practical issues arise with regard to hard material in drainage work, and this issue is discussed in 8.14.15.

8.14.14   Dealing with water

The item coverage for the excavation of drains and service ducts (including culverts <900 mm diameter) and filter drains are 'written short' to Series 600: *Earthworks*, Paragraphs 17 (topsoil), 18 (acceptable material) and 19 (unacceptable material). For narrow filter drains and fin drains, the item coverage is 'written short' to Series 500: *Drainage* Paragraph 22 which, in turn, refers to Series 600.

This means that the contractor shall allow in his rates and prices for drainage work for working within and below non-tidal open water or tidal water, for dealing with the existing flow of water, sewage and the like and for carrying out investigations in order to establish water boundaries, levels and fluctuations.

Risk Issue

The consequence of this is that the contractor is deemed to have included *keeping earthworks free of water* in his rates. In common with Series 600: *Earthworks*, this is a general obligation, with no limits, save for the test of what an experienced contractor could have expected having visited the site and perused the contract documentation.

8.14.15   Hard material

The excavation of hard material in drainage is measured as EO the excavation in which it occurs, with the unit of measurement in m^3, as illustrated in Table 8.21.

The measurement is the *volume of the voids formed by the removal of the Hard Material*, but this is qualified in Series 500 Paragraph 74 such that:

- For drains, service ducts and filter drains (except fin drains and narrow filter drains), the width measured is the internal diameter of the pipe plus 600 mm or, where there is no pipe, 600 mm.
- For fin drains and narrow filter drains, the width measured is 300 mm.
- For chambers, gullies and the like, the area measured is the horizontal area of the base slab or, where there is no base slab, the area of the bottom of the excavation.

The item coverage for excavation in hard material is 'written short', and Series 600 Paragraph 23 therefore applies.

However, a problem arises when it comes to the admeasurement of EO for excavation in hard material because the EO item coverage includes the extra cost of removal, such as blasting,

Table 8.21   Excavation of hard material.

					£	p
	SERIES 500 - DRAINAGE AND SERVICE DUCTS					
	Excavation in Hard Material		(NA)			
	Extra over excavation					
A	for excavation in hard material	900	m³			

Part 2

splitting, breaking, etc., and does not include for the additional costs associated with the void caused by the removal of hard material (i.e. 'overbreak and making good').

The additional excavation, disposal, pipe bedding and backfilling can be considerable, as is illustrated in Figure 8.12, because no one can accurately predict how rock will break or what sizes or dispositions the rock or artificial hard material will be in.

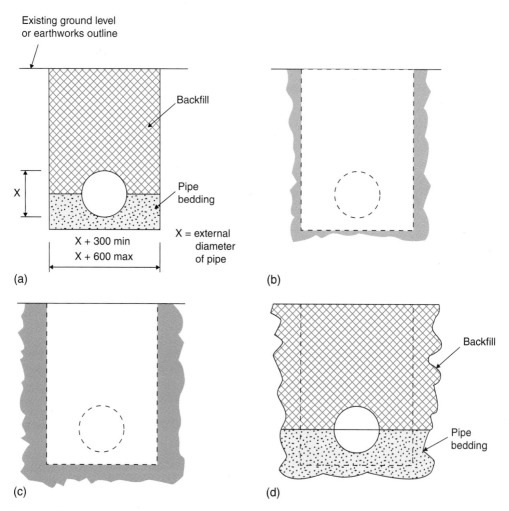

Figure 8.12 Overbreak in hard material. (a) Standard detail, (b) 'Normal' overbreak, (c) overbreak in hard material and (d) completed drain trench.

Consequently, reference must be made to Series 600 Paragraphs 18 and 19 (which is also written short) to find that it is in the relevant drainage item where the overbreak is deemed to be included.

From the borehole logs, superimposed on the long sections, tenderers may be able to interpolate where the hard material is likely to arise and, roughly, which drain runs are affected. In this way, the measured quantity of EO for excavation in hard material can be related to the length(s) of drain runs involved, and a suitable allowance can then be made in the relevant BQ rate(s).

Risk Issue

If the actual quantity of hard material is significantly more than that billed, the contractor could find that the additional cost of excavation, disposal, pipe bedding and backfilling, due to the overbreak, is not reflected in the rates for drain runs, as the 'extra over' item coverage excludes overbreak.

One way round this issue is for the contractor to price an enhanced rate for the extra over item to allow for the additional costs that would otherwise not be recovered, notwithstanding the item coverage. This, however, would increase the tender price, and consideration may have to be given to this issue in the contractor's 'commercial opportunity' and tender risk assessments.

8.15   Series 1600: Piling and embedded retaining walls

This series deals with:

- Precast concrete piles.
- Cast-in-place piles.
- Steel bearing piles.
- Steel sheet piles.
- Diaphragm walls.
- Secant pile walls.
- Contiguous bored pile walls.
- King post walling (e.g. steel 'H' beam sections, set into 'wet' cast-in-place piles, with precast concrete panels between).

There are some unusual features of this Series.

8.15.1   Piling plant

For precast concrete piles, bored cast-in-place piles, driven cast-in-place piles, steel bearing piles and steel sheet piles, Series 1600 Paragraph 4 requires a measured item for *establishment of piling plant* (item), and each type of pile, except steel sheet piles, requires an item for *moving piling plant* (number).

The establishment of piling plant is measured once only for each structure (e.g. a bridge) but shall not be admeasured to suit the contractor's method of working. The measurement of moving piling plant shall be measured once only per pile, but the moving of piling plant for steel sheet piling shall not be measured. This is illustrated in Table 8.22.

Table 8.22   Piling plant.

		Qty	Unit	Rate	£	p
	1600: Piling and Embedded Retaining Walls					
	Piling Plant					
A	Establishment of piling plant for 450 mm diameter; bored cast-in-place piles; in main piling; Bridge 16A		item			
B	Moving piling plant for 450 mm diameter; bored cast-in-place piles; in main piling; Bridge 16A	42	no			

The item coverages for 'establishment' and 'moving' piling plant include, *inter alia*, site preparation, levelling and access ramps, but, although the construction of piling 'mats' is not specified, it may be implied.

The provision for 'piling plant' seems somewhat out of step with the philosophy of the MMHW and a 'nod' in the direction of method-related charges. Whether this is to avoid arguments should a variation instruction require more piles, and subsequent moving, or even remobilisation, of the piling rig, is unclear. In the majority of instances, of course, piling will be a contractor-designed proprietary structural element, and so the problem as to whether the bill compiler should 'second-guess' the contractor's intentions as regards the number of piles raises its head again!

8.15.2   Cast-in-place piles

Where bored piles require casings or linings, these are included in the item coverage given in Series 1600 Paragraph 23 and are regarded as temporary *unless the Contract states specifically that they are to be left in place*. This is a provision of the *Notes for Guidance* and not the method of measurement itself.

Empty bores are only measured where the contract specifies a particular commencing level from which boring shall begin and shall be the length of empty bore measured from the finished level of the pile to the specified commencing level.

The measurement of reinforcement to cast-in-place piles requires the following itemisation:

- Bar or helical reinforcement.
- Nominal size 16 mm and under and 20 mm and over.
- Different types and grades of steel.
- Bars not exceeding 12 m in length.
- Bars exceeding 12 m in length but not exceeding 13.5 m and so on in steps of 1.5 m.

Consequently, rates are very much an 'average' in terms of bar diameter, and tenderers, therefore, need to undertake additional measurement at tender stage in order to determine the correct weightings.

8.15.3   Embedded retaining walls

For embedded retaining walls – diaphragm walls, secant pile retaining walls and contiguous bored pile walls – an item for the *establishment of embedded retaining wall plant* shall be given, and the unit of measurement shall be 'item'.

This measurable item shall be given only once for each embedded retaining wall.

As might be expected, the excavation of acceptable/unacceptable material is 'written short' in the item coverages, as per Series 600 Paragraphs 17–19, as is excavation in hard material (Series 600, Paragraph 23) and disposal (Series 600, Paragraph 39).

For complex-shaped diaphragm walls, the *Notes for Guidance* suggest that the wording of Series 1600, Paragraph 63 (i.e. the unit of measurement – m^2) may need to be amended as a departure from the method of measurement with, perhaps, an indication of the developed length given on a drawing for clarity.

Should an embedded retaining wall have a finishing thickness, such as brickwork, this should be separately measured in accordance with relevant series. Cleaning and treatment of the embedded retaining wall face are included in the item coverage in Paragraphs 65, 69 and 73 of Series 1600 and are not required to be measured separately.

8.16 Series 1700: Structural concrete

8.16.1 In situ concrete

In situ concrete to structures is measured in m^3 and itemised according to its class or design, if a designed mix is specified, as shown in Table 8.23.

Table 8.23 Structural concrete.

Group	Feature	
I	1	In situ concrete.
II	1	Different designs.
	2	Different classes.
III	1	Blinding concrete 75 mm or less in thickness.

In billing such items, it is important to follow the provisions of Chapter III: *Preparation of Bill of Quantities*, Table 1, which requires bridges or viaducts to be identified with a name or reference and for each structure to be separated into substructure end supports, substructure intermediate supports, superstructure and finishings.

Effectively, this provides all the locational information that the contractor needs to price the concreting items as illustrated in Table 8.24.

Table 8.24 Typical bill items for structural concrete.

Bridge 16A

			Qty	Unit	Rate	£	p
	1700: Structural Concrete						
	Substructure End Supports						
A	Insitu concrete; Class C40		108	m³			
B	In situ concrete; Class C10; in blinding 75 mm or less in thickness		5	m³			

Establishing the boundary between structural concrete in substructure and in superstructure is an interesting question as the MMHW does not provide a rule.

For the excavation of structural foundations, the boundary is the Earthworks Outline, but adopting this for the measurement of structural concrete would mean apportioning a bridge pier or abutment wall between substructure and superstructure as illustrated in Figure 8.13.

Measuring the foundation and pier or abutment wall together would be another option with 'superstructure' being limited to the bridge deck and support beams albeit that this would not be a 'strict' separation of substructure and superstructure.

8.16.2 Formwork

Series 1700 requires the measurement of (i) formwork (m^2) and (ii) void formers (linear metres). Group II features of the itemisation table provide for the separate measurement of horizontal, inclined,

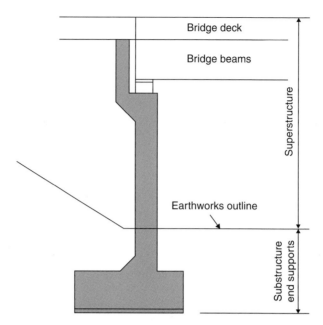

Figure 8.13   Structural concrete.

vertical and curved formwork and for void formers of different cross sections. Formwork of all types is measured as 300 mm wide or less and more than 300 mm wide, but all are measured in m².

Different classes of surface finish to concrete must be measured separately, but there is no classification given in the method of measurement. Surface finishes are determined by the SHW which specifies:

- Formed surfaces F1–F5.
- Unformed surfaces U1–U5.

Formed surfaces are those created by formwork, and the quality of the finish is made explicit in SHW Series 1708: *Concrete – Surface Finish*, Paragraph 4(i). The desired finish, and not the means of achieving it, is specified for each of the formed surfaces F1–F5, F5 being the most exacting standard. Unformed finishes do not require formwork.

The measurement of formwork is the area which is in contact with the finished concrete, measured over the face of openings of 1 m² or less and over certain listed features.

Formwork to concrete in structural foundations, other than blinding concrete, is measured to the sides of the foundations irrespective as to whether or not formwork is used, but, where stated on the drawings that the concrete is to be cast against the soil face, no formwork is measured.

Patterned profile formwork is accorded its own Group.

Risk Issue

An important point to remember with formwork under the MMHW is that the item coverage includes, *inter alia*, for falsework which is, therefore, deemed to be included in the rates per m².

Falsework is the temporary works needed to support the formwork itself and is defined as *any temporary structure used to support a permanent structure while it is not self-supporting* (British Standards Institution, 2011).

Such temporary structures can be significant cost items, but, as there is no separate method-related charge, a check on the quantities of formwork measured in the BQ would be prudent at tender stage should the contractor decide to price the falsework in with the formwork item.

8.17   Series 2700: Accommodation works, works for statutory undertakers, provisional sums and prime cost items

As with most other construction projects, highway works are never completely designed and detailed at tender stage, and some issues will remain that need to be resolved during the contract.

In order to cater for such work, the normal procedure is to include provisional quantities or provisional sums in the tender BQ so that instructions may be issued with regard to the quantum of such work or the expenditure of the provisional sums included. This is confirmed by the *Notes for Guidance* for Series 2700, Paragraph 1, which provides that:

- *Where accommodation works and works for privately and publicly owned services and supplies are known prior to tendering they should be billed in accordance with Chapter III and with the various Series of the MMHW.*
- *It might be appropriate to insert provisional quantities as provided for in Chapter I.*[15]
- *If neither of these options can be used and accommodation works and works for privately and publicly owned services and supplies are anticipated but cannot be defined then a Provisional Sum may be included in the Bill of Quantities.*

Such provisions in the contract bills ensure that instructions regarding the expenditure of provisional sums and the admeasurement of provisional quantities are covered by express contract terms and are not reliant on extra-contractual agreement with the contractor as would otherwise be the case.

8.17.1   Accommodation works

In view of the 'land take' required for a highway project, it is inevitable that land and property will be disturbed by the proposed works, especially in urban areas. Areas disturbed will have to be reinstated or modified by the contractor as part of the contract, and works may be required to divert watercourses; reconnect sewers and drains; reinstate gardens, paths and pavings and landscape disturbed areas; erect fencing to re-establish boundaries or to contain cattle or sheep; and so on.

Provision for such works is made in the contract bills by the inclusion of a bill of accommodation works. There is no provision in the MMHW as to where this bill shall appear except that the method of measurement *shall be in accordance with the various Series of this Method of Measurement.*

Problematically, at tender stage, the majority of accommodation works requirements will not have been resolved in sufficient detail to be measured, and, therefore, provisional quantities or a provisional sum will have to be included in the tender bills to allow for such works.

Remembering, however, that Highways England has adopted the ECC as its preferred form of contract, and that there is no mechanism in this contract for provisional sums as is the case with the ICC – Measurement Version form of contract, the bill compiler is faced with the problem as to how to include such provisions in the tender bills.

8.17.2   Works for statutory undertakers

Similar principles apply to works for statutory bodies as those for accommodation works with the additional problem that the whereabouts of statutory services may not be known or may not be where they were expected to be.

The traditional way of providing for investigative work is to include a daywork schedule in the contract which establishes competitive rates for time-based payments for work instructed by the contract administrator.

Again, however, the ECC form does not provide for daywork, and such matters are normally resolved by way of compensation events valued in accordance with the *Shorter Schedule of Cost Components*.

8.17.3   Provisional sums, etc.

The *Notes for Guidance* for Series 2700, Paragraph 2, draws the bill compiler's attention to the fact that *certain Forms of Contract do not support the inclusion of Provisional Quantities, Provisional Sums, Dayworks or Prime Cost Items. Reference should be made to the specific Form of Contract to be used before such items are included in the Bill of Quantities.*

This 'guidance' is no help whatsoever for contracts where the ECC form is to be used because, notwithstanding the use of a method of measurement that provides for such items, there must be an accompanying provision in the contract to empower the contract administrator to:

- Admeasure provisional quantities and adjust rates, where appropriate.
- Expend provisional sums, issue appropriate instructions and establish how the work shall be valued.
- Require the submission of daywork records, verify their correctness and value the work recorded.
- Expend prime cost sums for the carrying out of work or for the provision of goods, materials or services for the works.

It would appear that the ECC compensation event procedure, including the option to instruct the contractor to submit quotations for compensation events, is the only way to resolve the question of provisional or unforeseen work, together with the Clause 10.1 provision that the parties shall act *in a spirit of mutual trust and co-operation*.

8.18   Other works

Compared with CESMM, the MMHW is somewhat limited in the types of work covered.

True, there is a series for everything required for a standard highway project but not, it is submitted, for situations where tunnels, immersed tubes, railway trackwork and small building work may be required.

8.18.1   Building work

Where small building works, such as administrative and public buildings, public toilets and the like, are required, the MMHW lacks a suitable series for measurement purposes.

Creating a 'bespoke' series would be a considerable task as this would not only require the creation of a specific method of measurement but also a specification, standard construction details and notes for guidance.

Using other methods of measurement such as SMM7, NRM2 or CESMM (Class Z: *Simple building works incidental to civil engineering works*) is more attractive, but, in solving one problem, others may be created:

- BQ with items measured in the detail required of SMM7/NRM2 would create disproportionately lengthy documents and would be against the ethos of the MMHW.
- Introducing a second method of measurement into the contract would require careful attention with regard to drafting the contract appendix or works information and with regard to the amendment of Clause 57 of the ICC – Measurement Version form which states the default method of measurement.
- A designated outline to distinguish between works measured under the MMHW and another method of measurement might be needed.

8.18.2 Other civil engineering work

It is not uncommon to find highway projects with elements that are not included in the MMHW. Where a tunnel is required, for instance, there is no suitable series provided as a basis for measurement.

Amending the MMHW or adopting another method of measurement is an option to overcome this problem, but neither is ideal or without difficulties (see 8.18.1), neither is Series 2500: *Special Structures* **designed for** highway schemes incorporating large road tunnels and the like.

Fortunately, the availability and popularity of non-traditional procurement methods come to the rescue as such projects are often undertaken by consortia of engineers and contractors on a D&B basis.

Notes

1. https://www.gov.uk/government/organisations/highways-england (accessed on 6 April 2015).
2. http://www.dft.gov.uk/ha/standards/mchw/ (accessed on 6 April 2015).
3. http://www.constructionproducts.org.uk/publications/industry-affairs/display/view/construction-products-regulation/ (accessed 29 April 2015).
4. http://www.constructionenquirer.com/2013/10/24/race-starts-for-5bn-highway-agency-framework/ (accessed 29 April 2015).
5. https://www.gov.uk/government/organisations/highways-england/about/procurement (accessed 29 April 2015).
6. https://www.gov.uk/government/organisations/highways-england/about/procurement (accessed 29 April 2015).
7. Model Contract Document for Highway Works, Volume 0, Paragraph 1.11.
8. Model Contract Document for Highway Works (England), Chapter 4 *Instructions for Tendering*.
9. To be completed by compiler as appropriate.
10. Notes for Guidance on the Specification for Highway Works, NG 104 Paragraph 25 – *Statutory type approval is granted by the Secretary of State. Where the Contractor designs part of the works and makes application for approval, he should forward the information to the Overseeing Organisation in sufficient time for approval to be given, taking into account the programme for the works. Where statutory type approval is given, one copy of the approval certificate should be returned to the Contractor.*
11. Notes for Guidance on the Specification for Highway Works, NG 106 Paragraph 7.
12. Notes for Guidance on the Specification for Highway Works, NG 106 Paragraph 1.
13. http://www.ciht.org.uk/motorway/m6toll.htm (accessed on 6 April 2015).
14. Notes for Guidance on the Method of Measurement for Highway Works, Series 600, Paragraph 7.
15. Chapter 1: *Definitions*, Paragraph 1(d) – *items designated 'Provisional'*.

References

British Standards Institution (2011) *Code of practice for temporary works procedures and the permissible stress design of falsework* (BS 5975: 2008 + A1:2011), British Standards Institution.

Money B. and Hodgson G., (1992) *Manual of Contract Documents for Highway Works – A User's Guide and Commentary*, Thomas Telford, London.

Loots P., Charrett D., (2009) *Practical Guide to Engineering and Construction Contracts*, McPherson's Printing Group, Australia.

Mitchell H., (2014), *Managing with the MMHW*, Chartered Institution of Civil Engineering Surveyors.

Lee, S. Trench, W. Willis, A. (2014) *Willis's Elements of Quantity Surveying* Twelfth Edition, Wiley-Blackwell, Chichester.

National Audit Office (October 1992) *Department of Transport – Contracting for Roads (HC 226)*, HMSO.

Chapter 9
Principles of Measurement (International)[1]

Part 2

In the Foreword to the 2004 reprint of POM(I), Simon Cash, chairman of the RICS Construction Faculty, remarks that measurement-based procurement is still appropriate, despite the popularity of other 'newer' approaches, because circumstances still arise where there is design certainty before the contractor is appointed and where, by implication, the advantages of bills of quantities can benefit all concerned in reducing one-sided risk and ensuring a fair and balanced contract.

Simon Cash also points out that *measurement should be undertaken at the level at which design is carried out* and makes reference to the extent to which design decisions are being passed down the supply chain as a result of non-traditional procurement arrangements. Mr Cash also pleads for consistency in measurement, whoever does it, and underlines the consequent need for standard methods of measurement and appropriate measurement skills.

Mr Cash adds, importantly, that the Principles of Measurement (International) (POM(I)) *require a detailed specification and tender drawings to be provided* but hints that it is recognised by the RICS that they also *need to be flexible in order to accommodate* [local] *variations in practice and techniques.*

9.1 Introduction

Most standard methods of measurement are sector specific, focusing on either building or civil engineering work. CESMM does have a *simple building works* class, but no method of measurement offers the range of construction work that POM(I) does.

This might be criticised as the oversimplification of complex measurement issues, but there are merits to be found in an approach that links simple item descriptions with a well-developed design and clear and unequivocal specifications and contract conditions.

Despite being over 35 years old, POM(I) is, perhaps, more suited to modern procurement methods than other more sophisticated methods of measurement of more recent origin.

Managing Measurement Risk in Building and Civil Engineering, First Edition. Peter Williams.
© 2016 John Wiley & Sons, Ltd. Published 2016 by John Wiley & Sons, Ltd.

9.1.1   A word of warning

Users of POM(I), and especially contractors and subcontractors, should be not be misled or deceived by the simplicity of these principles of measurement.

POM(I) can be used for all sorts of building and engineering work, including demolitions, underpinning, piling, dredging, railway work and tunnelling, of any value. Because of its simplicity and brevity, however, very little provision is made for measuring, or otherwise describing, risk issues such as may be provided in other more complex methods of measurement.

For this reason, bill compilers, or contractors/subcontractors pricing bills of quantities based on POM(I), should be aware of four key provisions of Section GP: General Principles that condition the measured quantities provided:

1. POM(I) bills of quantities shall describe and represent the works to be carried out (GP2.2).
2. The principles of measurement are reliant on detailed conditions of contract, drawings and specifications being provided with the bill of quantities (Foreword and GP2.3).
3. There may be a need to provide more detailed description than is required by POM(I) in order to fully define the work to be carried out (GP1.1).
4. All BQ items shall be fully inclusive of the liabilities and obligations arising out of the contract (GP4.1).

Risk issue

Tenderers should be aware of the need to visit the site, to scrutinise the accompanying drawings and specification more carefully than ever and to fully understand the conditions of contract being used alongside POM(I).

It may be the case that risk issues not measured under POM(I), which might attract relief under other standard methods of measurement, may be provided for in the accompanying conditions of contract, but this is by no means certain.

If not, it might be necessary to turn to the common law of the land for relief in case of difficulty, but this is likely to result in a much less favourable conclusion than the provisions the JCT, ICC, FIDIC and NEC3 conditions would otherwise render.

9.1.2   POM(I) and computerised measurement

The item description requirements of POM(I) are so simple that a non-library-based software package could be used to create a bill of quantities.

There is no requirement to describe items of work in any specific way provided that the basic information required by each POM(I) clause is included. There are no additional description or supplementary information rules in POM(I), as there are in other methods of measurement.

The example shown in Figure 9.1 illustrates how easy it would be to create a library of item descriptions that can be saved as a template and then copied and exported into other projects.

It can be seen from Figure 9.1 that the software (Buildsoft Cubit) has produced a work breakdown structure, with headings referenced to POM(I) Section B: *Site Work* and Section C: *Concrete Work*. Item descriptions are simplicity itself, albeit some additional description might be thought necessary. Once the take-off has been completed, a variety of reports can be printed, such as individual trade mini-BQs or even a full bill of quantities. The software will also export the file into MS Excel for sending out trade enquiries to subcontractors who don't have Buildsoft Cubit.

Alternatively, an extensive library of standard item descriptions is available within the CATO take-off and billing software. It should be noted, however, that the authors of CATO appear to

	Work breakdown structure and take-off sheet							
	Description	**Result**	**Quantity**	**Unit**	**Rate**	**...**	**Total**	
1	◢ HOLLY FARM OVER-BRIDGE							
1.1								
2	◢ B - SITE WORK							
2.1	Trench excavation	🗑 Vo...		m3				
2.2	Disposal to tip	🗑 Vo...		m3				
2.3	Filling to excavation	🗑 Vo...		m3				
3	◢ C - CONCRETE WORK							
3.1	◢ Poured concrete							
3.1.1	Foundations	🗑 Vo...		m3				
3.1.2	Walls 300 mm thick	🗑 Vo...		m3				
3.2	◢ Reinforcement							
3.2.1	10 mm mild steel	▭ Le...		t				
3.2.2	12 mm high yield steel	▭ Le...		t				
3.3	◢ Shuttering							
3.3.1	Sides of foundations	▱ Area		m2				
3.3.2	Sides of walls	▱ Area		m2				
3.4	◢ Prestressed concrete							
3.4.1	Bridge beams type T	🔧 C...		each				
3.4.2	Bearings	🔧 C...		each				

Figure 9.1   Bespoke POM(I) library template. Produced with Buildsoft Cubit.

have taken the view that enhanced descriptive features are needed in some circumstances but, notwithstanding this, the software allows for the creation of 'rogue' descriptions.

This means that the bill compiler is at liberty to create bespoke item descriptions as need be. The functionality of CATO is a big help to bill compilers who would otherwise need to compile their own additional descriptions.

Examples of the approach to item description taken by CATO can be seen in Figure 9.2 which shows that trial holes are measured by number rather than depth (POM(I) Clause B2.1)

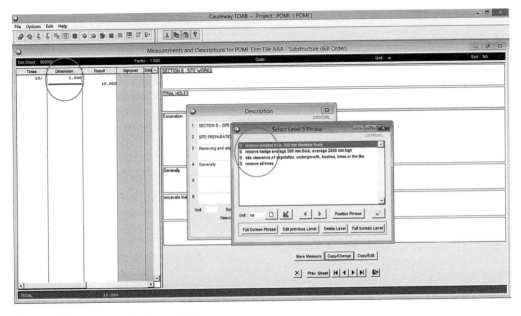

Figure 9.2   Item description – CATO.

and that additional descriptive features are provided for tree removal and the removal of hedges which are again supplementary to POM(I) rules.

9.1.3   Structure

POM(I) consists of 20 pages of measurement rules, comprising Section GP: General Principles, Section A: General Requirements and 15 measured Work Sections (B–R) together with an Appendix.

The measured Work Sections are subdivided into:

Section	Title
B	Site Works
C	Concrete Work
D	Masonry
E	Metalwork
F	Woodwork
G	Thermal and Moisture Protection
H	Doors and Windows
J	Finishes
K	Accessories
L	Equipment
M	Furnishings
N	Special Construction
P	Conveying Systems
Q	Mechanical Engineering Installations
R	Electrical Engineering Installations

The list of work types that are covered by POM(I) is impressive, especially when drilling down into individual Work Sections. Section B: *Site Works,* for instance, covers an extensive array of building and civil engineering work, including demolitions, underpinning, dredging, railway work and tunnelling.

The Appendix allows the body responsible for preparing the quantities to state any amendments to the principles of measurement where, for instance, additional measurement rules are needed for work items not covered in the standard document or where local conditions demand that certain items of work be measured in a specific or non-standard way.

Reading through the document will reveal one or two 'glitches' in the editing that may require a second look in order to fully understand the intended meaning.

9.2   Section GP: General Principles

Section GP contains 10 sets of general principles (GP1–GP10) which overarch the measured Work Sections B–R. It should be noted that some of these principles are optional (*may*) and some are compulsory (*shall*) which indicates that some of them are rules and some are not.

In some Work Sections, certain measurement rules are stated as being *subject to* specific GP clauses which tends to indicate, in such cases, that the general principle is to be observed.

9.2.1   GP1: Principles of measurement

GP1.1 requires that additional description *may* be given in order *to define the precise nature and extent of the work or the circumstances under which* the work is to be carried out. This is both an optional and 'either/or' clause. There is, therefore, no strict obligation to comply with it, but the bill compiler must consider whether not to do so would constitute a misrepresentation.

GP1.2 refers the bill compiler to the Appendix should it be necessary to amend the principles of measurement – that is, POM(I) – for local reasons or to include measured work not covered in the document.

GP1.3 states that the principles of measurement *may be applied equally* to the measurement of completed as well as proposed works. This contrasts with the certainty of CESMM4 that the same principles shall apply, and to the lack of any mention at all of completed work in NRM2. Certainty in a contract is to be encouraged, and an amendment to GP1.3 saying **will** could easily be listed in the Appendix, should this be desired.

9.2.2   GP2: Bills of quantities

GP2.1 simply describes the objectives of bills of quantities, and GP2.4 states that the way in which they are presented is not restricted by the headings and classifications used in POM(I).

GP2.2 is a rule – in fact, it is two rules:

- Bills of quantities *shall describe and represent the works to be carried out.*
- Work that cannot be measured accurately *shall be* either:
  - *Described as approximate.*
  - *Given in bills of approximate quantities.*

The first rule must be read in conjunction with GP1.1 which provides that more detailed information than is required by POM(I) *may be given* in order to accurately define the nature and circumstances of the work to be done.

Despite the brevity of item description required by the measurement rules, care must be taken not to misrepresent the work required by the contract, and the role of the conditions of contract, drawings and specification are of utmost importance in conveying to tenderers exactly what is required of them. POM(I) bills of quantities are reliant on these documents in the same way that a bill of quantities saying *build one house* or *construct one tunnel* would be.

It is, perhaps, tempting to assume that a document entitled *principles* of measurement contains no rules and that, therefore, there are no rules to be broken and no liability for misrepresentation can arise. This, in the author's view, would be a mistake because any principle that contains the words *shall be* can only be construed as a rule to be followed. Should the principle not be followed, therefore, a misrepresentation could occur unless supporting documents make the matter clear.

The first part of the second rule is standard practice where firm quantities cannot be given in what is otherwise a bill of firm quantities (i.e. the contract is a lump sum contract).

The second part of the second rule is less clear but, presumably, refers to situations where the entire works are to be remeasured on completion. If so, GP1.3 may need to be amended because, as it stands, there is no obligation to measure completed works in accordance with POM(I) rules. A further consideration might be that if POM(I) is used in conjunction with NEC3 ECC Option B, the complex rules for compensation events relating to differences between the estimated and final quantities would need to be pondered (ECC 60.4).

GP2.3 is a little unusual in that the conditions of contract *shall be provided with the bills of quantities* along with the drawings and specifications. Normal practice, for practical reasons as well as cost, is that the conditions of contract are either made available for inspection or a list of the clauses and amendments applicable is included in the bill of quantities with tenderers presumed to have access to the full printed conditions.

Part 2

9.2.3   GP3: Measurement

GP3.1 – measurements are net fixed in position to the nearest 10 mm (except dimensions stated in descriptions).

GP3.2 – no deduction of voids of less than 1 m² in superficial items.

GP3.3 – voids at the edges of areas deducted irrespective of size.

GP3.4 – amendments may be made to use other units of measurement with POM(I) which, presumably, refers to imperial rather than the default (but not expressly stated) metric units.

GP3.5 – separate minor buildings/structures may be enumerated.

9.2.4   GP4: Items to be fully inclusive

GP4.1 states that all items *shall be fully inclusive of all that is necessary to fulfil liabilities and obligations arising out of the contract* including:

- Labour and oncosts.
- Materials.
- Plant.
- Temporary works.
- Establishment charges, overheads and profit.

Attention needs to be paid to this clause, especially as regards temporary works.

Risk issue

Temporary works (except earthwork support) are not measured under POM(I), and therefore, the conditions of contract will, to a large extent, determine whether or not the contractor will be reimbursed if difficulties are encountered.

The ICC – Measurement Version and NEC3 ECC include provisions for situations that could not have been anticipated by an experienced contractor but JCT 2011 SBC/Q does not. Under the JCT conditions, below-ground risk issues are dealt with by SMM7 or NRM2, but this is not the case under POM(I) which does not have any similar provisions:

- No requirements to state/assume groundwater levels or existing services.
- No requirements to provide this information on drawings.
- No measured items for excavation below groundwater level or associated earthwork support.
- No requirement to measure excavation in instable ground.
- No requirement to measure earthwork support next to buildings or roadways.
- No relief to measure steel sheet piling if it becomes necessary.

9.2.5   GP5: Description of items

Most standard methods of measurement require additional description to be given in some circumstances. In most cases, additional description 'rules' are provided that clarify what additional description shall be given and to which items.

POM(I) is different in this respect, because there are no additional description **rules**. This is not to say that there are no additional description **requirements**, though, and GP1.1 states that additional description *may* be given in order *to define the precise nature and extent of the work or the circumstances under which* the work is to be carried out.

Risk issue

In order to avoid a possible misrepresentation, the bill compiler must decide whether it is necessary to convey additional information to tenderers relating to the nature and extent of the work, <u>or</u> whether the conditions under which it may be carried out need to be amplified in item descriptions, but not both.

GP5 instructs how items shall be described:

GP5.1 – enumerated items *shall be fully described*.

GP5.2 – linear items *shall state* the cross-sectional size and shape, girth or ranges of girths or other appropriate information, and the diameter of pipework items shall be stated as internal or external.

GP5.3 – items measured by area *shall state* the thickness or other appropriate information.

GP5.4 – items measured by weight *shall state* the material thickness and unit weight, if appropriate.

GP5.5 – if manufacturers measure items in a customary manner, then this is acceptable under POM(I).

GP5.6 – item descriptions may refer to other documents, drawings or published information, and reference to such information shall be deemed to fulfil the requirements of POM(I).

9.2.6   GP6: Work to be executed by a specialist nominated by the employer

Should the employer wish to have certain work carried out by a nominated specialist of his own choosing, GP6.1 requires that such work *shall be given as a sum* in the bill of quantities. This provision is subject to the important caveat *unless otherwise required by the conditions of contract* whose meaning is not entirely clear.

The phrase *otherwise required by the conditions of contract* could mean that:

- Such work shall not be given as <u>a sum</u> in the bill of quantities but shall be given as a <u>prime cost item</u> or <u>provisional sum</u>. Under the ICC – Measurement Version, such sums of money can be expended on work to be carried out by a nominated subcontractor.
- Certain conditions of contract do not allow the nomination of specialists or nominated subcontractors. This is the case under JCT and NEC3 contracts.
- A subcontractor can be named in the contract[2] as a nominated subcontractor or can be engaged by the contractor pursuant to the expenditure of a provisional sum. These are the arrangements under the FIDIC Conditions of Contract for Construction (1999).

It would appear, therefore, that GP6.1 recognises that the nomination of specialists can only be provided for in the bill of quantities if it is permitted by the conditions of contract and then only if the procedures in the conditions are followed.

GP6.1 also requires that an item *shall be given* for the addition of contractor's profit to each sum given in the bill of quantities in respect of such work.

GP6.2 states that an item *shall be given* in each instance for *assistance by the contractor*. In POM(I), 'assistance' may be taken as a synonym for 'attendance', but care should be taken that POM(I) makes no distinction between 'general' and 'special' assistance. In fact, GP6.2 lists in its item coverage for 'assistance' the usual 'general' items (1–6) and two items of what might be called 'special assistance' (7 and 8):

7. Scaffolding required by the specialist, *giving particulars* in the item.
8. Unloading, distributing, hoisting and placing in position items such as plant and machinery for which there is no requirement to state particulars.

Item 8 could involve considerable cost including cranes, tele-handlers, site transport and associated labour.

9.2.7 GP7: Goods, materials or services to be provided by a merchant or tradesman nominated by the employer

This is an unusual provision which does not appear in the mainstream forms of contract referred to Section 9.2.4. Under the ICC – Measurement Version, prime cost items may be used to order the supply of goods, materials or services by a nominated subcontractor, or by the contractor himself, but not by a supplier or tradesman.

It may be the case that this is an arrangement used in certain parts of the world using local conditions of contract in which case an item for profit shall be given with each such item as per GP7.1.

GP7.2 deals with the fixing of goods and materials supplied under GP7 arrangements which shall be given (i.e. measured or itemised) in accordance with the relevant section of POM(I), that is, Sections B–R. It should be noted that the item coverage for fixing includes *unloading, storing, distributing and hoisting* as well as the normal labour, plant and temporary works deemed to be included by Clause GP4.1.

9.2.8 GP8: work to be executed by a government or public authority

This is a similar arrangement to that used in GP7.1 where, *unless otherwise required by the conditions of contract*, such work *shall be given as a sum* in the bill of quantities with an accompanying item for profit.

GP8.2 adds the requirement to include an item *for assistance by the contractor* as per Clause GP6.2.

9.2.9 GP9: Dayworks

GP9 makes provision for the eventuality that some work may have to be carried out on the basis of time spent and resources used commonly known as 'daywork'. There are various ways of dealing with daywork, and POM(I) proposes two methods:

1. The provision of a sum money.
2. The provision of a schedule with provisional quantities of labour, materials and constructional plant as the case may be.

GP9.1 requires that any daywork schedule shall include *different categories of labour* with *a provisional quantity of hours for each category*. Whether a sum or schedule is given, the cost of labour is defined in GP9.2 as *wages, bonuses and all allowances to operatives* – including plant operators and drivers – *in accordance with the appropriate employment agreement or, where no such agreement exists, the actual payments made*.... GP9.7 further requires that an item shall be given for the addition of *establishment charges, overheads and profit* to the sum or schedule.

The same rules apply to the cost of materials (Clause GP9.3) and constructional plant (Clause GP9.5) except that:

■ The cost of materials shall be the net invoiced price, including delivery to site (GP9.4).

- The cost of constructional plant shall include fuel, consumable stores, repairs, maintenance and insurance of plant (GP9.6).

GP9.8 defines the meaning of *establishment charges, overheads and profit* in a seven-point list which *shall include*:

1. Labour employment costs.
2. Storage, handling and (storage) waste of materials.
3. Contractor's administration.
4. Constructional plant additional to that used on dayworks.
5. Contractor's facilities.
6. Temporary works.
7. Sundry items.

This list is less than clear. There is no mention of *profit* in the list and the distinction between *establishment charges* and *overheads* is not made either – normally, they mean one and the same thing.

It is probably incorrect to assume that *overheads* are represented by item 3 because a definition of 'administrative arrangements' is given in Section A, Clause A4.1, which resembles the usual labour 'oncosts' normally associated with daywork. Items 1 and 2 would also normally be daywork 'oncosts', and items 4–7 represent the usual 'preliminaries' or 'site oncosts' added to daywork rates. Contractors and subcontractors, especially, should note these anomalies and make sure that any daywork rates they quote include all necessary oncosts and overheads and profit.

A significant feature of GP9 generally is the absence of the phrase *unless otherwise required by the conditions of contract* which is used elsewhere in POM(I) in connection with nomination and provisional sums (which is what the BQ daywork provision is).

Different forms of contract deal with the issue of 'daywork' in different ways, and the POM(I) bill compiler should be aware of potential conflict with Clause GP9:

- FIDIC only refers to the provision of a daywork schedule.
- NEC3 ECC prefers the term 'compensation event' used in conjunction with the Shorter Schedule of Cost Components.
- The ICC – Measurement Version refers to a 'daywork schedule' or, alternatively, the Schedule of Daywork Carried Out Incidental to Contract Work published by the CECA.
- JCT 2011 SBC/Q employs daywork for the valuation of variations, where appropriate, using the Definition of Prime Cost of Daywork Carried Out under a Building Contract (RICS) together with percentage additions stated in the BQ by the contractor.

9.2.10   GP10: Contingencies

GP10.1 states that, *unless otherwise required by the conditions of contract*, contingencies *shall be given as a sum* with *no item* for the addition of profit.

Once again, potential conflict with the conditions of contract arises:

- ICC and FIDIC do not use the word 'contingency' and, instead, refer to the phrase 'provisional sums'.
- Neither 'provisional sums' nor 'contingencies' feature in NEC3 contracts.
- JCT 2011 refers to defined and undefined provisional sums consistent with SMM7 and NRM2.

9.3   Section A: General requirements

This section of POM(I) provides a set of rules to enable space to be created in the bill of quantities for the employer to state the conditions of contract that shall apply to the project, any limitations and restrictions that ought to be brought to the attention of tendering contractors and to give the contactor the chance to price the preliminaries, or general items, including facilities required for the employer or his representatives.

9.3.1   A1: Conditions of contract

POM(I) Clause A1.1 states that a schedule of contract clause headings *shall be set out in the bill of quantities*, and Clause A1.2 requires that, where there is an appendix to the conditions of contract, a schedule of the insertions made in the 'appendix' *shall be set out in the bill of quantities*.

Inexplicably, Clause A1.1 does not require a statement to be included in the bill of quantities as to which form of contract is to apply to the project, which is unusual.

As far as the appendix to the conditions of contract is concerned, tenderers need to be alerted to this important information because it has a direct impact on the tender price. In the appendix, the employer stipulates important information about the contract such as the commencement and completion dates, the rate of liquidated and ascertained damages and the rate of retention, etc. In some forms of contract, there is an appendix to the conditions, and in others, there is the equivalent but with a different name:

- ICC – Measurement Version – **Appendix Part 1**
- FIDIC (1999) Red Book – **Appendix to Tender**
- NEC3 ECC – **Contract Data Part 1**
- JCT 2011 SBC/Q – **Contract Particulars**

Notwithstanding POM(I) Clauses A1.1 and A1.2, the above information should be made clear in the bill of quantities, and it is common practice to include a completed 'appendix' with the tender documents as opposed to a *schedule of insertions*.

9.3.2   A2: Specification

Clause A2.1 refers to the relationship between the specification and POM(I) general requirements A3–A9, whereby the bill of quantities *shall make reference to* any specification clauses that are relevant to the general requirements. Such specification clauses may add detail to the general requirements that could impact on the tender price.

9.3.3   A3: Restrictions

Clause A3.1 lists five generic examples of possible restrictions that may be imposed on the contractor in the construction and completion of the contract. These include access restrictions, limitations on working space and requirements to complete the works in a specific order. The sixth item in the A3.1 list is *items of a like nature* which is an invitation for the employer/bill compiler to include any other restrictions that might be relevant to a particular project.

9.3.4   A4: Contractor's administrative arrangements

This is the first of six general requirements to be included in the bill of quantities that tenderers are likely to price. Clause A4.1 deals *inter alia* with the contractor's staffing and supervision of the project and would normally be priced in conjunction with Clauses A6 (employer's facilities) and A7 (site accommodation) as 'preliminaries'. Most contractors have a standard 'spreadsheet' for such items.

It should be carefully noted that Clause A4.1 only requires *an item* (i.e. a single item) to be given in the bill of quantities to cover all of the five items listed as 'administrative arrangements'. The words *shall include* means, in effect, that the items listed are 'understood to be included' (a POM(I) phrase) and also that it is not an exhaustive list.

Table 9.1   Contractor's administrative arrangements.

SECTION A - GENERAL REQUIREMENTS					
CONTRACTOR'S ADMINISTRATIVE ARRANGEMENTS					
Generally	(NA)				
The Contractor shall allow for all necessary administrative arrangements, including :					
A allow for site administration	ITEM				
B allow for site supervision	ITEM				
C allow for site security	ITEM				
D allow for the safety, health and welfare measures	ITEM				
E allow for transport of workpeople	ITEM				

CATO takes the view that it is reasonable to list items 1–5 as illustrated in Table 9.1. Whilst there is a certain logic to the CATO approach, there are also some drawbacks:

- This is not what the method of measurement says.
- The bill compiler must either:
  a) Select the items that he/she considers appropriate or
  b) Include them all.
- There is no guarantee that tenderers will price any of the items individually and could:
  a) Bracket the items and include a lump sum.
  b) Write 'included' against the list.

Contractors and subcontractors will have to consider the wisdom of disaggregating these items from other preliminaries and whether or not to 'write in' to the bill of quantities item those preliminaries which are fixed and those which are time related.

Risk issue

'Writing in' additional items in the bill of quantities may constitute a tender qualification, and this would have to be checked carefully with any tendering rules that might apply.

9.3.5   A5: Constructional plant

As with Clause A4.1, Clause A5.1 requires the provision of *an item* in the bill of quantities to cover a list of widely differing items of plant that the contractor may wish to price into the tender. The list ranges from small plant and tools to cranes and lifting equipment as well as scaffolding and site transport, as illustrated in Table 9.2.

Some of these items will have fixed costs – for example, erect/dismantle mobile crane (including crane pad), erect/dismantle scaffolding – and other costs will be time related such as crane and scaffold hire.

Part 2

Table 9.2 Constructional plant.

CONSTRUCTIONAL PLANT					
Generally	(NA)				
The Contractor shall allow for all necessary constructional plant including:					
F small plant and tools	ITEM				
G scaffolding	ITEM				
H cranes and lifting plant	ITEM				
J site transport	ITEM				
K plant required for specific trades	ITEM				

Risk issue

There is some danger in pricing fixed and time-related costs in a single item, particularly when the valuation of variations becomes an issue, and contractors and subcontractors may need to deal with this by 'writing in' appropriate items in the general requirements part of the bill of quantities (restrictions regarding tender qualifications permitting).

9.3.6   A6: Employer's facilities

There is no requirement in Clause A6.1 for a bill of quantities item to cover 'employer's facilities'. A6.1 simply states that *particulars shall be given* of any facilities required by the employer which *shall include* the seven items listed, such as temporary accommodation, telephones and the cost of calls which, paradoxically, *may be given as a sum*, any special requirements for programmes or progress charts and facilities such as progress photographs and signboards, etc.

The bill compiler may interpret A6.1 as meaning that a sum shall be given for these items, but if not, contractors and subcontractors should perhaps be careful to check that the employer's requirements are not hidden in a specification clause referred to in the bill of quantities pursuant to POM(I) Clause A2.1.

9.3.7   A7: Contractor's facilities

Clause A7.1 requires that *an item shall be given for facilities required by the contractor*, and a list is provided of the facilities which the item *shall include*. The list includes things like accommodation, welfare facilities, site fencing, water for the works (which might be supplied to the contractor), lighting and power, etc.

Unusually, in Clause A7.2, *particulars shall be given* if the *nature or extent* of the contractor's facilities is *not at the contractor's discretion*. In this context, *nature* may be taken to mean kind, type or sort and *extent* the amount or scope.

Table 9.3 Contractor's facilities.

			£	p
CONTRACTOR'S FACILITIES				
Generally	(NA)			
The Contractor shall allow for the following Contractor's Facilities				
A provide accommodation and buildings, including offices, compounds, stores, messrooms, laboratories and the like	ITEM			
B provide temporary fencing, hoardings, screens, roofs, guardrails and the like	ITEM			
C provide temporary roads, hardstandings,	ITEM			
D provide water for the works, as specified	ITEM			
E provide electric power for the works, as specified	ITEM			
F provide lighting for the works	ITEM			
G provide telephones for the works	ITEM			

Should the bill of quantities state that the *nature* of the contractor's site facilities is not at his discretion, this might be because the employer is running the site (say on a petrochemical works) and the contractor is required to use existing buildings or services provided by the employer. Alternatively, on a very large site, such facilities may be provided by a managing contractor.

Restrictions as regards the *extent* of the contractor's site facilities may be imposed should there be confined spaces on the site or limits may be imposed on the contractor's site set up under a cost reimbursement contract. In any event, the bill compiler must ensure that the contractor knows exactly what to price in order to comply with A7.2.

Sample items for 'contractor's facilities' are shown in Table 9.3 where it can be seen that CATO refers to the provision of water and power *as specified*. This may be an attempt to comply with A7.2, albeit this rule could equally apply to some of the other items billed.

9.3.8   A8: Temporary works

An item shall be given for temporary works in accordance with Clause A8.1, but in A8.2, *particulars shall be given* where the temporary works are *not at the discretion of the contractor*.

The temporary works item *shall include* any one, or more, of seven listed items such as traffic diversions, access roads, temporary bridges (e.g. Bailey bridges), cofferdams, pumping and dewatering and compressed air for tunnelling.

Table 9.4 provides an interpretation of such items where the bill compiler has taken the view that:

a)   A list should be provided for tenderers to price.
b)   All the items on the list in A8.1 are not relevant to the project in question.

Table 9.4 General requirements.

SECTION A - GENERAL REQUIREMENTS				£	p
TEMPORARY WORKS					
Generally		(NA)			
The Contractor shall allow for all necessary temporary works, including:					
A provide bridge; temporary; two-way traffic	ITEM				
B provide cofferdam; bridge pier A3	ITEM				
C provide dewatering; wellpoints	ITEM				

Risk issue

Bill compliers are playing a dangerous 'game' if they try to 'second guess' how a contractor is likely to go about constructing the works.

In Table 9.4, unless the provision of a temporary bridge is *not at the discretion of the contractor*, it may be unwise to create a specific item like this as the contractor may go about solving the problem in a completely different way.

If, for instance, the only access to the site is across a watercourse, the contractor could just as easily decide to divert the watercourse, or install a temporary culvert and ramp, as build a temporary bridge.

In any event, it seems, once again, that the contractor (or subcontractor) has to interpret what is required from the tender documents, and price everything into one item, as there is no provision for fixed and time-related charges.

Clause A8.2 states that *particulars shall be given* if *the nature or extent* of the temporary works is *not at the contractor's discretion*. Such particulars might be required for similar reasons to those given under Clause A7.2.

9.3.9 A9: Sundry items

Clause A9.1 requires that an item shall be given for sundry items which shall include those listed 1–9. Despite the designation 'sundry items', the list is fairly extensive ranging from the testing of materials to the control of noise and pollution. Removal of rubbish is included as is the drying out of the works and protection from inclement weather.

Clause A9.2 again provides for circumstances where the contractor has no discretion over the *nature or extent* of sundry items, and it can only be imagined that this might be the case should the employer or a managing contractor wish to control such items. In any event, particulars *shall be given* if this is the case.

9.4   Section B: Site work

POM(I) Section B takes five pages to go from site investigation to tunnelling, and so it is no surprise that the measurement rules are very brief!

Notwithstanding this, the Foreword to POM(I) recognises that local practice and contract-specific circumstances require a flexible set of measurement principles which, in conjunction with a well-developed design and specification, can nevertheless result in a meaningful and useful bill of quantities for the financial control and management of construction projects.

Risk issue

A well-developed design and specification are crucial to POM(I) because item descriptions are so brief and, even with additional description (GP1.1), they do not approach the level of detail offered by SMM7, NRM2 or CESMM4. There are no notes for guidance to bill compilers in POM(I), and there are no additional description rules of the sort found in other methods of measurement.

In keeping with the remainder of this book, the following subsections of POM(I) Section B identify the key risk issues concerned with the billing and pricing of items related to site work.

9.4.1   B1–B3: Site investigation

Clauses B1–B3 concern matters relevant to site investigation and, despite being brief, provide suitable measured items ranging from simple trial hole work to the complex soil sampling, laboratory testing and analysis that might be required on a large project:

- **B1: Site exploration generally**
  Record keeping of various tests and observations *shall be given as an item* (refer to Clauses B1.1–B1.3). No details are stated regarding the extent of records, tests and samples that might be required, but presumably, this will be stipulated elsewhere (e.g. in the specification) or may be given in the respective items as additional description in accordance with the provisions of Clause GP1.1.
- **B2: Trial holes**
  Clause B2.1 concerns the excavation of trial holes but requires no distinction to be made as to the purpose that trial holes might be required for (e.g. to determine subsoil conditions, to locate underground services). Unusually, trials holes *shall be measured by depth* stating *the number and the maximum depth below commencing level.* It would be more normal to measure by number stating the depth, and it would also be more usual to stipulate the length (or plan area) of the trial holes required.

  B2.2 confirms that earthwork support in connection with the excavation of trial holes is not measured unless ***not*** *at the discretion of the contractor* when it shall be measured by depth.
- **B3: Boreholes**
  B3 covers both boreholes for ground investigation and for driving test wells. B3.1 requires that boreholes *shall be measured by depth* stating *the number and the maximum depth below commencing level* with raking bore holes *so described.* Borehole linings are not normally measured (B3.2), but *cappings shall be enumerated* (B3.3).

Risk issue

It should be noted that the provision of additional description is not mandatory under GP1.1 and that, where there is any question as to item coverage, the drawings, specification and conditions of contract play a crucial role under POM(I).

Notwithstanding this, trial holes to discover existing services that could be live is both a health and safety issue and a cost issue for the contractor, and the bill compiler should consider including additional description where felt necessary.

A case for additional description could also be made regarding the length or plan size of trial holes to avoid unnecessary claims from contractors.

9.4.2   B4: Site preparation

Any work needed to prepare the site, including site clearance but not demolition work, is covered by Clause B4.

B4: Site preparation

Removing *isolated* trees *shall be enumerated* (B4.1), but there is no requirement to state girth.

Removing hedges *shall be measured by length* (B4.2).

Site clearance *shall be measured by area* (B4.3), but the removal of trees and the removal of hedges are both included in the coverage of the site clearance item (B4.3) which also includes the removal of vegetation, bushes and so on.

Risk issue

In view of Clauses B4.1 and B4.2, the bill compiler will need to exercise some judgement as to when to separately measure the removal of hedges and trees and when to include this work in the general site clearance item.

It would also seem sensible to provide additional description of the diameter of trees to be removed and the width and height of hedges to be grubbed up in order to avoid possible claims (refer to Table 9.5).

Table 9.5   Site preparation.

					£	p
	SECTION B - SITE WORKS					
	SITE PREPARATION					
	Removing and site clearance		(NA)			
	Generally					
A	remove isolated tree; 300 mm diameter trunk	5	no			
B	remove hedge average 500 mm thick; average 2000 mm high	68	m			
C	site clearance of vegetation, undergrowth, bushes, trees or the like	22800	m2			

9.4.3   B5: Demolitions and alterations

Demolition and alteration work is covered by Clauses B5.2–B5.4 which include removal of fittings and fixtures, demolition of structures, cutting openings and altering existing structures. B5.2 is effectively a soft stripping item, but to make this clear, additional description needs to be given.

Risk issue

Care should be taken when reading BQ items under B5.2 as this could include the removal of engineering installations which, in some instances, could be extensive.

Demolition of structures is *given as an item*, but this may be for single structures or parts thereof or for all structures on a site.

Forming openings in structures or carrying out alterations thereto is again *given as an item*, but there is also an item coverage rule that includes *making good all work damaged*. It should be noted that there is no requirement to describe the insertion of new work such as lintols which could either be referenced to a drawing or measured in accordance with Clause C5.

With each item of demolitions and alterations, locational information shall be given (Clause B5.1), and unless otherwise stated, all materials shall be cleared away by the contractor. The provision of temporary screens and roofs shall be *given as an item*, but any shoring requirements (and, presumably, needling and propping) shall be measured as per Clause B6.

9.4.4   B6: Shoring

Shoring may be temporary or left in place and may involve simple needling and propping or the provision of complex and extensive raking or flying shores to support existing buildings or structures. POM(I) Clause B6 distinguishes between:

1. Shoring *incidental to demolitions and alterations* (B6.1).
2. Shoring <u>not</u> *incidental to demolitions and alterations* (B6.2).
3. Shoring where the design of the shoring is not at the discretion of the contractor (B6.3).
4. Shoring required *by the specification* to be left in position (B6.4).

In the first case, any shoring required which is part and parcel of demolitions and alterations work is not itemised in the BQ and is *understood to be included* in the measured items. In the second case, any shoring that is required, such as the provision of shoring to buildings to be retained, is measurable and *shall be given as an item*. The design of any such shoring is the contractor's responsibility, irrespective of how extensive it is, as it is not part of the permanent works.

The third case provides for the situation where the design of the shoring is *not at the contractor's discretion*, in which case *particulars shall be given* (B6.3), but this is not a measurable item. There is no indication in B6.3 where the *particulars* shall be stated – this could be on a drawing or in the BQ item for shoring.

In the fourth case, shoring required to be left in position shall be identified in the specification and *shall be so described* in the bill of quantities.

Part 2

Risk issue

The issue of shoring, temporary or left in place, raises the issue of design liability and risk.

As with all temporary works, shoring is not usually identified and/or measured in the bill of quantities unless the work is of such significance that it is to be designed by the employer's engineer. In this event, design liability rests with the employer/engineer.

Under normal circumstances, the contractor would be responsible for the design of temporary works generally and shoring in particular, and its suitability in the context of his obligation to carry out and complete the works would also be down to the contractor.

However, under POM(I), all shoring, whatever the scale of work involved and whether it is a billed item or not, is the contractor's responsibility, unless otherwise stated, in which case the contractor carries the liability for the design of the shoring as well as for the installation.

Therefore, it would be dangerous to presume that design liability has shifted just because there is a billed item for the work. Design liability will only shift where clear particulars are given that *the design is not at the discretion of the contractor*, and this would have to be clear from the drawings, the specification and/or the billed item.

This important distinction is illustrated in Figure 9.3 where it is clear that, in one item description, the contractor is responsible for the design and, in the other, the shoring is 'as specified' and therefore the responsibility of the employer/engineer.

Where shoring is to be left in place – and this could be for several years – the question as to whether this is temporary or permanent works is a difficult one. Some shoring systems, such as RMD Kwikform 'Megashor' power shores,[3] are hired and therefore belong to the hire company, and it may not be feasible for the original contractor or subcontractor to remove them, maybe years later. A suitable contractual provision would have to be made for this eventuality and, perhaps, made clear in the BQ item description.

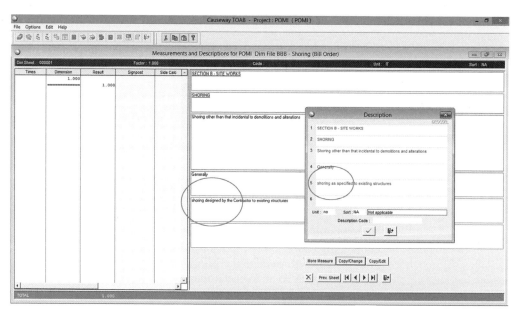

Figure 9.3   Shoring.

9.4.5   B7: Underpinning

Underpinning of existing structures can take many forms depending upon the nature of the problem, the condition of the existing building, the prevailing ground conditions and groundwater levels, access to the site, access to the interior and so on.

Anything beyond the traditional approach of excavating trenches or pits and building up walls or columns on new foundations underneath the existing foundation is likely to be specialist work because it will probably involve piling of some sort, cement–soil stabilisation (jet grouting) or resin injection, etc.

Risk issue

The treatment of underpinning in the bill of quantities would need to be considered in the light of the chosen procurement strategy and the form of contract to be used.

Traditional methods of underpinning and specialist piling work could be measured under relevant POM(I) rules (e.g. Piling B13–B15 and B18). Other methods of constructing underpinning could be measured according to other rules provided that they are stated as an amendment in the Appendix in accordance with Clause GP1.1.

Alternatively, any sort of underpinning work could be given as a sum under GP6 if a specialist contractor is to be nominated. Should the preference be for a contractor-designed solution, then a single BQ item could be given provided that the amended measurement rules applicable are stated in the Appendix (refer once again to Clause GP1.1).

POM(I) Clause B7 provides that underpinning work *shall be measured in accordance with appropriate* Work Sections (B7.2) under an *appropriate heading* that includes the location of the work (B7.1). The only measurement rules under B7 are:

- Temporary support to the existing structure *shall be given as an item* and particulars shall be given where the design of the temporary works is not at the discretion of the contractor (B7.3).
- Excavation *shall be measured by volume* (B7.4).
- Cutting away projecting foundations *shall be measured by length* (B7.5).

Clause B7.4 deserves some attention and this is dealt with as a case study in Chapter 17.

9.4.6   B8: Earthworks generally

Clause B8 deals with the general rules relating to all earthworks. This includes excavation of various sorts, dredging and tunnelling. In all cases, general rules apply that concern:

- The provision of ground information (B8.1).
- The proviso that quantities do not allow for bulking of materials (B8.2).
- That *multiple handling* is deemed included unless *required by the specification* and *described in the item of disposal* (B8.3).

There are also two measured items in B8:

1. Earthwork support which *shall be given as an item* (B8.4).
2. Rock which may be measured full value, and described as such, or extra over the excavation in which it occurs (B8.5).

In order to avoid duplication or ambiguity, B8.4 should be read in conjunction with Clause B26 (Tunnel support and stabilisation) wherein different methods of supporting tunnel excavations are measured in a variety of units.

Part 2

Risk issue

The measurement of earthwork support *as an item* is a strange one.

Normally, the choice is to measure such work superficially or not at all whereupon the earthwork support is deemed to be included in the rates. Creating an item in the bill of quantities is of no help to the contractor who, of course, would have to measure the area of earthwork support himself.

Perhaps the logic of POM(I) Clause B8.4 is to create an item for the contractor to insert a method-related charge for any temporary works needed in connection with the measured excavation items.

Clause B8.6 defines 'rock' as material that *can only be removed by means of wedges, special plant or explosives.*

Risk issue

The POM(I) definition of 'rock' follows that of SMM7 and NRM2, although the latter has also included *rock hammers* in the definition. This is a very old definition and not really fit for purpose, as splitting rock with wedges is rarely seen and the term 'special plant' is not defined.

The use of the term 'rock hammers' in NRM2 signals that rock hammers are not 'special plant' which can only be taken to mean plant specially brought to site for the purpose of digging or otherwise removing rock.

Arguments over what is rock and what is special plant have been going on since time immemorial, and weak definitions such as this give plenty of scope for claims experts who are well versed in the 'delay and disruption', 'compensation event' and 'experienced contractor' clauses in contracts!

Many years ago, Seeley (1965) recommended that a definition of rock be *drawn up to suit the local geological formation.* It is hard to disagree with the soundness of his logic.

9.4.7 B9: Excavation

The POM(I) classification of excavation is very broad and includes excavation for pits and trenches through to diaphragm walls.

All excavation items are measured on the basis of the volume of the void created by the permanent construction which is borrowed from the CESMM Class E Measurement Rule. This, of course, leaves the bill compiler with the dilemma of distinguishing between different classes of excavation in the same void (e.g. foundation excavation below cuttings) which is best solved by indicating a 'payment line' on the drawings.

The one exception to measurement by volume is trench excavation for service pipes, drains, cables and the like which shall *be measured by length, stating the average depth*, and shall include disposal and filling in the item coverage.

The role of the drawings is important in POM(I) bills of quantities in that no 'commencing' or 'excavated' surfaces or 'levels' are referred to in the measurement rules. Additional description could always be included in the measured item, in accordance with GP1.1, to overcome this issue.

Apart from the removal of topsoil, excavation items are not required to state the depth as is common in other methods of measurement. Consequently, the mini-BQ shown in Table 9.6 employs the phrase *any depth* which is exactly how the CATO software operates. The phrase is included to be illustrative, but it is clear that CATO intended additional description to be included and uses the phrase *any depth* simply because there is no POM(I) depth classification available.

In this example, the earthworks balance is made up of the following.

	m³	Total (m³)
Excavation		
Bulk excavation	48 000	
Excavation for pile caps	43	48 043
Disposal		
Landscaping	45 600	
Off-site disposal	2 443	48 043

Table 9.6 Excavation.

	EXCAVATION			
	Generally		(NA)	
	Oversite to remove top soil			
E	150 mm average depth	6000	m3	
	Reduce levels			
F	any depth	48000	m3	
	Pits to receive foundation bases			
G	any depth	43	m3	
	DISPOSAL			
	Disposal of material arising from excavations		(NA)	
	Generally			
A	backfilled over site in making up levels and contouring	45600	m3	
B	remove from site	2443	m3	
	LANDSCAPING			
	Soiling, seeding and turfing		(NA)	
	Vegetable soil selected from spoil heaps			
C	150 mm thick and levelling	38000	m2	

Excavation, disposal and topsoiling are measured in Sections B9, B11 and B22, respectively.

At this point, it is worth mentioning items of excavation for diaphragm walls (B9.1.7) which are required to state the width of the permanent construction and the type of support fluid used.

B9.1.7 should be read in conjunction with C2.1.10, where concrete to diaphragm walls appears. This is measured, in common with other concrete items, by volume.

Risk issue

The measurement of diaphragm walls is included in POM(I) at B9.1.7 (Excavation), B11.1* (Disposal) and C2.1.10 (Concrete).

No other items are measured elsewhere for such work, and tenderers must understand that there is no definition of Commencing Surface, nor any measurement of guide walls. Disposal is measured, either on or off-site, at B11.1*.

An example of a diaphragm wall 'mini-BQ' is illustrated in Table 9.7.

Table 9.7 Diaphragm wall.

				£	p
SECTION B - SITE WORKS					
EXCAVATION					
Average 8 m deep		(NA)			
Trenches for diaphragm walls; permanent construction					
A	1500 mm wide, support fluid Bentonite	1200	m3		
DISPOSAL					
Disposal of material arising from excavations		(NA)			
Generally					
B	remove from site	1200	m3		
SECTION C - CONCRETE WORK					
POURED CONCRETE					
Poured concrete; Grade 45; 20 mm aggregate		(NA)			
Diaphragm walls					
C	generally	975	m3		

9.4.8   B10: Dredging

Dredging is a special subset of earthworks that normally involves the removal and disposal of unwanted material submerged below a body of water. This may be required for a variety of reasons, including removal of silt from rivers and estuaries, cleaning of canals and ditches, the creation of trenches in the seabed to accommodate pipelines and the like and the removal of unacceptable material from within cofferdams for land reclamation and the construction of bridge piers and the like.

Dredging and land reclamation work can be carried out by a variety of means including:

- Shore-based long-reach backhoe or dragline.
- Barge or pontoon-mounted backhoe or grab.
- For large-scale works, bucket ladder, cutter suction and trailer suction hopper dredgers, etc.

Measurements can be taken from a survey launch or dredger using GPS and other horizontal positioning systems with depths measured using lead lines, echo sounding or bar sweeping (where acoustic soundings would be inaccurate). Some barges carry sophisticated equipment and hydrographic software capable of providing an accurate graphical profile of the depth of the seabed before, during and after completion of the work, and they also carry sophisticated data collecting instrumentation capable of recording, *inter alia*, cutter head coordinates and the mixture flow and density in a hydraulic dredger (Institution of Civil Engineers, 1995).

The scope of dredging and reclamation work can be relatively small scale, perhaps incidental to other civil engineering work, or it might involve extensive work at sea. Consideration, therefore, must be given to whether the work involved warrants a separate 'dredging-only' contract or whether it can be included with the civil works.

FIDIC publishes an internationally recognised standard *Form of Contract for Dredging and Reclamation Works* (2006) – the 'Dredgers Contract' or 'Blue Book' – that was developed in conjunction with the International Association of Dredging Companies (IADC). This contract can be used in conjunction with engineer-designed or contractor-designed projects, and it incorporates five different mechanisms for valuing the works according to the prevailing circumstances. The FIDIC form provides a legal framework for the specification and design of the work and for the documentation that describes and quantifies the work itself.

Issues that need to be considered when measuring and billing dredging work include:

- Method of measurement:
  - Horizontal positioning and soundings.
  - Half-sphere or centrifuge methods[4].
  - Hopper pressure method[3].
  - Tonnes of dry solids (TDS)[3].
- Calculation of volumes:
  - Average end area method.
  - Volume in barge.
  - Volume deposited as fill.
- Tolerance limits for over-depth dredging for payment purposes.
- The extent of work on side slopes or margins above or beyond the body of water.
- The presence of hard material, rock or artificial obstructions.
- Definition of rock.
- Whether blasting of hard material will be permitted for ecological or geological reasons.
- Disposal method:
  - Deposition of dredged material hydraulically, by hopper dredge or by self-dumping scow or barge.
  - Disposal in indicated fill areas.
- Responsibility for the removal of re-silting during the defects' correction period.

Other relevant issues include:

- Environmental protection limitations.
- Limitations of working hours.
- Interference with marine traffic.

POM(I) Clause B10.1 requires that dredging *shall be measured by volume* which, unless otherwise stated, will be assumed to be *taken from soundings*. Sounding is a method of determining

Part 2

the depth of a given point beneath the surface of a body of water which, nowadays, is carried out using ultrasonic echo sounders.

B10.1 further requires that items for dredging shall state the location and limits of the work with disposal measured as equal to the volume excavated in accordance with Clause B11.1.

This is illustrated in Table 9.8.

Table 9.8   Dredging.

					ʑ	p
SECTION B - SITE WORKS						
DREDGING						
Generally		(NA)				
Dredging within phase 2						
A within zone A and B; extending 1000 m	10000	m3				
DISPOSAL						
Disposal of material arising from dredging		(NA)				
Generally						
B remove from site	10000	m3				

9.4.9   B13–B18: Piling

In the absence of method-related charges in POM(I), except for the limited A5: *Constructional plant* provisions, tenderers should be aware that the mobilisation costs associated with piling work must be allowed for somewhere in the tender.

The measurement of piling distinguishes bored, driven and sheet piling, but there is some inconsistency in the rules that apply.

Risk issue

In bored piling and sheet piling, length measurements are taken from **the formation level of the ground** to the bottom of the pile hole or the bottom edge of the sheet piling when driven, whereas in driven piling, measurements are taken from where *the pile point* [is] *in contact with* **the ground** *when pitched*.

Formation level is not ground level, and a commencing surface should be identified in item descriptions.

Boring through rock is measured by length, *extra over* the boring item.

Disposal of piling arisings is measured by volume in accordance with B11. This rule, however, refers to 'volume' as being *equal to the volume of excavation*. As boring is measured by length, this makes little sense. Also, if there is the likelihood of an element of 'blind boring', perhaps for the contractor's convenience or to suit the method of working, should disposal of this element be measured or not?

CESMM deals with this issue quite simply by referencing calculations, both pre- and post-contract, to a 'Commencing Surface'.

In situ concrete is measured by volume with the piling items and not in Section C: *Concrete Work*. This is much clearer than Class P: *Piles* in CESMM4, which is, to say the least, vague about the subject of concrete in piles. On the other hand, there is no indication in POM(I) as to where the concrete is measured to and, perhaps, a cut-off level could be referred to in the item description as illustrated in Figure 9.4.

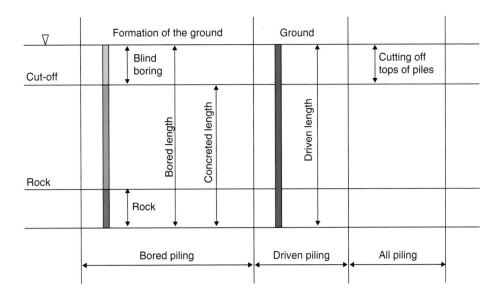

	Bored piling		
	Reinforced piling; as specified	(NA)	
	Boring for piles; 30nr		
C	450 mm diameter	360	m
	Extra over		
D	for boring through rock	24	m
	Disposal of material arising from excavation; Removed		
E	450 mm diameter	57	m3
	Poured concrete 30 N/mm2; 10 aggregate, unreinforced		
F	450 mm diameter	315	m
	Reinforcement, BS 4449: 1978, hot rolled plain round mild steel		
G	450 mm diameter	2.50	t
	Cutting off top of pile		
H	450 mm diameter	45	m

Figure 9.4   Piling.

Figure 9.4 demonstrates the billing of 30 nr bored piles, 450 mm diameter × 12 m long, with a cut-off level 1.5 m below Commencing Surface.

All items are measured by length in accordance with B15, with the exception of disposal, measured in accordance with B11.1, and reinforcement, which is measured by weight in accordance with Section C3. It will be noted that the number of piles is stated in the boring item (B15.1) and that the disposal item states 'removal'. This indicates removal off-site which includes providing a suitable tip (B11.1.4).

Sheet piling, conventionally measured only when expressly required in the contract, is measured in POM(I) according to the rules of B16. There is no indication of any rule as to 'if or when' sheet piling is measured, but reference to the drawings supplied with the tender documents should reveal whether such work is part of the design or not.

Supplying sheet piling is measured by area when in its final position, but unlike driving, which is measured from the 'formation level' to the bottom of the piling when driven, there is no indication of how the depth shall be determined. If sheet piling is measured in the BQ, B16.1 requires that the length is measured *along the centre line* and not along the developed length as is customary. This is illustrated in Figure 9.5.

Any strutting and walings required in connection with sheet piling is *understood to be included* according to B16.4.

Figure 9.5   Sheet piling.

One final point about piling concerns B17: *Performance designed piling* illustrated in Table 9.9.

The method of measurement says little about this type of piling, and so it must be assumed that such piles are to be designed by the contractor in response to a performance specification provided by the employer. The inclusion of such an item in the bill of quantities indicates a partial contractor design procurement route.

Table 9.9   Performance designed piles.

	SECTION B - SITE WORKS						
	PERFORMANCE DESIGNED PILING						
	Generally		(NA)				
	Pile to carry						
A	25KN with pile top 1.50 m below formation level of the ground	30	no				

Performance designed piles are to be enumerated, with reinforcement and disposal *understood to be included* (B17.1), and details are to be provided should the tops of the piles be required to terminate below the formation level of the ground (wherever that is!).

9.4.10   B19: Drainage

Drains are measured by length, over fittings, and thus, the B19.2 requirement to group and enumerate 'fittings' is effectively an extra over item. The diameter of pipes shall be stated as 'internal' or 'external' (refer to GP5.2).

Concrete beds and coverings for drain pipes are measured separately by length, including formwork, but there is no similar requirement to measure granular beds, haunchings and surrounds.

Chambers may be enumerated or measured in detail, according to relevant section of the method of measurement.

Risk issue

Important ancillaries to drainage work, such as excavation in rock, excavation and filling of soft spots, the provision of granular beds, haunchings and surrounds, and testing requirements are not included in B19.

Some of these items might come under the heading of 'required and indispensably necessary to complete the work', and others would give rise to claims under the contract, depending upon which form of contract is used.

It is hard to argue that granular beds and haunching should not be measured when there is a measured item for concrete beds and coverings. It is also a well-understood custom and practice to measure pipe supports and protection, even in CESMM.

The question is, 'would a contractor take the gamble' and risk having to fight through the dispute resolution process for payment, or would it be more sensible to raise the issue during the tender period or, if not, qualify the tender accordingly?

9.4.11   B20–B22: External works

Paving and surfacing, kerbing, fencing and landscaping are given short shrift by POM(I).

Road kerbing and edgings (B20.3), for instance, are measured by length, with curved work so described, but no mention is made of concrete beds and backings, formwork, excavation or disposal. As *curved work shall be so described*, the implication is that this is a full-value measured item and not an 'extra over'.

Paving and surfacing is measured by area (B20.1), but no distinction is made as to the composition of the paving or surfacing. Clearly, additional description is required here, especially to clarify whether or not excavation, disposal and filling (e.g. sub-base) have been measured elsewhere in the BQ.

9.4.12   B23: Railway work

Railway ballast is measured by volume, with no deduction for sleepers, in separate items for *top* and *bottom ballast*. This is normal practice, as is the method for measuring track – length along the centre line over all fittings. Sleepers are enumerated, and switches, turnouts and different types of crossings are separately classified and also enumerated.

There is no indication that the provision and laying of track should be separately measured, as is normal practice, but this could be clarified by the provision of additional description. There is, however, no provision for additional itemisation in POM(I), but if desired, this could be overcome via an amendment in the Appendix provided with the Method of Measurement pursuant to GP1.2.

Part 2

9.4.13   B24–B26: Tunnelling

The volume of excavation in tunnelling is measured as *the void which is to be occupied* including permanent linings. This includes the volume of <u>permanent</u> linings but not linings that would be classed as support (e.g. NATM linings). Consequently, the volume of overbreak and support linings must be allowed for in the BQ rates. Volumes may be grouped but must be separately classified as tunnels, shafts or other cavities stating whether they are straight or curved. Disposal is measured in accordance with B11: *Disposal*, and therefore, separate items must be measured both for disposal as on-site backfilling of different types and for disposal off-site to tips.

Table 9.10   Classification of tunnel linings and support and stabilisation.

Classification		Type		Unit
Linings	Lining	Poured concrete	Sprayed concrete	Area
			Cast concrete	Area
		Preformed segmental		Enumerated
	Secondary lining	Poured concrete	Sprayed concrete	Area
			Cast concrete	Area
		Preformed segmental		Enumerated
Supports and stabilisation	Timber			Volume
	Sprayed concrete			Area
	Rock bolts			Length
		Injection grout		Weight
		Face packers		Enumerated
	Metal arches			Weight

Section B - site works			
Tunnel excavation			
Excavation		(NA)	
Generally			
A Straight shaft	348	m³	
Disposal		(NA)	
Disposal of material arising from excavations			
B Removing from site	348	m³	
Tunnel linings			
Preformed segmental tunnel linings		(NA)	
Generally			
C 1500 × 200 × 450 mm	128	no	

Figure 9.6   Pumping station.

Tunnel *linings* and *support and stabilisation* are measured independently as summarised in Table 9.10 and typical bill items are shown in Figure 9.6. This illustrates a deep pumping station constructed of precast concrete segmental shaft rings.

It will be noted that there is no item for in-circle or cross caulking as this is not measurable in POM(I) Section B24.

9.5 Section C: Concrete work

Concrete work is the most extensive Work Section in POM(I).

General items C1.1–C1.4 establish that **reinforced** poured concrete and **plain** poured concrete shall be separately described and that horizontal surfaces are to be presumed *tamped* unless otherwise stated. Voids less than $1\,m^3$ shall not be deducted.

Concrete work is categorised according to Table 9.11.

Table 9.11 Concrete classification.

Reference	Subsection	Classification		Unit
C2	Poured concrete			Volume
C3		Reinforcement	Bar	Weight
			Fabric	Area
C4		Shuttering*		Area
C5	Precast concrete	Floor slabs, partitions, etc.		Area
		Lintels, sills, duct covers, etc.		Length
		Beams, stanchions and tunnel rings		Enumerated
C6	Prestressed concrete			Volume
		Reinforcement	Bar	Weight
		Shuttering*		Area
			Edges	Length
			Grooves, throats, etc.	Length
C7	Sundries	Surface finishes†		Area
		Expansion material		Area
		Designed joints		Length
		Sinkings, channels, etc.		Length
		Fixings, ties, inserts, etc.		Enumerated or by area
		Mortices, holes, etc.		Understood included

*Formwork.
†Understood included.

A further categorisation is provided for **poured concrete**, which shall be classified in the following groupings:

1. Foundations.
2. Pile caps.
3. Blinding.
4. Beds.
5. Suspended slabs.
6. Walls.
7. Columns.

8. Beams.
9. Staircases.
10. Diaphragm walls.
11. Other as may be appropriate.

Within each grouping of poured concrete are subsets, such that 'pile caps' include ground beams, 'beds' include roads and footpaths and 'columns' include stanchion casings. It is not clear whether the subsets of each grouping are to be separately measured, but for clarity, it would seem prudent to do so.

Times	Dimension	Result		SECTION C - CONCRETE WORK
30/	1.200			
	1.200			
	1.000			POURED CONCRETE
		43.200		
5/5/	8.800			Reinforced in situ concrete, Grade 20, 20 aggregate
	0.450			
	0.900			
		89.100		
6/4/	8.800			
	0.450			
	0.600			
		57.024		Pile caps, ground beams or the like
				generally

Times	Dimension	Result		SECTION C - CONCRETE WORK
30/4/	1.200			
	1.000			SHUTTERING
		144.000		
5/5/2/	8.800			Shuttering; generally
	0.900			
		396.000		
6/4/2/	8.800			
	0.600			
		253.440		Sides of foundations
				generally

SECTION C - CONCRETE WORK		
POURED CONCRETE		
Reinforced in situ concrete, Grade 20, 20 aggregate	(NA)	
Pile caps, ground beams or the like		
A generally	189	m3
SHUTTERING		
Shuttering; generally	(NA)	
Sides of foundations		
B generally	793	m2

Figure 9.7 Pile caps and ground beams.

Part 2

Notwithstanding the general requirement to measure concrete by volume, there are cases where a thickness needs to be stated in the item descriptions, such as beds, suspended slabs and attached columns. C2.3 provides an alternative for items where the thickness is to be stated, in that similar items of different thicknesses (e.g. beds) may be grouped within a range of dimensions of different thicknesses.

As far as shuttering is concerned, it appears that all surfaces requiring support will have an associated shuttering item measured. There is no exclusion for concrete cast against the earth. Certain classes of shuttering shall be identified in item descriptions, including that required to be left in position, curved, conical or spherical shuttering and shuttering requiring a special finish.

In Figure 9.7, it can be seen that items in connection with pile caps and ground beams have been measured.

The poured concrete and associated shuttering items follow the strict wording of C2.1.2 and C4.1.4 which respectively say that *pile caps … shall include ground beams* and that *sides of foundations* include *pile caps and ground beams*. This is to be contrasted with SMM7, which measures these items separately, and NRM2, which makes no distinction.

In this particular example, the contractor has decided to use blockwork as permanent shuttering. This is not *left in position* shuttering within the meaning of C4.5 but should nevertheless be included within the interim valuation for payment as if 'normal' shuttering had been used.

9.6   Section D: Masonry

The traditional UK work classification of 'brickwork and blockwork' is termed *masonry* in POM(I).

Measurement is by area, with sloping, battered, curved and reinforced work so described. Walls are classed as:

1. Walls.
2. Walls built against other construction.
3. Cavity walls.
4. Isolated piers.

For cavity walls, each skin and the cavity may be measured by area (D2.1.3) as illustrated in Table 9.12. Strictly speaking, no extra items are measured for forming and closing cavities or cavity ties, but in this example, the software adds this as an additional description.

Table 9.12 Masonry.

	SECTION D - MASONRY		
	WALLS AND PIERS		
	Blockwork: concrete blocks: Thermalite, 440 x 215; in cement mortar(1:4)		(NA)
	Cavity wall skins		
A	100 mm thick	1080	m2
	Facing brickwork: Spec A: coloured cement lime mortar (1:1:6); Flemish bond; flush pointing as the work proceeds		(NA)
	Cavity wall skins		
B	half brick thick; fair and pointed one side	1080	m2
	Forming cavity		
C	50 mm wide including ties type A, two per m2	1080	m2

Part 2

Faced or fair faced work may be so described in the item or, alternatively, may be measured 'extra over' the work concerned.

Cavity walls may also be measured as a composite item (both skins and cavity included) if desired (refer to D2.1.3).

Cills, copings and arches are measured separately by length, and reinforcement, concrete filling to cavities, joints and air bricks are all specifically measured in the masonry section.

9.7 Section E: Metalwork

Metalwork includes structural steelwork, measured by weight, and non-structural metalwork items whose units of measurement vary.

Structural metalwork is classified as:

- Grillages.
- Beams.
- Stanchions.
- Portal frames.
- Roof trusses.
- Support steelwork.
- Other as may be appropriate.

Fittings, grouped into caps, brackets, etc., and bolts, distance pieces, etc. are measured as items under E2.2 and E2.3.

There is no separation of fabrication and erection.

Risk issue

Protective treatments are given as an item and not measured by area as is conventional.

9.8 Section F: Woodwork

Woodwork comprises work that would conventionally be classed as 'carpentry' with 'joinery' being measured in Section H: *Doors and Windows*.

Sawn timbers must be distinguished from finished timber, but there is no requirement to identify timber requiring treatment or structural grading.

Structural timbers are measured by length in a classification that separates:

1. Floors and flat roofs.
2. Pitched roofs.
3. Walls.
4. Kerbs, bearers, etc.
5. Cleats, sprockets, etc.

Strutting, and the like, is measured by length, but there is no requirement to identify different types of strutting.

Boarding to floors, walls, ceilings, roofs, etc. is measured superficially with eaves, verges, trims and the like by length.

Grounds and battens are measured by length, but framing (e.g. stud partitions) is measured by area. Framing is stated to be taken [measured] overall, presumably meaning overall openings and the like. As an alternative, framing may be measured linearly.

Part 2

Composite items are enumerated. Exactly what is meant by this expression is not explained except that they may be, but not necessarily, fabricated off-site. The method of measurement makes no distinction. There is no requirement for a dimensioned diagram apart from the requirements of GP1.1 and GP2.3.

9.9   Section G: Thermal and moisture protection

This incongruously named section includes some substantial items of work including roofing, tanking and waterproof coverings.

Also included are eaves, ridges, flashings and roof lights, ventilators and special roofing sheets. Special roofing sheets, whilst not defined, may be measured 'extra over' other roof coverings.

Damp-proof courses and insulation are also measured in this section (G3), but no indication of any relationship with Section D: *Masonry* is made.

9.10   Section H: Doors and windows

Doors and windows are enumerated items, whereas jambs, mullions, transomes and the like are measured by length. Alternatively, frame and lining sets may be enumerated.

Screens are included in this section, leaving some doubt as to the meaning of 'composite items' in Section F: *Woodwork*. In Section H, it is presumed that 'screens' include entrance screens, shop fronts, etc., but there is no definition of the term. In common with Section F, there is no requirement for a dimensioned diagram, but the requirements of GP1.1 and GP2.3 nonetheless apply.

Tucked away in 'screens' is curtain walling – measured by area – and further down, in H6, patent glazing can be found, also measured by area. There is no extra detail provided on either.

Glass and glazed units appear under H5 with ironmongery in H4.

9.11   Section J: Finishes

This section makes the subtle, but important, distinction between 'finishes' and 'finishings'. Whilst not explained, it is fair to assume that 'finishes' is a collective noun that includes 'finishings'. In any event, J1.3 requires that *internal and external work shall each be so described* and, presumably, measured separately.

Section J is categorised as:

- Backgrounds (screeds).
- Finishings (floors, walls, ceilings, staircases).
- Sundries (inserts, dividing strips, mouldings, etc.).
- Suspended ceilings.
- Decorations.
- Signwriting.

Backgrounds are *floor, wall or ceiling finishes* (J2.1), and *each* [background] *shall be described*. Screeds and plasterboard backgrounds are specifically referred to in J2.2 and J2.3, respectively, but the method of measurement is silent on the subject of plastering. It is highly likely that plaster 'render' or 'backing' coats (i.e. the first of two-coat plaster work) will be included in 'backgrounds' on the simple logic that they are not 'finishings'.

Part 2

J3: *Finishings* makes no reference to types of finishings, but this is likely to include skim plaster (because it is not a 'background'), floor and wall tiling and, possibly, textured coatings, for example, Artex (if not part of J6: *Decorations*).

J6: *Decorations* is non-specific but is likely to include both painting and wallpapering. There is a large classification of items given in the subsection J6.3, all of which are measured by area. This includes items that would conventionally be measured as 'narrow widths'.

Decorations to small pipes (internal diameter <60 mm) are measured by length.

9.12   Section K: Accessories

'Accessories' includes partitions, doors and glazed units and cubicles or the like.

It is not clear how Section K relates to Section H: Doors and Windows, which includes H3: *Screens*.

9.13   Section L: Equipment

'Equipment' is defined as *specialist equipment related solely to the function of a building or department*, examples of which are provided in L1.1 (food preparation, laboratory equipment, etc.).

Such items are synonymous with 'fit-out' contracts that are normally separate from the main construction contract. However, in an international context, where the popularity of turnkey procurement is greater than in the United Kingdom, it is a sensible inclusion in this particular method of measurement.

Such items are to be itemised in the bill of quantities but will, in all probability, be given initially as a 'nominated' or 'prime cost' item.

9.14   Section M: Furnishings

The same comments as Section L apply.

9.15   Section N: Special construction

Work in this section includes *air supported or geodetic structures, prefabricated buildings or ... radiation protection installations* (N1.1).

Such items shall be described as:

1. Enclosures of a specialist construction.
2. Installations of a specialist nature.

Air-supported structures would come under the first classification and radiation protection under the second.

At a more prosaic level, prefabricated buildings would also be classed as *enclosures of a specialist construction*. In view of the popularity of off-site prefabrication of bathroom pods and the like and the provision of a special Work Section 2: *Off-site manufactured materials, components or buildings* in NRM2, this is a far-seeing inclusion in a method of measurement that is some 35 years old!

9.16   Section P: Conveying systems

Section P is concerned with lifts, hoists, conveyors, escalators, etc.

Each item shall be enumerated, but separate measured items shall also be given for supports, identification, testing and commissioning, tools and spares and documents, including operation and maintenance manuals.

Incidental work – conventionally referred to as 'builder's work' – is to be provided as an item (P3.1) as is each of removing protective coverings and cleaning and polishing exposed surfaces.

9.17   Section Q: Mechanical engineering installations

A separate heading in the bill of quantities is to be given for each 'installation', classed by function, such as hot and cold water installations, heating, air conditioning and so on.

Under Q2, 'pipework and gutterwork' are measured by length which, whilst not defined, are likely to be equivalent to Work Section 33: *Drainage above ground* (NRM2) or R10: *Rainwater pipework/gutters* (SMM7).

The remainder of this section deals with the measurement of Q3: *Ductwork*, Q4: *Equipment*, Q5: *Automatic controls*, Q6: *Connections to supply mains* and Q7: *Insulation* in connection with mechanical engineering installations.

Builder's work in connection with mechanical engineering installations is measured in accordance with P2: *Sundries* in Section P: Conveying Systems (Q8.1 refers).

9.18   Section R: Electrical engineering installations

The POM(I) approach to electrical work is refreshingly pragmatic and is divided into:

- R2: *Main circuits.*
- R3: *Sub-main circuits.*
- R4: *Final sub-circuits.*

Accessories (R5), control gear (R6) and equipment (R7) are each measured separately as are R8: *Connections to supply mains* and R9: *Sundries*.

Installations are classed by function and given separately under an appropriate heading. The Section R measurement rules may be followed or the work *may be enumerated on a locational basis* (e.g. lighting installation to ground floor).

The 'main' circuit is the incoming supply to the main distribution board, and the 'sub-main' is the supply which runs from the main distribution board to a sub-main distribution board. Final sub-circuits include lighting points, socket outlets, telephone outlets and so on. Ceiling pendants, light switches and other 'final fix' items are included in 'Accessories' (see R5.1).

Builder's work in connection with electrical engineering installations is measured in accordance with P2: *Sundries* in Section P: Conveying Systems (R9.1 refers).

Notes

1. *The headings in this chapter generally follow those of POM(I).*
2. That is, in any of the various documents that make up the contract – for example, contract agreement, letter of acceptance, letter of tender, contract conditions, specification, drawings, schedules (including bills of quantities), etc.

Part 2

3. http://www.rmdkwikform.com/uk?gclid=CNCa6uqL3bwCFSEHwwoda0UACg (accessed April 29, 2015).
4. Methods of payment rather than methods of measuring the absolute net volume of solids removed.

References

Institution of Civil Engineers (1995) *Dredging: ICE Design and Practice Guides*, Thomas Telford Ltd, London.

Seeley, I.H. (1965) *Civil Engineering Quantities*, Macmillan, London.

PART 3

Measurement Risk in Contract Control

Part 3

Managing Measurement Risk in Building and Civil Engineering, First Edition. Peter Williams.
© 2016 John Wiley & Sons, Ltd. Published 2016 by John Wiley & Sons, Ltd.

Chapter 10
Contract Control Strategies

As discussed in Chapter 4, many different types of pricing documents are employed in construction:

- Some are formal – issued by the employing authority.
- Some are informal – prepared internally by the contractor or subcontractor.
- Some become formal – when contractors and/or subcontractors issue their internal documents for tendering purposes and/or for incorporation as a contract document (e.g. a contractor's contract sum analysis or activity schedule).

Irrespective of the pricing document, or its formality or informality, and regardless of whether it is included in the contract documents, the pricing document forms the basis for contractor's/subcontractor's financial control of the project.

10.1   Financial control

Morris (1999), Burke (2013) and others consider that project management concerns the management of change.

Looking at the job of construction site managers, this is certainly true, as most of their time is consumed dealing with unforeseen events, unexpected design changes and incidents that arise 'out of the blue'.

The same could be said of the site quantity surveyor (QS), who has to deal with the financial consequences of what happens on-site. Added to this, upwards of 50% of site QS's time is spent preparing the monthly cost-value reconciliation, which compares actual cost and value with planned, and most of the remaining time is spent dealing with dayworks and claims and finding ways to make money out of loss-making activities.

Managing Measurement Risk in Building and Civil Engineering, First Edition. Peter Williams.
© 2016 John Wiley & Sons, Ltd. Published 2016 by John Wiley & Sons, Ltd.

10.1.1   The role of measurement

Change has to be managed and this requires a strategy for control. Measurement is central to the financial management of construction projects and is integral to the role played by the pricing document in the contractor's financial control strategy. The pricing document:

- Establishes **the contractor's budget** for the contract.
- Forms the basis for **subcontract enquiries** and the subsequent **subcontract agreement**.
- Provides a means for making **main contract payment applications** to the employer.
- Provides a means for making **subcontract payment applications** to the main contractor.
- Establishes a starting point for **valuing variations** to the contract.
- Creates the basis for **the contractor's internal valuation** which, in turn, enables the contractor's monthly **cost-value reconciliation** to be prepared.
- Forms the basis of the contract **final account**.

Both the contractor's QS and that of the employer perform many measurement iterations during the course of a contract, even on lump sum contracts. The PQS is mostly involved with measuring and valuing variations to the contract and work in connection with the expenditure of provisional sums.

10.1.2   Measurement risk

Risk is the chance of something happening that will have an impact upon the outcomes of the contract. It cannot be completely eradicated, but it can be managed.

Measurement might be considered as comprising three parts:

1. Magnitude.
2. Dimensions (units).
3. Uncertainty[1].

The use of the word 'uncertainty' is interesting because it also introduces the notion of 'risk' simply because the two words are often considered synonymous. However, in a useful discussion, Ross and Williams (2013) suggest that 'uncertainty' has a separate and distinct meaning from 'risk'. They quote a number of authorities who concur in this view amongst whom is Winch (2010) who perhaps makes the clearest distinction:

- Uncertainty is *the absence of information required* for decision making.
- Risk is *the condition where information is still missing but a probability distribution can be assigned to the occurrence* of a particular event.

Transposing this logic into a measurement context would seem to suggest, therefore, that:

- Uncertainty arises in circumstances where there is insufficient information available to measure the work or object in question.
- Risk is where a judgement is needed to assess the likely consequences arising from the occurrence of a specific event or circumstance in relation to an item of work or object that can be measured.

Table 10.1 provides some examples of each.

Consideration of risk and uncertainty in measurement is not, however, confined to measurable items. The procurement and contractual arrangements for a project can also have a considerable impact on measurement issues, whether in relation to the main contract between employer and contractor or a subcontract, or sub-subcontract, further down the supply chain.

The measurement process doesn't end when the bill of quantities, or other pricing document, is completed, and a great deal of measurement goes on during the contract whilst work is progressing in order to:

Table 10.1 Measurement risk and uncertainty.

Measurement	
Risk	**Uncertainty**
• A measured item for a drain trench excavation, 3–4 m deep, where the contractor has to include for an earthwork support system in his price • Where an item is included in a bill of quantities for the construction of a structure which the contractor is required to design • A provisional quantity in a bill of quantities where the quantity stated is artificially inflated to provide a hidden contingency	• A provisional quantity in a bill of quantities where the amount of work in the item can only be estimated • A prime cost sum is included in a measured item of brickwork for facing bricks that the designer has yet to specify • The inclusion of a provisional sum in a bill of quantities where the item in question is known about, or anticipated, but cannot be quantified

- Measure the quantities of work done according to the type of contract being used.
- Deal with provisional quantities and provisional sums.
- Measure variations to the contract.
- Prepare the contractor's cost-value reconciliation.

10.2   Measuring the quantities of work done

Post-contract measurement is conducted by different people for different reasons during the currency of a construction contract.

The accuracy, reliability and contractual validity of such measurements can be extremely variable and much depends upon who is doing the measuring.

An experienced PQS is usually qualified and very able when it comes to measurement, but contractors' QSs come in all shapes and sizes and levels of competence in measurement, whether qualified, partially qualified or non-qualified. Some subcontractors are competent in measuring their particular trade, whilst others may have little or no QS background and may struggle with the finer points of methods of measurement and contract conditions.

Measurement may be undertaken:

- For or on behalf of the employer, usually in conjunction with the contractor, is a valid process under standard forms of contract, and the results of the measurement process will find their way into interim payments and into the final account.
- By the contractor, and submitted to the employer, may or may not be accepted and it is for the contract administrator to decide whether the measurements are valid or not.
- By subcontractors, and submitted to the contractor, may be accepted if they are capable of being accepted by the employer.
- By subcontractors, and submitted to the contractor, will be rejected if they are not capable of being accepted by the employer unless the terms of the subcontract are different to those of the main contract.

10.2.1   Admeasurement and remeasurement

'Admeasurement' is often taken as a synonym for 'remeasurement', but the words have different, if subtle, meanings. The term *remeasurement* is used in NRM2, Paragraph 3.3.8.2, whilst *admeasurement* is used in several places in CESMM4, but neither term is defined.

Part 3

According to the Oxford Dictionary:

- **Admeasurement** is *the act of ascertaining and apportioning.*
- **Remeasurement** means to *measure again.*

In terms of construction, *admeasurement* originated from Clause 56 of the ICE Conditions of Contract (now ICC – Measurement Version), meaning to establish <u>the difference</u> between a final quantity and an original quantity of work, whether more or less. It is this difference that determines, for instance, whether or not any of the rates and prices in the contract bills are *rendered unreasonable or inapplicable in consequence of* the change in quantity.

Whilst the term 'admeasurement' is peculiar to the ICC – Measurement Version, other standard contracts use words and phrases that essentially mean the same thing. Therefore, in a measure and value contract, the admeasurement process will take place, if not in name, in order to identify differences in the estimated and final quantities of work for the purpose of arriving at a fair valuation of work done.

The dictionary definition of 'remeasurement' infers the measurement of something that has already been measured but is to be measured again. This process results in a fresh set of quantities that replace the original rather than establishing the difference between the two as in admeasurement.

Remeasurement is common in lump sum contracts where the basis for determining the final account is the omission of one item from the contract sum and the substitution of another should a change or variation have occurred.

Consider a bill of quantities item for *mass concrete in foundations* in a lump sum contract, where the original quantity is measured as $93\,m^3$. Under the JCT 2011 contract, a change in this quantity would be a variation. If the final quantity is measured as $136\,m^3$, the original item would be omitted from the contract sum and a new item, with the same description but a remeasured quantity of $136\,m^3$, would replace it. This adjustment would be made in the 'variation account' of the final account as illustrated in Table 10.2.

Table 10.2   Remeasurement.

Omit			Add		
'In-situ' concrete works			**'In-situ' concrete works**		
Plain in-situ concrete			Plain in-situ concrete		
A Mass concrete; Class C20; Any thickness; In trench filling; Poured on or against earth or unblinded hardcore	93 m3		A Mass concrete; Class C20; Any thickness; In trench filling; Poured on or against earth or unblinded hardcore	136 m3	

The same item in a measure and value contract would be dealt with differently as measure and value contracts do not have a contract sum. Consequently, any difference in the original quantities would be admeasured (i.e. adjusted ±) to reflect the final quantities of work carried out, and the bill of quantities would be altered accordingly.

The extent to which work executed is remeasured or admeasured depends on the type of contract.

10.2.2   Lump sum contracts

Lump sum contracts are based on the concept of a fixed and agreed price to do a job which establishes the contract sum agreed by the parties (e.g. employer–main contractor and main contractor–subcontractor). The contract sum in a lump sum contract may only be adjusted if the contract terms contain express provisions to do so.

Under the JCT SBC/Q 2011, for instance, the contract sum may only be adjusted for specified reasons:

- Variations issued by the architect/contract administrator.
- Adjustment of provisional sums.
- Adjustment of approximate quantities.
- Payments for direct loss and expense.
- Insurance payments.

Consequently, the quantities given in a lump sum contract are, to all intents and purposes, fixed, agreed and not subject to adjustment unless there is a legitimate reason why the works should be remeasured.

Subcontractors, particularly, often fail to understand this concept. Sometimes, this stems from ignorance of contract law, sometimes from lack of information from the main contractor and sometimes because they instinctively remeasure the work they have done in the belief that this is the means whereby maximum payment can be guaranteed.

Additionally, subcontractors are often unaware that they are signed up to a lump sum contract despite the main contract being a measure and value contract and vice versa. There is no question that main contractors can often make extra margin in such circumstances by either:

- Profiting from an under-measured bill of approximate quantities whilst at the same time subcontracting the works as a series of lump sums.
- Tying subcontractors to a measure and value arrangement whilst the main contract is a lump sum contract with a generously measured bill of quantities or in the belief that it will be possible to 'pull the wool over the eyes' of the subcontractors when it comes to admeasuring their work.

10.2.3   Measure and value contracts

Under measure and value contracts, there is no contract sum at the outset. The only thing fixed and agreed by the parties beforehand are the rates and prices that will apply to the contract, and it is these rates and prices that will be used in order to value the work carried out by the contractor (or subcontractor).

The prices will consist of the contractor's preliminaries, including any fixed or time-related charges or method-related charges, and the rates will be for individual items of measured work such as earthworks, concrete work, wall and floor finishes, drainage etc.

Measure and value contracts are often classified as 'remeasurement' contracts, but this is not strictly correct. They are really 'admeasurement' contracts because the difference between the original quantities and the final quantities is what determines the final payment and whether any change is needed in the rates and prices to reflect the consequences of the change in quantities. In practice, both terms are used interchangeably.

Measure and value contracts arise out of uncertainty:

- With the scope of works required.
- With the design or parts of the design.

There is an element of risk for the contractor (or subcontractor) with measure and value contracts in that the eventual admeasure may equally be less than envisaged as more, and as such, the risk is that margin may be lost as well as turnover. Ross and Williams (2013) illustrate this point with a worked example which shows that a 10% reduction in measured quantities could lead to a 24% loss of profit on the contract.

With a measure and value contract, the original pricing document may be a bill of approximate quantities or possibly a schedule without quantities. In both cases, the work carried out will be measured according to the rules of measurement stated in the contract documents, and the value of this work will be established by applying the agreed rates and prices to the quantities.

The JCT SBC/AQ 2011 is an example of a measure and value contract whereby the contractor's rates and prices are used in the first instance to establish a tender total (i.e. *not* a contract sum) and, thence, following admeasurement of the completed work, to determine an ascertained final sum which is the amount to be paid by the employer to the contractor.

There may be a presumption that a measure and value contract will be admeasured on the basis of a physical on-site measure of the completed work. This may be far from the case because:

- The original drawings upon which the approximate quantities were based may have been less inaccurate or uncertain than at first envisaged.
- Revised drawings may have been prepared during the course of the contract which may be much more representative of what was actually constructed.
- 'As-built' drawings may have been prepared which exactly depict what was constructed.

On a measure and value contract, it is usual practice to admeasure from the drawings, and it is only where these are uncertain or inaccurate that actual site measurements are taken. Subcontractors often have difficulties with this in the mistaken belief that nothing can be more accurate than measuring that which is actually built. This is not the case, and site measures are notoriously inaccurate because of:

- The natural tendency to 'over-measure' (in order to make more money).
- Failure to follow precisely the standard method of measurement.
- Failure to measure 'net' as depicted on the drawings or in standard construction details.
- The idea that 'pinching' a few metres here and there will not be noticed.

Subcontractors are often at a distinct disadvantage when it comes to admeasures because:

- Subcontract enquiries rarely contain full and complete information.
- Very often, the work gets underway and the subcontractor has little or no drawn information to work from.
- Main contractors are frequently reluctant to provide subcontractors with full or up-to-date drawings from which to measure.
- Main contractors invariably fail to pass on contract administrator's instructions so that subcontractors find difficulty in distinguishing variations from measured work.
- At final account stage, subcontractors either have out-of-date drawings or no drawings at all.

As a consequence, subcontractors regularly have to resort to the time and expense of a physical measure which (i) will undoubtedly be far less reliable than measuring from drawings and (ii) is more than likely to be disputed by the main contractor.

10.2.4   Cost reimbursement contracts

By their very nature, there is no remeasurement/admeasurement involved with cost reimbursement contracts.

The idea with this sort of contract is that the contractor is paid on the basis of the actual cost of carrying out the work – that is, the cost of labour, materials, plant and subcontractors to which is added a pre-agreed percentage, or sometimes a fixed fee, to cover preliminaries, overheads and profit.

Paradoxically, there is often a cost reimbursable element to both lump sum and measure and value contracts in the form of **daywork**. This is a means of valuing work done on the basis of the time spent and the materials and plant used together with a percentage addition to cover oncosts, overheads and profit.

Where payment is made on the basis of time spent, it could be argued that the very act of recording the time expended by various classes of labour, and types of plant and equipment, is a form of measurement with the unit of measurement being **the hour**.

10.3   Provisional quantities and provisional sums

Provisional sums are to be distinguished from provisional quantities because they each arise, and are dealt with, differently both contractually and in terms of measurement. The measurement of these items is conducted by, or on behalf of, the contract administrator.

Provisional quantities arise in lump sum contracts. They are the estimated quantities of items of work to be carried out by the contractor where accurate quantities cannot be measured. Provisional quantities do not arise in measure and value contracts because all the quantities in such contracts are estimated.

In SMM7, work that can be described in accordance with the rules of measurement but cannot be measured accurately is to be identified as *approximate* (General Rule 10.1). The same rule applies in Paragraph 3.3.8.1 of NRM2, but such quantities are to be identified as *provisional*. This gives rise to a measured item as illustrated in Table 10.3.

Table 10.3   Provisional quantity.

Excavating and filling		
Excavations		
Excavation		
A	Bulk excavation; Not exceeding 2m deep [Provisional]	12500 m3

It will be noted from Table 10.3 that the provisional quantity ($12\,500\,m^3$) is a 'round' figure. This should put the contractor 'on warning' of the approximate nature of the quantities even if the word 'provisional' doesn't! The large quantity billed may indicate the possibility of a hidden contingency.

Care should be exercised with approximate/provisional quantities because a lump sum contract, by definition of its nature, requires a mechanism for adjusting any quantities in the contract bills, whether provisional or not.

Where SMM7 and NRM2 are used in conjunction with the JCT 2011 SBC/Q contract, this is taken care of in Clause 5.1.1 which defines changes in quantity as a variation to the contract and confers legitimate powers to deal with them on the contract administrator.

Should it not be possible to create an approximate/provisional quantity, SMM7 (General Rule 10.2) and NRM2 (Paragraph 2.9.1.1) provide that a provisional sum be given as 'defined' or 'undefined' work.

An issue for contractors is that approximate/provisional quantities are not distinguished from other quantities in JCT 2011, and the contract administrator can vary such items with seeming impunity. However, JCT 2011 does redress the balance somewhat via the valuation of variations Clause 5.6.2, which states that a fair allowance shall be made to the rates and prices where the change in quantity is significant.

The word 'significant' is not defined in the contract, but Paragraph 3.3.8.2 of NRM2 does introduce the idea of a threshold of change (the 20% rule) despite it being of questionable contractual merit.

Provisional sums are described, but not measured, in the bill of quantities and represent a sum of money to be expended, as required, by the contract administrator. If, in the fullness of time, the work described is not required, the provisional sum(s) will be omitted from the contract sum calculation at the final account stage.

Provisional sums are to be identified as 'defined' or 'undefined' work pursuant to NRM2 Paragraph 2.9.1.1, depending upon the level of detail available at the time the bill of quantities is prepared. In CESMM4, there is only one classification – provisional sum for defined work.

Part 3

In the event that a particular provisional sum is expended, the employer's QS, or equivalent, will normally measure the work involved. This does not constitute a remeasure or admeasure because no quantified allowance has been made in the bill of quantities.

Such work is treated as a variation to the contract, under both the JCT 2011 and ICC – Measurement Version forms of contract, unless the work is to be carried out by a nominated subcontractor under the ICC form. The work is, therefore, measured in accordance with the rules of measurement applicable to the contract and dealt with in the provisional and prime cost sum part of the final account as shown in Table 10.4.

Table 10.4   Dealing with provisional sums in the final account.

Provisional sum in BQ						Measurement of work in connection with Provisional sum in final account		
Omit						**Add**		
Ref	Item	Quantity	Unit	Rate	£	p	General joinery		
	Provisional Sums								
	Allow the Provisional Sum of £3 000 for sundry joinery work in Workroom A. (Undefined work)				3 000	00	Unframed isolated trims, skirtings and sundry items **Skirtings,**		
							A Softwood; 169 mm x 20 mm; Torus; Fixed to blockwork	69	m
							Architraves and the like		
							B Softwood; 56mm x 17 mm; Torus; Fixed to blockwork	33	m
							Window boards		
							C Softwood; 219 mm x 33 mm; Rebated and rounded one edge; Fixed to blockwork	9	m
							Pinboards		
							D 1200 mm x 1200 mm x 12 mm; Unframed; Fire-retardant felt; Burgundy; Fixed with 4 nr brass plates to blockwork	3	nr

10.4   Measuring variations to the contract

Where empowered to do so, contract administrators may issue variation instructions where changes to the design are needed or where additional or less work than envisaged is required. Under some forms of contract (e.g. JCT 2011), changes in quantity qualify as a variation, whilst in others (ICC – Measurement Version), they don't. Changes to the Works Information under the ECC are 'compensation events'.

On occasion, certifiers will pay the contractor 'on account' for variations whilst awaiting appropriate written instructions from the architect or engineer (often one and the same person!), but strictly speaking, no payment should be made until the variation has been measured and valued in accordance with the contract conditions.

The measurement of variations follows the rules of measurement in the contract and normally involves deleting one BQ item and substituting it with another in the 'variation account' part of the final account for a lump sum contract. Work may equally be omitted altogether, changed in quantity or quality or simply changed.

Variations are usually measured from revised drawings but, where necessary, can be made on-site. In this event, the contractor normally has a right to be present when the measurements are taken. This acts as a check on what the PQS is measuring, which is not needed if measuring from drawings, as the contractor will have a copy.

Attention should be paid to the distinction between remeasurement/admeasurement and the measurement and valuation of variations.

If measured work in the bill of quantities is omitted or changed in some way, it is normally within the power of the contract administrator to value this as a variation, and this power extends, where appropriate, to the power to decide the appropriateness of BQ rates and prices if considered necessary.

If this process indicates that the BQ rates should be revisited in the circumstances, then method-related charges will be subject to scrutiny and adjustment, if the contract administrator thinks fit, in order that a fair valuation is determined.

The form of contract must be carefully examined, however, because different standard forms define variations in different ways. In some contracts, a change in quantity is a variation but not in others.

10.5   Preparing the contractor's cost-value reconciliation

The biggest monthly task for the contractor's QS is to accurately measure the work carried out on-site (the internal valuation) so that this can be compared with the valuation of work in progress determined for interim payment purposes (the external valuation).

The **external valuation**, whether conducted by the PQS or by the contractor, is a 'theoretical' measure of work done simply because of the way that the valuation is carried out.

Ross and Williams (2013) suggest that, most commonly, external valuations are based on **inspection**, which is a judgement of the percentage of work completed to date. The external valuation is prone to inaccuracy because the work may well be over- or under-measured due to the lack of precision in establishing the amount of work completed.

This contrasts with the contractor's **internal valuation**, which is a measured valuation that accurately establishes how much work has really been done.

The internal valuation needs to be accurate because it is compared with actual cost in order to establish the profitability, or otherwise, of the contract. The purpose of the exercise is to determine the real value of work in progress, and profit or loss, so that this can be reported to management for control purposes and for inclusion in the quarterly, six-monthly and annual accounts of the company. The cost-value reconciliation process is examined in detail by Ross and Williams (2013).

The accuracy of the internal valuation depends on how able the site QS is at measurement and upon the reliability of site records, diaries, daywork and time sheets and measurements taken by non-QS personnel.

10.6   Physical measurement

The physical measurement of construction work on-site is arguably the most difficult task facing the QS, measurement engineer, site engineer or subcontractor. Physical measurement poses a number of problems for the measurer, and the eventual output from the site measurement process can often provide a fruitful area for argument and dispute. Site measurement may be required for a number of reasons, but it frequently lacks the precision of measurement from paper drawings or computer-generated drawings and models – hence the arguments!

Another problem is that the QS frequently has to rely upon measurement data prepared by others. For example, a foreman or ganger might record the dimensions of a soft spot which is then filled with stone or concrete, or a subcontractor might remeasure some extra work that has been carried out. At some stage, the site QS will receive a copy of this information, but there may be problems:

- The measurements taken may have been exaggerated (with the best of intentions!) but are nevertheless inaccurate.
- The dimensions may not have been agreed or verified by the employer's representative (which a QS would normally do as a matter of course).
- The work may have been recorded as daywork with no dimensions taken.
- Records of the materials used may not have been kept.

Part 3

Where a site engineer takes a grid of levels during earthworks operations, this information will be passed over to the QS at some stage. On the face of it, this sounds fine but the QS might have preferred three grids of levels: one favourable to the contractor for presentation to the employer, another favourable to the contractor and less favourable for the earthworks subcontractor and a final accurate set for internal costing or CVR purposes. This may sound like questionable practice or 'cooking the books', but it is simply a matter of making money by placing the levelling staff in high spots or low spots as appropriate.

It is always a good idea for the QS to ask site staff to take digital photographs of work that they have measured and to make sure that there is an object in the photo that will give an idea of scale – a measuring tape or levelling staff could be used, for instance. This is easier said than done because site staff are normally under pressure and may not have the time for QS 'niceties'.

Measurement data prepared by non-QSs may also suffer from deficiencies vis-à-vis the standard method of measurement:

- Measurable items may be missed.
- Claims opportunities may be overlooked.
- The dimensions taken may be inaccurate.
- Dimensions may be written down in a way that is difficult to follow (remembering that QSs have their own conventions for setting down dimensions and side casts).

There may be a number of reasons why it is necessary to measure construction work physically on-site:

- The quantities given in the bills of quantities are approximate, and no 'as-built' drawings have been prepared from which to measure the completed work.
- There is no grid of levels for the site, or the levels given on the drawings are either insufficiently detailed for accurate measurement or have been proved to be inaccurate when checked physically.
- The bills of quantities provided – either by the employer to the main contractor or by the main contractor to a subcontractor – are inaccurate because either the quantities are incorrect or the standard method of measurement has not been followed, or there is a conflict with the drawings and/or specification or for any combination of such reasons.
- The contract may be such that no quantities were provided at the outset, there are few if any drawings and the work must necessarily be measured from the finished items of work (a typical circumstance in refurbishment, repair and maintenance work).
- Instructions for variations to the contract may have been issued by the contract administrator, but no revised drawing has been issued to reflect the work involved (irrespective of whether the variation is an omission or an addition to the contract).
- Circumstances may have arisen on-site where unplanned or unexpected work has been carried out that needs to be measured and recorded before the work is covered over or excavations are backfilled and the data is lost.
- A subcontractor may have carried out additional work on the basis of a verbal instruction and has not been issued with any drawings representing the work involved.
- A subcontractor may have been requested by the main contractor to return to site in order to rectify completed work that has been damaged by another subcontractor, and there is no alternative but to physically measure the remedial work needed.

For these, and perhaps many other, reasons, physical measurement on-site may be necessary, but there are traps for the unwary, and it is vital to understand the contractual arrangements for the project in question before jumping to the conclusion that work done should necessarily be remeasured.

Site measurements are often taken when one trade has finished its work but the following trade has yet to commence or is only at a preliminary stage.

For instance, plastering to block walls may have been completed, but skirting boards and suspended ceilings have yet to be done. For practical reasons, the plastering will normally extend behind where the skirting is to go and beyond the finished level of the suspended ceiling, and it is, therefore, tempting to measure the plastering that has actually been applied to the walls. This is to forget the concept that:

a)   Standard methods of measurement prescribe that measurements shall be net.
b)   The tops of skirting boards and the finished line of suspended ceilings denote a **payment line** for measurement of the plastering work.

10.6.1   Conducting a site measure

Site measures are often undertaken by QSs working alone and with no one available to hold 'the other end of the tape'.

At one time, this posed a problem and most QSs will be familiar with the difficulty of finding a convenient point for 'hooking' the end of the tape to in order to take a measurement. Invariably, the tape will fall off at the last minute, and the QS is then obliged to trudge back through the mud and start again – frustrating, time-consuming and tiring work!

Measuring floor to ceiling heights, other than in domestic-scale buildings, is also problematic when working alone, particularly when the scaffolding has been dismantled or if there are dormer-style roof lights or atria to measure.

The advent of a modern generation of measuring devices has helped to solve such problems and make site measuring much quicker and more accurate. With some devices, measurement data can be downloaded directly to a PC or laptop and thence into a spreadsheet or measurement software package.

Basic QS equipment for site measures includes a retractable tape, long tape, measuring staff and dimension book.

10.6.2   Site measurement of specialist work

Some types of construction work cannot be measured on-site in conventional ways due to the nature of the work concerned. Piling work, embedded retaining walls and the like cannot be physically measured in the same way as 'normal' construction work. Imagine, for example, the problem of physically measuring the depth of a rotary bored pile 300 mm in diameter. A small person dangling on a rope would not be allowed for health and safety reasons!

An additional difficulty is the contract administration procedures and protocols required to agree such measurements with the main contractor and with the employer's representative so that payment without dispute is assured.

The Federation of Piling Specialists guidance is that piling work and embedded retaining walls, etc. *should always be billed as 'provisional' and measured and valued as executed*, but no guidance is given on how this should be done. The solution to the problem with regard to piling depends on the type of pile.

Driven piling, for instance, requires the development of a static design from the geotechnical information supplied with the contract documents or from a subsequent subsoil survey. This determines an estimated depth for the pile which is used for estimating purposes and as the basis for an order or contract. If billed, this work would have to identified as provisional in a lump sum contract.

For driven piles, a 'pile set' calculation is performed according to the 'industry standard' *Hiley Formula*,[2] which takes into account the weight of the piling hammer, the efficiency of the rig, the drop height of the hammer and the safe working load of the pile. The set calculation ensures that the pile is driven to the correct depth to carry the required loads. The pile set is a measurement of how much the pile moves for every 10 blows of the piling hammer. The site operative makes a mark on the pile section, hits the pile 10 times with the hammer and then marks the pile again and measures the distance between the two marks.

As the ground can vary from the geotechnical information supplied, there will invariably be variations in pile depths for each of the driven piles supplied. The desired length of pile will be made up as required from standard lengths (typically 3, 4 or 6 m) in order to minimise any waste pile protruding from the ground.

The driven piles are measured to the 'pitched length' which is the depth of the pile in the ground including any pile length protruding above the ground. The lengths of piles are calculated from the sum of the lengths of pile sections installed as noted on a 'pile record sheet' by the piling operative. The sheet is then signed off as a true record by the operative, the contractor/employer representative and the piling engineer.

In the case of driven cast in situ piles (Cementation Skanska's FRANKI® pile), the hydraulic hammer rigs have digital instruments in cab which provide a continuous display of depth, driving resistance and set. The data is recorded for each pile and can be saved to a PC for analysis and measurement purposes. Site printouts can be made from the in-cab printer which produces a record on-site in graphical form for verification and signing off, if required.

In the case of augered (not rotary bored) piles, the design process is the same as for driven piles, but the measurement process is different.

The augered pile is installed to the design depth, taking into account the borehole commencement level, the piling platform level and the cut-off level for the pile. Within the auger rig cab is a computer system which the operator uses to measure the depth of the pile, torque, rotation speed, concrete pressure and a few other things. This information is manually written onto a pile record sheet. The difference with augered piles is that the computer also produces a pile synthesis which states the depth of the pile and, as they are installed to the designed depth, there is very rarely any variation on pile length.

Notes

1. http://en.wikipedia.org/wiki/Measurement (accessed 29 April 2015).
2. http://anbeal.co.uk/hiley.html (accessed 29 April 2015).

References

Burke, R. (2013) *Project Management: Planning and control techniques*, 5th edition, John Wiley & Sons, Inc., Hoboken, NJ.
Federation of Piling Specialists, *Measurement of Piling and Embedded Retaining Wall Work* 2007.
Morris, P.W.G. (1999) *The Management of Projects*, Thomas Telford, London.
Ross, A. and Williams, P. (2013) *Financial Management in Construction Contracting*, John Wiley & Sons, Inc., Hoboken, NJ.
Winch, G.M. (2010) *Managing Construction Projects*, 2nd edition, John Wiley & Sons, Ltd, Chichester.

Chapter 11
Measurement Claims

Whilst the subject of 'claims' is not strictly within the ambit of this book, it is nevertheless an important issue in the context of measurement for the simple reason that various sorts of claims have their origin in matters concerning how work is to be measured and paid for.

This might arise from events on-site, from differences of opinion between the employer and the contractor or between the contractor and the subcontractor(s) or from disputes where, for example, the correctness of quantities or item descriptions, the application of measurement rules or the interpretation of those rules comes into question.

Disputes may also arise where there is a question regarding the basis of the contract, as it is not always clear whether a contract is for a lump sum (fixed) or a lump sum (adjustable) or a measure and value contract. It may seem an obvious thing to check, but it is surprising how often contracts, and especially subcontracts, are entered into without a basic understanding of the basis of the tender, and thence the contract.

11.1 Claims

The UK construction industry is notorious for being 'claims conscious' and has an unenviable reputation for conflict, disputes and litigation that has been highlighted in numerous industry reports dating from the 1944 Simon Report to the more recent Latham (1994) and Egan (1998) reports.

Despite changes in procurement methods and increased emphasis on partnering and developing better supply chain relationships, contractors and subcontractors remain conscious of their contractual entitlement when the occasion arises, and construction industry claims remain a fruitful area for lawyers and claims experts.

A claim arises in construction work when conditions on-site, and/or the circumstances in which work is carried out, are not as envisaged in the contract. Claims may also arise when the documentation upon which the contract is based is flawed in some way.

Managing Measurement Risk in Building and Civil Engineering, First Edition. Peter Williams.
© 2016 John Wiley & Sons, Ltd. Published 2016 by John Wiley & Sons, Ltd.

Part 3

11.1.1   Definition of 'claims'

The word 'claim' is, to some extent, an idiomatic term that is widely used in the construction industry to signify a demand for additional payment, but it is also a term that appears in some standard contracts, notably the Infrastructure Conditions of Contract.

Other less emotive terms tend to be preferred in some contracts – JCT contracts refer to 'loss and expense' and the NEC Engineering and Construction Contract uses 'compensation event', for example. Claims submitted to the employer tend to reflect badly on the employer's professional team, whereas claims submitted to contractors by subcontractors tend to reflect badly on the contractor's profit margin!

A definition, therefore, may be useful to clarify exactly what is meant by the word 'claim', and the following, taken from the excellent book by Hughes and Barber (1992), now sadly out of print, serves the purpose most eloquently:

> **Claim:** *a request, demand, application for payment or notification of presumed entitlement to which the contractor, rightly or wrongly at that stage, considers himself entitled and in respect of which agreement has not yet been reached.*

Notwithstanding the above definition, which might be considered to reflect 'active' claims – that is, claims actively pursued by the contractor – certain claims may be considered as 'passive' to the extent that they are initiated by the contract administrator under an express duty in the contract. For example, where there is a variation to the contract, the contractor may automatically receive an enhanced BQ rate, or additional preliminaries, as a result of an instruction issued by the contract administrator. The contractor may not agree with the valuation, but this will then become an 'active' claim.

Both 'active' and 'passive' claims arise for a number of reasons, and there are several types of claims that may be classified in different ways.

11.1.2   Classification of claims

A glance at any of the standard texts on construction claims will reveal that claims may be classified under a variety of headings. Chappell (2011), for instance, prefers a 'legalistic' classification:

- Contractual claims.
- Common law claims.
- *Quantum meruit* claims:
  - Where there is a contract.
  - Where there is no contract.
- *Ex gratia* claims.

Hewitt (2011), on the other hand, classifies claims more pragmatically:

- Claims for variations.
- Claims for extensions of time.
- Prolongation claims.
- Acceleration and disruption claims.
- Claims for damages under the law.
- Interim and final claims.

Hughes and Barber (1992) propose a classification by subject which is perhaps more pertinent to this book:

- Claims concerning the existence or applicability of the contract.
- Claims concerning contract documentation.

- Claims concerning the execution of the work.
- Claims concerning payment.
- Claims concerning prolongation (delay and disruption).
- Claims concerning default, determination, forfeiture, etc.

Chappell (2011) reminds us that claims are not just a one-way street and that employers may have legitimate claims against contractors for a variety of reasons. These might include the payment of liquidated and ascertained damages for late completion, the payment of a debt due on a final certificate, or by way of set-off, against sums due (e.g. where the employer has paid others to rectify the contractor's faulty work), or in circumstances where the contractor's employment under the contract has been determined.

However, none of the standard classifications of claims refer specifically to 'measurement claims' albeit that practitioners in the industry are very familiar with the phrase.

11.1.3   Measurement claims

Hughes and Barber (1992) come closest to the phrase 'measurement claims' under their heading of **claims concerning contract documentation**. In the context of this book, therefore, **measurement claims** are taken to mean contractual, common law, *quantum meruit* or *ex gratia* claims submitted by contractors or subcontractors under the following heads of claim:

- Where there is an error in a quantity.
- Where there is an error in an item description.
- Where there is a discrepancy between a pricing document (e.g. BQ, schedule of rates, etc.) and any other contract document(s), such as drawings or specifications.
- Where there is a departure from the rules of a method of measurement.
- Where there has been an omission, or alleged omission, to measure something required by the method of measurement.

Where quantities are provided by one party to a contract (e.g. the employer) to another party (e.g. contractors) for the purpose of submitting a tender, errors in the quantities provided will inevitably influence the tender figure (lump sum contracts) or tender total (measure and value contracts) submitted. This may mean that the tender figure is inflated as a consequence or that the tender figure is lower than it otherwise would be. More often than not, such errors will be adjusted pursuant to express contractual provisions, but each contract, especially non-standard or amended standard contracts, should be carefully scrutinised to be sure of the contractor's entitlement.

The same principle applies to descriptive errors, departures from the method of measurement and failure to measure something required by the method of measurement:

- Different contracts treat errors in quantities and other errors, departures and omissions in different ways.
- In some cases, changes in quantities are not necessarily reflected in changes to the rates and prices.
- Some contracts may provide for adjustments to rates and prices but not to preliminaries.
- There may be no provision in the contract to adjust errors.
- There may be a specific term in the contract excluding any express or implied warranty as to the accuracy of the quantities.
- Where items of work are obviously required to complete a lump sum contract (e.g. floor boards in a house), the contractor's price may be deemed to include them even though the items were erroneously omitted from contract specification. This follows judgements in *Williams v Fitzmaurice (1858)* and *Patman and Fotheringham v Pilditch (1904)*.

Part 3

- Faced with uncertainties in the BQ, a prudent contractor may feel that a qualified tender may be the best way to avoid problems later on, although the invitation to tender may preclude such a tactic.

11.2 Extra work

Margins are notoriously slim in construction, and scrutiny of the annual accounts of contracting companies will typically reveal 2–3% profit on turnover. It is not surprising, therefore, that contractors are always looking for 'extras' on their contracts in order to generate more margin. Such 'extras' may result from:

- Extra work ordered by the employer that was not included in the original contract.
- Expenditure of provisional sums for defined or undefined work.
- Increased quantities of work on top of that billed in the contract documents resulting from design changes.
- Additional work arising as a result of unexpected events on-site (e.g. bad ground conditions, unknown services).
- Additional quantities of work due to errors in the contract documents:
  - Incorrect measurement.
  - Omission to measure something required by the method of measurement.
  - Departures from the method of measurement.

11.2.1 Are quantities included in the contract?

In order to determine whether or not the contractor should be paid for 'extra work', it must first be established whether or not the quantities given in the contract documents form part of the contract. This may seem obvious, but it is not always clear whether quantities are included in the contract or not; this may be due to oversight, mistake or misrepresentation or simply that documents that should have been incorporated into the contract have been omitted for some reason.

In some standard forms of contract, it is clear that the bill of quantities is a contract document (e.g. JCT 2011 SBC/Q, ICC – Measurement Version), but not all construction contracts are based on standard forms, not all standard forms provide for a bill of quantities to be included in the contract, and not all contracts are formulated using 'bills of quantities' in the normally accepted meaning of the phrase.

Fortunately, Ramsey and Furst (2015) provide authority as to whether quantities are part of a contract or not and suggest that, where a bill of quantities forms part of a contract for a lump sum:

- The quantities *are introduced into the contract as part of the description of the contract work* (*Patman and Fotheringham v Pilditch*, 1904).

Ramsey and Furst (2015) also remark, however, that it is sometimes difficult to determine whether, in fact, the quantities do form part of the contract and they quote a number of cases that have been brought before the courts:

- In *Young v Blake* (1887), it was held that quantities do not form part of the contract where there is an express power in the contract for the architect to rectify any mistakes in the quantities.
- In *Sharpe v San Paulo Railways* (1873), a contract to build a railway for a fixed price lump sum according to a specification, the judgement was that the contractor was not entitled to be paid for additional work where quantities were included in a schedule to the contract.
- In *Re Ford v Bembrose* (1902), it was similarly concluded that, where the contract was to construct certain buildings according to plans and a specification which included quantities, the contractor was not entitled to be paid extra.

- In the case of *Williams v Fitzmaurice* (1858), floor boards were omitted from the specification to build a house which the contractor had undertaken to complete in its entirety ready for occupation. In this particular case, *the language of the specification* clearly inferred that it was the contractor's obligation to complete the work omitted from the specification without further recompense.

In a further case, however, where a block of flats was to be built for a lump sum according to plans, invitation to tender, specification and bill of quantities, Ramsey and Furst (2015) report that it was held that the quantities did form part of the contract and that the contractor was entitled to be paid for items that were omitted from, or understated in, the bill of quantities. The judge qualified this judgement, however, with respect to *things that everybody must understand are to be done, but which happen to be omitted from the quantities* and which would not, therefore, qualify as 'extra work' following the judgement in *Williams v Fitzmaurice*.

11.2.2 Lump sum versus measure and value contracts

Ramsey and Furst (2015) make the point that it is a matter of construction as to what is included in the contract in each case and they add two important distinctions with respect to lump sum contracts:

1. Lump sum contracts may be broadly classified into:
   a) Those in which the contractor's obligation is broadly defined (e.g. to build a house).
   b) Those in which the contractor's obligation is precisely defined (e.g. to execute so many cubic metres of digging).
2. Where a contractor is to complete a whole, specific or entire work (e.g. a house, a railway from A to B), *the courts readily infer a promise on his part to provide everything indispensably necessary to complete the whole work* on the basis that *necessary works are not extras but are impliedly included in the lump sum.*

In the case of measure and value contracts (as opposed to lump sum), Ramsey and Furst (2015) argue that *it is usually immaterial whether any particular item of work that the contractor has to do is in the contract or not, because the contractor is entitled to be paid for it at the contract rate if it is applicable, or at a reasonable price if it is not* (Re Walton-on-the-Naze Urban District Council v Moran, 1905).

However, where a measure and value contract includes a specified sum for a specified item of work, Ramsey and Furst (2015) say that *it is a question of construction to determine what work is impliedly included in that item of work*. If the question arises as to whether or not the contractor could claim for extra work in such circumstances, Ramsey and Furst (2015) conclude that *the principles of construction applicable to lumps sum contracts* would apply to each item. Nonetheless, they also remind us that *it is important to determine whether the work is of the type contemplated by the contract* and is therefore governed by its conditions or whether the work *is outside the contract*.

In all this, lump sum contracts have to be contrasted with measure and value contracts.

In the case of lump sum contracts, the contract sum must be adjusted for variations by means of additions to or deductions from the lump sum and not by remeasurement. This is how JCT 2011 works.

However, a bill of quantities can form part of a measure and value contract where it does not define the work for which a lump sum is payable but merely constitutes a schedule of rates and quantities by which the actual work done is measured and paid for (this is how the ICC – Measurement Version works).

The case law that applies to 'extra work' is very old and predates the first standard method of measurement (1922 SMM of Building Works) and the first 'recognisable' standard forms of contract (1903 RIBA form), albeit that the first 'standard forms of contract' were probably developed by public corporations (Thomas, 2001) in the nineteenth century. Nonetheless, there is no reason to suppose that a modern court would not follow precedent but would clearly do so in the light of prevailing conditions of contract and standard methods of measurement.

11.3   Departures from the method of measurement

Where there are departures from the method of measurement stated in the contract, Hughes and Barber (1992) argue that the BQ description should prevail, provided that it is clear and unambiguous, on the legal axiom that the particular overrides the general.

This may well be the case with contract conditions that construe the contract documents equally. Under the ICC – Measurement Version Clause 5, for instance, there is no particular priority of documents, and so any departure from the stated method of measurement should not be problematic.

11.3.1   Priority of documents

Where there is priority of documents, however, the situation is more tricky.

Under the JCT2011 SBC/Q, for instance, Clause 1.3 states that *nothing contained in the Contract Bills or CPD documents, nor anything in any Framework Agreement, shall override or modify the Agreement or these Conditions*, but **unstated** departures from the method of measurement are, nevertheless, correctable (Clause 2.14.1 refers).

In the Engineering and Construction Contract, there is, strictly speaking, no priority of documents. Broome (2013) suggests, however, that, in practice, some documents *sit above* others, but the priority will depend upon individual circumstances. He also points out that *many employers* include an *order of precedence* in the articles of agreement to be referred to in the event of a conflict of documents or dispute. Broome (2013) proposes that the contract should sit above the Z clauses and that bills of quantities, or employer-written activity schedules, should sit above the Works Information, the Site Information and the accepted programme. On the bottom of the list are contractor-written activity schedules.

The ICC – Measurement Version Clause 5 states that *The several documents forming the Contract are to be taken as mutually explanatory of one another*. This indicates no priority of documents, but if the clause were to read, *The several documents forming the Contract are listed in the Contract Agreement*, the Contract Agreement may list the several documents in order of precedence, and in this case, there would be priority.

In the FIDIC conditions (Clause 1.5), documents are to be taken as mutually explanatory, but for the purposes of interpretation, follow an (a–h) sequence with the contract at the top and schedules (which includes bills of quantities and the like, if any) at the bottom.

The decision of Akenhead J. in 2013 in the Technology and Construction Court (TCC)[1] may be *read as a strong discouragement to place reliance on order of precedence provisions for every apparent discrepancy in the language of the contract without first properly analysing the contract and applying a commercial interpretation* (Weston, 2013).

11.3.2   Non-compliant item descriptions

Should a non-method of measurement compliant item description be included in the bill of quantities – perhaps where the method of measurement does not provide a suitable description for the item in question or is inadequate in some way – there should be a statement of derogation in the BQ or in a preamble. If not, the item would be correctable under the JCT conditions and (a suitably amended) ICC – Measurement Version.

However, if the item is nevertheless clear and unambiguous, it would not make sense for the method of measurement to take precedence over the item description, albeit this is effectively what some contracts say. Hopefully, common sense would prevail in such circumstances, but the issue is nevertheless open to question and potential dispute.

Where an item is omitted from the bill of quantities entirely or an item is incorrectly described, it could be argued that the item was *contingently and indispensably necessary* to complete the work and the contractor should therefore have allowed for it in his price.

A case in point would be where formwork to support concrete should have been measured but wasn't. Whilst the formwork is clearly contingently and indispensably necessary to complete the work, the principle can only be taken on a narrow construction if the method of measurement is stated in the contract and this states that formwork should be measured.

Most PQSs would accept this as an error of omission, measure the item and agree a rate, but other circumstances may be less clear.

This is a 'grey area', as it cannot be hoped that any BQ item will be scrupulously complete, despite the presence of item coverage and additional description rules in the method of measurement. Each bill compiler will have an individual way of interpreting the method of measurement, and indeed, some methods of measurement actively encourage additional description at the discretion of the person writing the item (e.g. CESMM4 Paragraph 5.11).

11.4   Errors in bills of quantities

Where errors are found in bills of quantities or other pricing document, this may well give rise to a head of claim depending on the extent to which, if at all, there is a contractual provision to deal with the issues arising:

- Quantities:
  - Wrong quantity.
  - Misleading provisional quantity.
- Descriptions:
  - Errors in descriptions.
  - Omission of information.
  - Discrepancies between related item descriptions.
  - Discrepancies between the bill of quantities and other contract documents.
  - Ambiguities or inconsistencies.
- Measurement rules:
  - Departures from the rules for item descriptions.
  - Departures from the rules for division of the work into items.
  - Items not measured.

Different standard forms of contract have different arrangements to deal with errors. This issue will now be considered in the context of the following standard forms of contract:

- JCT 2011 SBC/Q (JCT 2011).
- Infrastructure Conditions of Contract (ICC) – Measurement Version (ICC).
- NEC3 Engineering and Construction Contract (ECC).
- FIDIC Conditions of Contract for Construction 1999 (FIDIC).

11.4.1   JCT 2011

JCT SBC/Q 2011 is a lump sum contract with the option to enable parts of the work to be designed by the contractor. This is called the Contractor's Designed Portion (CDP).

Part 3

As a result, the tender BQ will contain both the measured items of work that are to be priced and carried out in the usual way and the separately identified 'contractor designed works' which will have their own priced breakdown of the work involved (CDP Analysis). Errors in the measured items of work are treated differently to errors in the CPD Analysis unless, of course, the CDP has been measured by the PQS and included in the tender BQ.

For normally measured bill of quantities items, unstated departures from the method of measurement, errors in descriptions, omission of items and an error or omission of information in a provisional sum for defined work are corrected according to JCT 2011 Clause 2.14.1 and treated as a variation pursuant to Clause 2.14.3.

In the case of errors in the Contractor's Proposals or CDP Analysis, these shall be corrected, but no addition to the contract sum is made unless there is an error in the Employer's Requirements.

Errors treated as a variation to the contract are dealt with under the variation rules of Clause 5.6, and therefore, for errors resulting in additional work (i.e. where the BQ item has been under-measured), the validity of the BQ rates and prices for such work may come into question:

- For work of a similar character, executed under similar conditions and with no significant change in quantity, the BQ rates shall apply.
- Where the contrary is the case, the BQ rates shall form the basis of the valuation, but a *fair allowance* shall be made for differences in the nature of the work, the conditions in which it is executed and/or the differences in quantity.
- In the unlikely event that a difference in quantity creates additional work of an entirely different nature, the valuation shall be based on *fair rates and prices*.

Should the error in quantities result in a reduction in the amount of work required (i.e. where the BQ item has been overmeasured), the BQ rates shall apply to the valuation of the work concerned with no adjustment. In the case of both under-measurement and overmeasurement, the contract provides that there shall be an addition to or a reduction of preliminary items albeit that the contract is silent on how this shall be valued.

It is not uncommon to find that the employer's quantity surveyor has included inflated quantities in the BQ so as to create a 'hidden' contingency. Where this is the case, the contractor is faced with the 'double whammy' of suffering a reduction in the quantity of work to be carried out, and subsequent loss of turnover and profit, and a possible reduction in the value of preliminaries due to the reduced quantity of work required.

Prudent contractors will always scrutinise significant items in the BQ and, where it is suspected that the quantities are incorrect, make an appropriate adjustment:

- For under-measured quantities, enhance the rate by moving money from elsewhere in the priced BQ and thus make money when the quantity is increased.
- For overmeasured quantities, reduce the BQ rate and move the money elsewhere in the BQ in order to avoid losing money when the quantity is reduced.

JCT 2011 SBC/Q does not specify any 'trigger point' for the adjustment of BQ rates in the event of an error in quantities, and it is for the quantity surveyor to decide whether the error is significant enough to warrant a change in the rate and to decide what is fair in the circumstances. There is no clarification on this matter in SMM7 should this method of measurement be used in conjunction with the contract.

NRM2, however, provides a 'rule of thumb' for the adjustment of rates where the quantities are inaccurate, but this only applies to any provisional quantities that may have been included in the BQ. Such adjustments are intended to compensate the contractor only where the provisional quantities may have been misleading and are not intended for the correction of errors.

11.4.2 Engineering and Construction Contract

Under the ECC, ambiguities or inconsistencies in or between documents are resolved pursuant to the project manager's instructions. Errors in Works Information are treated as compensations events, but the pricing documents (Option A – Activity schedule; Option B – Bill of quantities) are not Works Information, and errors in these documents are treated differently.

ECC **Option A** is a lump sum contract based on an activity schedule. This may be prepared by the employer, in which case errors would be corrected, but, more usually, is prepared by the contractor, in which case the contractor will have to stand by any error.

The activity schedule does not have the status of Works Information or Site Information (Clause 54.1), but where there are changes to the Works Information which affect the contractor's prices, the project manager will assess a compensation event in accordance with Clause 60.1(1). There is no such relief for changes to the Site Information.

Option B – priced contract with bill of quantities – is a measure and value contract.

Under this option, the difference between the final total quantity of work done and the quantity stated in the BQ is a compensation event if the difference <u>does not</u> result from a change to the Works Information (Clause 60.4). This rule does not specifically refer to errors in quantities, but as this is a measure and value contract, they would be adjusted as normal by the process of admeasurement.

It would appear that Clause 60.4 is aimed mainly at circumstances where the contractor simply does more or less work than the bill of quantities states, that is, not an error, not a change to the Works Information, just a change in quantity. However, if the difference in quantity *is* the result (or partial result) of an error, the same valuation rule would apply where the contractor has simply done more or less work.

The Clause 60.4 valuation rule is complicated and Broome (2013) would prefer that it was changed. A compensation event arises if:

- The difference does not result from a change to the Works Information.
- The difference in quantity causes the Defined Cost per unit of quantity to change AND
- The rate in the BQ at the Contract Date multiplied by the final total quantity of work done is more than 0.5% of BQ total at the Contract Date.

Once the quantity for an item of work has been admeasured, the contractor can do the sums and decide whether or not to notify a compensation event to the project manager. This fits with the early warning ethos of the NEC3 contract and also ensures that, should the compensation event be validated, the contractor will be paid at the next opportunity. If the Defined Cost per unit reduces, then the affected rate is reduced.

Where a difference in quantity delays completion, this results in a compensation event under Clause 60.5.

Under Option B, should there be departures from the rules of measurement for item descriptions in the bill of quantities or for the division of the work into items (e.g. incorrect classification of items), the project manager will make the necessary corrections according to Clause 60.6 and each correction shall be a compensation event.

Where there are ambiguities, or inconsistencies, in the bill of quantities, Clause 60.6 also applies. In both instances where mistakes are corrected under Clause 60.6, there is the caveat that the correction of mistakes in the BQ may lead to reduced prices.

A novel provision in the ECC is in the assessment of compensation events resulting from the correction of inconsistencies between the bill of quantities and another document. In such circumstances, the contractor is assumed to have taken the BQ as correct, and this is the starting point for the assessment (refer to Clause 60.7).

Part 3

11.4.3   Infrastructure Conditions of Contract

The ICC – Measurement Version is a measure and value or admeasurement contract which means that the quantities stated in the BQ at tender stage are estimated quantities (refer to Clause 55).

This also means that the priced BQ submitted at tender stage is purely a schedule of rates whose purpose is to facilitate the valuation of the actual quantity of work carried out. Consequently, the tender total submitted by the contractor is simply a total figure which enables the various tenders received to be compared.

As this is a measure and value contract, there is no contract sum to be adjusted at final account stage, nor is there any requirement for a written instruction in the event that the actual quantities vary from those stated in the BQ as is normal with a lump sum contract (refer to Clause 51(4)). There is no need for a mechanism to correct errors in quantities under the ICC form as they are, or should be, automatically picked up in the admeasurement process.

There may, however, be an error in an item description and this does require to be corrected. Clause 55(2) deals with this issue.

The decision to admeasure any part or parts of the work rests with the engineer under Clause 56(3), and therefore, should the engineer not require such admeasurement, it will not happen unless the engineer is prompted by the contractor. In this situation, the contractor will need to consider:

- Whether any such admeasure may result in an increase in quantities in which case it is in his interests to bring the matter to the engineer's attention.
- Whether a significant increase in quantities may result in a decrease in the BQ rate as is the prerogative of the engineer under Clause 56(2).
- Whether any admeasure may result in a decrease in quantities in which case it may be prudent to keep quiet.
- Whether it is likely that a significant decrease in quantities may result in an increase in the BQ rate under Clause 56(2).

Should the contractor decide to alert the engineer to a change in quantities, he may do so:

- Informally in the first instance, perhaps via the engineer's representative, which would be a courtesy to the engineer.
- Formally in a letter to the engineer.
- Formally in the contractor's monthly application for payment under Clause 60(1)(a) which states *the estimated contract value of the Permanent Works carried out up to the end of that month*.
- Formally in the contractor's statement of final account under Clause 66(4) which is required to show *the value in accordance with the Contract of the Works carried out*, albeit that this might fall foul of the contractor's obligation to give early warning of potential claims under Clause 12(2) in respect of adverse physical conditions and artificial obstructions.

However, as stated earlier, the ICC – Measurement Version does have a mechanism to reflect the consequence of a difference between the actual quantities of work carried out and those stated in the bills of quantities under which the engineer has the power to increase or decrease the appropriate BQ rates.

Therefore, if, in the opinion of the engineer and after consultation with the contractor, the BQ rates are rendered unreasonable or inapplicable by reason of the change in quantities, the engineer may increase or decrease the rates accordingly. Should the contractor be dissatisfied with the outcome, he has recourse to Clause 66(2)(b) to make his views known before possibly embarking on the contract dispute resolution procedure.

Item descriptions in the bill of quantities must, of course, comply with the specified standard method of measurement, unless there is a specific statement to the contrary somewhere in the contract documents; if they do not, then the engineer must take action to put matters right.

It must be noted that *No error in description in the Bill of Quantities or omission therefrom shall vitiate* [invalidate] *the Contract* – Clause 55(2) – nor do they release the contractor from his obligations to carry out the works in accordance with the drawings and specification or from any of his other obligations or liabilities under the contract. The error(s) must simply be corrected by the engineer, and the work actually carried out must be valued. Because the original item in the bill of quantities will be changed, this constitutes a variation to the contract, and thus, the valuation of that change must be in accordance with Clause 52(2) or (3) – Valuation of ordered variations.

The correction of errors in the 'rates and prices' contained in the bill of quantities is expressly excluded by Clause 55(2) as they are at the contractor's risk. This exclusion also applies to 'descriptions' inserted by the contractor. Where there is partial contractor design, CESMM4 should be amended pursuant to paragraph 5.4 (i.e. in a Preamble) but any descriptions, rates and prices inserted by the contractor would similarly not be subject to correction for errors or 'wrong estimates'.

11.4.4   FIDIC

The old FIDIC 'Red Book' was always regarded as a 'rebranded' version of the old ICE Conditions – in other words a measure and value/admeasurement contract.

The first edition of FIDIC Conditions of Contract for Construction 1999, however, has been extensively redrafted and now has its own personality and idiosyncrasies.

The standard Letter of Tender gives the idea that the new FIDIC Red Book is a lump sum contract because the tenderer offers to execute the works for a specified sum of money or such other sum as shall be determined in accordance with the conditions of contract. This is not the language of a measure and value contract.

However, turning to Clause 12.3 reveals the procedure for determining the contract price which is by *applying the measurement agreed* to the appropriate rates or prices in the bill of quantities or other schedule. Measurement shall be *the net actual quantity of each item of the permanent works* (Clause 12.2). This now sounds like a measure and value contract, especially when read in conjunction with Clause 14.1(c) which says that *any quantities which may be set out in the bill of quantities or other schedule are estimated quantities and are not to be taken as the actual and correct quantities.*

FIDIC is, in fact, written on the basis of a measure and value contract with provision for a lump sum if desired. Suggestions for appropriate amendments to the contract in order to create a lump sum arrangement are made in the *Guidance for the Preparation of Particular Conditions* bound into the Red Book. Additionally, the term *contract price* is replaced by *accepted contract amount* in a lump sum contract arrangement.

For reasons discussed earlier, the distinction between lump sum and measure and value contracts is important, especially as regards quantities and changes to the quantities.

There is not necessarily a bill of quantities with FIDIC, and there could quite easily be a schedule of rates (Clause 1.1.1.7: *Schedules*) or, possibly, an activity schedule, although the latter would presumably be an employer-drafted document.

Where there is an ambiguity or discrepancy in the documents that make up the contract, they are to be clarified by the engineer by way of an instruction (Clause 1.5), but such instructions may not necessarily result in a variation (Clause 3.3).

Technical errors in documents are notifiable by one party to the other (Clause 1.8), but there is no contractual remedy available. However, this clause will undoubtably be caught by Clause 3.3 which empowers the engineer to issue instructions *necessary for the execution of the works.*

Part 3

Unusually, the method of measurement is not stated in the Appendix to Tender but is referred to in Clause 12.2 as being *in accordance with the Bill of Quantities or other applicable Schedules*. Presumably, this means that the method of measurement would be stated in these documents, failing which there would be no method of measurement. A bill of quantities prepared to 'any rules you like' is not a method of measurement!

On the assumption that a sensible method of measurement is used, changes in the measured quantities attract an adjustment to the appropriate BQ rates when the change in quantity *is more than 10%* of that stated in the bill of quantities/schedule (Clause 12.3(a)(i) refers) and:

- The change in quantity × the rate is more than 0.01% of the accepted contract amount.
- The change directly changes the cost per unit of the item by more than 1%.
- The rate in question is not a 'fixed rate item' (e.g. a fixed charge or a lump sum).

11.5   Procurement issues

Issues relating to entitlement claims from a measurement perspective are inextricably linked to the method of procurement employed for a particular project, to the form of contract between the employer and the contractor or between the contractor and any subcontractor and to the method of measurement (if any) used to quantify the work if, indeed, there has been any quantification carried out.

So-called measurement claims need to be considered in the context of the pricing documentation employed for the project in hand, whether or not a recognised standard method of measurement has been used and whether or not the specification is incorporated in the pricing document or whether it is separately bound.

A further consideration is whether the documentation has been prepared formally by, or on behalf of, the employer or whether it has been prepared by a contractor or subcontractor informally.

Note

1.   RWE Npower Renewables Ltd v J N Bentley Ltd [2013] EWHC 978 (TCC)

References

Broome, J. (2013) *NEC3: A User's Guide*, ICE Publishing, London.
Chappell, D. (2011) *Building Contract Claims*, 5th edition, Wiley-Blackwell, Oxford.
Egan, J. (1998) *Rethinking Construction: Report from Construction Task Force*, UK Department of Transport and the Regions, London.
Hewitt, A. (2011) *Construction Claims and Responses*, Wiley-Blackwell, Oxford.
Hughes, G.A. and Barber, J.N. (1992) *Building and Civil Engineering Claims in Perspective*, 3rd edition, Longman, Essex.
Latham, M. (1994) *Constructing the Team: Joint Review of the Procurement and Contractual Arrangements in the UK Construction Industry*, HMSO, London.
Patman and Fotheringham v Pilditch (1904) Hudson's Building Cases, 4th edition, p. 368.
Ramsey, V. and Furst, S. (2015) *Keating on Building Contracts*, 9th edition, Sweet and Maxwell, London.
Re Walton-on-the-Naze Urban District Council v Moran (1905) Hudson, Building Contracts, 4th edition, p. 376.
The Simon Committee Report 1944.
Thomas, R. (2001) *Construction Contract Claims*, Palgrave, Basingstoke.
Weston, A. (2013) Building, http://www.building.co.uk (accessed on 27 August 2013).

Chapter 12
Final Accounts

The term 'final account' is one of the many euphemisms that abound in the construction industry. It is used to represent an accounting process that determines the amount, if any, that is owed by one party to another at the conclusion of a construction contract.

'Final account' is a term that everyone recognises and vaguely understands. However, not everyone is familiar with the final account process or the procedures required by the various forms of contract used in the industry or the role and status of the pricing document (bill of quantities, schedule, etc.) when preparing of the final account.

12.1 Purpose

In practice, several 'final accounts' are necessary in order to determine amounts owing, if any, and to conclude the contract:

a) The 'official' final account between the employer to the contractor.
b) Final accounts between the contractor and the many subcontractors who are likely to have been engaged to carry out the work.

Good practice suggests that final accounts, whilst settled at the end of the contract, are started much earlier, as it is always best to conduct site measures at the time and to measure and agree other work, variations and extras to the contract whilst minds are fresh and evidence is easy to establish.

Some of the tasks involved in preparing final accounts include:

- The **remeasurement** of approximate quantities in a lump sum contract.
- The **admeasurement** of work in a measure and value contract.
- The adjustment of the rates or prices for a variety of reasons.
- The **measurement** and valuation of variations to the contract.
- The evaluation of claims and contractual entitlement.

In some cases, where the original contract has changed beyond recognition, the entire project will need to be remeasured as a basis for agreeing the quantum of work carried out and the final amount owing.

Managing Measurement Risk in Building and Civil Engineering, First Edition. Peter Williams.
© 2016 John Wiley & Sons, Ltd. Published 2016 by John Wiley & Sons, Ltd.

The four standard forms of contract referred to in this book take different approaches to the 'final account'.

12.2 Forms of contract

Few of the usual standard forms of contract use the term 'final account' – ICC contracts do – but most employ terms such as *the amount due* (ECC Clause 50), the *final payment* (JCT 2011 SBC/Q Clause 4.15) or *the amount which is finally due* (FIDIC Clause 14.13) instead.

Each contract has slightly different procedures which must be respected.

12.2.1   JCT 2011 SBC/Q

- The contractor provides all necessary documents (Clause 4.5.1).
- The quantity surveyor (for the employer) prepares a **statement** of all adjustments to be made to the contract sum (Clause 4.5.2.2).
- The contract administrator sends a copy of the **statement** to the contractor (Clause 4.5.2).

12.2.2   Engineering and Construction Contract

The ECC does not have a final account or final statement process as such and merely provides that the project manager shall *assess the amount due* (Clause 50).

However, this doesn't alter the fact that, generally speaking, the process of preparing the final account for an ECC contract pretty much follows well-established industry 'custom and practice'.

12.2.3   ICC: Measurement Version

- The contractor submits a **statement of final account** to the engineer along with detailed supporting documentation (Clause 60(4)).

12.2.4   FIDIC 1999

- After the works have been taken over (substantial completion), the contractor submits a **statement at completion**, with supporting documents, to the engineer (Clause 14.10).
- After the performance certificate is issued (thereby completing the defects correction period), the contractor submits to the engineer a **draft final statement** with supporting documents (Clause 14.11).
- Once discussions are concluded, and notwithstanding any possible dispute, the contractor submits a **final statement** (Clause 14.11).

12.2.5   Final account statements

Generally speaking, once the amount owing has been calculated, a final certificate, or equivalent, can be issued which signals that all obligations under the contract have been discharged and that the contract is concluded. The amount is calculated according to express contractual procedures, and although different terminology may be used, they all mean the same thing:

- Collect all relevant information needed to calculate the amount due.
- Prepare a final account statement.
- Certify the amount, if any, owing.

Preparation of the final account statement is done by the quantity surveyor for the employer on some contracts (e.g. JCT), by the project manager (ECC, although there is no 'official' statement) and by the contractor on others (e.g. ICC, FIDIC). Similar procedures apply to subcontract accounts.

Notwithstanding the final certification arrangements in contracts, it is common practice in the industry for the employer (or the employer's representative) and the contractor to sign the final account statement in order to signify that the amount claimed in the final account statement is in full and final settlement of all claims, actions, liabilities, costs or demands under the contract. In some cases, interim final account statements are submitted as well as a final one following FIDIC practice. The idea is to expedite the payment of monies owing to the contractor, thereby helping with cash flow, and to give certainty to the final certification process.

However, where one of the parties decides that it does not wish to be bound by the terms of the final account statement, even when it has been signed, the decisions in *Hurst Stores and Interiors Ltd v ML Europe Property Ltd. [2004] EWCA Civ 490* and in *YJL London Limited v Roswin Estates LLP [2009] EWHC 3174 (TCC)* have both held that the signed final account statement is not binding and that adjustments can still be made to the final account in accordance with the provisions of the contract.

12.2.6   Final certificate

Once a construction project is completed and patent defects have been corrected, it is normal practice for the contract administrator to issue a final certificate.

Usually, the final certificate is the last act in any construction project unless, of course, there are ongoing legal proceedings or an arbitration or adjudication has arisen out of the contract. In most cases, however, the final certificate determines what sums (if any) are due to the contractor (or employer) and brings to an end the contractor's obligations under the contract.

In order to facilitate issue of the final certificate, most standard contracts require the contractor to supply information to the contract administrator so that the amount to be stated in the final certificate may be calculated. The volume of information is often very substantial, and numbers of lever-arch folders may be required. Subcontract final accounts tend to be much slimmer, often because the subcontractor doesn't understand the process and often due to a lack of information, records, signed instructions, signed daywork sheets, etc.

Included within the information supplied with the final account may be details of claims for the payment of monies to which the contractor/subcontractor believes there is an entitlement.

12.3   Lump sum contracts

12.3.1   The contract bills

A normal lump sum contract with interim payment, such as JCT 2011, is based on an accepted contract sum. The contract sum is the accepted tender figure submitted by the successful contractor which is calculated by totalling the amounts included in the bill of quantities (the contract bills) for:

- Measured items of work.
- Priced preliminaries.

Part 3

- Provisional sums for defined and undefined work.
- Risk allowances (or contingencies).
- An adjustment item (under NRM2, for last minute changes to the tender figure).

The contract bills are not touched when preparing the final account, and this document is used solely as a point of reference for:

- Rates and prices.
- Provisional quantities.
- Provisional sums.
- Risk allowances.
- Adjustment items.

In some contracts, the contract bills may be replaced by an activity schedule, but the same principle applies – it is a reference document to assist with establishing rates for variations and for the valuation of work executed under defined and/or undefined provisional sums.

12.3.2   The contract sum

The starting point for calculating the final account is the contract sum.

Various 'accounts' are then prepared so that the contract sum may be adjusted to cater for everything that has happened from a measurement and financial perspective over the course of the contract. The various accounts include:

- Variation account.
- Daywork account (may be included with the variation account).
- Provisional sums.
- Provisional quantities.

The total of each of these accounts is then added to the contract sum, and the corresponding amount included in the contract bills is deducted. By the process of omitting from and adding to the contract sum, the final total of the contact works is determined.

12.3.3   Changes in quantities

A good deal of measurement is required in the final account process, but there is an important distinction to remember about lump sum contracts:

- Changes in 'firm' quantities are treated as variations to the contract and are valued according to the valuation rules for variations contained in the contract.
- Changes in provisional quantities are remeasured but are not measured as variations unless the work concerned has been altered in some way other than in quantity.
- If an item with provisional quantities is changed, other than in quantity – for example, the specification of the item is altered – it will be valued as a variation and the variation rules will apply to its valuation.

Under NEC3, Options B and D, the prevailing method of measurement is stated in the Contract Data, but differences between the actual and billed quantities are a compensation event unless the work is otherwise varied. There is no undertaking in NEC3 that actual quantities shall be measured in accordance with the stated method of measurement nor is this a stated compensation event.

In NRM2 Paragraph 3.3.1, there is no statement that *the rules apply to measurement of proposed work **and** executed work* such as may be found, for example, in General Rule 1.2 of

SMM7. This may be an oversight, or it may have been a conscious decision by the authors of NRM2 that the JCT 2011 SBC/Q would be the default contract to use with NRM2.

12.3.4   Variations

Each variation is measured according to the applicable rules of measurement and priced according to the rates and prices in the contract bills.

JCT 2011 Clause 5.6.1.3 states that the measurement of variations *shall be in accordance with the same principles as those governing the preparation of the Contract Bills* referred in Clause 2.13. Clause 2.13 concerns preparation of the contract bills, and Clause 2.13.1 refers expressly to the *measurement rules* (i.e. NRM2) pursuant to the JCT August 2012 NRM Update.

Failing this, fair rates may be established to account for changes in the type or nature of work concerned.

12.3.5   Daywork

Daywork is not measured but is priced up at contract rates according to the agreed number of hours spent by each trade on the work, together with the cost of recorded plant used and materials consumed.

12.3.6   Provisional sums

Provisional sums are deducted from the contract sum, and the work carried out is measured from 'scratch' according to the rules of measurement and the contract rates.

This process will result in a 'mini-bill of quantities' for each provisional sum with items priced at contract rates, where possible. There may be an issue with the valuation of a defined provisional sum which was incorrectly described in the bill of quantities.

12.3.7   Provisional quantities

Provisional quantities are a bit different in that the relevant item is remeasured in order to establish the correct and final quantities of work done. Unless there has been a change in the nature or specification of the work, the final account is calculated by deducting the provisional quantity and adding back the final 'remeasured' quantity.

Should the provisional quantity not be a reasonably accurate forecast of the amount of work needed, usual practice is that the contract rates may no longer be applicable and a 'star rate', making a 'fair' allowance for the difference, is agreed. This is the arrangement in the JCT 2011 contract, but where this contract is used in conjunction with NRM2, there are some added complications:

- JCT 2011 refers to 'approximate' not 'provisional' quantities.
- NRM2 includes Paragraph 3.3.8 which deals with 'provisional quantities':
    3.3.8.1 Where work can be described and given in a BQ item in accordance with the tabulated rules of measurement, but the quantity cannot be accurately determined, *an estimate of the quantity shall be given* and identified as a 'provisional quantity'.
    3.3.8.2 Provisional quantities:
    a) *shall be subject to remeasurement* when the work has been completed.
    b) *The 'approximate quantity' shall be substituted by the 'firm quantity' measured.*

    c)   *The total price for that item **shall be** adjusted to reflect the change in quantity.*

    d)   Where the variance between the provisional and firm quantities is <20%, *the rate tendered by the contractor **shall not** be subject to review.*

    e)   Where the variance is ≥20%, *the rate **can be** reviewed to ensure that it is fair and reasonable* to the parties.

Consequent to Paragraph 3.3.8.2, the use of the words *can be* indicates that the contractor has no <u>right</u> under the rules of measurement to a re-rate for significant changes in provisional quantities and that the possibility of a re-rate is purely discretionary. As a consequence, the contractor could be at risk of financial loss, and it might, therefore, be in the contractor's interest to seek relief via the conditions of contract in such circumstances.

Whether the Paragraph 3.3.8.2. rule overrides Clauses 5.6.1.4 and 5.6.2.5 of the JCT conditions is debatable to say the least.

As discussed in Chapter 11, the threshold for revising contract rates is different, and more complex, in the FIDIC 1999 contract, but the general idea is the same as NRM2.

In the ECC, lump sum contracts are based on (usually) contractor-prepared activity schedules, and changes in the Works Information, failure to provide access to the site and instructions to stop work are dealt with via the compensation event procedure. This relies on the accepted programme and 'Defined Cost' to establish the impact of changed resources on the contract sum; there is no 'valuation of variations' arrangement in the ECC.

Subcontract accounts for lump sum contracts should follow the same pattern as the main contract, but if the subcontractor has prepared the final account, there may be a lot of adjustment required to bring the account into line with correct practise. Complications may also arise where the basis of the subcontract is different to that of the main contract (e.g. main contract – lump sum, subcontract – measure and value).

For partial contractor design items and for full contractor design, the contractor is 'stuck' with the quantities that he has prepared at tender stage unless there has been a variation to the contractor design element of the BQ or a change in the employer's requirement in a full design and build contract. In either event, the variation provisions of the relevant contract will take effect.

Provisional, or approximate, quantities in 'firm' bills of quantities are a particular bone of contention with contractors.

This is because some PQSs are prone to inflate these quantities in order to add extra 'contingency' into the bill of quantities.

In Table 12.1, it can be seen that a provisional quantity of 650 m³ has been allowed for extra over for excavation in rock. If this quantity is later reduced substantially, the net effect will be:

- The contractor will lose turnover and therefore potential profit and overhead recovery.
- The PQS will have a sum of money available that can either be spent on something else by the architect or given back to the employer as a saving on the contract sum.

Standard forms of contract cater for this situation in different ways in order to provide relief for the contractor and avoid extra-contractual claims against the employer:

- By classifying changes in quantity as a variation, if there has been an accompanying change in the nature or quality of the work, and thereby opening up the possibility of revaluing the relevant bill of quantities rate(s) to a more equivalent or fair rate.
- By providing a mechanism whereby the contractor can request a re-rate which would then be considered at the contract administrator's discretion.
- By establishing a threshold percentage (say, ±20%) against which changes in quantities can be judged:
  - Changes in quantity above the threshold would be considered for rerating:
    - 20% of 650 m³ = 130 m³
      i.e. 'final' quantities of 520 m³ (or less) or 780 m³ (or more).

○ Changes in quantity below the threshold (i.e. 19% change or less) would not be considered for a new rate:

- $19\% \times 650\,\mathrm{m}^3 = 123.5\,\mathrm{m}^3$
  i.e. 'final' quantities of no less than $527\,\mathrm{m}^3$ or no more than $773\,\mathrm{m}^3$.

Contractors would probably argue that such protocols are 'hit and miss' and that full recovery of loss or expense is never achieved. They might also argue that any system that relies on the contract administrator's judgement or discretion would always be construed in favour of the employer and not the contractor.

Table 12.1   Provisional quantities.

		Qty	Unit	Rate	£	p
	Excavating and filling					
	Site Clearance / preparation					
	Site preparation					
A	Remove topsoil: depth 150mm	1287	m2			
	Excavations					
	Excavation; commencing level 150mm below original ground level					
B	Bulk excavation; over 2m not exceeding 4m deep	3458	m3			
	Extra over all types of excavation irrespective of depth					
C	Breaking up; rock [Provisional quantity]	650	m3			
	Disposal					
D	Excavated material off site	3458	m3			
	Retaining excavated material on site					
E	Top soil; to temporary spoil heaps; average distance 100m	193	m3			

12.4   Measure and value contracts

With measure and value contracts, the final account process determines the contract sum, whereas the total of the bill of quantities at tender stage is purely a 'tender total' for the purpose of comparing tenders received.

Where there is an unquantified schedule of rates, there is no tender total and tenders are compared by comparing respective rates, perhaps by using some 'nominal' quantities.

The contract bills (or schedule of rates) are central to the process of determining the final account figure in a measure and value contract as it is the total of the quantities multiplied by the contract rates that determines the contract sum.

Preparation of the final account for a measure and value contract differs in several keys ways to that for lump sum contracts:

1. Quantities included in the contract bills at tender stage are admeasured for the final account, that is, increased or decreased according to the final and correct quantities of work done.
2. Where there is a schedule of rates, all the work undertaken is measured (not remeasured or admeasured) according to the rules of measurement stated in the contract.
3. The total of the admeasured bill of quantities or schedule, together with other sums determined in accordance with the contract (e.g. variations), becomes the contact sum or contract price.
4. There are no provisional quantities in a measure and value contract as all quantities at tender stage (if any) are estimated.
5. An increase or decrease in quantities is not a variation unless the work is varied in some other way as well (e.g. change in the kind or quality of work).

In all other respects, measure and value contracts follow equivalent procedures to those for lump sum contracts with regard to:

▪ The measurement of work in connection with provisional sums.
▪ The adjustment of rates where changes in quantity invalidate the original rates.
▪ Variations.
▪ Daywork.

12.5   Daywork accounts

The daywork method of valuation is a form of cost reimbursement, except that the costs reimbursed are not the actual costs incurred by the contractor (see Chapter 4).

Daywork is also an alternative to the contractor making a 'claim' for damages. This might be a 'contractual' claim based on an express or implied term of the contract or an 'extra-contractual' claim for damages in law. This might arise where, for no fault of his own, the contractor or subcontractor is required to:

▪ Disrupt normal working in order to investigate something not described in the contract (e.g. existing services).
▪ Disrupt normal working in order to correct something that would otherwise be a defect in the finished work (e.g. dealing with a soft spot in an excavation, repainting a completed wall damaged by another trade).
▪ Carry out a minor item of work which is not a measurable item under the standard method of measurement.
▪ Etc.

It is often misunderstood that:

▪ Daywork is a method of valuing variations when all else fails.
▪ Applications for payment on a daywork basis do not necessarily result in payment on this basis.
▪ Variations are valued according to the valuation rules written into the contract, and daywork is just one of several possible methods of valuation which include the use of BQ rates, adjusted BQ rates and 'fair' rates.

Accordingly, the PQS or the contractor (*vis-à-vis* a subcontract claim) must assess how work should be valued in accordance with the rules written into the contract. Daywork is the 'backstop' to those rules. The rule of thumb is that, where the work involved in a variation cannot be properly valued by measurement, daywork should be used as the means of valuation.

In contracts where the word 'daywork' is not used (e.g. NEC3 contracts), the contract administrator/project manager must nevertheless assess the circumstances of an event, and if the contractor deserves compensation for the event, a fair means of payment must be found.

Under NEC3 ECC Option A (priced contract with activity schedule) and Option B (priced contract with bill of quantities), the backstop is the Shorter Schedule of Cost Components (SSCC) which provides a means of valuing work on the basis of 'Defined Cost' (i.e. the resources used to do the work) plus a fee (see Chapter 4.13.2).

Daywork is treated as a provisional sum, and consequently, the final account is adjusted by omitting the provisional sum and adding back the 'daywork account' compiled by the contractor or by the PQS.

12.6   'Final accounts' under the ECC

For the purpose of determining the final value of the contract works, arrangements under the ECC depend on the Main Option chosen and upon whether the contract is the main ECC or the Short Contract.

12.6.1   ECC Options A and B

- For normal work, the contractor is reimbursed according to the items listed in the activity schedule or bill of quantities, respectively.
- For compensation events, the contractor is paid according to the change in resources resulting from the event compared with those indicated in the accepted programme and accompanying method statements provided that the resource in question is listed in the SSCC.

12.6.2   ECC Options C and D

- For normal work, the contractor is reimbursed according to the items listed in the full Schedule of Cost Components (SCC) for his own work and in accordance with the appropriate subcontract for subcontracted work.
- For compensation events under Options C and D, the contractor is paid according to the change in the target price resulting from the event.

12.6.3   ECC Option E

- For normal work, the contractor is reimbursed according to the items listed in the full SCC for his own work and in accordance with the appropriate subcontract for subcontracted work.
- For compensation events, the contractor is paid according to the change in estimated final cost resulting from the event.

12.6.4   ECC Short Contract

- For normal work and for changes in quantities, the contractor is reimbursed according to the rates listed in the Price List.
- For compensation events, other than changes in quantities, the contractor is paid according to the impact of the compensation event on Defined Cost; this may be forecast or actual depending upon whether the event has already occurred or not.

Part 3

12.6.5   Defined Cost

There is an added complication in the ECC 'final account' arrangements, because the definition of 'Defined Cost' varies according to which contract, and which Main Option, is chosen:

- Under the full ECC, 'Defined Cost' is variously expressed as:
  - *the cost of components in the Shorter Schedule of Cost Components whether work is subcontracted or not* (Options A and B: Clause 11.2(22)).
  - *the amount of payments due to Subcontractors for work which is subcontracted … and the cost of components in the Schedule of Cost Components for other work* (Options C, D and E: Clause 11.2(23)).
- Under the ECC Short Contract, Clause 11.2(5) states that *Defined Cost is the amount paid by the contractor for:*
  - *people employed by the Contractor.*
  - *Plant and Materials.*
  - *work subcontracted by the Contractor and*
  - *Equipment* (i.e. construction plant).
- Under Options A and B, Defined Cost is used uniquely for assessing compensation events.
- Under Options C, D and E, Defined Cost is used both for the assessment of compensation events and for reimbursing the contractor for work in progress; the difference is because Options C, D and E are cost reimbursement contracts, whereas Options A and B are priced contracts.

12.6.6   SCC or SSCC

In order to administer this arrangement, the full ECC contract uses either the SCC or the SSCC or both, as appropriate:

- For Options A and B, the shorter schedule only is used.
- For Options C, D and E, the full schedule is normally used, but the short schedule may be used by agreement for assessing straightforward compensation events.

Both the SCC and the SSCC appear at the back of the full ECC – the so-called 'Black Book'. In each case, definitions of what is included in 'Defined Cost' are categorised as follows.

Resource heading	Non-NEC3 equivalent
1. People	Labour
2. Equipment	Plant
3. Plant and materials	Fixed equipment such as HVAC plant and normal building materials delivered to site
4. Charges	Site overheads or preliminaries
5. Manufacture and fabrication	Plant and materials manufactured off-site such as prefabricated units, structures or structural elements
6. Design	The hourly cost of employees providing design services off-site (i.e. it is not a fee percentage)
7. Insurance	Refers to deductions from cost where the contractor has failed to insure in accordance with the contract or has been otherwise reimbursed

There are some differences in the above definitions as between the full and short schedules, but this is at a detailed level which is beyond the scope of this book.

As far as the ECC Short Contract is concerned, neither the SCC nor the SSCC applies. In this case, Defined Cost is the amount paid out by the contractor in accordance with Clause 11.2(5), and the contractor is reimbursed for compensation events:

- By the contractor submitting a quotation or revised quotation.
- By the employer assessing the compensation event and notifying the contractor accordingly.
- By assessing the changes to the contractor's prices on the basis of forecast changes in Defined Cost due to the compensation event (pre-event).
- By assessing the changes to the contractor's prices on the basis of Defined Cost incurred due to the compensation event (post-event).

Under the ECC, data for the calculation of Defined Cost is submitted by contractors at tender stage. This data is written into the Contract Data Part 2 and includes:

- Percentage for people overheads.
- Time charges for equipment (i.e. construction plant) according to a stated list (e.g. RICS or CECA published lists of hourly/weekly rates).
- Percentage adjustments ± on the rates in the published list.
- Rates for equipment not listed in the published list.
- Hourly rates and overhead percentage for design work.
- And so on.

12.6.7   The fee

In all cases, the contractor is entitled to a fee which is added to Defined Cost. There are two fees that tenderers quote in the Contract Data Part 2:

- *The direct fee percentage.*
- *The subcontracted fee percentage.*

The fee is defined in ECC Clause 52.1 as including *all the contractor's costs which are not included in Defined Cost....* Fees for Options C, D and E also include Disallowed Costs determined by the project manager according to ECC Clause 11.2(25). There is no fee as such under the ECC Short Contract, but the Contract Data refers to a *percentage for overheads and profit* to be added to the Defined Cost for people and a further percentage to be added to other Defined Cost.

12.6.8   Compensation events

The various fees, rates, lists of equipment hire, etc. quoted by contractors at tender stage appear in the Contract Data (Part 2 where the ECC applies), but unlike traditional contracts, there is no provision under the ECC for provisional sums or for estimates of Defined Cost in the bill of quantities against which the contractor prices competitive percentages.

Therefore, bills of quantities prepared under ECC Options B or D do not include provisions for the cost of compensation events in the tender figure as they do in traditional contracts (e.g. contingencies or risk allowances) nor do they appear in the activity schedules prepared by contractors under Options A and C.

The implications of this are twofold:

- Tender figures do not represent the project budget, and employers must understand that compensation events may add considerably to the accepted tender/target figure by the time the final account is prepared.

Part 3

- Assessing the 'competitiveness' of the tendered fee percentages must be done by separate analysis of each of the tenders submitted, and tenderers must understand that their tender figure will be adjusted commensurate to the estimated impact of compensation events.

In this regard, a model tender assessment sheet for ECC Options A and B appears in the ECC *guidance notes* which illustrates how a tender figure may be adjusted in order to compare tenders received. This is done by applying the various tenderers' fees and other percentages to the difference between the estimated out-turn cost of the project, less the relevant tender figure received, suitably apportioned between the various elements of defined cost (people, equipment, subcontract, etc.).

This analysis could be done by using a separate calculation sheet for each tenderer as suggested in the ECC *guidance notes* or by devising a spreadsheet similar to that shown in Table 12.2. Here, it can be seen that the estimated final account figure, less tender figures received, provides a forecast of compensation events for the contract. The estimated final account figure is exclusive of tendered overhead and other fee percentages.

Table 12.2   ECC tender comparison spreadsheet.

TENDER COMPARISON SHEET

			A		B		C		D	
							Tender			
Forecast final account			1890000		1890000		1890000		1890000	
Tender Sum			1725898		1709225		1759850		1770956	
Forecast compensation events			164102		180775		130150		119044	
People element	35%		57436		63271		45553		41665	
Equipment element	25%		41026		45194		32538		29761	
Subcontract element	40%		65641		72310		52060		47618	
Tender adjustment										
Tender Sum				1725898		1709225		1759850		1770956
People overheads		55%	31590		75%	47453	60%	27332	45%	18749
Equipment adjustment		-15%	-6154		-5%	-2260	-10%	-3254	-10%	-2976
Additional direct cost			25436			45194		24078		15773
Direct fee percentage		7.5%	1908	27343	10%	4519	49713 12.5%	3010	27087 8%	1262	17035
Subcontracted fee percentage	12.5%	65641	8205	15.0%	72310	10847 12.5%	52060	6508	10%	47618	4762
ADJUSTED TENDER TOTAL				1761447		1769785		1793445		1792753
Tender ranking										
Original				2		1		3		4
Adjusted				1		2		4		3

This forecast provides a means of allocating percentages in order to establish the forecast direct cost for people, equipment and subcontractors. From these figures, the various tendered percentages for people and equipment overheads can be applied which can be summated in order to add the tendered direct fee percentage to the forecast direct cost (i.e. the contractor's own costs). For the subcontracted element of the forecast of compensation events (i.e. the indirect costs), the tendered subcontracted fee percentage is added to give a total for this part of the calculation.

Once the calculations have been completed, tenders can be compared to see if the tendered overhead percentages and other fees have influenced the tender rankings. In Table 12.2, it can be seen that the original lowest tender (Tender B) is now ranked second and the lowest adjusted tender is Tender A.

With regard to changes in quantity under ECC Option B, a compensation event arises under Clause 60.4 if:

- The difference does not result from a change to the Works Information.
- The difference in quantity causes the Defined Cost per unit of quantity to change **and** (note the 'and').
- The rate in the BQ at the Contract Date multiplied by the final total quantity of work done is more than 0.5% of BQ total at the Contract Date.

As touched on in Chapter 11, there are complications with this clause, but in principle, Table 12.3 illustrates the intended mechanism in relation to an item measured under CESMM4.

Table 12.3   ECC Option B: Change in quantity.

Reference	Item	Quantity	Unit	Rate (£)	Total (£)
E.3.2.5	Excavation for foundations; material other than topsoil, rock or artificial hard material; maximum depth 2–5 m	6900	m³	3.50	
	Final total quantity	8300	m³		
	Difference	+1400	m³	3.50	4900.00

BQ total at Contract Date			£5 000 000
%	$= \dfrac{£4900 \times 100}{£5\,000\,000}$		**=0.098**

NB:

1. In order to warrant a compensation event, there would need to be a demonstrable change in actual cost to the contractor for performing this item of work *and* the percentage would have to be above the 0.5% threshold.
2. Being below the 0.5% threshold, there would be no rate adjustment for this item irrespective as to whether the actual cost per unit quantity had changed.

PART 4

Measurement Case Studies

Managing Measurement Risk in Building and Civil Engineering, First Edition. Peter Williams.
© 2016 John Wiley & Sons, Ltd. Published 2016 by John Wiley & Sons, Ltd.

Part 4

Chapter 13
New Rules of Measurement: NRM1

This chapter concerns the application of the NRM1 rules to the preparation of the works estimate part of an order of cost estimate for a proposed new crematorium, which is to replace an existing facility that no longer meets modern emission standards.

The design is to be developed by the client to RIBA Plan of Work 2013 Stage 3. The contract is to be awarded on a design and build, 70/30 quality/price cost reimbursement basis, with preliminaries and profit as per a Framework Agreement. A 0/100 'pain/gain' arrangement weighted in favour of the client organisation will be included in the contract.

13.1 Project details

Figure 13.1 shows the site plan and general arrangement of the crematorium.

The design concept for the main building is based on three interlinking 'ellipses'. To the east there is an open utility area enclosed with a curved boundary wall and access gates and to the east there is a separate Book of Remembrance building linked to the main building with a partially covered walkway.

The roof is timber framed, with a plywood-boarded insulated deck and single ply roofing membrane, designed to mimic a standing seam roof.

There is a new one-way road system, car parking for 55 vehicles and the site is to be extensively landscaping and planted.

13.2 Accommodation

The main building consists of:

1. Ground floor:
    - Area 1:
        ○ Entrance.
        ○ Waiting area.
        ○ Public toilets.
        ○ Offices and ancillary rooms.

Managing Measurement Risk in Building and Civil Engineering, First Edition. Peter Williams.
© 2016 John Wiley & Sons, Ltd. Published 2016 by John Wiley & Sons, Ltd.

Part 4

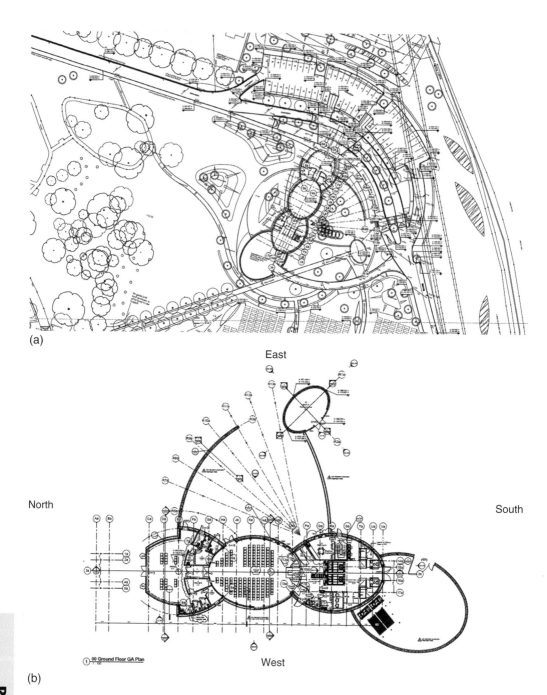

(a)

(b)

Figure 13.1   (a) Site plan and (b) general arrangement. Reproduced with the kind permission of Kier Construction.

- Area 2:
  - Chapel.
- Area 3:
  - Committal hall and cremator hall.
  - Staff rooms and toilets.
2. First floor:
   - Area 3:
     - Air handling plant and other M&E installations.

13.3 Gross internal floor area

At the early design stages, NRM1 Paragraph 2.6 suggests that either the cost per square metre of gross internal floor area (GIFA) or cost per functional unit would be appropriate estimating methods. NRM1 Appendix B suggests commonly used units of measurement for order of cost estimating, but being fairly uncommon, there is no suggested unit for crematoria.

NRM1 Paragraph 2.3 suggests that the accuracy of an order of cost estimate is dependent on the quality of information provided to the quantity surveyor/cost manager and Paragraph 2.3.2(b), in particular, requires that a schedule of GIFAs and an accommodation schedule should be provided by the architect.

This information is often provided on the drawings but should be regarded with suspicion. Designers are prone not to indicate all floor areas on the drawings, and there is never any guarantee that schedules of accommodation are complete or that they include all corridors and the like.

In this particular case study, there is some variance between the stated floor areas, especially for ground floor area 3 and the floor above, although they are identical.

13.4 Calculating GIFA

More often than not, even just as a basic check, it is good practice to carry out a simple floor area take-off.

This is quickly performed using on-screen measurement software such as the Causeway CATO suite.

For the purposes of this case study, Buildsoft Cubit has been used, and Figure 13.2 shows the floor plans and polyline area measurements obtained. Table 13.1 provides a work breakdown structure for the main building together with respective GIFAs. The GIFA total of 662 m² represents a difference of +32 m² compared to the areas stated on the drawings.

On-screen measurement is especially useful where complex or unorthodox shapes are to be measured, as in this case, and accurate answers can be obtained quickly at the click of a mouse.

13.5 Special design features

Crematoria are not 'run-of-the-mill' buildings, and new ones are both few and far between and often designed imaginatively and sensitively, as in this case study. Finding comparable elemental cost analyses for suitable historic cost data is difficult, therefore, and even if one can be found, a good deal of interpretation would be needed on the part of the cost planner.

Comparative cost analysis data, and indicative costs per square metre, always need to be viewed in the light of the individuality of the design, and this is especially true with regard to this case study. Amongst the design features that need to be accounted for are:

- The ellipsoid floor plans.
- Curved wall construction.
- Complex interface between walls and roofs.
- Special roof configuration.
- High quality of internal finishes.
- Under floor heating.
- Extensive boundary walling.
- Extensive roadworks and landscaping.

Part 4

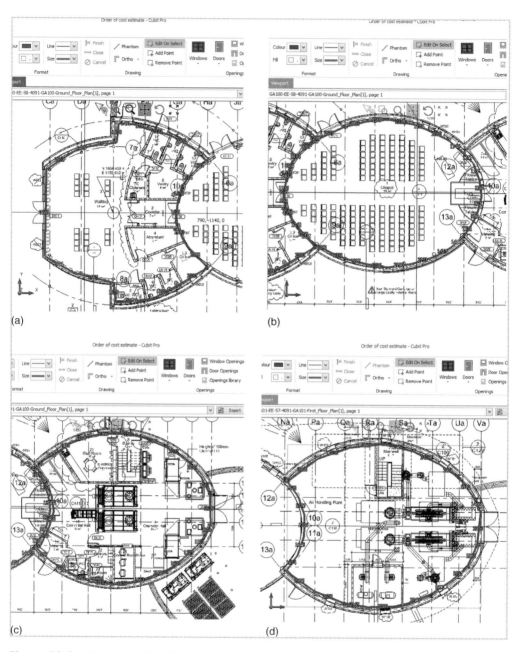

Figure 13.2 On-screen take-off – gross internal floor area. (a) Ground floor: Area A, (b) ground floor: Area B, (c) ground floor: Area C, (d) first floor: Plant room and (e) GIFAs in Buildsoft Cubit estimate screen. Reproduced with the kind permission of Kier Construction.

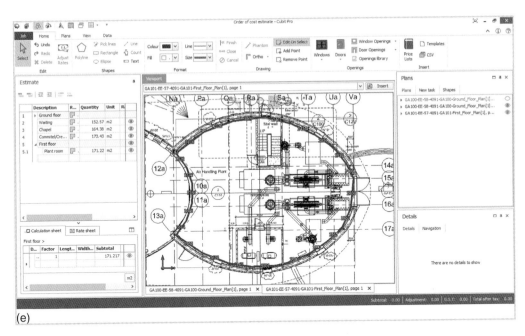

(e)

Figure 13.2   (*Continued*)

Table 13.1   Gross internal floor area (GIFA).

	Description	R...	Quantity	Unit	Rate	...	Total
1	◢ Ground floor						
1.1	Waiting	▢ .	152.57	m2			
1.2	Chapel	▢ .	164.38	m2			
1.3	Committal/Cremator	▢ .	173.43	m2			
2	◢ First floor						
2.1	Plant room	▢ .	171.22	m2			
3	TOTAL	▢ .	661.60	m2			

13.6   GIFA measurement rules

When using the GIFA as the basis of calculation, the measurement rules for order of cost estimating are those laid down in the RICS *Code of Measuring Practice* as stipulated in NRM1 Paragraph 2.6.1(a)(ii).

The code is reproduced in the NRM1 appendices and is discussed in Chapter 5, but GIFA is the sum of the areas at each floor level measured to the internal face of the perimeter walls.

13.7   Roof

The list of inclusions and exclusions in the *Code of Measuring Practice* is of particular interest to this case study and, especially, with regard to the crematorium roof which is shown in Figure 13.3. Here, it can be seen that the roof cantilevers beyond the building line at the north end and effectively forms a canopy over the front entrance to the crematorium.

Part 4

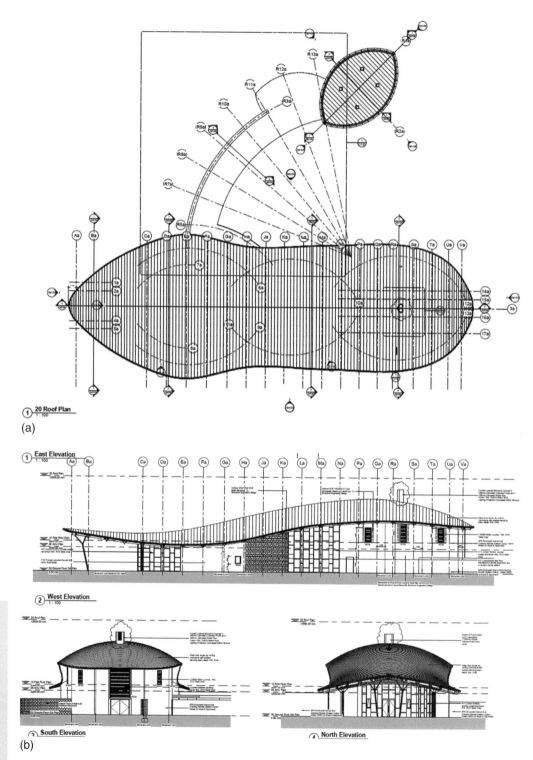

Figure 13.3   Crematorium roof. (a) Roof plan and (b) Elevations. Reproduced with the kind permission of Kier Construction.

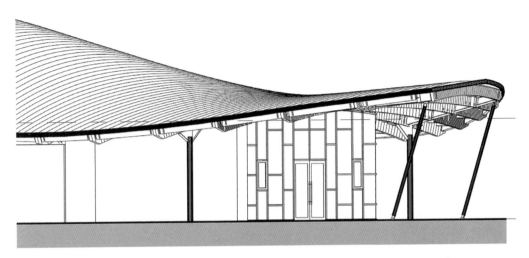

Figure 13.4   Entrance canopy. Reproduced with the kind permission of Kier Construction.

The *Code of Measuring Practice* specifically excludes canopies from the GIFA calculation, and a decision must, therefore, be made whether to:

- Adjust the chosen range of costs per square metre for the order of cost estimate to include for the cost of the additional roof area or
- Treat this particular section of the roof as a canopy, measured separately in the order of cost estimate

It can be seen from the elevations in Figure 13.3, and from Figure 13.4, that the roof over the entrance to the crematorium has vertical supports outside the footprint of the building and, it must be imagined, that this support has a foundation.

Some clues as to how this item may be dealt with can be found in NRM1 Part 4: *Tabulated rules of measurement for elemental cost planning*:

- Element 2.3.1 Roof structure:
    - Note 9 – canopies to <u>external **areas**</u> are included in sub-element 8.8.2: *Ancillary buildings and structures*.
    - Note 10 – canopies to <u>external **doors**</u> are to be included in sub-element 2.6.2: *External doors*.

The question now arises, is the crematorium canopy to an external area or to an external door?

In effect, it could be construed as both, but Measurement Rule C1 of Element 2.3.1.1 states that *the area measured for pitched roofs is the area of the roof on plan, to the extremities of the eaves*. This implies that the roof over the main entrance is not a canopy, but this doesn't necessarily mean that the structure supporting the cantilevered roof is part of the frame of the building.

Element 2.1.1: *Steel frames* is silent on the issue, except for Note 3 which says that *roof trusses which can be separated from the frame* shall be included in sub-element 2.3.1: *Roof structure*, which brings us full circle back to the canopy issue.

The conclusion from a measurement point of view is, probably, that:

a)   The roof is part of Element 2.3: *Roof*.
b)   The canopy support structure is not part of the frame but is either:
   (i)   Part of 2.6.2: *External doors* on the basis of Note 9 (providing protection to external doors) or
   (ii)   Part of 8.8.2: *Ancillary buildings and structures* as a canopy to an external area

Having exhausted the logic of NRM1, common sense must prevail on the basis that (b)i is the likelier option, as it is hard to be convinced that the canopy structure is an ancillary building or structure. This issue will assume even greater importance when it comes to the cost plan.

Part 4

13.8   Works cost estimate

Taking the foregoing into account, the works cost estimate part of the order of cost estimate can now be put together.

When compiling the works cost estimate, a historic elemental cost analysis needs to be chosen where the costs of preliminaries and external works are not distributed amongst the elements. These costs need to be stripped out of the analysis so that contract specific figures can be calculated and added back into the estimate.

It will be seen in Table 13.2 that such adjustments have been made to the building works estimate as well as for:

- The ellipsoid floor plans and curved wall construction.
- The complex interface between walls and roofs.
- The special roof design and cantilever.
- The high specification.
- The canopy support structure.

Table 13.2   Works cost estimate.

					Subtotals (£)	Subtotals (£)	Total (£)
Reference	**Constituent**	**Quantity**	**Unit**	**Rate (£)**			
a	Facilitating works estimate						
b	Building works estimate	662	m²	4000		2648000	
	Adjustments						
	Curved work	7.5	%	2648000	198600		
	Complex roof	662	m²	100	66200		
	Specification	5	%	2648000	132400		
	Canopy structure				10000		
	External works and landscaping				300000	707200	
						3355200	
c	Main contractor's preliminaries estimate	11	%			369072	
d	Subtotal					3724272	
e	Main contractor's overheads and profit estimate	7	%			260699	
f	**Works cost estimate**						3984971

Heading of table: **Order of cost estimate for new crematorium**

Chapter 14
New Rules of Measurement: NRM2

14.1 Excavation in unstable water-bearing ground

Further to Chapter 6, Paragraph 6.10.8, this case study relates to the measurement of suitable items for the excavation of a deep basement.

The basement is 45.750 m × 25.750 m on plan as indicated in Figure 14.1 and is of reinforced concrete construction, comprising:

- 300 mm diameter augered piles (60 nr perimeter; 32 nr internal).
- 1.5 m² × 1.0 m deep pile caps.
- Integral perimeter ground beams 1.5 m × 1.0 m in cross section.
- Inner ground beams 0.8 m × 0.6 m in cross section.
- 300 mm thick reinforced concrete walls.

The borehole information in Figure 14.1 shows that the basement is to be constructed in unstable soil and that the water table is 1.5 m below existing ground level.

In order to deal with the measurement issues raised by NRM2, it might be informative to begin with SMM7!

14.1.1 SMM7 rules

The measurement of excavation in water-bearing ground under SMM7 is straightforward. If water is present in the ground, the measurable items are:

- Extra over excavation D20.3.1.
- Earthwork support D20.7.*.*.2 or D20.7.*.*.3 if the ground is unstable and water bearing.
- Disposal D20.8.2.
- If conditions on-site are worse than envisaged at tender stage, omit the earthwork support item D20.7.*.*.2 and add sheet piling D32.2.*.

The contractor then decides how to carry out the work, short of sheet piling, and must include in his tender sum for a dewatering system to remove groundwater (e.g. pumping, well-points, etc.).

Part 4

Managing Measurement Risk in Building and Civil Engineering, First Edition. Peter Williams.
© 2016 John Wiley & Sons, Ltd. Published 2016 by John Wiley & Sons, Ltd.

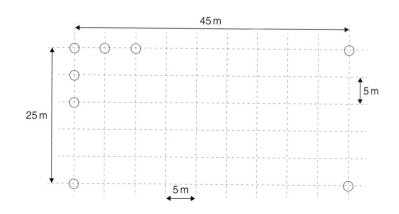

(a)

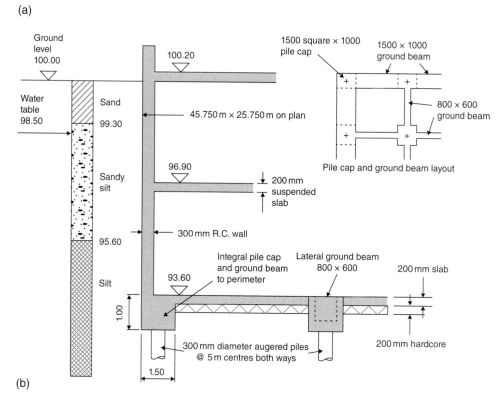

(b)

Figure 14.1   Deep basement. (a) Pile layout and (b) Cross section of basement.

14.1.2   NRM2 rules

NRM2 Work Section 5: *Excavation and filling* is the starting point for measuring the excavation items for the basement. As discussed in Chapter 6, the rules are more complex than SMM7:

- Basement excavation is measured as bulk excavation under 5.6.1.* (refer to Note 1).
- Excavation below **groundwater** is measured as *extra over all types of excavation irrespective of depth* (5.7.1.3).
- *Excavating in unstable ground* is also measured as *extra over all types of excavation* under 5.7.1.5.
- Disposal of groundwater is measured under 5.9.1.*.
- Disposal of excavated material is measured under 5.5.2.*.

Groundwater is defined in Work Section 5.7.1, Note 2, as:

- *Any water encountered below the established water table level*, excluding water arising from streams, broken drains, culverts or surface flooding, and also excludes running water from springs, streams or rivers.

Unstable ground is defined as:

- *Running silt, running sand, loose ground and the like* (refer to Note 4 of 5.7.1.5).

An added complication to the above is that NRM2 Work Section 6: *Ground remediation and soil stabilisation* provides a measurable item for *site dewatering*, whether or not the choice of dewatering method is at the contractor's discretion (refer to 6.1.*.*).

14.1.3   NRM2 measurement issues

In water-bearing ground, excavation item 5.7.1.3 would be measured, but where the ground is unstable, item 5.7.1.5 (excavating in unstable ground) also comes into consideration.

There is no earthwork support item to measure, unless specified in the contract documents, but an item for disposal of groundwater is needed (5.9.1.*). A further item for site dewatering is also required under NRM2 6.1.*.

Under NRM2, the quantity surveyor/cost manager has a dilemma:

- How to describe unstable water-bearing ground.
- Whether to measure an item of support to excavations not at the contractor's discretion (e.g. sheet piling) and/or make a suggestion to the architect/engineer.
- Whether to measure a site dewatering item as well as an item for disposal of groundwater.

Tenderers also have a dilemma:

- If there <u>is no measured item</u> for site dewatering, what is the item coverage for the disposal of groundwater item (e.g. normal pumping or well-point dewatering)?
- If there <u>is a measured item</u> for site dewatering, what is the disposal of groundwater item for?
- If there is an item for excavating in water-bearing ground, but no measured item for site dewatering or support to excavations not at the contractor's discretion, should the tender price include for sheet piling?

14.1.4   Possible approaches to NRM2

Under SMM7, there was no problem if both water-bearing ground and unstable ground were present because there was no item measurable for unstable ground – only for excavating below groundwater level.

In NRM2, however, there is no rule as to how these items should be measured and several approaches might be taken:

<u>Method 1</u>	Measure the entire excavation below water table as 'below groundwater level' and include a further item for excavating in unstable ground.
Problem:	This would create an 'extra over an extra over', that is, double counting.
<u>Method 2</u>	Measure the two volumes separately.
Problem:	This would avoid double counting but might give tenderers the impression that the two volumes are in different excavations or in different parts of the site, which could well be misleading. Tenderers could also be misled because unstable ground is not necessarily water bearing and thus the item description would not be complete.
<u>Method 3</u>	Measure the entire volume as an extra over item for excavating below groundwater level and ignore the unstable ground item.
Problem:	This might appear a better idea (consistent with SMM7), but would not comply with NRM2 as tenderers would be denied pricing an unstable ground item; this could have implications for the choice of earthwork support, for overbreak, for additional disposal and filling requirements and could also lead to a measurement claim.

Part 4

Method 4   Measure an item for extra over all types of excavation for excavating below
                            groundwater level and include reference to unstable ground in the item
                            description.
        Problem: This is not how the library-based software systems work and creation of an item
                            description like this is effectively a 'rogue item'.

14.1.5   Site dewatering

The next question is whether an item for site dewatering should be measured <u>in addition to</u> items for extra
over excavation for excavating below water table and in (water-bearing) unstable ground.

This was not the case with SMM7 as there was no provision for measuring site dewatering.

Within the rules of NRM2, it would appear that if excavations in water-bearing ground require site
dewatering, then an appropriate item is measurable. This would be a matter for the quantity surveyor/
cost manager's judgement when preparing the tender documents.

If, however, the quantity surveyor/cost manager decides not to include an item for site dewatering
in the bill of quantities, but site dewatering is nonetheless required on-site, then the question arises as
to whether the contractor would be entitled to a measured item by default and extra payment as
a result.

Should the provisions of SMM7 for measuring steel sheet piling (if required on-site) be taken as a precedent, it would appear that a site dewatering item should be measured under NRM2. Under the JCT 2011
SBC/Q, this would constitute a variation and would thus be subject to the valuation rules in the
contract.

It would be a fair assumption on the part of the contractor, therefore, that the extra over items
measured under NRM2 Work Section 5.7.1.* exclude the cost of site dewatering, as this is a measurable
item.

The 'extra over' excavation item would consequently represent only the additional degree of difficulty in excavating below the water table, and possibly in unstable ground, together with the cost of
additional overbreak, disposal and backfill, but not the cost of site dewatering.

14.1.6   Earthwork support

Whilst earthwork support is not measured in Work Section 5: *Excavating and filling*, <u>it is measured</u> where
not at the contractor's discretion.

Earthwork support would be measured in two circumstances with regard to the basement excavations
following Note 1 under NRM2 5.8.1.1:

- If it is felt sensible at tender stage, a bill of quantities item could be included for, say, steel sheet piling in
  order to reduce the contractor's risk and avoid a potential claim if the ground conditions are not exactly
  as indicated in the borehole logs.
- Earthwork support could be measured pursuant to a contract administrator's instruction should
  prevailing site conditions warrant it.

14.1.7   Worked example

A worked example of how the issues discussed above may be resolved is demonstrated in the bill of
quantities presented in Table 14.1. This assumes that Method 4, described in Paragraph 14.1.4, is
adopted as the most sensible approach for describing the work concerned.

The various Dim Sheets that accompany the bill of quantities items are included in Table 14.2, and the
relevant side casts are shown in Figure 14.2.

Table 14.1   Bill of quantities – deep basement excavation.

			£	p
5 EXCAVATING AND FILLING				
Excavation				
Bulk excavation				
A over 6 m not exceeding 8 m deep	8011	m3		
Foundation excavation; commencing 6.8 below original ground level				
B over 6 m not exceeding 8 m deep	212	m3		
Extra over all types of excavation irrespective of depth				
Excavating in				
C below ground water level; unstable ground	6455	m3		
Disposal				
Excavated material off site				
D generally	8223	m3		

Excavation items

6 GROUND REMEDIATION AND SOIL STABILISATION				
Site dewatering; wellpoint dewatering				
[Description]				
E area of site to be dewatered 1178 m2; pre-contract water level 98.50 m; water level to be lowered to 92.60 m	ITEM			

Site dewatering item

14.2   NRM2 Director's adjustment

Chapter 6 identifies a number of issues concerning the Director's Adjustment item that is provided on the pricing summary sheet of the bill of quantities under NRM2.

The adjustment is intended as a means of changing the tender sum prior to submission of the tender rather than having to make wholesale changes to the rates and prices in the bill of quantities. This is a sensible arrangement (following CESMM) as the tender figure is invariably decided at the last minute when changes to the priced bills would be difficult to make.

Unlike CESMM, NRM2 provides no rules for dealing with the Director's Adjustment item at the post-contract stage, and this may lead to problems when negotiating the final account.

Part 4

Table 14.3 illustrates three ways of dealing with the Director's Adjustment item at final account stage based upon a tender sum of £1 183 000 for a lump sum contract, assuming that JCT 2011 standard conditions apply:

1. A lump sum adjustment.
2. An adjustment in proportion to the net value of measured work carried out divided by the net value of measured work at tender stage.
3. An adjustment in proportion to the gross value of measured work carried out divided by the net value of measured work at tender stage.

Net value is the value of work excluding the contractor's overheads and profit margin, and gross value is where the contractor's overheads and profit are included. Self-evidently, the tender allowances for provisional sums and risks have been omitted from the contract sum in the final account figures. This is because the actual value of any work instructed would be added to the contract sum, as is normal practice when preparing a final account for a lump sum contract.

It can be seen that each method of adjusting the contract sum results in a different answer leading to the conclusion that:

a) There will be an argument about this when attempting to settle the final account.
b) The prudent approach would be to include a preamble in the bill of quantities detailing exactly how the Director's Adjustment should be dealt with post-contract.

Table 14.2   Dim Sheets.

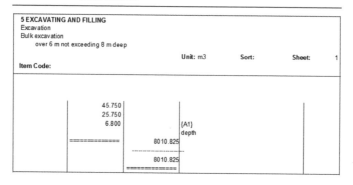

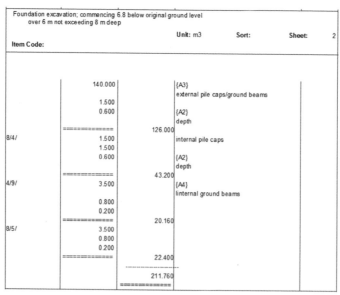

Table 14.2   (*Continued*)

Extra over all types of excavation irrespective of depth
Excavating in
   below ground water level; unstable ground

Unit: m3   Sort: Sheet: 5

Item Code:

Item Code		Dim	Total	Description
		45.750		Anded-on from file AAA sheet 000001 - START
		25.750		
		6.800		depth
		=============	8010.825	Anded-on from file AAA sheet 000001 - END
		(45.750)		
		25.750		
		1.500		{A5} excavation above water table
		=============	(1767.094)	
		140.000		Anded-on from file AAA sheet 000002 - START - external pile caps/ground beams
		1.500		
		0.600		depth
		=============	126.000	Anded-on from file AAA sheet 000002 - END
8/4/		1.500		Anded-on from file AAA sheet 000002 - START - internal pile caps
		1.500		
		0.600		depth
		=============	43.200	Anded-on from file AAA sheet 000002 - END
4/9/		3.500		Anded-on from file AAA sheet 000002 - START - linternal ground beams
		0.800		
		0.200		
		=============	20.160	Anded-on from file AAA sheet 000002 - END
8/5/		3.500		Anded-on from file AAA sheet 000002 - START
		0.800		
		0.200		
		=============	22.400	Anded-on from file AAA sheet 000002 - END

			6455.491	
			=============	

Disposal
Excavated material off site
   generally

Unit: m3   Sort: Sheet: 4

Item Code:

Item Code		Dim	Total	Description
		45.750		Anded-on from file AAA sheet 000001 - START
		25.750		
		6.800		depth
		=============	8010.825	Anded-on from file AAA sheet 000001 - END
		140.000		Anded-on from file AAA sheet 000002 - START - external pile caps/ground beams
		1.500		
		0.600		depth
		=============	126.000	Anded-on from file AAA sheet 000002 - END
8/4/		1.500		Anded-on from file AAA sheet 000002 - START - internal pile caps
		1.500		
		0.600		depth
		=============	43.200	Anded-on from file AAA sheet 000002 - END
4/9/		3.500		Anded-on from file AAA sheet 000002 - START - linternal ground beams
		0.800		
		0.200		
		=============	20.160	Anded-on from file AAA sheet 000002 - END
8/5/		3.500		Anded-on from file AAA sheet 000002 - START
		0.800		
		0.200		
		=============	22.400	Anded-on from file AAA sheet 000002 - END

			8222.585	
			=============	

Table 14.2 (*Continued*)

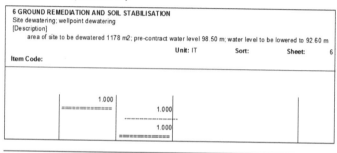

Figure 14.2 Side casts.

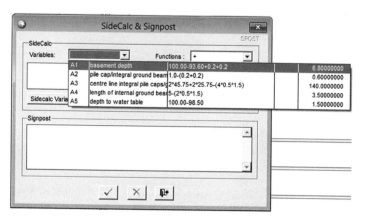

Table 14.3 Director's Adjustment pre- and post-contract.

Item	Tender	A (lump sum adjustment)	B (adjustment in proportion to net* measured work value)		C (adjustment in proportion to gross** value)	
			Final account			
Measured work	1 000 000	1 200 000		1 200 000		1 200 000
Prov sums (defined/undefined)	50 000	0		0		0
Risks	30 000	0		0		0
	1 080 000	1 200 000		1 200 000		1 200 000
OH&P 10%	108 000	120 000		120 000		120 000
Subtotal	1 188 000	1 320 000		1 320 000		1 320 000
Director's Adjustment ±	(15 000)	(15 000)	15 000 × 1.2/1.0	(18 000)	15 000 × 1.32/1.0	19 800
Subtotal	1 173 000	1 305 000		1 302 000		1 300 200
Dayworks	10 000	12 000		12 000		12 000
Total	**1 183 000**	**1 317 000**		**1 314 000**		**1 312 200**
Variance		0		3 000		4 800

*Excluding OH&P.
**Including OH&P.

Chapter 15
Civil Engineering Standard Method of Measurement

15.1 Canal aqueduct

An aqueduct is required to enable a canal to be carried over a new road which is to be constructed beneath. The abutment walls of the aqueduct, and the adjacent wing walls, require the installation of 30 nr ground anchors. The work involves drilling into ground which includes some rock.

15.1.1 Construction method

The aqueduct is to be constructed using 'top-down' construction in the following (simplified) sequence:

- Stank-off existing canal during autumn–spring possession period.
- Remove section of existing canal.
- Install two rows of secant piles to form the abutment walls to carry the aqueduct over a new road to be constructed below.
- Construct capping beams.
- Install ground anchors.
- Construct reinforced concrete aqueduct.
- Commission canal and open to traffic.
- Excavate beneath aqueduct and construct new road.
- Face up secant pile walls with brickwork.

This is illustrated in Figure 15.1.

15.2 Ground anchors

Figure 15.2 shows a typical cross section at an anchor point where the Commencing Surface is below the Original Surface.

For simplicity, the anchors are assumed to be 100 mm diameter and 15 m long. The length drilled into rock at each anchor is 3 m.

15.2.1 Commencing surface

It can be seen from Figure 15.1(a) and (d) that the Commencing Surface for the ground anchor work is at an indeterminate level, below the deck of the new aqueduct. This is the level at which there is a

Managing Measurement Risk in Building and Civil Engineering, First Edition. Peter Williams.
© 2016 John Wiley & Sons, Ltd. Published 2016 by John Wiley & Sons, Ltd.

(a)

(b)

(c)

(d)

Figure 15.1 Construction details. (a) Stanking to existing canal and ground anchor locations, (b) Access to bridge deck, (c) Bridge deck construction and (d) Abutment wall and ground anchors.

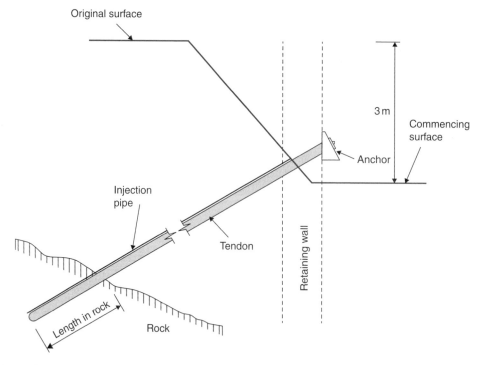

Figure 15.2 Typical cross section at anchor position.

hardstanding for the falsework which supports the deck construction, but it is not formation level for the road which has yet to be excavated.

Clearly, the Original Surface is not the Commencing Surface for the ground anchors, and so in accordance with Measurement Rule M1, a Commencing Surface needs to be chosen for use when preparing the bill of quantities.

In making the decision, the bill compiler needs to ask:

- Is the 'top-down' method of construction the only method that could be adopted?
- Is the 'top-down' method expressly required in the tender documents or is this a contractor choice?
- In any event, should the Commencing Surface be related to:
  - A depth/level below the Original Surface?
  - The underside of the aqueduct deck?
  - The pile capping beam (which is the bearing for the ends of the aqueduct)?
  - Somewhere else?

CESMM4 Section 1 Paragraph 1.10 defines *Commencing Surface* as *the surface of the ground before any work covered by* [an] *item has been carried out.*

In this particular example, the Commencing Surface has been chosen as the reduced level below the Original Surface. A strict interpretation could conclude that the inclined surface is where the ground anchor work commences (see Figure 15.2) but it is difficult to describe where work commences on a slope. Measurement Rule M1 comes to the rescue to some extent by stating that whichever Commencing Surface is adopted for preparing the bill of quantities, this will be adopted for the admeasurement of the completed work.

It would be easy to conjecture that the Commencing Surface could be shown on a drawing but in practice, where ground anchors are to be installed in disparate locations, this might prove to be more difficult than imagined.

Figure 15.3 illustrates the ground anchor installation process for the aqueduct project.

15.2.2 Measurement

The completed bill of quantities items for the Stage 1 Earthworks are illustrated in Table 15.1, where it will be noted that the Excavated Surface is 3 m below existing towpath level.

The Excavated Surface for the Stage 1 Earthworks is the Commencing Surface for the ground anchor installation as illustrated in Table 15.2, which shows the completed bill of quantities items for the ground anchor work.

The principal dimension sheets for the earthworks and ground anchors are given in Table 15.3.

NB: Some technical details have been omitted from the item descriptions for clarity.

Figure 15.3   Ground anchor installation. (a) Drilling (Class C 2 3 3), (b) Pressure test, (c) Tendons, (d) Taping injection pipe to tendon, (e) Tendon strands, (f) Cement and grout pump and (g) Anchor head.

Table 15.1   Bill of quantities – Earthworks.

CLASS E: EARTHWORKS				
Excavation for foundations				
Material other than topsoil, rock or artificial hard material				
maximum depth : 2-5 m				
E Stage 1 excavation; Excavated surface 3m below existing towpath		818	m3	
Excavation ancillaries				
Disposal of excavated material				
F material other than topsoil, rock or artificial hard material		819	m3	

Table 15.2   Bill of quantities – Anchors.

Page 1

			£	p
CLASS C: GEOTECHNICAL AND OTHER SPECIALIST PROCESSES				
Drilling for grout holes through material other than rock or artificial hard material	(NA)			
Horizontally or downwards at an angle less than 45 degrees to the horizontal				
in holes of depth : 10-20 m				
A	100 mm diameter	360	m	
Drilling for grout holes through rock or artificial hard material				
Horizontally or downwards at an angle less than 45 degrees to the horizontal				
in holes of depth : 10-20 m				
B	100 mm diameter	90	m	
Driving injection pipes for grout holes diameter 100 mm				
Horizontally or downwards at an angle less than 45 degrees to the horizontal				
C	in holes of depth : 10-20 m	450	m	
Grout holes materials and injection				
Number of holes				
D	cement and sand	30	nr	
Single water pressure tests				
E	cement and sand	30	nr	
Materials				
F	cement	0.57	t	
G	sand	3.71	t	

Page 2

			£	p
Injection				
A	number of injections	30	nr	
B	cement and sand filler grout	7.42	t	
Ground reinforcement				
Number in material which includes rock or artificial hard material to a [stated] maximum depth				
C	permanent with double corrosion protection	30	nr	
Total length of tendons which includes rock or artificial hard material				
D	permanent with double corrosion protection	30	m	

Part 4

Table 15.3   Principal Dim Sheets.

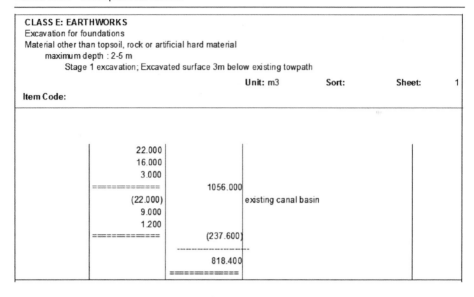

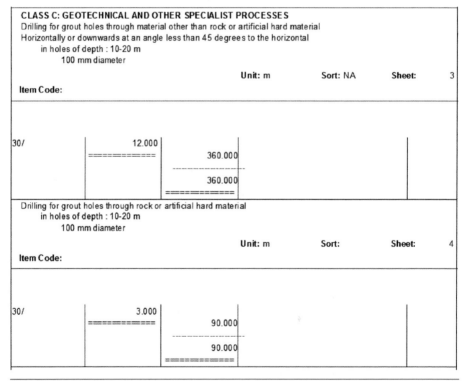

Chapter 16
Method of Measurement for Highway Works

16.1 Measurement and billing of proprietary manufactured structures

Each proprietary structure to be designed by the contractor is billed according to Paragraphs 15 and 16 of the MMHW Chapter III: *Preambles to Bill of Quantities* and to Series 2500: *Special Structures* of Chapter IV of the MMHW.

16.1.1 Billing of items

Guidance is provided in the *Notes for Guidance on the Method of Measurement for Highway Works* Series 2500: *Special Structures*. As a consequence, the BQ items for proprietary structures to be designed by the contractor would appear in a separate Bill No. 5 as illustrated in Table 16.1.

16.1.2 Measurement

It can be seen that, within Bill No. 5: *Structures Designed by the Contractor*, a separate 'sub-BQ' is provided for each structure (e.g. Bill No. 5.2: Highmore Lane Underbridge).

The take-off for this item would be generated as shown in Figure 16.1 which shows the item description in the upper grid and the 'base item' or 'root narrative', from which the item description is derived, in the lower grid.

The base item is taken from the *Library of Standard Item Descriptions for Highway Works* (1*2* **designed by the Contractor 3***), and this is edited as shown in the upper grid.

The item would appear in the final bill of quantities as illustrated in Table 16.2 pursuant to the requirement of Series 2500: *Special Structures* that contractor-designed structures are to be billed as separate single items, with an associated name or reference.

16.1.3 Tender stage

At tender stage, the contractor prices the item by means of a single lump sum that is deemed to include for design, approvals, submissions and resubmissions and *everything necessary for the completion of the Works within the Designated Outlines, as shown in the relevant item coverages in the Chapters and Series of the Method of Measurement, with the exception of those works scheduled as not to be included.*

Managing Measurement Risk in Building and Civil Engineering, First Edition. Peter Williams.
© 2016 John Wiley & Sons, Ltd. Published 2016 by John Wiley & Sons, Ltd.

Part 4

Table 16.1   Bill No. 5 – structures designed by the contractor.

Bill no.	Title	Name/reference
1	Preliminaries	
2	Roadworks	
3	Structures	
4	Structures where a choice of designs is offered	
5	**Structures designed by the contractor**	
5.1		Elton Brook Culvert
5.2		Highmore Lane Underbridge
5.3		Grassington Road Gabion Wall
6	Service areas	
7	Maintenance compounds	
8	Accommodation works	
9	Works for statutory or other bodies	
10	Daywork	
11	PC and provisional sums	

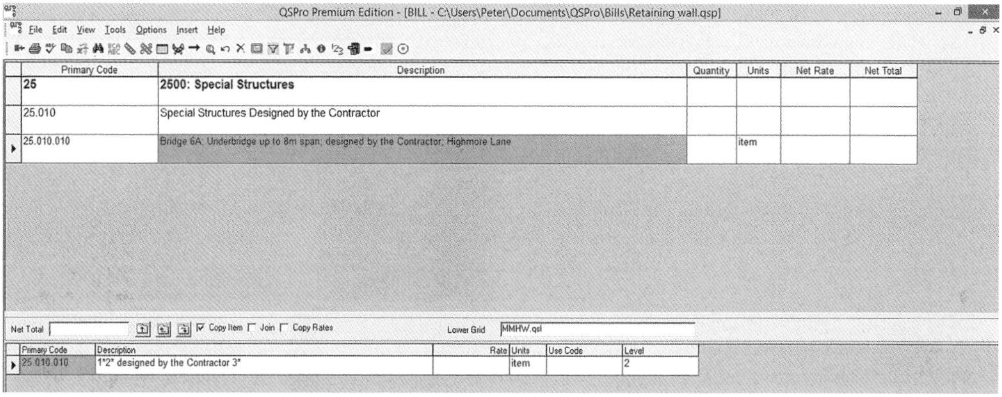

Figure 16.1   Highmore Lane Underbridge – 1.

Table 16.2   Highmore Lane Underbridge – 2.

		Qty	Unit	Rate	£	p
	2500: Special Structures					
	Special Structures Designed by the Contractor					
A	Bridge 6A; Underbridge up to 8m span; designed by the Contractor; Highmore Lane		item			

This means that not only does the item coverage for Series 2500: *Special Structures* apply, but also other item coverages for other relevant Series pertinent to the work involved in constructing the proprietary structure.

The tender documentation must clearly show the extent of work excluded from within the designated outline as this work would be measured in the Roadworks or other Bills by the overseeing organisation. Where this cannot be shown clearly, a schedule of exclusions must be provided.

> **Risk issue**
>
> Tenderers must decide how much effort to put into designing proprietary structures at tender stage.
>   Clearly, designs need to be sufficiently developed so that an accurate price can be derived but not so detailed as to incur design fees disproportionate to the cost of tendering and to the chances of winning the contract.

Exclusions from the contractor's lump sum include common items of work such as drainage, kerbing and safety barriers, etc. Such items will be billed by the overseeing organisation in other Bills, unless impracticable to do so, in which case the contractor's attention must be drawn to the need to price these items into the lump sum price for the proprietary structure. Earthworks within designated outlines are excluded but, if this is not scheduled on the drawings, the contractor could be caught out by the provisions of Series 600: *Earthworks* Paragraph 11 which states that:

- *Earthworks within Designated Outlines shall not be measured in this Series.*

> **Risk issue**
>
> Contractors should be aware that earthworks volumes within designated outlines will not be included in the earthworks schedules prepared by the overseeing organisation and that no claims will be entertained should the contractor not take this work into account in the tender price.
>   The consequence of this is that tenderers must calculate earthworks volumes within designated outlines for themselves and include these quantities in the pricing of the lump sum items for proprietary structures to be designed by the contractor.

Once the contractor's tender is accepted, detailed design will follow and, once accepted by the overseeing organisation, the contractor must submit a schedule of rates totalling the tendered lump sum. The purpose of the schedule of rates is to provide a means of valuation of work in progress and in order to value variations to the contract. The schedule of rates must be itemised in accordance with the relevant Chapters and Series of the MMHW pursuant to the requirements of Paragraph 16 of the MMHW Chapter III: *Preambles to Bill of Quantities*.

16.2   Measurement and billing of structures where there is a choice of designs

In accordance with MMHW Chapter III: *Preambles to Bill of Quantities*, Paragraph 6, the measurement and billing of structures where there is a choice of designs requires <u>two</u> bills of quantities to be included in the contract documents as part of the main BQ so that the contractor is able to choose which of the options to price.

16.2.1   Billing of items

Where a choice of designs is offered, each bill (e.g. Bill No. 4.1 – Hough Farm Box Culvert) will comprise two bills of quantities as illustrated in Table 16.3:

1. One bill of quantities quantifying the design prepared by the overseeing organisation.
2. A second bill of quantities for the contractor's design.

Part 4

Table 16.3   Bill No. 4 – structures where a choice of designs is offered.

Bill no	Title	Name/reference
1	Preliminaries	
2	Roadworks	
3	Structures	
4	**Structures where a choice of designs is offered**	
4.1A	Structure designed by the overseeing organisation	Hough Farm Box Culvert
4.1B	Structure designed by the contractor	
4.2A	Structure designed by the overseeing organisation	Tarrant Road Retaining Wall
4.2B	Structure designed by the contractor	
4.3A	Structure designed by the overseeing organisation	Wavin Lane Footbridge
4.3B	Structure designed by the contractor	
5	Structures designed by the contractor	
6	Service areas	
7	Maintenance compounds	
8	Accommodation works	
9	Works for statutory or other bodies	
10	Daywork	
11	PC and provisional sums	

Each structure must be billed in this way so that:

- The measured work in the engineer-based design is kept separate from other measured work in the contract to avoid confusion.
- The BQ for the engineer-based design may be left unpriced and omitted from the contract where the contractor design option is preferred by the tenderer.
- It is clear that there is no duplication in the eventual tender total.

The structure of the full tender bill of quantities is illustrated in Table 16.3, where it can be seen that there are three structures where a choice of designs is offered.

Table 16.3 shows that:

- **Structures Where a Choice of Designs is Offered** is Bill No. 4.
- Bill No. 4 is quite separate from Bill No. 5 Structures Designed by the Contractor which deals with proprietary structures.
- Bill No. 4 is subdivided as necessary into separate bills for each structure (e.g. Bill No. 4.1, 4.2, etc.).
- Each bill is then subdivided into separate bills according to how they are measured (e.g. Bill No. 4.1A and 4.1B).

16.2.2   Measurement

Taking Bill No. 4.1, Hough Farm Box Culvert, as an example, the work would be measured as follows:

- Bill No. 4.1A is measured in detail, according to the various appropriate Series of the MMHW, because it is based on a design by the overseeing organisation.
- Bill No. 4.1B is measured according to Series 2500 (i.e. a single item) because the design is to be carried out by the contractor.

Bill No. 4.1B would be created as a single item under Series 2500 as illustrated in Figure 16.2 which shows the 'base' item or 'root narrative' in the lower grid.

The final billed item would appear as illustrated in Table 16.4.

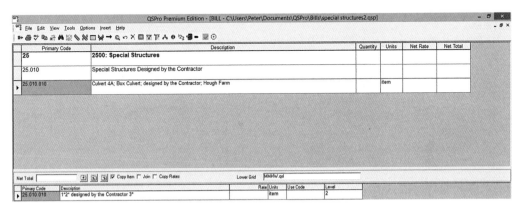

Figure 16.2   Hough Farm Box Culvert – 1.

Table 16.4   Hough Farm Box Culvert – 2.

		Qty	Unit	Rate	£	p
	2500: Special Structures					
	Special Structures Designed by the Contractor					
A	Culvert 4A; Box Culvert; designed by the Contractor; Hough Farm		item			

16.2.3   Tender stage

At tender stage, tenderers have a choice:

1. Price Bill A – The bill of quantities for the engineer-based design:
   - This is measured, in detail, according to the appropriate Chapters and Series of the MMHW.
   - If priced by tenderers, this bill is brought to a distinctly separate total to the remainder of the measured work in the tender bills.
   - The design and quantity risks are retained by the Overseeing Organisation.
2. Price Bill B – The contractor design option:
   - This is again prepared in accordance with the MMHW but this time uniquely under the provisions of Series 2500: *Special Structures*.

Part 4

- This BQ will comprise only one item and the price, if submitted by the contractor, will be a single lump sum.
- The contractor takes both the design risk and the quantity risk for this item.

Once the tenderer has decided on the appropriate strategy, the tender is finalised by pricing either Bill A or Bill B (but not both) and totalling this with the remaining bills in contract (i.e. Preliminaries, Roadworks, etc.).

16.3 Measurement of proprietary manufactured structural elements

Figures 16.3 and 16.4 illustrate the construction of an engineer-designed railway bridge where the contractor is to be responsible for designing the proprietary structural elements. These consist of bridge bearings and piles.

The bridge is required to carry the railway line over a proposed highway. In order to keep the railway line running, the bridge is to be slid into position during a line possession thereby reducing disruption to a minimum. Piles, capping beams and service ducts can be seen in Figure 16.3 which have been installed in previous possession periods. In a further possession, the railway line embankment, ballast and track have been removed, and the bridge is ready to be slid into position.

A simulation of the process can be seen at https://www.youtube.com/watch?v=Nw4luhVNjsU.

16.3.1 Billing of items

In Figure 16.4, the bearings for the central pier can be seen highlighted, and it is this item of work that the contractor is to design as a proprietary structural element, along with the piles.

Figure 16.3 Bridge slide – 1.

Figure 16.4   Bridge slide – 2.

As there are no notes for guidance for measuring these items in the *Notes for Guidance on the Method of Measurement for Highway Works*, the bill compiler is on his/her own when it comes to describing and billing the work.

16.3.2   Measurement

In this particular case study, the number and length of piles and the number of bridge bearings have to be measured. These items are classed as proprietary manufactured structural elements to be designed by the contractor, and the MMHW requires that they be billed in specific units of measurement.

As discussed in Section 8.10.4, however, the bill compiler has a dilemma because these elements have yet to be designed but a quantity is, nonetheless, required to complete the BQ.

After the *Notes for Guidance*, the first place to look for ideas is the root narrative for the item concerned – for example, bridge bearings.

Figure 16.5 illustrates the root narrative for bridge bearings where it can be seen that the type of bearing must be known in order to complete the two descriptions – supply and installation – and a quantity must be known to complete the items.

Primary Code	Description	Quantity	Units	Net Rate	Net Total
21	**2100: Bridge Bearings**				
21.010	Bearings				
21.010.010	Bearing 1"		no		
21.010.020	Installation of 1" bearing		no		

Figure 16.5   Root narrative – bridge bearings.

However, without knowing what the contractor's design will be, it is impossible to say what type the bearing is or what the quantities should be, leaving the bill compiler to come up with quantities for items that are unknowns!

Obviously, the method of measurement must be respected but, on the other hand, reality must be faced and a solution found. Therefore, purely as a personal suggestion, a method of billing the bridge bearings is illustrated in Table 16.5.

Table 16.5   Suggested billing of bridge bearings.

Rail Bridge 118A						
		Qty	Unit	Rate	£	p
	2100: Bridge Bearings					
	Bearings to be designed by the Contractor					
A	Bearing Type [ ]*; quantity to be inserted by the Contractor * to be completed by the Contractor		no			
B	Installation of bearing Type [ ]* ; quantity to be inserted by the Contractor * to be completed by the Contractor		no			

Once individual bill compilers have decided on their own plan of action, it may be thought appropriate to amend the Method of Measurement. This may be done by including additional pages in the Preambles to the Bill of Quantities and turning to MMHW Chapter III Paragraph 20 for a suitable form of words:

- *For the purposes of the Contract the Method of Measurement for Highway Works is amended in accordance with the pages immediately following.*

16.3.3   Tender stage

Faced with the problem of billing proprietary structural elements, tenderers are in a better position than bill compilers because, at some point, they will have a design from which quantities may be derived, enabling them to come up with a price.

However, faced with uncertainty in the method of measurement, contractors are prone to overprice or load rates or both as a reaction to risk and, therefore, need a clear and unambiguous item to encourage a competitive figure.

This requires not only a suitable item description – notwithstanding the method of measurement – but an appropriate item coverage that deals with the interface between the contractor's design and that of the engineer.

Chapter 17
Principles of Measurement (International)

17.1 Underpinning

Application of the POM(I) measurement rules is considered in this case study in relation to the underpinning of a $1^1/_2$ brick thick masonry wall as detailed in Figure 17.1.

The underpinning is to be constructed with C25 mass concrete filling using the traditional alternate bay method of construction.

17.1.1 Measurement rules

Underpinning is measured under POM(I) clause B7, which provides five rules for such work:

B7.1. Underpinning shall be given under an *appropriate heading* that includes the location of the work.
B7.2. Work *shall be measured in accordance with the appropriate* Work Sections.
B7.3. Temporary support *shall be given as an item*, and particulars shall be given where the design of the temporary works is not at the discretion of the contractor.
B7.4. Excavation *shall be measured by volume*.
B7.5. Cutting away projecting foundations *shall be measured by length*.

17.1.2 Itemisation

Excavation in underpinning is required to be given in two separate items:

1. Excavation in preliminary trenches.
2. Excavation below the base of the existing foundation.

B7.4.1 is clear that the depth of excavation for the preliminary trench is measured to the base of the existing foundation and the depth of the underpinning pit is the depth below that (B7.4.2).

Whilst POM(I) rules require excavation in underpinning to be billed in two items, there is, however, no rule governing the width of the preliminary trench, or the underpinning pit.

B7.4 merely states that *excavation shall ... be taken to the outside line of the projecting foundations or to the outside line of the new foundation (whichever is the greater)*.

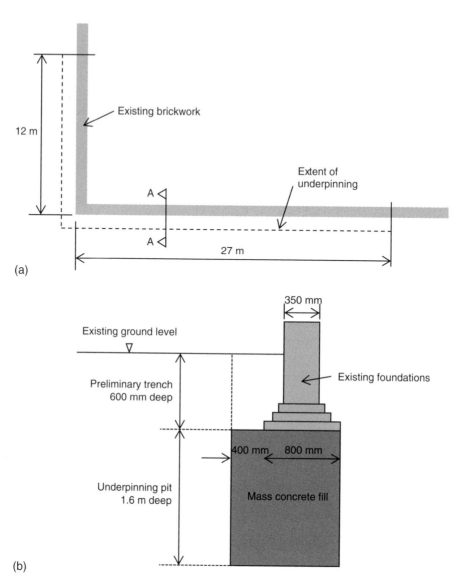

(a)

(b)

Figure 17.1   Underpinning details. (a) Plan and (b) Section A–A.

17.1.3   Rule B7.4

This rule is illustrated in Figure 17.2 where it can be seen that the width for the measurement of excavation of both the preliminary trench and the underpinning pit is given by the greater of dimensions (a) or (c).

This width shall apply to the volume calculation of <u>both</u> excavations.

The approach taken under POM(I) is to be contrasted with that of SMM7, wherein a *width allowance* is added to the calculation of the width of each excavation as illustrated in Figure 17.3. The width allowance is considerable – a minimum of 1 m – and this reflects the working space required for operatives to work in an underpinning trench which, because of its short length in the alternate bay system, is effectively a confined space from a health and safety standpoint. The consequence of this, for the case study in question, is that a formwork (shuttering) item would be needed for the concrete as well as the obligatory earthwork support item required under SMM7 D50.4.

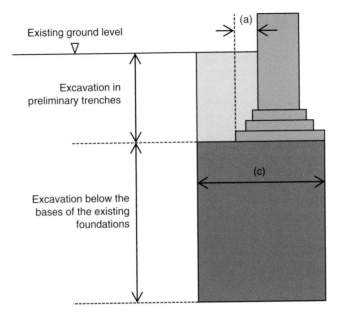

Figure 17.2   POM(I) trench widths.

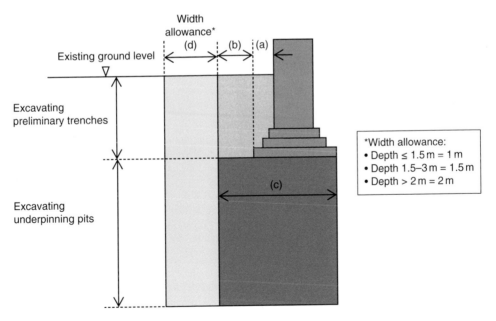

Figure 17.3   SMM7 trench widths.

Risk issue

Following clause B7.4, contractors and subcontractors should be alive to the fact that the volume of excavation measured in the bill of quantities will probably be significantly less than that necessary to carry out the work.

In this event, the rate stated for the work will need to allow for the extra excavation and backfilling required as well as for earthwork support, which is not measured at all.

Consequently, under POM(I) rules, the volume measured is the *net* volume with no allowance for working space or earthwork support.

17.1.4   Working space

Working space is not measured in underpinning, or elsewhere in POM(I), but temporary support for underpinning shall be *given as an item* (refer to B7.3).

Exactly what is meant by the phrase *temporary support* is not clear as this could be taken to mean:

- Temporary support to excavations.
- Temporary support to the wall to be underpinned (e.g. raking shores).
- Temporary support to the foundation prior to concreting (e.g. needles inserted through the wall at ground level).

In common with much of POM(I), the intention here may be to give the bill compiler the latitude to refer to details on drawings or to add additional description to items. The contractor also has the freedom to price the work as seen fit and to price a lump sum (or method-related charge) that can be shown separately from the measured work in the bill of quantities.

In any event, the contractor is obliged to measure his own quantities for the 'support' item from the drawings.

17.1.5   Dimensions

Apart from the support item (B7.3), none of the underpinning work is measured under B7. Consequently, reference must be made to other appropriate sections of the method of measurement in order to find suitable items. In this case, Sections B: *Site Work* and C: *Concrete Work* are relevant.

The software used for this case study is CATO Take-off and Bills, which is one of several modules in the Causeway CATO suite.

The basis for measurement in CATO is the dim sheet which is illustrated in Figure 17.4. Here, it can be seen that the first dimension, the trench centre line (39.025 m), is signposted as such and there is a reference {A4} to the side cast for this dimension. Reference to Figure 17.5 shows how the basic side casts have been calculated in the CATO software. The side casts are as follows:

A1. The existing foundation overhang which is needed to work out the width of the preliminary trench.
A2. The width of the preliminary trench.
A3. The centre line of the preliminary trench, allowing for the corner.
A4. The centre line of the mass filling, again allowing for the change of direction, which serves both the excavation and concrete items.

Tables 17.1 and 17.2, respectively, are the dimension sheets for the excavation and concrete work. The dimensions are presented in the traditional vertical format.

Taking Table 17.2 as an example, it can be seen that the dimensions have been referenced to both the centre line signpost that appears in the corresponding dim sheet (Figure 17.4) and to the side cast {A4} for this dimension. CATO, therefore, provides a clear audit trail for the dimensions and for the side casts used to create those dimensions.

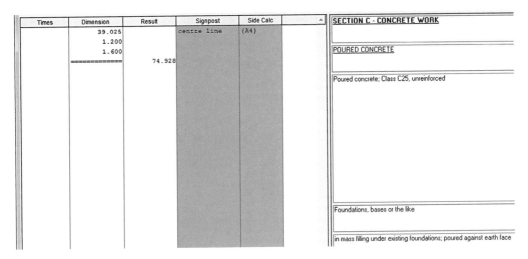

Figure 17.4   CATO dim sheet (part).

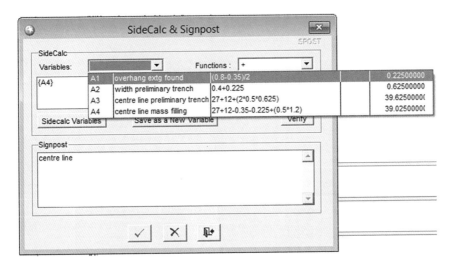

Figure 17.5   Side casts.

17.1.6   Billing

The dimension sheets in Tables 17.1 and 17.2 can now be related to the quantities billed as illustrated in Table 17.3. In the final BQ, it can be seen that:

- An appropriate heading has been created for the underpinning work which is described under Sections B and C of the method of measurement.
- Items A, D and G contain additional description which has been included pursuant to Paragraph GP1.1 of POM(I).
- All item descriptions have remained faithful to the 'brevity principle' of POM(I) which encourages tenderers to pay close scrutiny to the drawings and specification that accompany the bill of quantities.

Table 17.1   Dim sheets – excavation work.

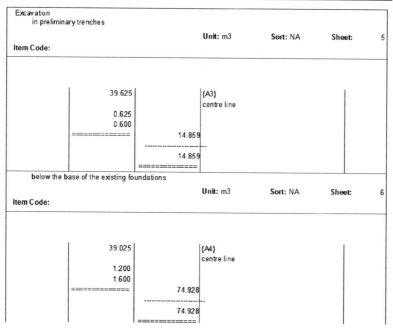

Excavation
    in preliminary trenches

Unit: m3   Sort: NA   Sheet: 5

Item Code:

39.625	{A3} centre line
0.625	
0.600	
============	14.859

	14.859
	============

below the base of the existing foundations

Unit: m3   Sort: NA   Sheet: 6

Item Code:

39.025	{A4} centre line
1.200	
1.600	
============	74.928

	74.928
	============

DISPOSAL
Disposal of material arising from excavations
Generally
    backfilled into excavations

Unit: m3   Sort: NA   Sheet: 9

Item Code:

39.625	Anded-on from file UUU sheet 000005 - START - centre line
0.625	
0.600	
============	14.859 Anded-on from file UUU sheet 000005 - END

	14.859
	============

remove from site

Unit: m3   Sort: NA   Sheet: 10

Item Code:

39.025	Anded-on from file UUU sheet 000006 - START - centre line
1.200	
1.600	
============	74.928 Anded-on from file UUU sheet 000006 - END

	74.928
	============

Table 17.2   Dim sheet – concrete work.

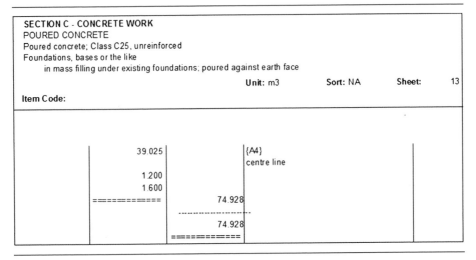

SECTION C - CONCRETE WORK
POURED CONCRETE
Poured concrete; Class C25, unreinforced
Foundations, bases or the like
        in mass filling under existing foundations; poured against earth face
                                                            Unit: m3   Sort: NA   Sheet: 13
Item Code:

                    39.025   {A4}
                                                 centre line
                     1.200
                     1.600
                =============   74.928

                                     74.928
                                    =============

17.1.7   Footnote

It should be noted that the BQ for the underpinning has been prepared strictly in accordance with POM(I) rules. However, POM(I) allows the bill compiler considerable latitude to not only include additional description but also to amend the rules of measurement as seen fit (clause GP1.2).

In this particular case study, it would seem appropriate to amend the method of measurement for several reasons:

- There is a clear need for a measured item in respect of working space for excavation work under the existing foundation.
- Consequently, an earthwork support item would be appropriate.
- A shuttering item should be measured for the mass concrete fill.

Table 17.3   Bill of quantities – underpinning.

				£	p
SECTION B - SITE WORKS					
UNDERPINNING					
North and east elevation; Howard building		(NA)			
Temporary support					
A to sides of excavation	ITEM				
Excavation					
B in preliminary trenches	15	m3			
C below the base of the existing foundations	75	m3			
Cutting away projecting foundations					
D cutting away projecting brickwork foundation; one side	39	m			

Excavation items

DISPOSAL					
Disposal of material arising from excavations		(NA)			
Generally					
E backfilled into excavations	15	m3			
F remove from site	75	m3			

Disposal items

SECTION C - CONCRETE WORK					
POURED CONCRETE					
Poured concrete; Class C25, unreinforced		(NA)			
Foundations, bases or the like					
G in mass filling under existing foundations; poured against earth face	75	m3			

Concrete item

Chapter 18
Builders' Quantities

18.1 Lift pit

Figure 18.1a–d shows the plans and cross sections for a lift pit which is part of a proposed new school. The project is to be procured on a design and build basis, and builders' quantities are needed to price this particular item of work. On-screen measurement from a PDF file will be employed to prepare the builders' quantities using the Buildsoft Cubit non-SMM-based software.

18.1.1 Take-off list

Having imported PDF drawings into the Buildsoft Cubit Viewport, it is good practice to prepare a take-off list which can then be entered into the Estimate sheet of the software. Once this has been done, quantities can be prepared.

The take-off list and the quantities will be quite different from the equivalent for an NRM2 take-off and will reflect the estimator's view of site practicalities and how the work will be carried out:

- **Excavation:**
  - The pad foundation for the lift pit is a complex shape and will, in all likelihood, be excavated as a rectangle.
  - There will, therefore, be additional excavation, disposal and backfilling compared to the 'net' measurement required by NRM2.
  - Site excavations never resemble the clean lines on the drawing, and there will inevitably be some 'overbreak' leading to an even greater volume of dig, disposal and backfill. Bearing in mind the depth of the excavation (over 2.5 m), earthwork support would be needed for safety reasons; this is not a measurable item under NRM2 (unless specified).
  - The estimator might feel that a 'battered' excavation would be more practical; this would lead to extra excavation, disposal, concrete and backfill.

- **Concrete to lift base:**
  - As the concrete pad foundation is un-reinforced mass concrete, NRM2 would describe the item as being *poured on or against earth or unblinded hardcore* (11.1.1.3.1), with no formwork measured; the estimator would thus need to consider:
    - The need for some rough shuttering (formwork).
    - Making an allowance for additional concrete to fill the excavation.
    - Whether some rough shuttering would avoid a lot of wasted concrete; if so, some working space would be needed which is not measurable under NRM2.

Managing Measurement Risk in Building and Civil Engineering, First Edition. Peter Williams.
© 2016 John Wiley & Sons, Ltd. Published 2016 by John Wiley & Sons, Ltd.

Part 4

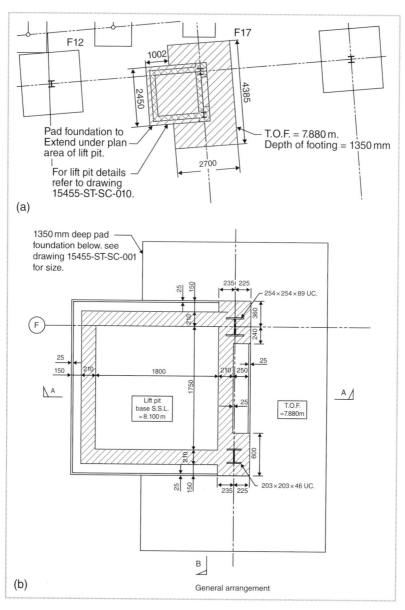

Figure 18.1   Lift Pit. (a) Plan – 1. (b) Plan – 2. (c) Section A–A. (d) Section B–B. Reproduced with the kind permission of Ramboll UK Ltd.

- **Concrete to steelwork:**
  - ○ Two steel columns are to be fixed to the pad foundation, and these are to be encased in concrete as an integral part of the lift pit wall; this creates a complex shape for both formwork and concrete.
  - ○ As a result, the estimator may decide that the concrete casing to the columns would be cast first, and the wall formwork and concreting would follow as a subsequent activity.
  - ○ This would mean measuring formwork differently to NRM2 which would measure formwork as an attached column (11.21.1) but only to the outer faces; the inner face would be measured with the wall formwork.

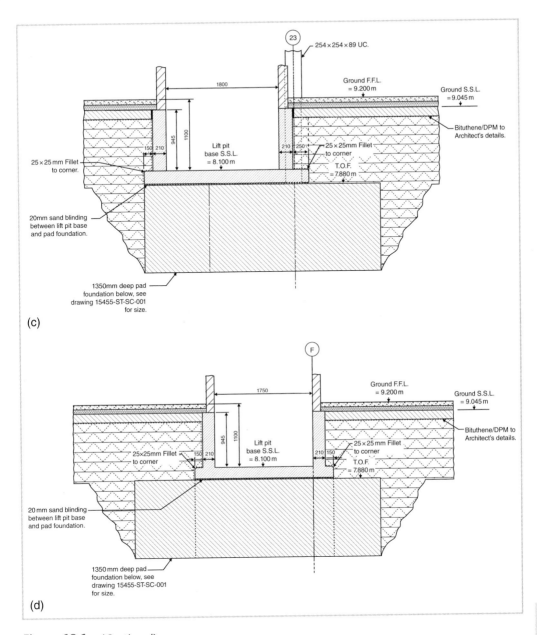

Figure 18.1 (*Continued*)

Table 18.1 shows the take-off list and compares the NRM2 rules of measurement with how builders' quantities might be approached for the lift pit take-off.

18.1.2   Preparing the quantities

Once the estimator has decided on the preferred take-off approach, the take-off list can be entered into the Buildsoft Cubit software 'Estimate' screen as a list of items to be measured on-screen.

Part 4

Table 18.1   NRM2 and builders' quantities compared.

Reference	Item	NRM2	Unit	Builders' quantities	Unit
A	Excavate pad foundation	Measured 'net'	m³	Measured 'gross'	m³
B	Disposal off-site	Ditto	m³	Ditto	m³
C	Working space	Not measured	—	Optional – could be measured or not	m²
D	Earthwork support	Not measured	—	Should be measured as it will be needed	m²
E	Mass concrete to lift base	Measured 'net' as plain in situ 'mass' concrete in lift pit base (11.1.1.3.1)	m³	Measured 'gross'; rough shuttering could be measured instead to limit the amount of waste	m³
F	Sand blinding	Imported filling (5.12.1.1.1)	m³	Could be measured or included with base slab	m²
G	Concrete base slab	Horizontal work ≤ 300 thick (11.2.1.2.2)	m³	Could be measured in m³ but m² might be more practical as the surface finish and sand blinding could be included in the rate	m²
H	Surface finish	Trowelled finish (11.1.1)	m²	Could be measured or included with base slab	m²
I	Formwork	Plain formwork to sides of foundations and bases (11.13.1)	m	Measured	m
J	Wall kickers	Suspended kickers measured on centre line – both sides deemed included (11.32.2)	m	Would not be measured but allowed for in the rate	—
K	Rebar	Measured (but not in this exercise)	t	Measured or an allowance made (but not in this exercise)	t/%
L	Concrete casing to columns	Vertical work > 300 thick in structures (un-reinforced) (11.5.2.1)	m³	Measured	m³
M	Formwork	Sides of attached columns; regular shape (11.21.1)	m²	Measured but on all four sides	m²
N	Concrete to lift base walls	Vertical work ≤ 300 thick in structures (reinforced) (11.5.1.1.2)	m³	Could be measured in m³ but could also be measured in m² so that the estimator could include both concrete and formwork both sides in the rate	m³/ m²
O	Formwork	Plain formwork to faces of walls; vertical (11.22.1)	m²	Could be measured or included with rate for concrete in walls	m²
P	Rebar	Measured (but not in this exercise)	t	Measured or an allowance made (but not in this exercise)	t/%
Q	Backfill with suitable material	Imported filling in beds > 500 thick but only above void created by the concrete base (5.12.3.1.1)	m³	Could be measured or included in rate for working space or earthwork support	m³

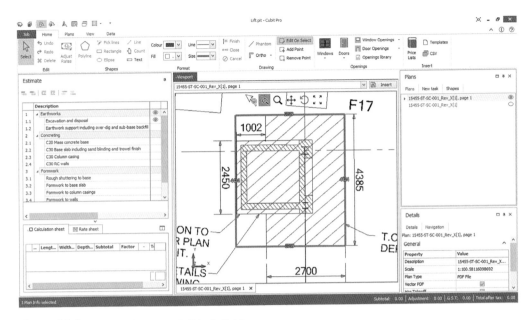

Figure 18.2   Take-off list using Buildsoft Cubit.

This can be seen in Figure 18.2 which shows the take-off list in the Estimate screen and the imported PDF file in the Viewport. Figure 18.2 also shows a rectangle around the lift pit base indicating the limits of the excavations to be measured as opposed to the 'net' dimensions shown on the drawing.

The waste calculation for the depth of excavation is calculated as follows:

Ground level	9.045	
Less		
Top of foundation slab	7.880	
	1.165	
Add		
Foundation thickness	1.350	**2.515**

Following on-screen measurement of the lift pit using Buildsoft Cubit, a completed take-off can be exported into Excel, and edited as required, as shown in Table 18.2.

18.1.3   *Quantities comparison*

It makes an interesting contrast to compare the builders' quantities take-off with the equivalent using NRM2.

Table 18.3 has been prepared with CATO, according to the NRM2 rules of measurement, and it is immediately obvious that there are 13 measured items compared to only 10 in Table 18.2. Numerically, there is very little difference, but extrapolated for a complete bill of quantities, there would be 30% more items in the NRM2 take-off than the builders' quantities. It goes without saying that the item descriptions are much more extensive in the NRM2 take-off (Table 18.3).

Some of the quantities vary quite a bit, but this is largely explained by the fact that the builders' quantities have been taken off as the estimator sees the work being done operationally, and not *net as fixed in position* in accordance with NRM2 Paragraph 3.3.2(1)(a) requirements.

Table 18.2 Builders' quantities for lift pit.

Take-off for lift pit				
Description	**Quantity**	**Unit**	**Rate**	**Total**
Earthworks				
Excavation and disposal	40	m³		
Earthwork support including over-dig and sub-base backfill	40	m²		
Total for Earthworks				
Concreting				
C20 Mass concrete base	22	m³		
C30 Base slab including sand blinding and trowel finish	6	m²		
C30 Column casing	1	m³		
C30 RC walls	2	m³		
Total for Concreting				
Formwork				
Rough shuttering to base	22	m²		
Formwork to base slab	10	m		
Formwork to column casings	4	m²		
Formwork to walls	8	m²		
Total for Formwork				
Total				

18.1.4   Software

Guidance on getting started with Buildsoft Cubit may be found at http://www.youtube.com/watch?v=Ek3rxdX_D14 which explains how to:

- Set up projects (e.g. school) and jobs within projects (e.g. lift pit).
- Enter items and simple descriptions.
- Enter and manipulate dimensions.
- Import PDF files and scaling.
- Carry out on-screen measurement.
- Etc.

For guidance using CATO, visit Tim Cook's excellent YouTube channel at http://www.youtube.com/watch?v=ur2jcCPw6Ag.

Table 18.3   NRM2 bill of quantities for lift pit (CATO).

					£ p
	Lift Pit Base				
	Page 1				
	5 EXCAVATING AND FILLING				
	Excavation				
	Foundation excavation				
A	over 2 m not exceeding 4 m deep	36	m3		
	Disposal				
	Excavated material off site				
B	generally	36	m3		
	Imported filling				
C	Sand blinding; 20 thick; Level	1	m3		
D	Granular sub-base; Beds exceeding 500 thick	2	m3		
	11 IN SITU CONCRETE WORKS				
	Plain in situ concrete; grade C20				
	Mass concrete				
E	in Lift pit base; poured on or against earth or unblinded hardcore	19	m3		
	Plain in situ concrete; grade C30				
	Vertical work				
F	exceeding 300 thick; In structures; Column casings	1	m3		
	Reinforced in situ concrete; grade C30				
	Horizontal work; in base slab				
G	not exceeding 300 thick; reinforced over 5%	1	m3		
	Vertical work; in structures; walls				
H	not exceeding 300 thick; reinforced over 5%	1	m3		
	Surface finishes to in situ concrete				
	Trowelling				
I	to top surfaces	6	m2		
	Page 2				£ p
	Formwork; plain				
	Sides of attached columns				
A	regular; rectangular	3	m2		
	Faces of walls and other vertical work				
B	vertical	13	m2		
	Sides of foundations and bases				
C	200 high	10	m		
	Wall kickers				
D	suspended	10	m		

Index

activity schedules, 36, 77, 82–5, 87–94, 96, 293, 364, 370–372, 374–5, 487, 507, 509, 514, 519

adjustment item
  tender, 293–5, 514
  Director's, 171, 179–80, 198, 539–40, 542
  drainage, 435–436,

admeasurement, 68, 265, 291, 293, 299, 304–6, 308–9, 314, 324, 346, 372, 385, 403, 424–7, 431, 439, 445, 489–92, 494, 507–9

appendix, to contract/tender, 271, 293, 378, 387, 391, 458, 467, 510

barrier to trade, 332, 397–398, 401, 404–5, 407

bills of quantities
  decline in use of, 15, 39, 43, 63–4, 66, 166
  formal, 5, 15–16, 18, 36, 39, 43–4, 63, 66, 68, 79, 81, 90
  informal, 5, 15–16, 18, 40, 64, 66, 78–9, 487
  operational, 36, 63, 77–8

BIM *see* building information modelling

budget, 25, 98, 109, 113, 118, 132–3, 142, 196, 521

builders' quantities, 16, 19, 24, 51, 78–82, 137, 139, 141, 565, 567–9
  risk 16

building information modelling (BIM)
  Bew-Richards Maturity Wedge, 28, 33
  3D BIM, 9, 30
  4D BIM, 31
  5D BIM, 31, 35–6, 61
  design/design liability, 6, 28–9, 31, 244
  extra over, 60
  government strategy, 9, 21, 29

IFC, 30–31, 58, 60
  implementation, 6, 9, 14, 29
  information, 3, 7–8, 9
  measurement/quantities, 6, 9, 14, 18, 26–7, 31–6, 46, 57–61, 80
  RIBA plan of work, 23
  uniclass, 13

CAWS *see* Work Sections (Coordinated Arrangement of)

CESMM *see* civil engineering standard method of measurement

civil engineering standard method of measurement (CESMM)
  CESMM4, 16
  characteristics, 14
  CIRIA report, 13
  item coverage, 272, 303, 312, 324, 344, 350, 355
  origin, 13

cloud, 29, 31, 52, 55, 59

commercial opportunity, 99, 190, 196, 306, 441

communication
  information, 6, 8
  Tavistock Report, 8–9

conditions of contract,
  BIM 9

construction industry
  business models, 9
  constructing excellence, 8
  medieval builders, 8, 22
  reports, 8, 499
  structure, 9–10

Managing Measurement Risk in Building and Civil Engineering, First Edition. Peter Williams.
© 2016 John Wiley & Sons, Ltd. Published 2016 by John Wiley & Sons, Ltd.

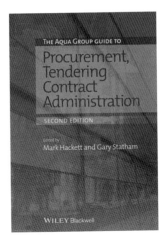
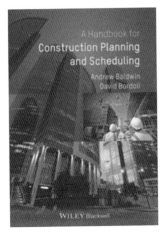

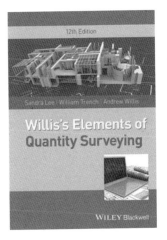